Lithostratigraphic Analysis
of Sedimentary Basins

Lithostratigraphic
Analysis
of Sedimentary Basins

C. E. B. CONYBEARE

Centre for Resource and Environmental Studies
Australian National University
Canberra, Australia

1979

ACADEMIC PRESS

A Subsidiary of Harcourt Brace Jovanovich, Publishers

New York   London   Toronto   Sydney   San Francisco

ACADEMIC PRESS, INC.
111 Fifth Avenue, New York, New York 10003

United Kingdom Edition published by
ACADEMIC PRESS, INC. (LONDON) LTD.
24/28 Oval Road, London NW1 7DX

Library of Congress Cataloging in Publication Data

Conybeare, C. E. B
    Lithostratigraphic analysis of sedimentary basins.

    Bibliography: p.
    1. Rocks, Sedimentary. 2. Geology, Stratigraphic.
3. Geology, Structural. I. Title.
QE471.C72   551.7   79–6951
ISBN 0–12–186050–7

PRINTED IN THE UNITED STATES OF AMERICA

79 80 81 82   9 8 7 6 5 4 3 2 1

Contents

Preface   vii
Acknowledgments   xi

Chapter 1 **Introduction**

Classification of Sedimentary Basins   1
Methods of Study   30

Chapter 2 **Evolution and Development**

Concepts   37

Chapter 3 **Stratigraphic and Structural Concepts**

Sedimentation   130
Consolidation of Sediments   167
Stratigraphic Framework of Basins   173
Structural Framework of Basins   183

Chapter 4 **Depositional Environments**

Geomorphic and Sedimentary Associations   197
Lithofacies and Stratigraphic Associations   282

Chapter 5 **Analysis of Outcrops, Drill Core, and Cuttings**

Macrostratigraphy   294
Microstratigraphy   320
Physicochemical Properties   333

Chapter 6 **Construction of Maps and Sections**

Sedimentary Basin Configuration   351
Lithofacies   391
Biofacies   412
Collation and Representation   414

Chapter 7 **Application of Basin Analysis to Economic Geology**

Introduction   438
Exploration for Oil and Gas   439
Exploration for Coal   464
Exploration for Mineral Deposits   468
Exploration for Aquifers and Storage Reservoirs   487

References   498

Index   533

Preface

Lithostratigraphic analysis of sedimentary basins concerns fundamental concepts of tectonics and sedimentation, and it requires the application of many methods and techniques designed to elucidate various aspects of the geological history of sedimentary layers within a basin. These methods may be direct, as in the case of surface geological mapping and the examination of subsurface drill core and cuttings, or indirect, as in the case of geophysical surveys and petrophysical logging of drill holes. They are concerned with lithology, fossil and mineral content, grain characteristics, physical and chemical properties, and the geometry of stratigraphic bodies of rock. Several books and a great number of papers have been written on subjects pertaining to sedimentary basin analysis, particularly with reference to specific aspects of sedimentology, stratigraphy, paleontology, and tectonism; but there is a need for a general review and commentary on the evolution and development of sedimentary basins with special reference to lithostratigraphy. This book is intended to fill that gap and to serve as a general reference to the concepts and methodology of lithostratigraphic analysis.

Implicit in any consideration of the history of a sedimentary basin is an understanding of the concepts regarding depositional environments, paleogeographic settings, structural framework, and present solid geometry of basins. These concepts involve principles of stratigraphy and sedimentation and the influence of tectonism. Their relevance to various methods employed in the analysis of sedimentary basins is discussed in this book, although methods used in biostratigraphic, geophysical, tectonic, and structural analysis are referred to only in the context of lithostratigraphic analysis.

Of paramount importance to any analysis is the quantity and quality of the data available. The results of various analyses may differ considerably in response to each of these factors, and the inferences and conclusions which can be drawn will consequently vary in validity. It is therefore essential to compare the results of various analyses in order to ascertain in which respects they affirm or negate each other and to evaluate the validity of each individual analysis in the context of the composite results. Some analyses are based on data derived from direct methods of observation of rock outcrops and samples. Others are based on data derived from indirect methods of observation such as seismic, gravimetric, and magnetic surveys on the surface, or electrical, sonic, and radiometric surveys of drill holes penetrating sedimentary layers in the subsurface.

Certain lithological features, including sedimentary structures, textures, and assemblages of features that are considered to be diagnostic or indicative of particular depositional environments, are discussed in this book. The implications of these features to interpretations of the geologic history of a sedimentary basin are reviewed on the basis of both macrostratigraphic and microstratigraphic criteria. Other lithologic analyses that are mentioned relate to petrophysical properties such as porosity and permeability and to chemical properties such as trace element, organic, and hydrocarbon content. Methods employed in the examination of outcrops and rock samples are discussed mainly from the viewpoint of the rationale determining their use and of their relevance to the gathering of information that may have a bearing on concepts of basin analysis.

The interpretation of geophysical survey records is essentially the provenance of geophysicists who normally work in close cooperation with geologists familiar with the stratigraphy and physical principles involved in the survey. The construction and interpretation of geophysical contour maps, derived from seismic, gravimetric, and magnetic records, are ideally the result of cooperation between geophysicists and geologists. The various records obtained from drill-hole logging are interpreted by geophysicists, geologists, and engineers whose fields of expertise and interest tend to merge in practice. In each field specific methods and techniques tend to become specialities which attract individuals, many of whom have published extensively in their areas of interest. This book does not purport to explain the geophysical methods employed, but where appropriate briefly discusses the implications of geophysical data to structural and lithostratigraphic aspects of basin analysis.

Presentation of data in the form of contoured or colored maps is common practice in basin analysis and must be undertaken with imagination but within the constraints allowed by the data. It may be desirable to emphasize certain features or anomalies revealed by the interpretation, and this is permissible provided the data are not thereby invalidated. It is at this stage that comparison of the various preliminary maps is essential in order to come to a decision as to the best illustrative methods of presenting the final maps and sections. This book comments on some of the criteria involved in the construction of various maps and sections and emphasizes the considerations entailed in consolidating separate and possibly disparate views into a comprehensive interpretation.

In the final chapter some examples are given of the application of sedimentary basin analysis to concepts of exploration for hydrocarbons, coal, minerals, subsurface water resources, potential storage reservoirs for natural gas, and strata suitable for the disposal of chemical waste. For some geologists the study of a sedimentary basin, or of a stratigraphic sequence within a basin, will be undertaken purely for academic interest. Others will pay particular attention to certain formations or beds which may contain hydrocarbons, coal, or mineral deposits of industrial interest. Many geologists will at some stage of their careers be employed as part of a team, including engineers and geophysicists, engaged in studies on the economic potential of particular sedimentary basins. Whatever the reason for the study, it is essential that it be undertaken in a systematic manner in order to maximize interpretation of available data within the constraints of validity and the known geologic history of the sedimentary basin.

Lithostratigraphy covers a vast field of information, and the choice of material for the text and illustrations in this book has been based on the availability of literature and suitability of subject for a general review and commentary. The book has been written with these reservations in mind and with the knowledge that concepts and methods involving sedimentary basin analysis are themselves continually under review.

Acknowledgments

The author is most grateful for the facilities offered him by Professor F. J. Fenner, Director, Centre for Resource and Environmental Studies, The Australian National University, and by Professor T. A. Oliver and Professor J. E. Klovan of the Department of Geology, The University of Calgary. He also wishes to acknowledge the assistance of Mrs. Linda Kind of the University of Calgary in the preparation of the manuscript.

Publication of this book would not have been possible without reference to the work of a great number of earth scientists whose research has appeared in journals of the following institutions, societies, and organizations:

The Geological Society of America
The Geological Society of Australia
Canadian Society of Petroleum Geologists
The Royal Geological and Mining Society of the Netherlands
The German Society for Geological Sciences
The Geological Society of London
The Geological Society of Switzerland
The Geological Society of South Africa
Society of Economic Paleontologists and Mineralogists
Société Géologique de Belgique
The West Texas Geological Society

Société Nationale des Pétroles d'Aquitaine
Société Géologique de France
The American Association of Petroleum Geologists
Australian Petroleum Exploration Association
The Rocky Mountain Association of Geologists
International Association of Sedimentologists
The American Geological Institute
The South African National Institute for Metallurgy
Geological Institute of Poland

Israel Geological Survey
The United States Geological Survey
Yale University
The University of Uppsala
Brigham Young University
University of Cape Town
The National Research Council of Canada
United Nations Economic Commission for Asia and the Far East (ECAFE)

United Nations Economic and Social Commission for Asia and the Pacific (ESCAP)
Elsevier Scientific Publishing Company
Cambridge University Press
The University of Chicago Press
Blackwell Scientific Publications
Australian National University Press
Gulf Publishing Company
The Economic Geology Publishing Company

CHAPTER

1

Introduction

CLASSIFICATION OF SEDIMENTARY BASINS

The term sedimentary basin is in constant use by students of geology, many of whom accept the concept of a lenticular or wedge-shaped accumulation of strata thousands of feet thick but are uncertain of the parameters by which sedimentary basins may be classified. Indeed when one attempts to do so, the fundamental question arises: What is a sedimentary basin?

It is apparent that a basin comprises a pile of layers of rock which are predominantly of sedimentary origin but which may also include volcanics. It is also apparent that this predominantly sedimentary pile commonly occupies an area that is roughly oval to trough-shaped, hence the term basin, and that the pile thins considerably or pinches out completely along the edge of this area. In fact, most sedimentary basins mapped today show themselves to be only erosional remnants of former accumulations of strata which may have been much thicker and more widely distributed. The term sedimentary basin in itself does not tell us much about the nature of the strata or the depositional environments and tectonic framework in which they were formed. In order to determine the nature of a sedimentary basin and to attempt to answer the question "what is a sedimentary basin?" it is necessary to define the parameters of the basin

1

as it exists today and thereby consolidate a basis on which certain conclusions may be drawn as to its origin and geological history. Having stated the parameters, as they appear to be, the next step is to arrange them in a systematic order by which a classification may be constructed, depending on the emphasis one wishes to place on particular aspects of the basin's history. But herein lies a difficulty, for many sedimentary basins do not readily fall into a rigid set of pigeon holes, and one is often forced to ask whether the criteria used to set up a classification are valid.

The purpose of any classification is to systematize information into a form that can be readily understood and utilized and also to give a better understanding of the nature and relevance of the information. Inherent in any process of classification is a bias, which may be intentional or otherwise. Whatever the bias, it must be recognized and assessed, or it may unduly influence the basis on which the classification is constructed and, consequently, the conclusions that may be drawn from it.

The criteria by which a sedimentary basin is classified will fall within several categories which determine the nature of the classification. The weight that is given to each criterion may determine the category to which it is allocated. Consequently, the categories within a classification of sedimentary basins may be mutually inclusive. Furthermore, different stratigraphic intervals within the same basin may be placed in different categories. For example, a sedimentary basin may be classified in accordance with its lithologic and stratigraphic characteristics, the geometry of the basin, the tectonic framework of the basin, the paleogeographic setting of the basin, or the depositional environment of the basin. Each of these and other categories which may be included in a classification of sedimentary basins must be clearly defined so that users of the classification may know as precisely as possible its scope and limitations.

Depositional Environments

A comprehensive classification by Le Blanc (1972), based on studies by Bernard and Le Blanc (1965), is shown in Figs. 1-1-1-3. This classification is based on a sequence of geomorphologic and sedimentary features which commonly are found within a particular segment of a range extending from intracratonic through epicontinental to neritic, bathyal, and abyssal depositional environments. In any such classification one is confronted with the problem of defining terms and avoiding unintended connotations; but for lack of precise nomenclature within a particular area of reference, terms connoting genesis, sedimentary processes, and geomorphologic features are commonly interrelated within the same classification. This prac-

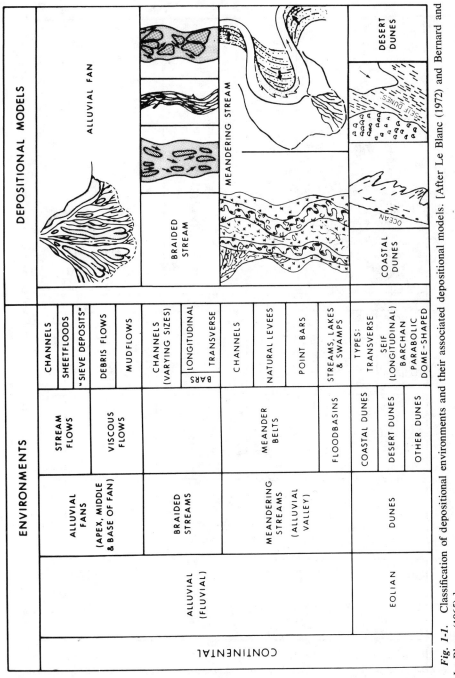

Fig. 1-1. Classification of depositional environments and their associated depositional models. [After Le Blanc (1972) and Bernard and Le Blanc (1965).]

3

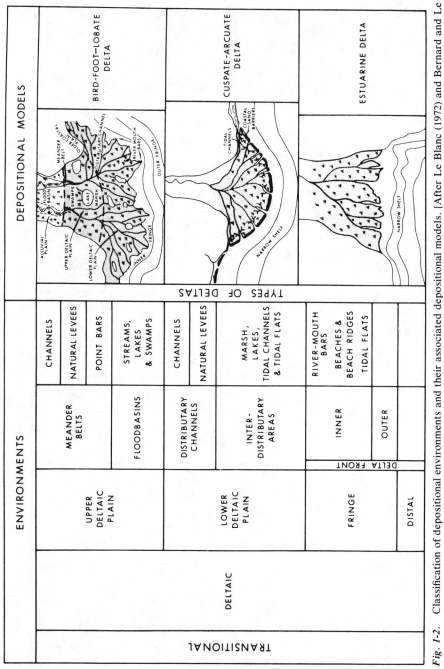

Fig. 1-2. Classification of depositional environments and their associated depositional models. [After Le Blanc (1972) and Bernard and Le Blanc (1965).]

4

tice, although not always unavoidable, has the possible consequence of being misinterpreted to some extent. In short, the reader may not form the precise conceptual image intended by the writer.

In Figs. 1-1-1-3 the main divisions are alluvial (fluvial), eolian, deltaic, coastal interdeltaic, shallow marine, and deep marine. The first two divisions connote sedimentary processes, the third and fourth refer to coastal geomorphologic features and geographic distribution, and the last two imply ranges of depth to the sea floor. These terms are used concomitantly with the presumption that the reader understands the scope and limitations of the definitions and appreciates the difficulties of trying to devise too rigid or specific a classification. The fact that Le Blanc defines the deltaic and coastal interdeltaic categories as transitional between the continental and marine environments attests to the difficulties involved in classification. These difficulties may become particularly apparent in cases where the stratigraphic interval, such as a zone of sandstone bodies, can be traced continuously from a continental to a deep marine environment, as evidenced by Bloomer (1977).

The recognition of depositional environments will, to a significant extent, depend upon qualitative and quantitative analyses of the rock layers comprising the stratigraphic section within the sedimentary basin, the type of sequence of sedimentary structures within these layers, and the sequence of the layers themselves. In other words, recognition will in large part depend on a number of studies that collectively are termed lithostratigraphic analysis. Also of fundamental significance to the recognition of depositional environments are studies of the fossil content of the rock layers and the paleoecological interpretations placed on them. These paleobiological studies are collectively referred to as biostratigraphic analysis.

Normally, where fossils are available, both lithostratigraphic and biostratigraphic analyses are undertaken. The relative contributions made by both types of analysis to the interpretation of the geological history of a stratigraphic sequence will vary from basin to basin and from section to section within a basin. Obviously, it is essential that fossil material be available for study. In the field it may be difficult to secure good specimens of both macro- and microfossils, particularly where the surface exposures of rock are badly weathered. In the subsurface it may be impossible to rely on the hit-or-miss chance that a core taken in a borehole will collect macrofossils, and the main reliance is placed on microfossils which are commonly more liberally distributed within the rock layer. Even so, cores are normally taken only at particular stratigraphic intervals which are expected to yield significant stratigraphic or petrophysical information. Smaller samples of sidewall cores may be taken at more fre-

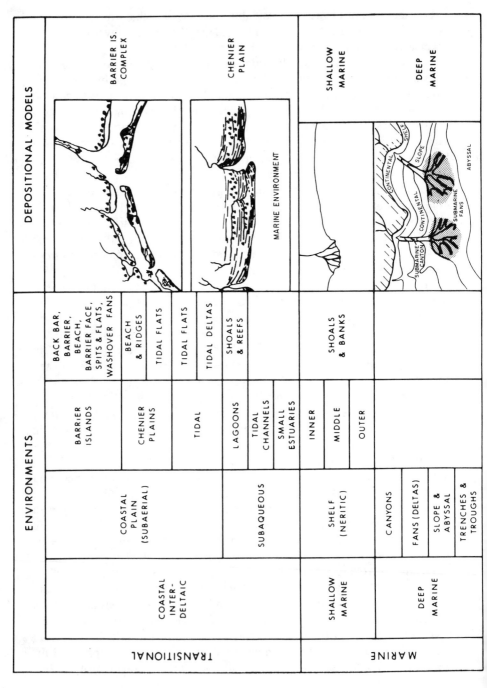

ENVIRONMENTS | DEPOSITIONAL MODELS

TRANSITIONAL

- COASTAL INTER-DELTAIC
  - COASTAL PLAIN (SUBAERIAL)
    - BARRIER ISLANDS — BACK BAR, BARRIER, BEACH, BARRIER FACE, SPITS & FLATS, WASHOVER FANS
    - CHENIER PLAINS — BEACH & RIDGES / TIDAL FLATS
  - SUBAQUEOUS
    - TIDAL — TIDAL FLATS / TIDAL DELTAS
    - LAGOONS — SHOALS & REEFS
    - TIDAL CHANNELS
    - SMALL ESTUARIES

BARRIER IS. COMPLEX

CHENIER PLAIN

MARINE ENVIRONMENT

MARINE

- SHALLOW MARINE
  - SHELF (NERITIC) — INNER / MIDDLE — SHOALS & BANKS / OUTER
- DEEP MARINE
  - CANYONS
  - FANS (DELTAS)
  - SLOPE & ABYSSAL
  - TRENCHES & TROUGHS

SHALLOW MARINE

DEEP MARINE

CONTINENTAL SHELF — CONTINENTAL SLOPE — ABYSSAL — SUBMARINE FANS — SUBMARINE CANYON quent intervals. Otherwise, the search for microfossils will depend on the examination of drill cuttings from known depths.

Biostratigraphic analysis encompasses a large field of paleontological studies which are beyond the scope of this book and which will only be referred to in the context of lithostratigraphic analysis. But it must be borne in mind that the determination of paleodepositional environments in sedimentary basins commonly requires a multifaceted approach in which lithostratigraphic analysis is only one facet.

Intracratonic–epicontinental depositional environments are essentially nonmarine although epicontinental deposits may interfinger with marine shoreline deposits. This category obviously includes a number of specific depositional environments associated with rivers, fresh-water lakes, glaciers, alluvial fans, and deserts. Of these, river deposits of the delta distributary type commonly interfinger with marine shoreline deposits. Fresh-water lake deposits may be closely underlain or overlain by marine shoreline deposits but would seldom be regarded as having an interfingering relationship. Glacier deposits may interfinger with marine shoreline deposits or may be dumped by floating ice as glaciomarine deposits. Alluvial fans are commonly intracratonic deposits, well inland from any shoreline; but in some areas of the world, such as along parts of the Red Sea coastline, alluvial fans spread down to the sea from actively eroding mountains and may be transgressed periodically by marine shoreline deposits. This stratigraphic relationship is particularly common in regions of active tectonism, in particular along graben or rift structures open to the sea and flanked by eroding terrain. Desert deposits may also interfinger with marine shoreline deposits where the desert extends to the coast, although the relationship may be masked by the reworking from wave action of desert sand deposits along the coast. Also the marine and desert deposits may be partly separated by the incorporation within the interfingering sequence of sabkha and tidal flat deposits lying between the desert and the open sea. Such exceptions and reservations impose restrictions on the use of any classification and in many cases require a somewhat arbitrary approach to the determination or assignment of particular parameters. On a statistical basis, weighing the balance between those parameters that fit the pattern and those that appear to be exceptions, a decision can usually be made with a reasonable degree of confidence as to the appropriate category. But circumspection is essential.

Fig. 1-3. Classification of depositional environments and their associated depositional models. [After Le Blanc (1972) and Bernard and Le Blanc (1965).]

Neritic–bathyal depositional environments are marine environments within the limits of the continental shelf and slope. Where the shelf is broad, it may be expedient to divide the neritic environment into outer, middle, and inner zones. The seaward limit of the outer zone bordering on the bathyal environment is generally considered to be the 100-fm contour from sea level to sea floor, but divisions separating the outer, middle, and inner zones are purely arbitrary and can be applied only in situations where the continental shelf is sufficiently broad. A classic and well-studied example of such a broad shelf is that of the Gulf of Mexico in the vicinity of the Mississippi Delta. In other regions such a division of zones could not be applied. A case in point is the region off the coast of southern California where the true continental shelf is very narrow, although relatively shallow areas, separated from each other and from the shelf by deep trenches, are present for up to 150 km from shore. The differences in the physiography of the sea floors in the two regions result from differences in the tectonic framework of the Gulf of Mexico and the offshore region of southern California. The former is tectonically fairly stable, although mild earth tremors do occur in the region, and very minor displacement of surface features such as roads has been recorded. Throughout a long period of time the peripheral regions of the Gulf of Mexico have been filling in with subsiding wedges of sediment, subject to the effects of compaction and growth faults but comparatively stable from a tectonic point of view. This infilling has progressed since the Upper Cretaceous and has resulted in a regression of the sea evidenced by the fact that, in general, the older the shoreline facies the deeper it lies and the farther inland it is situated. On the other hand, the offshore region of southern California is a drowned topography which is a continuation of the basin and range topography found on land. This topography, controlled by a graben and horst type of tectonic framework, is subject in a marine environment to comparatively slow rates of deposition on the shallow plateaus and to periodic floods of turbidites in the deep trenches.

In regions similar to the one off the coast of southern California, there is no continental slope shelving seaward from a broad continental shelf but rather a group of trenches and plateaus extending seaward to the edge of the continental mass where the sea floor slopes rapidly downward to abyssal depths of more than 4000 m. This outer slope could be defined as a continental slope but should be regarded as a particular type where placed in a classification. Sediments deposited on the slopes or on the floors of trenches lying between the outer slope and the mainland will have the characteristics of both bathyal and abyssal deposits, which may result in some confusion where a classification of depositional environments implies some geographical relationship to the mainland. Similar reservations must be kept in mind in dealing with sedimentary sequences deposited in submarine valleys and canyons which cut into both the continental shelf and slope.

Abyssal depositional environments are perhaps the least well understood for the simple reason that until a comparatively few years ago they were inaccessible to direct observation by submarine, and photographs of the sea floor taken by means of a camera lowered on a cable were not of the quality obtained today. Also until the development of deep-sea drilling techniques and the availability of drilling equipment, it was not possible to obtain cores from deeper than a few meters below the sea floor. Another, and very important factor in our understanding of the stratigraphic sequences underlying the abyssal and bathyal zones, in particular, is the development of marine seismic survey techniques.

Seismic surveys, using a variety of equipment and different sources of shock generation, can be designed to determine the topography of the sea floor, the internal structure of sea-floor features, and the thickness and structural configuration of shallow strata beneath the sea floor. They can also be designed to penetrate the strata lying hundreds of meters deep beneath the sea floor to the basaltic layer or thousands of meters deep in areas underlain by large submarine fans such as the Bengal Fan (Curray and Moore, 1971). In these great piles of sediment accumulated beneath the abyssal and bathyal sea floor, the depths to acoustical basement as well as the variations in thickness of layers having different ranges of acoustic velocity and, consequently, of density can be determined.

Abyssal environments range from the base of the continental slope to depths of 12,000 m or more. They may be characterized by abyssal plains where the normal rate of deposition is exceedingly slow, of the order of 2–3 cm in a thousand years, or by abyssal troughs subject to periodic flows of turbidity currents which rapidly deposit successive layers of turbidites. Each layer represents a distinct flow and has a thickness up to one meter or more. In these troughs the rates of accumulation can be rapid, amounting to many meters in a thousand years. The rates of accumulation of sediments and intercalated or intermixed volcanics, such as submarine basaltic flows and pyroclastic material, will normally be much more rapid in areas of active earthquakes and volcanism. Obviously, ranges of depositional rates and ranges of depth to sea floor cannot be precisely related although, in general, the shallower ranges of depth, particularly where submarine fans are developed, have higher rates of deposition. Exceptions may be found where deep submarine trenches are flanked by active submarine and subaerial volcanoes. These generalizations may seem to be self-evident but are emphasized in order to show the limitations of any classification of abyssal depositional environments and, consequently, the limitations of any such application to the classification of sedimentary basins. Certain characteristics of shallow-water sediments, both carbonate and noncarbonate, are well known and can be applied to the interpretation of depositional environments of particular strata and, therefore, to the classification of the basin or to that part of the basin which contains the strata. But definitive characteristics of deep-sea sediments and associated volcanics are more elusive, and the evidence for depth ranges within abyssal environments or even for an abyssal environment itself is somewhat equivocal.

Most of the great piles of sediment accumulating today as delta complexes and submarine fans are being deposited on continental shelves and continental slopes or rises. The physiographic and depth demarcations between bathyal and abyssal environments must be somewhat arbitrary, and, consequently, some submarine fans or the lower slopes of some fans may be ascribed to an abyssal environment. But on the whole, the bulk of sedimentation today is taking place in neritic and bathyal environments, and it can fairly be presumed that such has been the case in the past. On the basis of this assumption it will become apparent that comparatively few sedimentary basins have been deposited, in whole or in part, in the deeper parts of the sea within the limits normally defined as an abyssal environment.

Lithology

Classification of sedimentary basins on the basis of the stratigraphic sequence of rock types and other lithologic characteristics is a logical approach to understanding the geological history of any basin. Some basins, in particular those which consist predominantly of a particular assemblage of rock types, can be readily classified. Others may fall within two categories representing different stages in their genesis. A classification based on the gross nature of the stratigraphic sequence of rock types may, or may not, indicate the depositional environment or tectonic framework in which the basin was formed. For example, a sequence of sandstone, siltstone, and shale may have been deposited in the shallow-water environment of a delta complex or in the deep-water environment of a submarine fan. In order to differentiate these and other environments, it is necessary to classify the internal and external features of the rock types themselves by macroscopic and microscopic examination. Such features include composition, roundness of the grains, grain gradation within the layer of rock, and sedimentary structures. It is also necessary to classify the stratigraphic relationships, such as the sequence or cyclical repetition of the various layers of rock. Although stratigraphic sequences of sandstone, siltstone, and shale may be deposited within a wide range of depth below sea level, sequences of carbonate rocks are commonly or exclusively deposited in a shallow-water neritic environment.

In general, sedimentary basins can be classified as consisting predominantly of a carbonate sequence of limestones with associated shales or of a noncarbonate sequence which may or may not include volcanics or tuffaceous sediments. But sandstone and volcanic beds may be intercalated in an otherwise predominantly carbonate sequence, and carbonate beds may be found in a sequence of noncarbonate rocks. So classification of sedimentary basins on the basis of the general lithologic characteristics and assemblage of rock types may in some cases be somewhat arbitrary. Nevertheless, it is obviously the first approach to any sedimentary basin, as evidenced from direct observation by field mapping of the nature and sequence of the layers of rock, supported by indirect observation from geophysical surveys of the petrophysical characteristics and sequence of beds in the subsurface.

The importance of qualifying and quantifying the lithologic characteristics of individual rock types within a stratigraphic sequence, and of the sequence itself, cannot be overemphasized. A limestone is not just a limestone, nor a shale just a shale. The macroscopic and microscopic parameters by which the rock is described must be clearly and unequivocally defined if they are to be understood by others. Similarly, a stratigraphic sequence cannot yield all the information it may contain unless it is analyzed on the basis of both qualitative and quantitative parameters.

Classifications of sedimentary basins based on depositional environments will depend to a significant extent on the lithologic characteristics of the basin and will, consequently, be related to classifications based on predominant rock types and sequences in the basin. Classifications based on tectonic characteristics of the basin will, to a lesser degree, be related to the lithologic characteristics, although it is evident that the presence within a stratigraphic section of a substantial thickness of volcanic flows or pyroclastics indicates crustal instability. There may also be a general relationship between the lithologic characteristics of a basin and the shape and structural framework of the basin in that carbonates are commonly found in nonlinear basins. The relationships between classifications based on depositional environments and lithologic characteristics of a basin are evidently close, as are the relationships between classifications based on shape and structural framework of a basin and on the regional tectonic movements that influenced the development of the basin. But it is obvious that no two classifications are mutually exclusive.

Carbonate sequences are always of marine origin, although nonmarine carbonate beds may occur in a noncarbonate sequence. They may be several hundred to a few thousand meters thick, and in some regions, notably in the area of the Bahamas, off the east coast of Florida, the thickness of the carbonate sequence probably exceeds 10,000 m. In aerial extent carbonate sequences may underlie tens of thousands of square kilometers where they have not been highly folded. Where fold belts occur, palinspastic adjustments are needed to unravel the folds and realign the beds displaced by thrust faults. When this is done, the original extent of the area underlain by the carbonate sequence can be estimated. These geometric parameters are primary factors in the classification of carbonate sequences.

Such sequences commonly include beds of shale and sandstone in various volumetric proportions. They may also include beds of evaporites, in particular anhydrite, or more rarely of volcanics. These intercalations may be distributed fairly evenly throughout the carbonate sequence, or they may be predominantly within a particular section of the carbonate sequence, in which case it may be desirable to place each section in a different category within a classification. Some beds may present a problem in defining them as carbonates, where the clay and silt content is high. In such cases they may be referred to as calcareous shales or siltstones. The reverse is also true. Particularly where the rock is dark gray or bituminous, it may be called a shale whereas it is, in fact, a limestone. Such divisions of nomenclature, based on qualitative rather than quantitative characteristics, are likely to be prone to the influence of subjective interpretation in the field. Every effort must be made to be objective in describing such rocks, and where a quantitative analysis is required, the rock unit should not be designated other than as a tentative means of identification until the analysis is done. In some cases it may be thought insufficient to merely analyze by weight or volume the proportions of clay, silt, and carbonate. It may be deemed necessary to determine the nature of the clay mineral content. Such additional parameters may be of significance in the classification of limestone types and, consequently, in the classification of carbonate sequences in sedimentary basins. Other lithologic characteristics may be relevant to the classification of carbonate sequences, particularly with respect to the paleogeographic setting of their depositional environments. For example, rounded, pitted, and frosted grains of quartz disseminated within a stratigraphic interval of a carbonate sequence may be interpreted as sand grains blown into a carbonate sediment from a nearby desert. In evidence, the frosting on the grains may have resulted from abrasion in sand dunes or from replacement by carbonate after the grains were buried in the sediment. Further, the sand grains may not have been blown in but may have been dropped in the sediment with the excreta from pelagic organisms. Or the grains may have been transported in some other way. The point is that classifications must be objective and should, insofar as possible, avoid the use of categories based on features which, although objective in themselves (e.g., frosted and pitted grains of quartz), can fall prey to subjective interpretations. This is not to say that a subjective interpretation may not be correct but to caution the user of classifications based on features which have equivocal origins.

Other features of equal importance to the classification of carbonate rock types are observed either with a hand lens or in a thin section under the microscope. These features are the physical composition of the rock, the proportions of carbonate grains or fragments of various sizes, shapes and derivations, and the proportions of fine-grained indeterminate carbonate matter and noncarbonate matter such as glauconite. These featues, defined quantitatively as parameters, are the bases for various classifications devised in particular to indicate the depositional environments of carbonate rocks. Studies of present-day environments in areas of carbonate sedimentation have given many examples that can be duplicated in carbonate rocks, so that the environments and processes of sedimentation that controlled the deposition of many ancient carbonate sequences are well known.

To the uninformed layman, a rock is a rock; to the informed layman, some rocks are of limestone. But the geologist recognizes a variety of limestone types, each of which tells the story of its history when examined in detail and classified according to its chemical and physical characteristics. For example, a limestone composed largely of the fossilized remains of colonial organisms such as algae, polyzoans, crinoids, rudistids, and corals has originated as an accumulation of the debris of these organisms which either grew in situ to form a reef or when dead were rolled by currents and washed into piles or banks. In some cases the precise part of such a reef can be identified, such as the forereef or backreef, and consequently the general seaward direction and paleogeographic setting. For example, an outcrop of coarse limestone rubble composed of fragments of colonial coral having no preferred orientation of the coral organisms can be interpreted as part of the forereef. Additional mapping and sampling may disclose other reef facies. Where borehole samples are the only source of information, it is particularly important to record the qualitative and quantitative characteristics of the samples which may subsequently deteriorate or become lost and which in any case may not always remain as accessible to examination as an outcrop. Whatever the source of the sample, classification is essential to an understanding of its origin.

Associated with reef limestones and the rubble formed from them are other facies characterized by grain size or particle size and by fossil content. Formed in a shallow, high-energy zone of a neritic environment, some of the debris from the reefs, calcareous algae, and shells become ground to sand grain size and form underwater banks or intertidal beaches of carbonate sand. Other parts of the reef complex are subject to the chemical precipitation of calcium carbonate which rains down on the sea floor as tiny crystals of aragonite to form a carbonate mud. When lithified, the carbonate sand is called calcarenite, and the carbonate mud is called calcilutite. A particular type of carbonate sand, composed of rounded, concretionary grains, is commonly associated with carbonate mud and has been formed in areas where currents stir the bottom, rolling nuclei of shell or other fragments in the carbonate mud. Like rolled snowballs, the nuclei gather fine particulate matter in the mud and grow to form concretions commonly ranging in size from that of a pin head to a pea. The larger type is called pisoliths and the smaller ooliths. The lithified sediments are referred to as pisolitic or oolitic, and in the case of a rock composed entirely of ooliths, it is called an oolite.

It may seem to the reader that some very elementary and self-evident points are being unnecessarily belabored, but the crux of the matter is that although it is essential to classify limestone types by their physical or chemical characteristics, it is advisable to exercise some reservation in making interpretations as to their genesis that may be implied by the classification. With many examples, the physical characteristics or properties do indeed indicate the genesis of the rock, and the terms applied to some rocks carry the connotation of genesis or depositional environment. This is true of reef limestones formed from colonial corals whose present-day habitat is well known. It may not be so true of chalk composed largely of the microscopic plates of coccoliths, calcareous planktonic algae, and forams or of harder and more massive limestones formed mainly of algal remains, the depositional environments of which are somewhat speculative. And it certainly is not necessarily true of descriptions, based on drill cuttings, of thinly interbedded limestone and anhydrite. Where the beds are too thin to be satisfactorily differentiated by means of electrical logs and where core is not available, caution and experience are required in logging the drill cuttings from such sections, which may be interbedded limestone and anhydrite or anhydritic limestone. The former relationship may indicate a coastal mud flat or sabkha type of depositional environment, whereas the latter may be of diagenetic origin and related only to the subsurface movement of solutions through the preconsolidated carbonate sediment or subsequently through the permeable limestone.

Classifications of limestones can be entirely objective, or they may be in part subjective, particularly in the case of classifications using terms that have some connotation as to the genesis of the rock. In this respect it must be noted that some confusion may arise where different geologists place a somewhat different connotation on a particular term. Classifications that are entirely objective may be constructed on the basis of a number of chemical, physical, or biologic parameters such as the ratio of calcite to dolomite, of carbonate to phosphate or other noncarbonate matter, porosity and permeability, particle or grain size, texture, or type of organic remains. Each classification is devised to serve a particular purpose and does not in itself provide the answer to all questions that arise. Cross reference from one classification to another may be helpful in providing a better understanding of stratigraphic relationships and of the geological history of carbonate sequences.

Noncarbonate sequences may be of marine or nonmarine origin and may be classified as volcanic or nonvolcanic sequences. The differentiation is somewhat arbitrary as many sedimentary sequences are found to include substantial thicknesses of water-laid sediments derived in large part from volcanic ejecta, but with only minor beds of pyroclastics or intercalated volcanic flows. In such cases the differentiation may depend on the relative proportions of volcanic flows and pyroclastics to water-laid sediments which themselves may consist largely of reworked volcanic material. There is often a problem with lithic sandstones in determining whether the content of volcanic rock fragments and grains was airborne to the site of deposition, washed in from adjacent ash falls, or reworked from distant volcanic deposits and transported by rivers, coastal currents, or turbidity currents. In describing such sediments, the geologist must be careful to define precisely what he means by the terms he uses. For example, the term tuffaceous is commonly used for a lithic sandstone containing variable but substantial volumetric proportions of volcanic rock fragments. The term tuffaceous indicates the origin of part of the rock content but at the same time carries a connotation as to the origin of the sediment from which the rock was derived. The connotation of the term tuffaceous may not carry the same implication to all geologists and must be qualified. Although used, particularly in the field, as a descriptive term intended to be objective, it may in fact be decidedly subjective.

Volcanic sequences may also be either nonmarine or marine. Non-marine volcanic sequences may be several thousand meters thick, consisting of basaltic or andesitic flows, ash flows or falls, reworked volcanic rubble, mud flows, and interbedded fluvial and lacustrine deposits. These deposits accumulate in a tectonic environment characterized not only by volcanic activity but commonly by earthquakes and faulting. The accumulations of volcanics and terrestrial sediments fill valleys and other topographic depressions which commonly are underlain by actively down-sinking wedges of the earth's crust dislocated by complex graben structures. In such areas of deposition thick interfingering fans of volcanic flows, reworked volcanic ejecta, and alluvial material can build up.

Marine volcanic sequences are commonly associated with volcanic arcs which flank deep submarine trenches overlying, although not necessarily directly, zones of crustal subduction. These sequences may also be several thousand meters thick where they have accumulated in down-sinking submarine trenches and commonly comprise an interbedded sequence of submarine volcanic flows, volcanic ash and other ejecta, turbidites, and thin layers of clay deposited during periods of quiescence. Such sequences are generally classified as indicative of a tectonic–depositional environment termed eugeosynclinal. But it must be kept in mind that the classification of so-called geosynclines is not entirely unequivocal, and the definitive characteristics by which sedimentary basins may be separated from geosynclines are subject to various interpretations. Nevertheless, and regardless of whether a stratigraphic pile of volcanics and sediments is referred to as a geosyncline or sedimentary basin, the lithologic nature of the pile is indicative of the tectonic and depositional environment and can in whole or in part be defined in a useful classification.

Another type of marine volcanic sequence is associated with rifts in the earth's crust, up which well periodic flows of basaltic lava. Normally, sediments consist of muds, including siliceous or calcareous oozes formed from the rain of tests of dead plankton, and rates of sedimentation are relatively slow. Lithification of such a sequence may form a section comprising claystone, cherty claystone, and pillow lavas; and subsequent metamorphism may alter this section to layers of argillite, chert, and greenstone. Here again, the lithologic character of the sequence may indicate the tectonic–depositional environment, but commonly with a greater degree of uncertainty than in the case of a thick sequence of volcanics and turbidites. For example, the chert layers in such a sequence can be objectively described, both chemically and physically, but the elucidation of its origin, as possibly a radiolarite, or a chemical precipitate of silica resulting from submarine activity, may be open to various interpretations in the absence of any evidence of organic remains.

Nonvolcanic sequences of noncarbonate rocks may also be of either marine or nonmarine origin. As a generalization, nonmarine sequences are much thinner than marine sequences and range in thickness up to a few thousand meters. They commonly consist of fluvial, floodplain, backswamp, and lacustrine sediments but in some cases comprise alluvial fans or fluvioglacial deposits. In rare situations thin beds of limestone formed in fresh-water lakes by the chemical precipitation of calcium carbonate, and by the accumulation of ostracod tests and the remains of calcareous algae, occur in these sequences. Notable examples are the chalky beds, a few meters thick, formed in Pleistocene lakes in Alberta.

Where deposition has clearly taken place inland, and not along a coastal plain or within a delta complex, the sequence may comprise part of an intracratonic sedimentary basin. It may, on the other hand, merely form a thin but possibly widespread veneer over an erosional surface. This type of sequence is not so commonly preserved in the geological record because it is often removed by subsequent erosion. Thicker sequences formed in topographic depressions underlain by down-sinking fault blocks may be preserved entirely, or the lower part of the section only may be preserved.

Classification of such nonmarine sections on the basis of lithology depends on the sequence, thickness, and sedimentary structures of the various rock layers. The apparent absence of marine fossils is negative evidence and helpful as such, but certainly not diagnostic. In fact, the commonly used term "unfossiliferous" is subjective. Beds described as such may be sparcely fossiliferous insofar as megafossils are concerned but abundantly fossiliferous in microfossils. The ability to find fossils is not a gift granted equally to all geologists, as any worker in the field will know. The sequence of sedimentary structures and other internal features of various rock layers may be indicative of the sedimentological processes by which the deposits were formed. For example, a gravelly and gritty sand may be deposited along a beach front or on a river point bar. In both cases the gravel layers are lenticular and may show some degree of imbrication of the pebbles. Cross bedding within the sand is evident in both cases but notably of different types. Whereas the river point bar deposit is characterized by cross beds having angles of dip generally exceeding 20°, the beach deposit cross beds have much lower angles of dip reflecting changes in direction of the dip slope of the beach. Fluvial-type cross beds formed in river deposits are also characterized by their planar or festoon arrangement, depending on the angle at which they are viewed and on the scale of the scour-and-fill structures. Marine shoreline sands may locally have high-angle cross beds where they are formed by tidal currents but can be distinguished on the basis of other criteria such as grain gradation.

In beach and offshore bars the grain gradation is from finer below to coarser above, whereas in point bar deposits the reverse is true, provided there is a range of grain sizes available. The gradations can be observed in specimens and are also reflected, although with variable definitive character, in the self-potential curves of electrical logs, the point bar deposits exhibiting a bell-shaped curve and the beach or bar deposits exhibiting a funnel-shaped curve. These characteristics, although indicated by a geophysical method that measures the petrophysical parameter of permeability, can be recorded as also indicating the lithologic feature of grain gradation and can be classified as such.

Other nonmarine sequences may be geographically and stratigraphically closely associated with marine sediments, as in the case of river channel, floodplain, and backswamp deposits in a large delta complex. Where the river system changes course to some other part of the delta, the nonmarine sediments may be overlain by marine sediments deposited by a locally transgressive sea. Transgression results from local sinking by compaction of that part of the delta temporarily abandoned by the river system. Coal beds commonly occur with such nonmarine sequences originating in backswamps of a river floodplain or in coastal marshes. Particularly where the coal beds are formed in coastal marshes adjacent to coastal bays open to the sea, and consequently to marine fauna, shaly beds immediately underlying or overlying the coal bed may contain forams, a relationship which refutes the notion that coal beds necessarily indicate nonmarine environments. Coal beds are commonly formed by inland swamps but are in themselves not diagnostic of distance from the sea, except to the extent that coal beds formed in a coastal environment tend to be very much thinner than those formed by inland swamps. The thickness and sequence of coal beds are measurable parameters that can be used in classification. In this connection it is of interest to note that although most coal beds have thicknesses within a range of less than one meter to a few meters, some are very thick indeed. The lignitic brown coal section of the Eocene Latrobe Formation in the Gippsland Basin of Victoria, Australia, has a thickness of about 100 m. The coal seams thin seaward into the basin.

Nonvolcanic sequences of marine origin are currently being deposited in all known depths of the sea. But it cannot be presumed that the present-day sediments deposited in a particular locality are necessarily characteristic of the underlying sedimentary sequence. In neritic environments major changes of sea level may have a marked influence on the processes of sedimentation and consequently on the nature and thickness of sedimentary sequences deposited on the continental shelf. In bathyal environments the effects of sea level changes are much less pronounced,

dimentary basin has been deposited may considerably alter or modify
riginal structural framework. Further, structural dislocations within a
mentary sequence may be penecontemporaneous folds and growth
ts formed by compaction and gravitational sliding or slumping of the
rs of unconsolidated sediment.

he present structural framework of a basin may have resulted from
vements which have also considerably altered the basin's original
pe. This is particularly true of fold belts. Unraveling the folds within a
limentary basin and restoring the layers to their approximate original
ographical outline involve calculations referred to as palinspastic ad-
tment and are of particular application in the construction of
leogeographic and facies maps which will be discussed later.

Classifications of sedimentary basins based on structural framework are
herently arbitrary to some degree, particularly so as the classification
tegory depends on the interpretation of structural features observed or
ferred in the basin. An example is given in Fig. 1-4 modified from an
ustration in a paper by Olenin (1967) who discusses the classification
ith reference to the occurrence of oil and gas accumulations. The dia-
rams show, in a simplified manner, some of the structural relationships
etween sedimentary basins and the basement massif. No vertical scale is
iven, but it must be presumed that the vertical exaggeration is of the
rder of three or more. The uppermost category, referred to as down
warpings or synclises, is probably inaccurately illustrated to the extent
that such depressions of the basement are commonly effected by a multiple pattern of faults. Intracratonic basins are commonly of this type or of
the type shown to be controlled by grabens (1). The other category (2)
adjacent to the illustration of grabens is typical of many offshore basins in
various parts of the world, where continual movement of adjacent basement blocks results in depositional thinning or thickening, and of deformation of the overlying layer of sediment. The category illustrating activated
platform basins is related to categories comprising both the downwarping
of basement and graben structures. The difference is mainly one of degree,
the basement of the activated platform being much more intensely dislocated by faults and the overlying sedimentary pile being much thicker and
locally folded. Olenin states that most examples of this type of basin are
known from China. The category of interjacent basins is well known and
often shown as a classic example of the relationship between a structural
platform and a miogeosyncline. Referring to this type of sedimentary
basin, Olenin (1967, p. 42) says, "They arise and develop between platforms and geosynclines. That is the reason for their strong asymmetrical
structure. In each such basin, one of its slopes is a gently dipping platform
shelf, and the opposite is a folded and faulted frontal part of the

and in abyssal environments such changes have no influence whatsoever.
In general, changes of sea level do have a considerable influence on the
deposition of the bulk of nonvolcanic marine sequences because, on a
volumetric basis, a very large proportion of these is deposited on continental shelves in water depths of less than 100 meters. Large volumes of
sediment are also deposited in bathyal environments where they are swept
off the slopes of continental shelves and deposited as submarine fans on
the continental rises flanking abyssal plains.

The classification of nonvolcanic marine sequences on a lithologic basis
depends largely on the stratigraphic relationships of the beds within the
sequence, on the nature and sequence of sedimentary structures within
the beds, and on the thickness and proportion of various sediment types.
The gross aspects of sediment types in themselves are not commonly
diagnostic, as sands, silts, and clay muds are deposited in all depths of the
sea. Although sands are generally characteristic of shallower depths, they
are frequently transported into deeper water by turbidity currents. In this
respect, the importance of reworking or secondary deposition in marine
environments should be recognized. For example, during the Pleistocene
much of the continental shelf of eastern North America and of Europe was
exposed. As the glaciers melted, the sea levels rose with respect to the
land by up to 100 m, reworking the terrestrial sediments to form neritic
deposits.

To reiterate, it is stated above that the gross aspects of sediment types
are not diagnostic of depth or environment of deposition. In other words,
grain size is not diagnostic; although an analysis based on proportion of
volume would show a higher content of sand deposited in shallower
depths, particularly in proximity to deltas where the sands have been
transported by river distributary systems. The degree of sorting of grain
size and grain shape, on the other hand, is diagnostic of the energy level of
waves and ocean currents winnowing the sand.

Another diagnostic feature of sandstone is composition. Quartzose
sand, in particular sand composed of well-rounded and well-sorted grains,
is indicative of deposition in a high energy environment and is characteristic of beach and other offshore sand bodies. Sandstones of this type do not
occur in sedimentary sequences deposited in bathyal environments. Lithic
sands composed of fragments of rock with variable but minor proportions
of quartz grains are deposited in a wide range of water depths and indicate
relative immaturity in their sedimentological history. Commonly they
form thicker beds than the quartzose sands, interstratified with beds of silt
and clay muds. Such a sequence indicates rapid deposition with consequent lack of winnowing and sorting by waves and currents. Deposition of
this type may occur near the mouth of a delta distributary in shallow water

or on a submarine fan in deep water. Lithic sandstones that form wide-spread layers of fairly uniform thickness interbedded with claystones and siltstones, particularly where they show graded bedding, are characteristic of turbidite sequences deposited in relatively deep water on bathyal submarine fans or on the floor of abyssal plains. Within such a section the stratigraphic sequence of beds and the sedimentary structures are also important criteria in recognizing a depositional history in which normal deep-sea muds are periodically swept over, scoured, and overlain by poorly sorted floods of lithic sand and silt. As in all cases where stratigraphic sequences are examined in order to interpret their geological history and depositional environment, the use of paleontological methods is essential. In turbidite sequences there is commonly a paucity of megafauna, although trace fossils such as crustacean, worm, and gastropod trails and burrows may be abundant. Microfauna, on the other hand, may be abundant but poorly preserved. Some siliceous forms, such as Radiolaria, may be fairly well preserved in siliceous bands of rock, showing in hand specimen as tiny specks. Where very abundant, the rock may be designated as a radiolarite and fitted into a lithologic as well as a paleontological classification.

Shape and Structural Framework

Classifications of sedimentary basins based on the shape and structural framework of a basin must clearly distinguish the original shape and framework from what is apparent today. Most sedimentary basins have, during one or more periods of uplift, been beveled to some extent. Their depositional margins have, in most cases, disappeared entirely or in part. The uplift that resulted in erosion of the basin margins may also have resulted in structural dislocations of the basin. This is particularly true in areas that are now within or adjacent to fold belts or which are underlain by the graben and horst structures that feature basin and range topography.

Another aspect which requires clarification is differentiation between the geometry of a sediment pile deposited during a particular interval of time and the geometry of the total pile within the whole basin. A sedimentary basin consists of the sum of its parts, each of which may have been deposited on a fairly flat to gentle slope and may have its own depositional outline. For example, a sedimentary basin described as linear in shape may, in fact, comprise several sedimentary piles deposited at various intervals of time. Each of these piles of sediment may have formed a

sequence of prograding wedges or lenses which thic... mately pinch out as the sediment load is deposited. successive piles may account for the overall linea sedimentary pile within the basin. This concept must the possibility that sediment was transported mainl basin and then moved along the axis of the basin by a ocean currents. Care must be taken in the construction based on shape so that the terms used do not have to the genesis or depositional pattern of the basin. equivocal terms, or terms that may connote some asp history, is unavoidable, it should be noted.

In defining the shape of a sedimentary basin, the ter somewhat misleading as it carries the connotation of a pression in which deposition is taking place. There basin-shaped sedimentary basins, many of which are i nonmarine. Conversely, the Bass Basin of South Austral shaped and marine. But by and large the shapes of se within a basin, or of the accumulation of such piles comp itself, are not markedly basin-shaped unless one thinks dinted, battered, or crushed basin. On the other hand, linea result of deformation within a fold belt, with consequent depositional edges of the basin. In general, intracratonic an tal sedimentary basins deposited within the limits of a land nental shelf tend to be nonlinear in shape, although sedim such as formed by reefs or shoreline sand bodies may be ma There are exceptions to this generalization, particularly in sedimentary basins formed by the accumulation of sediments the elongate graben structures of a continually active rift. marine sedimentary basins formed in the bathyal to abyssa continental rises and deep-sea trenches generally show a di trend.

A point which is essential to remember in classifying the sedimentary basin is that in constructing cross sections of a ba of vertical exaggeration will distort the true shape of the basi with no vertical exaggeration, most sedimentary basins appea infillings of minor depressions in the earth's crust.

Classifications based on the structural framework of sedimenta must also clearly distinguish the present framework from th existed during the period in which the basin was deposited. Basi framework may be the same, as old deep-seated faults tend to b nated by periodic movements. But structural dislocations occurr

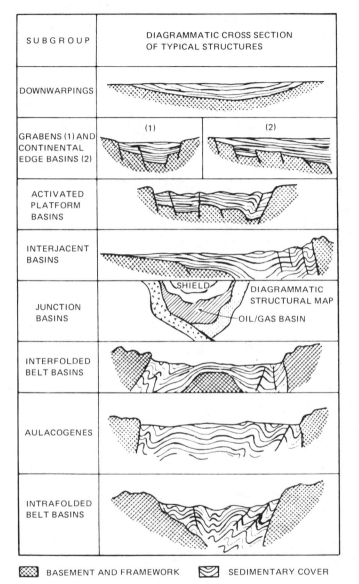

SUBGROUP	DIAGRAMMATIC CROSS SECTION OF TYPICAL STRUCTURES
DOWNWARPINGS	
GRABENS (1) AND CONTINENTAL EDGE BASINS (2)	(1) (2)
ACTIVATED PLATFORM BASINS	
INTERJACENT BASINS	
JUNCTION BASINS	SHIELD DIAGRAMMATIC STRUCTURAL MAP — OIL/GAS BASIN
INTERFOLDED BELT BASINS	
AULACOGENES	
INTRAFOLDED BELT BASINS	

▨ BASEMENT AND FRAMEWORK   ▽ SEDIMENTARY COVER

Fig. 1-4. Classification of sedimentary basins based on structural framework. [After Olenin (1967).]

geosyncline belt." It should be noted here that Olenin does not differentiate between the thicker flank of the sedimentary basin and the frontal flank of the geosyncline. Indeed the distinction between some sedimentary basins and some geosynclines is commonly tenuous and may hinge on semantics rather than on geological characteristics. The category referred to as junction basins includes examples of complex basins formed by the superposition of two interjacent basins. Olenin states that the resulting thickness of strata in the basin can exceed 18,000 m. This category is possibly superfluous if the two superimposed basins are considered separately. There are many examples of superimposed basins, including the Mesozoic Surat Basin which overlies part of the Permian Bowen Basin in Queensland, Australia, and such basins are not necessarily of the same type. For this reason Olenin's definition of junction basins appears to cover a special case. Another special category is that referred to as interfolded belt basins. The category is similar to that which includes activated platform basins; but whereas the latter are formed by intense graben-type downfaulting, the former includes examples where sedimentary basins have been interfolded within the basement. Such interfolding results in belts of metamorphosed sedimentary rocks including argillite, siliceous siltstone, quartzite, and crystalline dolomite. The category termed aulacogenes appears to be a special case within the category of interfolded belt basins and is characterized by a sedimentary basin in a long and narrow depression formed by major faults within a young, unstable platform adjacent to an active geosyncline. The sedimentary sequence in such basins may be very thick and strongly folded. The lowermost category includes intrafolded belt basins formed in fault-controlled depressions within belts of active folding. In this sense the category has some affinity with category (1) which includes basins formed in depressions caused by grabens. The difference lies in the stipulation that intrafolded belt basins are developed in active fold belts.

   With all of these categories it should be noted that the genesis of the sedimentary basin is implied, and care must be taken in formulating any classification based on structural framework that the categories are separated on the basis of structural and tectonic considerations. It is also important that the terms used to describe the categories are clearly defined.

   It is axiomatic that sedimentary basins develop where they do, not necessarily because they are filling topographic depressions but because the underlying basement is subsiding at a rate compatible with the rate of sedimentation and the degree of compaction of the sediments. To what extent the weight of the pile of sediments causes the basement to sink, commonly by means of faulting, is arguable. The fact is, the relationship obtains, signifying the importance of structural framework to the growth of sedimentary basins.

Tectonics

The relationships between major zones of crustal deformation or disruption and the origin, nature, and solid geometry of sedimentary basins are of considerable interest as they reveal not only the probable geological history of the earth's sedimentary layers but also of some belts of metamorphic rocks that lie within the Precambrian shields. These belts have at times been described as predominantly occupying the peripheral zones of the shields, and, geographically speaking, this relationship is probably valid. But it must be remembered that the Precambrian shields may extend under a cover of Paleozoic and Mesozoic rocks to the continental margins. In this respect it has been postulated that the shields themselves have grown by the accretion of sedimentary basins on their flanks and by the subsequent downwarping and metamorphism of these basins. Many hypotheses have been put forward to explain the mechanics of such a growth process. The commonly held view is that where continental masses have impinged during the Precambrian, the resulting subduction has caused buckling of the igneous massifs with consequent rapid erosion and deposition of sediments and volcanics in linear basins and subsequent downwarping of the basin to depths where the pressure–temperature conditions are sufficiently high to metamorphose the strata. Subduction during later periods has involved not only the igneous and metamorphic massifs but also the sedimentary and volcanic sequences of basins themselves.

It has been suggested by Anhaeusser *et al.* (1969) that the greenstone belts found within the Precambrian shields were formed very early in the earth's history by the outpouring of basic lava into linear depressions caused by tension fractures in the crust. Associated with these lavas there would probably be pyroclastic material. At some stage, when bodies of water were able to form, water-laid tuffs and sediments would also be deposited in these linear depressions which, when subsequently incorporated into the mass of a Precambrian shield, would constitute a greenstone belt. Some of these belts may prove to be the oldest volcano-sedimentary basins in existence.

The greenstone belts described above are characterized by a comparatively low degree of metamorphism which is compatible with their probable origin as infilling within a tension fracture zone of the crust. Belts of metamorphic rock within the Precambrian shields may, on the other hand, have originated as mobile belts of volcano-sedimentary basins. The linearity of these belts is not in itself a criterion of the tectono-depositional environments of the basins from which they are inferred to have developed, a point which must be remembered in constructing any classification based on basin shapes.

Classifications based on tectonic activity generally separate sedimentary basins on the basis of depositional environment. Intracratonic, epicontinental, and neritic environments are considered as relatively stable, whereas bathyal and abyssal environments are considered as relatively unstable. But there are so many reservations that this generality is hardly valid. For example, the thick deposits of alluvial fans formed in valleys in a basin and range topography shaped by active faulting can hardly be considered as having formed in a tectonically stable area. Even areas such as the Gulf of Mexico, regarded as tectonically stable, are subject to faulting and minor earth tremors. On the other hand, some abyssal environments are tectonically stable; although the deep trenches of the ocean floors are prime examples of major zones of dislocation in the crust and probably are loci for the accumulation of volcanic and sedimentary deposits that may in the future become mobile belts of metamorphic rock. Deposits formed in such environments are typically of the eugeosynclinal type. Flanking such deposits there may be thick sequences of neritic sediments of the miogeosyncline type. This juxtaposition is a classical relationship and is illustrated by Cobbing (1976) in Fig. 1-5.

The miogeosyncline to the east comprises about 6000 m of shale, sandstone, and limestone; the eugeosyncline comprises about 7000 m of water-laid andesitic tuff, pillow lava, and minor shale. Both geosynclines developed on a block-faulted basement of Precambrian crystalline rocks

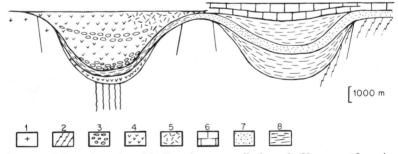

Fig. 1-5.   Schematic section of the Peruvian geosynclinal couple (Uppermost Jurassic and Cretaceous), showing the juxtaposition of a miogeosyncline comprising shales (8), sandstones (7), and limestones (6) with a eugeosyncline comprising subaerial pyroclastics (5), water-laid tuffs (4), and pillow lavas (3). Basement rock consists of Precambrian gneiss (1) and pre-Ordovician schist (2). [After Cobbing (1976).]

and pre-Ordovician schist. This geosynclinal couple flanks the continental margin beneath which is believed to be a zone of subduction. A similar case can be made for the Cretaceous sequence in Alberta and British Columbia, the Alberta miogeosyncline lying adjacent to somewhat fragmentary remnants of a sequence of sediments, tuffs, pyroclastics, and volcanic flows preserved along the flanks of the Coast Range batholiths.

Fig. 1-6 is a geological sketch map showing the geographical relationship of the geosynclinal couple to the Precambrian granitoid massif to the west, and to the pre-Ordovician schist belt and Ordovician to Devonian flysch trough to the east. The axis of the deep submarine trench trending parallel and in proximity to the Peruvian coastline is also shown. This trench is the surficial evidence of deep-seated dislocation of the earth's crust and is part of a system of similar trenches throughout the Pacific region. These trenches are flanked by volcanic arcs and coincide with the trends of earthquake epicenters. They are believed to be the surficial reflection of subsidence along zones of subduction where one continental mass is impinging upon another.

Classifications based on tectonic activity generally separate sedimentary basins formed in intracratonic, miogeosynclinal, and eugeosynclinal environments. Where sedimentary basins are linear and exceptionally thick, they may be referred to as geosynclines, and in such cases the terminology is arbitrary. Sedimentary basins formed in a eugeosynclinal environment are generally thicker than those formed in a miogeosynclinal environment, and intracratonic sedimentary basins are the thinnest. For example, intracratonic basins commonly have thicknesses of up to 2000 m, basins of the miogeosyncline type up to 8000 m, and basins of the eugeosyncline type up to 15,000 m. There are exceptions to this statement which is valid only on the basis of average thicknesses. Classifications based on tectonic activity also indicate a relationship with classifications based on lithology.

Intracratonic basins have formed from sediments deposited as alluvial fans in river valleys and in lakes and swamps. These deposits may comprise sediments which become boulder conglomerate, poorly sorted lithic sandstone, siltstone, and shale. Coal seams also form from accumulations of plant remains in lakes and swamps. The nature and volume of these various rock types may indicate the possible degree of tectonic activity that obtained during the period of deposition. For example, a section of boulder conglomerate showing fluvial features and having a thickness of 500 m or more suggests rapid accumulation, possibly on the upper part of an alluvial fan, by high-energy but possibly periodic flows of water. This type of deposit can form along the flanks of rapidly rising mountains, particularly in arid to semiarid regions subject to periodic rainfall.

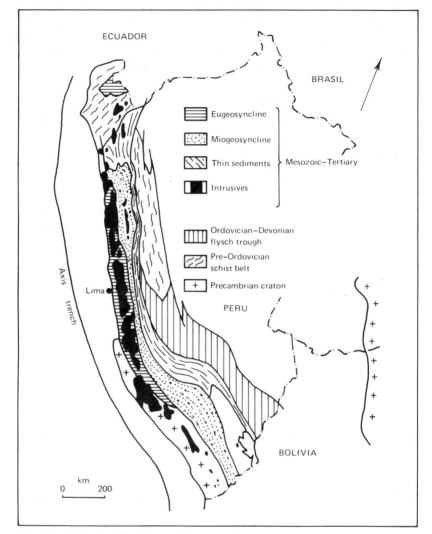

Fig. 1-6. Geological sketch map of the Peruvian geosynclinal couple (Uppermost Jurassic and Cretaceous), adjacent flysch trough (Ordovician to Devonian), and basement rock (Precambrian to pre-Ordovician). [After Cobbing (1976).]

Basins formed in a miogeosynclinal environment largely comprise sequences deposited in delta complexes, along interdeltaic coasts, and on continental shelves and slopes. These sediments consist predominantly of sand, silt, clay mud, and carbonate mud. Evaporite deposits of salts and gypsum may be associated with carbonate muds where these have been deposited on intertidal mud flats or in lagoons having the restricted circulation of a sabkha environment. Peaty deposits may also be deposited in coastal bays in association with muds and silts. These subsequently form coal seams interbedded with shales which may contain marine to brackish-water forams. Sand deposits along marine coastlines are commonly well sorted as they have been transported a long way by rivers and possibly undergone several stages of reworking. Furthermore, they have been subject to winnowing action in a high-energy beach environment of waves and currents. Such deposits tend to be quartzose, and some are almost entirely of quartz sand. These and other lithologic features, including rock types and their sequences, indicate a miogeosynclinal type of environment.

In contrast, the eugeosynclinal type of sedimentary basin is characterized by rapid accumulation of sediments and volcanics consisting commonly of graded-bedded lithic sand forming adjacent layers or interbedded with layers of silt and clay mud, of falls of volcanic ash and pyroclastics, and of flows of basic lava. These lithified sequences may include beds or lenses of chert which may have formed from collodal silica precipitated on the sea floor by volcanic effusions; but in many cases their origin is not clear. Pillow structures may also be observed in the basic lavas, a feature indicating their underwater origin on the sea floor.

In general, classifications of sedimentary basins based on tectonic activity are closely related to classifications based on shape, structural framework, and lithology. Cross reference from one to the other is essential in any study involving the analysis of a sedimentary basin. Also the parameters of movement of the earth's crust must always be borne in mind in dealing with the tectonic aspects of the development of sedimentary basins. In abyssal depths the rates of deposition in a period of 1000 years may range from 1 cm to 1 m or more, in a large and active delta the rate may be 10 m or more, and in an alluvial fan the rate may be considerably greater. In contrast, orogenic movements may be of the order of a few millimeters a year. But given an interval of 10 million years and a rate of uplift of 5 mm–1 cm in a thousand years, the total uplift would amount to 50–100 m. Over a period of 40 million years, the approximate time span of the Early Cretaceous, uplift would amount to 200–400 m. In some cases these estimates of rates may be grossly conservative. In any case it can reasonably be presumed that a stated rate of uplift is merely an estimate of an average rate of movement based on measurements taken over an extremely limited period of time in geological terms, or on somewhat subjective inferences based on stratigraphic and structural considerations. One is on somewhat firmer ground in calculating the horizontal movement of continental masses where, for example, they are moving apart from great fissures in the earth's crust such as are marked by the Atlantic Ridge. Measurements taken in Iceland during recent years appear to confirm an east–west drift of a few centimeters a year, and Steiner (1977) quotes rates of sea-floor spreading within the past 5 million years in the range 4–10 cm/yr.

The point to keep in mind is that classifications of sedimentary basins based on tectonic considerations such as rates of orogenic uplift cannot be quantitative and depend on relative terms such as active or moderate.

METHODS OF STUDY

The Basin Studies Group

Analysis of a sedimentary basin involves so many aspects that it is ideally carried out by a group of experts in various fields such as stratigraphy, paleontology, structural geology, and geophysics. Each applies his special knowledge to the problem as a whole, the project being coordinated in such a way that the group works as a team. In any such undertaking it is essential to state firmly the terms of reference, to determine the priorities involved, and to come to a decision on the scope and limitations of various aspects of the study. This procedure is normally followed in any such undertaking or research project as limitations of time and funds preclude a rambling or extended approach to the project. The people involved in such basin studies groups must keep themselves closely in touch with the work of their colleagues and must keep firmly in mind the purpose of the study. This may not be easy to do in that interesting sidelines of research, some of which may prove important, often turn up and lead the researcher along a path that does not essentially contribute to the main object of the study. Set aside for the time being, such subsidiary avenues of research can be later followed up, particularly in cases where the research worker is free to follow where the path may lead. But as a member of a team, his inquiries must be somewhat circumscribed, and to this extent the personality of the worker may be as much involved as his expertise.

Analysis implies the dissection of the whole into its parts, in other words, the division of the complete stratigraphic sequence in the sedimentary basin into its lithologic, time-stratigraphic, biostratigraphic, and ecologic units. It further implies an understanding of depositional and climatic environments, of sedimentologic conditions that obtained during the formation of each unit, and of the paleogeographic relationships of the units. It also implies an understanding of the influence exerted by tectonism on the genesis of the basin, on the structural framework that existed during the formation of the basin, and on the present structural configuration of the basin.

The selection of a sedimentary basin for study normally involves some rationale related to the requirements of the organization employing the study group. This organization may be a research unit within a university, a geological survey, or a company engaged in exploration for hydrocarbons, coal, or minerals. Having selected a basin, a decision must be made as to the targets within the basin, that is to say, the stratigraphic units of most interest with reference to the purpose of the study. Investigations will be focused on these units but not necessarily to the exclusion of other units or of the geological history of the basin as a whole. In fact, it is essential that studies of particular stratigraphic units not be undertaken out of context with the geological record of the entire basin, although priorities and limitations of time may preclude the undertaking of more comprehensive studies.

The collation and interpretation of the data gathered should be a continuing process, otherwise the sheer mass of information may prove to be somewhat daunting. In this respect it is necessary for every member of the study group to always keep in mind the purpose of the study so that he will not become so lost in details that he cannot see the woods for the trees. Not all the findings will fall into place along the way, and some may tend to be misleading until further information becomes available. But the construction of an interpretation by the assemblage of blocks of information from all members of the study group, and by the continual review of this assemblage, is critical to the orderly expedition of sedimentary basin analyses.

Having determined the organization of the basin studies group, selected the personnel, and decided on the basin or stratigraphic interval within the basin to be studied, the next task is to proceed with the study. This must be done in an orderly and systematic way in order to be most effective. The procedure must also be sufficiently flexible to accommodate the inclusion of additional personnel in the group and of new information or other methods of approach. The first prerequisite for the undertaking of a research or review project is the assimilation of information relating to the project. This requires the collation of all available published and unpublished findings, some of which will be pertinent, some not relevant. The gathering and sifting of information can be a daunting task, but it is essential to the preliminary planning of a study. Other workers may have carried out similar investigations, and it is well to know what they thought. This is not to say that the same ground cannot be gone over again. Seen with different eyes and possibly with additional information, the previous interpretation will probably be modified. For although the data used in a study are accepted as valid and the methods of employment of this data are scientific, the interpretations arrived at by different geologists may not be the same. There is inevitably a certain element of subjective and imaginative thinking in the formulation of interpretations, and were it not so, the earth sciences would stagnate and the discovery of hydrocarbon and mineral deposits would be retarded. Be this as it may, a basin studies group must be apprised of the available information relating to a project before it is undertaken. Unfortunately, because many companies employ such study groups and because the reports issued by these groups are considered to be confidential by the companies, a great deal of repetitious work has been done without the benefit of knowledge derived from previous studies.

This is particularly frustrating to academic and scientific workers in universities and geological surveys who must publish their work with the knowledge that other professional geologists in industry may be better informed on the subject than themselves. Eventually, ideas and concepts become generally known, in part because the companies think it in their interest to release information or because the information is no longer of particular economic value to a competitor and in part because geological personnel move about from industry to academic and scientific institutions.

Having gathered and digested the available published and unpublished geological and geophysical information relating to a project, the next step is to decide how the study will be carried out in detail and the particular aspects of coordination that will be necessary. Geologists specializing in lithostratigraphic aspects of the study will work closely with geophysicists, on the one hand, and paleontologists, on the other. They will also coordinate their work with that of geochemists and mineralogists engaged in age-dating, trace element or clay mineral studies, and with that of petroleum engineers evaluating the petrophysical characteristics and possible economic potential of particular stratigraphic zones. Frequent consultation between individuals and with the basin studies group as a whole is essential to the effective conduct of the study. In this regard mention should be made of the limitations of time. Usually one never has enough time to undertake a study and complete a report entirely to one's satisfaction. Most projects must be completed within a definite period of time, although some may be subject to extensions. Time must be carefully budgeted and the progress of a study watched and adjusted regularly. This applies not only to projects undertaken by a basin studies group but to projects carried out by individuals who must not only organize and coordinate their own work but must apply the necessary self-discipline to carry it out.

Having embarked on a study, the basin studies group should preferably proceed without interruptions resulting from other assignments. It is important that the continuity of thought be maintained. Discontinuities in the course of a study commonly result in a tendency to fragmentation in the presentation of a final report which should comprise contributions from all members of the group. In the event of a significant difference of interpretation between members of the group, the alternative opinion should be stated either in the body of the report or in a footnote. In most cases such differences are resolved by consultation prior to the final preparation of the report. In this regard it is timely to suggest that the report have a particular theme rather than consist of a loose assemblage of separate reports on various aspects of the study. This theme should relate to the terms of reference of the study. It is also advisable to clearly and concisely summarize the text and illustrative material in a form that can be easily and quickly understood by others who do not have the time to assimilate the bulk of the report. Although this should be self-evident to those who prepare such reports, in many examples the writers are essentially addressing others of similar expertise and either do not include in the report a layman's summary or fail to adequately express themselves in writing it. In these days of specialization it is important that earth scientists in various fields communicate their ideas and findings in terms that are meaningful and unequivocal to other earth scientists who may themselves be laymen in certain specialized fields of geology, geochemistry, or geophysics.

The final report can take a number of forms which possibly include several appendixes containing the bulk of data and information of particular relevance to special fields. The salient features may, if time allows, be presented in the form of an atlas containing maps, sections, and diagrams.

Selection of a Stratigraphic Objective

Most studies of sedimentary basins are undertaken with a particular stratigraphic objective in mind. The objective may be of economic or of purely scientific significance. It may be of considerable interest in both respects. Whatever the emphasis placed on this objective, it must be related to the stratigraphic sequence within the basin and preferably to some particular stratigraphic sequence within the basin and preferably to some particular stratigraphic unit that can serve as a key zone in reconstructing the geological history of the basin. In other words, although the major efforts of the study are concentrated on a particular stratigraphic unit or feature, this must not be considered out of context with the stratigraphic sequence and geological history as a whole. It is necessary to establish a system of regional and local stratigraphic correlations in order to formulate an overall view of the geological events that have shaped or influenced the development of the basin. This must be regarded as a prerequisite for the undertaking of studies aimed specifically at particular stratigraphic units.

In order to construct the basic framework of a basin, it is necessary to construct a series of master cross sections. These usually include both structural and stratigraphic sections not only of the total section within the basin but also of selected stratigraphic intervals within the section. Obviously, the construction and accuracy of such sections depend on the quantity, geographical distribution, reliability, and availability of both surface and subsurface data. In this respect the element of time may be of prime consideration, and the methods employed may depend on the information readily available. In cases where information is not readily available, a decision must be made as to whether the possible or probable usefulness of the information warrants the expenditure of effort and time necessary to obtain it. This is not an easy decision to make and one which depends largely on the exigencies at the time.

In cases where the stratigraphic objective has been determined on the basis of economic or scientific interest, it is necessary to select a stratigraphic unit, above or below the objective, that can be established as a key unit for purposes of correlation. This unit should be related in time to the objective as closely as possible; in other words, the stratigraphic interval between the objective and the key unit should be minimal. It must not be separated from the objective by an unconformity either but must preferably be part of the sedimentary sequence within a major geological event. In some basins it may be possible to relate the objective to a widespread unit such as a bentonitic bed that formed as an air-fall bed of volcanic ash. Such a unit, deposited in the same short interval of time throughout its entire distribution, represents a time-marker and as such is an excellent datum for the construction of stratigraphic sections. Other thin stratigraphic beds may be widespread (a necessary feature in selecting a unit on which to hinge the stratigraphic framework of a basin) but diachronous. This states in general terms the rationale behind the methods of study employed; but it must be emphasized that the selection of key stratigraphic units and stratigraphic data is a matter of judgment and very much depends on the nature of the sedimentary basin and the availability of data.

It is emphasized that the study of individual stratigraphic zones must be seen in context with the geological history of the basin as a whole. The preexisting tectonic setting, structural framework, paleogeographic relationships, and sedimentary environments are all related to and influence the subsequent geological history of the basin. With this course of events in mind, it would appear sensible to begin a study of any sedimentary basin by looking at the basement and overlying strata. The basement is commonly the prime mover in geological events, such as beveling at unconformities, deformation of strata, localization of subbasins, and depositional thinning that occurs from time to time throughout the period of deposition of the entire stratigraphic section within the basin. Information on present-day basement structure is derived mainly from seismic surveys which have been of inestimable value in understanding the structural configuration of basins situated, in particular, within the block-faulted crust of continental shelves.

Although it is preferable to begin the study of any sedimentary basin by investigating the nature and structure of the basement and of the beds immediately overlying the basement, it is not always possible to do so because of lack of information or time. Seismic data may not be available, and subsurface information may be from scattered boreholes that have not penetrated the section to basement. Surface investigations may be profitable provided the outcropping sections overlie basement, can be correlated, and have continuity in the subsurface. All of these factors may in varying degrees preclude the study from beginning with an investigation of the basal section of the basin. But where sufficient information is available to construct cross sections of a sedimentary basin from the basement upward, much can be inferred of the basin's early history by constructing stratigraphic sections, unit by unit. Such constructions depend on the selection of a stratigraphic datum plane within each unit or on the use of either the upper surface in the case of a shale bed or the lower surface in the case of a bioherm, as a datum plane. A further assumption may be made as to the probable slope or undulating configuration of a datum plane, or it may be illustrated in sections as the trace of a horizontal plane. This technique of reconstructing in sectional view the probable geometry of successive units comprising the stratigraphic sequence can be applied to any unit within the section, without reference to basement or the basal section of the basin. But the growth of the stratigraphic sequence, the influence of compaction on the shape of buried layers, the effect of underlying layers on the deposition and consequent geometry of overlying layers, and the possible movement of basement blocks to accommodate deposition of variable thicknesses in the sedimentary pile are all demonstrated with most effect where they are shown in upward sequence from the basement.

Most studies of sedimentary basins are undertaken with a view to assisting directly or indirectly in the furthering of knowledge about a particular section within the basin. The interest in such a section commonly derives from a pragmatic approach to exploration for hydrocarbons, coal, or minerals. The study may be oriented specifically to a particular exploration project, or it may be designed to furnish a general background of information for use as reference by those employed in more specific investigations. In all cases the planning of a basin study requires the selection of a stratigraphic objective that may include a sequence of several units or may be confined to a particular unit. The selection of this objective is not a random process, and although it implies the requirement to gain as much information as possible within the bounds of constraints applied to the study, it also implies a preconceived conception of the possible application of information to other investigations such as exploration programs. In other words, where the basin study leads to interpretations that may give a clue to the possible origin, distribution, or location of hydrocarbon or other deposits, it should be clearly stated and demonstrated by means of maps or sections. This is not to imply that such interpretations should be forced on the basis of insufficient information, or information taken out of context, but rather that the investigator should think positively about the purpose for which the study was undertaken and in particular the reasons for selecting the stratigraphic objective. Otherwise the final report on the study may consist of a mass of information all of which is relevant but poorly related in the text and visual presentation of the report.

CHAPTER

2

Evolution and Development

CONCEPTS

In the beginning of the earth's sedimentary history, when the outer crust consisted entirely of volcanics, sediments must have accumulated as the products of erosion in topographic depressions and along the shores of the primeval oceans. The sequence of events that gave rise to the origin of the oceans and that shaped the development of landmasses can only be surmised, but it seems not unreasonable to assume, as Stewart (1972) has done, that during the Precambrian zones of tension within the crust produced widening fractures which were filled along their margins with wedges of sediment. A similar concept was expressed by Anhaeusser *et al.* (1969) to explain the origin of greenstone belts within Precambrian shields. Both concepts conceive the development of linear depressions in the crust formed by tensional rifts. In one case the margins of the depression are filled with sediment of igneous origin; in the other case the depression itself is filled with basaltic and andesitic lavas. It is probable that both sedimentation and volcanic extrusion of basic lava flows occurred simultaneously. The Precambrian occupies most of the earth's history, and whereas one may generalize on the tectonic, sedimentary, and volcanic history of the Late Precambrian, one can only guess as to the events that took place during the preceding interval of perhaps 2000 million years

37

when sedimentation was taking place. The history of these early sedimentary events has long since vanished, the early accumulations of sediments having been eroded, metamorphosed to gneiss and schist, or completely assimilated into the igneous massif. The early sediments, derived entirely or in large part from basic volcanic rocks, would also become metamorphosed to greenstones, probably indistinguishable from greenstones formed by the alteration of basic volcanics. Such greenstone belts may be the earliest record we have of sedimentation. Later in the earth's history, as granitic rocks became exposed through orogeny along belts of batholithic intrusions, quartzose and feldspathic detritus was added to the predominantly ferromagnesian content of the sediments. Subsequently, reworking of the sedimentary deposits themselves occurred.

The concept of Stewart (1972) with reference to the origin of the earliest sedimentary deposits in the Cordilleran Geosyncline of western North America is illustrated by Figs. 2-1 and 2-2. These Late Precambrian to

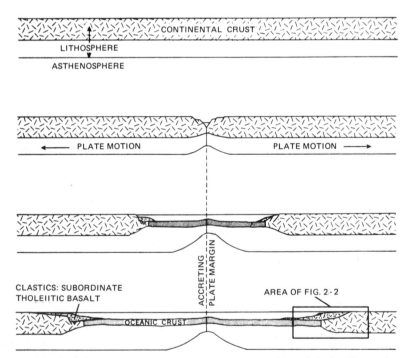

Fig. 2-1. Diagram showing inferred Late Precambrian rifting of the earth's continental crust, the infilling of basic lavas to form an oceanic crust, and the development of wedges of clastic sediments and basalt flows (Fig. 2-2) along the margins of the rift. [After Stewart (1972).]

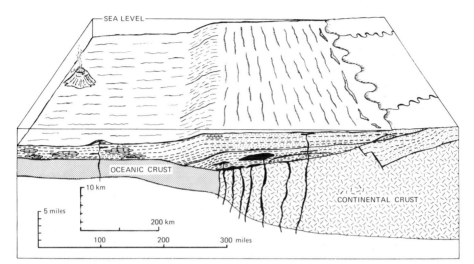

Fig. 2-2. Diagram showing a generalized section of Late Precambrian to Early Cambrian strata in the Cordilleran Geosyncline, Nevada and Utah, and a paleogeographic reconstruction illustrating depositional and tectonic relationships. See Fig. 2-1 for location. [After Stewart (1972).]

Early Cambrian deposits form a narrow belt extending from Alaska to Mexico. The belt comprises up to 8000 m of predominantly siltstone, shale, argillite, quartzite, and conglomerate, with minor beds of chert in what is presumed to be a deeper water facies. The basal unit is commonly a conglomeratic mudstone, interpreted as a diamictite, and the lower part of the entire sequence contains intercalated flows of basalt interpreted as tholeiitic. The upper half of the sequence, as shown in Fig. 2-2, appears to be a molasse facies forming a thickening wedge from the landmass and grading seaward into a deeper water possibly flysch facies.

This Cordilleran belt of Late Precambrian rocks extends for approximately 4000 km, and Stewart's concept consequently implies a long, somewhat sinuous coastline. Deposition along this coastline may not have been continuous, nor at a constant rate from place to place. Furthermore, it is likely that sections representing part of the paralic and much of the terrestrial record were removed by erosion prior to the Cambrian, although some parts of this record, possibly including the upper part of the Beltian, appear to have been preserved.

It is not until the Paleozoic that we have stratigraphic sequences which exhibit features of sedimentary environments and processes identifiable with depositional sequences and sedimentary structures seen in geomorphologic features of basins forming today. Structures such as grooved pavement indicating glacial activity, varved siltstones and claystones indicating lacustrine and possible glacial lakes, algallike structures in limestones suggesting a carbonate mud flat environment, varicolored thin-bedded siltstones and shales suggesting deposition on river floodplains, and other examples are present in Late Precambrian rocks; but the record of sedimentary basin history during the Precambrian is fragmentary and incomplete.

With some exceptions, the best and most complete records of depositional environment and development of sedimentary basins are found in Mesozoic and Cenozoic sequences, particularly in the Tertiary. This is to be expected as the younger sequences have commonly, but by no means consistently, been less affected by metamorphism, erosion, or structural dislocation. Studies of present-day sedimentary processes, geomorphologic features, distribution of sedimentary, biological, and ecological facies, rates of sedimentation and compaction, growth and geometry of strata, and of the internal and external structures that feature a growing accumulation of strata such as underlie a delta, can readily be applied to the study of many Mesozoic and Cenozoic sedimentary basins, particularly with reference to the interpretation of depositional environments. Interpretations with respect to the stratigraphic framework of a basin depend on the preservation of sequences representing various cross sections of the basin and also on reconstruction of the original shape of the basin, a process that may entail palinspastic adjustment of folded and faulted strata. Interpretations with respect to the structural framework of a basin commonly depend on relationships mapped in the field and on seismic surveys of strata in the subsurface. Tectonic control can in places be demonstrated, particularly where the trends of earthquake epicenters and volcanic activity coincide with depositional and structural trends such as submarine trenches, where known and active faults result in rapidly growing alluvial fans, or in the subsidence of a deltaic sequence. In other places the influence of tectonics may only be inferred.

The growth of some sedimentary basins is also influenced by growth features which may be organic or inorganic, such as bioherms, submarine volcanic cinder cones, and salt domes. The internal configuration of a basin may also be influenced by solution features where thick layers of salt or other evaporite have been removed, causing collapse of the overlying layers and disruption or deformation of the internal structural configuration of the basin. In addition, if solution within the stratigraphic section is penecontemporaneous with deposition at the top of the section, subsidence will have some influence on the paleogeography and consequently on the depositional patterns reflected as lithofacies of the strata within the basin.

Subsequent uplift of a sedimentary basin. by block faulting or orogenic buckling and folding, results in erosion and beveling of the basin edges. As a depositional unit the sedimentary basin becomes incomplete and presents only part of its history, commonly that part revealed by the thicker sections of the basin. These may be entirely marine, whereas the thinner sections removed by erosion may have been of paralic or terrestrial origin. Other stratigraphic relationships may also obtain. The point is that sedimentary basins are seldom entirely what they seem to be; they commonly present only certain facets of their original constitution, key sections having disappeared by erosion of the basin's margins. It is consequently of importance, in reconstructing the original configuration of a sedimentary basin by means of maps or sections, to clearly define erosional and possible or inferred depositional edges.

This involves a concept of the present and possible original solid geometry of a basin, which hinges on all of the previous considerations. Hopefully, at the conclusion of a basin study, one can define the three-dimensional shape of the basin, demonstrate its internal structure and stratigraphic arrangement of beds, show the facies relationships layer by layer, and draw certain inferences as to the paleogeographic setting, depositional environments, and tectonic background in which the basin developed.

Depositional Environments and Stratigraphic Framework

Depositional environments comprise the overall influence of geographical situation, physical processes, and biological factors that obtain in an area of sedimentation. Obviously, there are numerous facets to each of these factors that can be taken into account, with varying degrees of significance, in constructing any classification of depositional environments. Concepts inherent in the classification of sedimentary environments must be clearly sorted out, particularly with regard to geographical scale. For example, in defining a deltaic environment, a distinction must be made between the comparatively small area of deposition of a river flowing into a lake and comparatively very large areas, including floodplains, river distributaries, coastal marsh, bays, sand bars, and intertidal mud flats such as the delta complex of the Mississippi River. Within the area of such a delta complex, distributary branches flowing into bays may deposit fans of sediment referred to as deltas. These may exhibit, where cored, the typical textbook illustrations of topset, foreset, and bottomset bedding which are commonly thought to characterize the internal structure of deltas. The very large deltas, such as the Mississippi Delta, also show an internal structure formed by prograding bodies of sediment. In such great deltas the sediments deposited on the continental shelf may be compared to topset beds, whereas those deposited on the continental slope may be compared to foreset beds. To carry the comparison further, although it would only be valid in a limited number of deltas, the sediments deposited on the continental rise, seaward from the base of the continental slope, may be compared to bottomset beds.

Other parameters that must be examined in the context of concepts of depositional environment include those that come under the general heading of physical processes. For example, the category defined as a high-energy environment is commonly thought of as characterizing a paralic environment where beaches of sand and pebbles are subject to strong current and wave action. Some beaches, for example those of Ghana and Nigeria which are subject to large tidal fluctuation, are constantly pounded by great breakers, whereas in areas of low tidal fluctuation, particularly where the sea floor slopes gently seaward for many kilometers, the beaches are winnowed by comparatively gentle wave and current action. Also, the category may equally apply to some terrestrial environments where chaotic mud flows or imbricated cobble and gravel deposits are formed in alluvial fans. It may also apply to deep-sea deposits of turbidites which form fans at the base of submarine canyons. In some case the physical processes produced by high-energy conditions are constant, in others they are periodic, although they may be frequent.

The influence of biological factors on depositional environments is purposely excluded from the scope of this book, as previously explained. It is an important, indeed in some cases a prime factor in the development of an environment in which a particular set of sedimentological conditions may operate to produce a particular sequence of sedimentary beds. A good example is the development of a chain of organic reefs which may tend to cut off circulation of seawater from a large body of shallow water lying between the reef chain and extensive carbonate mud flats. The resulting higher salinity in the enclosed body of seawater will contribute to the deposition of magnesia-rich carbonate muds and gypsum and affect the previously existing ecological balance. Certain fauna may disappear, whereas algal growth, preserved in the sediments and subsequent limestone as wavy laminations or stromatolitic-type structures, may become much more abundant. In areas such as the carbonate platform of the Bahamas, where tens of thousands of square kilometers are overlain by less than 10 m of water, the sedimentological processes forming the carbonate sands and muds are directly or indirectly influenced by the chain of reefs that fringe the edge of the platform. This chain of reefs and its predecessors have in large part been responsible for the continuation, over a period of time extending back into the Tertiary, of the depositional environment we see today. The result has been the buildup of many thousands of meters of limestone deposited by physical and chemical processes in depositional environments similar to those now observed in the Bahamas.

Concepts of depositional environments must also be defined in terms of regional environments, local environments, and microenvironments. The boundaries between these categories are somewhat arbitrary and not always easy to define, but should where possible be stated. Statements of definition will hinge primarily on geographic situation but will refer also to physical processes and biological factors.

Geographical situation relates to the classification of depositional environments outlined in Chapter 1, in which the intracratonicepicontinental, neritic–bathyal, and abyssal environments are discussed. These categories define geographical situations ranging from within a continent (intracratonic), to the continental shelf and slope (neritic–bathyal), to oceanic trenches and plains (abyssal). They also define ranges of depth below sea level, with the exception of the intracratonic category which has no reference to depth but implies deposition in a body of water either above sea level or, in the case of an inland sea, at depths comparable to a shallow neritic environment. These categories of depositional environments not only relate to geographical situations and ranges of depth below sea level but also have a regional implication. Without quantifying the definitions, they suggest deposition over a wide area and imply a regional depositional environment. An example of a regional environment is the lower portion of the Mississippi Delta, an area of river distributaries, bays, and coastal marsh covering several thousand square kilometers.

Local environments can be defined as those in which deposition of sediments is related to a particular geomorphic feature such as a river system, lake, or a chain of barrier islands. Each of these features has its own facies. For example, within a river system deposition takes place in point bars, on floodplains, and in the backswamp areas of the system; in lakes the deposits of mud and sand may be fringed by swamp; and along a chain of barrier islands the seaward side forms a sloping beach of sand, whereas the lagoon side may be a coastal marsh or intertidal mud flat. Each of these facies may, in turn, be referred to as a microenvironment from a sedimentological point of view, although from a biological point of view the division may be finer still.

Considered on a regional geographical basis, keeping in mind the implications and limitations previously discussed, depositional environments are broadly classified with reference to their position on or adjacent to the continental crust. Intracratonic environments are within a continent, neritic–bathyal environments are on the fringe of a continent, and abyssal environments may be either on the fringe of the continental crust or overlying the oceanic crust.

Fig. 2-3 illustrates an example of an Early Proterozoic intracratonic sedimentary basin that developed on the Archean basement of Western Australia. To the north the depositional edge has long since been eroded, but the remnant of the Hamersley Basin as seen today covers approximately 100,000 square kilometers. The sedimentary sequence within the basin contains carbonates and chert beds, exhibits cyclic sedimentation, and is noted for its cherty and banded iron formation described by Trendall (1973). The sequence, which is about 2000 million years old, has been folded and metamorphosed. Its depositional environment is problematical, but it is significant that the sequence lies directly on Archean basement which suggests that deposition occurred within an inland body of water which may have formed a restricted embayment of the sea. Some caution is needed in the interpretation of depositional environments of

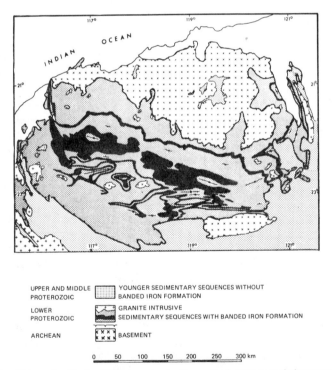

Fig. 2-3. Hamersley Basin of Western Australia, an Early Proterozoic intracratonic basin resting on Archean basement and noted for its banded iron formation. [After Trendall, (1973).]

Precambrian sediments because the conditions of atmosphere and ocean salinity that obtain today possibly did not exist so far back in the earth's history. Physical and chemical principles remain the same, but physical and chemical processes with respect to erosion and sedimentation are subject to change. One can only surmise the conditions under which these Proterozoic beds were laid down. Since then more than twice the interval of time spanning the Cambrian to the present has elapsed.

The preservation of such ancient sedimentary basins lying directly on Archean basement is uncommon. Uplifted parts of the Archean shown by de Villiers and Simpson (1974) to form cratonic nuclei (Fig. 2-4) within the continental masses have periodically been uplifted and eroded, resulting in the removal of older sedimentary sequences. The erosional products have, from time to time and during a period of many hundreds of millions

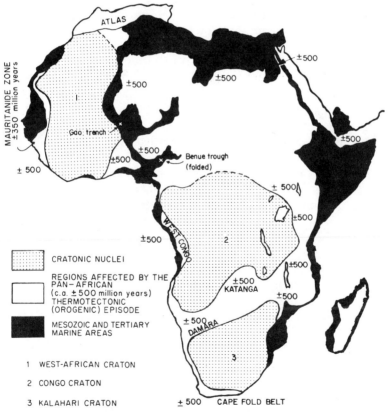

Fig. 2-4. Distribution of African Mesozoic and Tertiary sedimentary basins in relation to cratonic nuclei of the Precambrian basement and to areas affected by Proterozoic–Early Paleozoic orogeny. [After de Villiers and Simpson (1974) and Kennedy (1964).]

of years, been transported to and deposited along the exposed edges of these cratonic nuclei. The process has not been continuous, as periodic transgressions of the sea have laid down sedimentary sequences on the Archean shields; but the overall trend of the cratonic nuclei has been one of uplift with attendant accretion of successive sedimentary basins or remnants of basins along their periphery. This accretion has resulted in the deep burial and metamorphism of older basins which have become metamorphic belts of schist, gneiss, and granitized rock incorporated in the Archean shield which thereby grows marginally.

The distribution of cratonic nuclei in Africa is shown in Fig. 2-4. Also shown is the distribution of Mesozoic and Tertiary sedimentary basins formed around the edges of the uplifted areas, the erosion of which has contributed to the load of sediment deposited in these basins. On a mega-scale of continental dimensions the figure illustrates the geographical relationships of the younger basins to the areas underlain by older deformed basins and to the uplifted areas or cratons of the Precambrian shield. These younger marine basins deposited on the margins of the continental mass are in large part of epicontinental to neritic origin.

In this context, an example of a regional depositional environment comprising epicontinental and neritic features is the delta complex of the Ganges–Brahmaputra river system which occupies a large part of Bangladesh. Much of the sediment forming this delta has been derived from the Himalayan orogenic belt to the north during a period of uplift and erosion extending from the Miocene to the Recent. The delta complex, which has numerous distributary mouths emptying into the Bay of Bengal along a coastal stretch of approximately 400 km, is presently prograding on to a continental shelf that extends east-west for 500 km and seaward to the south for 200 km. This is one of the largest deltas in the world, and, like the Mississippi Delta, it has been growing since well back in the Tertiary. Subsurface studies of such deltas show that the present depositional environments are in large part a recapitulation of the previous history of the delta and that insofar as the Tertiary is concerned, the present is very much the key to the past. Present-day studies of depositional environments can certainly be applied to the Tertiary and with a reasonable degree of certainity to the Mesozoic. They can also be applied with a fair degree of confidence to Paleozoic strata extending back in age to the Devonian. An example is the considerable accumulation of information published on the Devonian carbonate reefs of North America, based in large part on studies of carbonate deposition and reef growth in the Bahamas, the Yucatan area of Mexico, the Persian Gulf, the Great Barrier Reef of Australia, and in coral atolls of the Pacific. But interpretations of depositional environments arising from studies of older Paleozoic and of

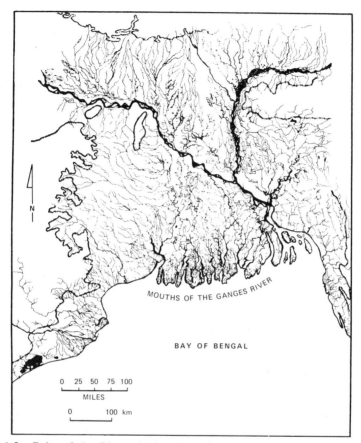

Fig. 2-5. Delta of the Ganges–Brahmaputra river system, Bangladesh (Ganges and Mahandi deltas).

Proterozoic strata become more problematical. Sedimentary structures and other features are commonly not as well preserved as in the younger beds. Also one cannot be certain that on a quantitative basis sedimentary processes operated as they do today or as they appear to have done during the Tertiary. These processes are reflected in the nature, thickness, and distribution of strata within the sedimentary framework of a delta, an example of which is given later with reference to the Ganges–Brahmaputra Delta.

Another large delta that has not been formed from the erosional material of an orogenic belt but from the erosion of a comparatively stable area of Precambrian and Paleozoic rocks (Fig. 2-4) is the Niger Delta shown in Fig. 2-6. This delta has been growing since the Late Cretaceous and is

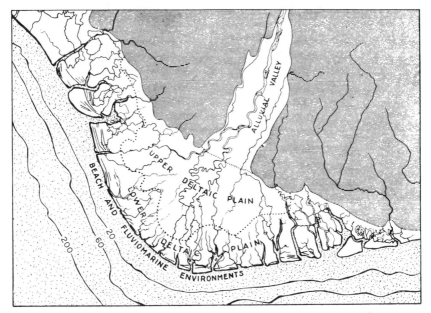

Fig. 2-6. Delta of the Niger river system, Nigeria. Bathymetric contours in meters.

prograding on to a narrow continental shelf. The depositional environment of the lower coastal plain is, like that of the Ganges–Brahmaputra environment, characterized by a coastline indented with estuaries and lagoons. A marked difference, and one that arises from the high tidal variations and very strong wave action along the Nigerian coast, is the truncated configuration of the Niger Delta marked by straight and commonly steeply dipping beaches. Internally, as shown in studies by Weber (1971), the delta exhibits a complex of sand bodies developed in a cyclical sequence of offlapping sedimentary beds, grading upward from marine clays to fluviomarine, interlaminated silts and sands overlain by barrier bar and distributary channel–fill sands. Each cycle, ranging in thickness up to 100 m, is terminated by a marine transgression which eroded part of the underlying sequence, leaving a thin layer of fossiliferous and glauconitic coarse sand. These transgressions of the sea may not have resulted from changes in the absolute sea level, or from downward movements by faulting of parts of the delta itself, but more probably by local compaction in areas deprived of sediment as the result of lateral shifts in the main river course. Similar events have been noted on the surface and recorded in sequences of subsurface beds in the Mississippi Delta. Where lateral shifts of a distributary cut off the supply of sediment to an area, that area begins to sink by compaction. The sea invades the coastal marshland which is extremely flat and barely above sea level, and the result is a drowned topography. Subsequent shifts of the river system may cause a main distributary to again debouch its load of sediment in the drowned area, producing a prograding wedge or fan of sediment which once more raises the area above sea level. This concept explains in part the complexity and relationships of sedimentary facies developed in local sedimentary environments of a delta complex.

As a delta continues to prograde seaward, any particular location marked by a borehole will commonly migrate landward with respect to its geographical situation on the surface of the delta. But if faulting occurs along a trend between the location and the landmass, then the location will sink and, with respect to facies indicated by depth of water, will migrate seaward. Faulting in such cases may be of tectonic origin and have its source in the basement rock. On the other hand, in great piles of sediment underlying the continental slope, it may arise within the pile of sediment itself as a result of gravitational slumping. These slumps occur periodically during the growth of the pile of sediments and are known as growth faults. They have an important effect not only on the facies within the sedimentary pile but also on the growth of fold structures and the consequent development of potential reservoir traps for oil and gas.

An example of a submarine plateau which is thought to be a downfaulted part of the continental shelf is recorded by von der Borch (1967) in sections C–D and E–F of Figs. 2-7 and 2-8. It lies off the coast of South Australia and Victoria and is probably the result of fault block movements within the Precambrian basement which is overlain over much of the area of the Great Australian Bight by Mesozoic arenites and lutites, and by Cenozoic limestones having a cumulative thickness of less than 1500 m. According to van der Borch, the plateau has acted as a local sediment trap, preventing the prograding Tertiary shelf sediments from slumping down to form submarine fans at the base of the continental slope. This area has not been drilled, but if the plateau described by van der Borch is indeed a sunken part of the continental shelf, then from top to bottom the sedimentary sequence formed during the period of downward movement should show facies variations reflecting a change in the depositional environment from deeper to shallower water provided the rate of sinking exceeds the rate of deposition.

Fig. 2-7 also shows clusters of submarine canyons cut into the continental slope, and section G–H of Fig. 2-8 shows the relative depths of the canyons in one cluster. The origin of these canyons is problematical; it seems likely that they are not drowned topographic features but have been formed under the sea. They serve as chutes for the runoff of sediment from the upper slopes of the continental shelf and terminate in broad fans

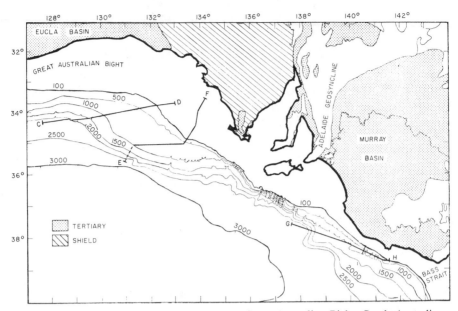

Fig. 2-7. Bathymetry of the Region of the Great Australian Bight, South Australia, showing submarine canyons and plateaus. Depths in fathoms. [After von der Borch (1967).]

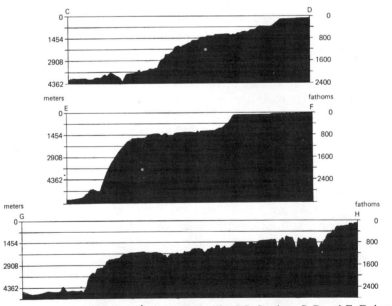

Fig. 2-8. Depth profiles of sections shown in Fig. 2-7. Sections C–D and E–F show a submarine plateau, believed to be a downfaulted part of the continental shelf. Section G–H shows the relative depths of a cluster of submarine canyons. [After von der Borch (1967).]

of sediment which spread out on the relatively flat portions of the sea floor
in abyssal depths at the base of the canyons.

A very much longer canyon, also of problematical origin, is shown in
Fig. 2-9. This is the Northwest Atlantic Mid-Ocean Canyon described by
Heezen *et al.* (1969). The canyon does not have any major tributaries but
extends in an arc southward from the northern tip of Labrador to south of

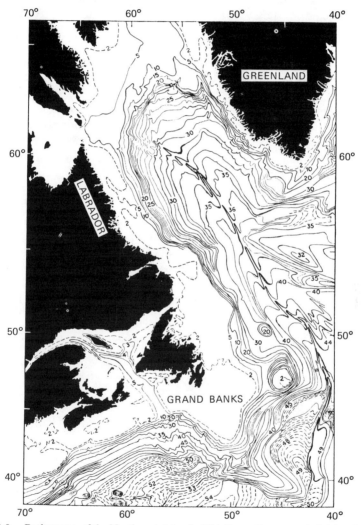

Fig. 2-9. Bathymetry of the Northwest Atlantic Mid-Ocean Canyon and vicinity showing
the arcuate canyon extending southward from the northern tip of Labrador to east of the
Grand Banks. Contour interval is 500 m. [After Heezen *et al.* (1969).]

the Southeast Newfoundland Ridge. The submarine depression in which it is enclosed is flanked to the northwest and west by the broad continental shelf of North America, of which the Grand Banks forms a notable part. To the west of the Grand Banks the ancient and submerged valley of the ancestral St. Lawrence River cuts across the shelf to the edge of the continental slope. The bathymetry of the area suggests that sediment is transported from the mouth of this valley down the slope to build a complex of submarine fans. It is believed that the Northwest Atlantic Mid-Ocean Canyon was formed during the Quaternary and that it is not a tectonic feature. But the sedimentary processes that shaped this giant canyon have not been established, and the canyon is tentatively referred to by Heezen *et al.* (1969) as a depositional–erosional linear feature.

Another type of linear feature developed in the abyssal environment and forming the deepest known depressions of the ocean floor is the submarine trench. These features are commonly flanked by arcs of volcanoes, are known to overlie trends of earthquake epicenters, and are believed to reflect zones of subduction along trends of contact between impinging plates of the earth's crust. As such, submarine trenches are of tectonic origin. An example, illustrated in Fig. 2-10, shows part of the Middle America Trench flanking the Pacific coast of El Salvador in Central America. This trench, according to Fisher (1961), is continuous at depths greater than 4400 m for a distance of more than 2000 km. Off the coast of Guatemala it is deeper than 5500 m for 600 km. For part of its course the trench is generally U-shaped with a steeper shoreward flank and a flat bottom suggesting sedimentary fill; elsewhere, according to Fisher, the trench is V-shaped, bordered by a chain of volcanoes along the coast, and has very little fill (Fig. 2-10). As suggested by Fisher, the trench

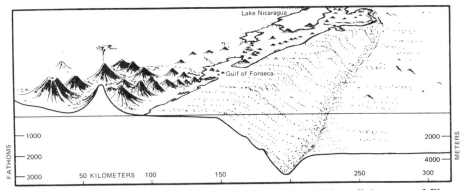

Fig. 2-10. Sketch of a part of the Middle America Trench lying off the coast of El Salvador, Central America. View looking southeast from La Libertad. Vertical exaggeration 10×. [After Fisher (1961).]

is probably not of the same age along its entire length. Also the tectonic influence that has shaped it has probably been active at different times along its length. The relationship between subsidence of the trench floor caused by subduction and rates of sedimentation may have considerable control over the shape of the trench floor. Where the rate of sedimentation exceeds the rate of subsidence, the floor of the canyon will build up and become flat. But where there is little sedimentation from the continental shelf and slope and where subsidence is rapid, the canyon will tend to become V-shaped. Such a canyon may be underlain by thick deposits of volcanoclastics ejected from the adjacent chain of volcanoes and rapidly buried in the canyon by turbidity currents.

An association of submarine trenches, basins, and plateaus is shown in Fig. 2-11 which outlines the general features of the bathymetry of the New Hebrides region in the southwest Pacific. Thick accumulations of turbidities, volcanoclastics, and submarine volcanic flows are formed within such abyssal basins and trenches. The configuration of a basin will change with time, and centers of optimum sedimentation within the basin will shift laterally; but where tectonism and accompanying volcanism are active over long periods of time, the resulting pile of sediments and volcanics will eventually constitute what is referred to as a eugeosyncline. The depths of accumulation are certainly abyssal. As shown in Fig. 2-11, the New Caledonia Basin is more than 2 km deep, the South Fiji Basin more than 4 km deep, and the New Hebrides Trench more than 6 km deep.

As with the large deltas of the world whose history goes back into the Tertiary and whose growth features are sedimentary processes today are a recapitulation of the past, so it is with areas of volcanic activity and tectonic disruption such as the southwest Pacific shown in Fig. 2-11. According to Mitchell (1970), the Early Miocene sequence on Malekula Island, New Hebrides, consists of volcanoclastic rocks, detrital limestones and pelagic sediments, are rare lava flows. Much of the volcano-sedimentary pile was transported by mass slumping from slopes into a basin, as indicated by sedimentary structures and facies relationships. Texture and composition of clasts in the volcanoclastic rocks indicate that some were derived from subaerial eruptions and others from submarine eruptions. The clastic limestones consist largely of algal and coral fragments carried into the basin from shallow depths and of the tests of forams. Mitchell concludes that the paleogeography during the Early Miocene in the area of Malekula Island must have been similar to what is seen today, and that carbonate detritus derived from reefs fringing the volcanic islands, and pyroclastic debris were washed into troughs and basins of deep water bordering the islands.

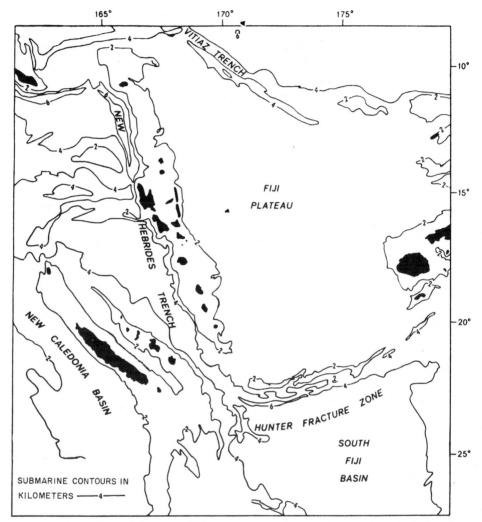

Fig. 2-11. General features of the bathymetry of submarine trenches, basins, and plateaus in the New Hebrides region, southwest Pacific. Contours in kilometers. [After Mitchell (1970).]

Stratigraphic framework of a sedimentary basin refers to the external and internal geometry of a sedimentary basin as well as to the nature of the accumulation of beds which compose the basin. The external geometry of a basin must be considered in its true perspective without vertical exaggeration. Viewed in this way, the basin may appear to be remarkably thin and flat, resembling a section cut through a pancake of uneven thickness. It may also show a marked degree of thickening along one flank of the basin or along a trend in the middle of the basin and so indicate asymmetry or symmetry respectively. The aerial configuration of the basin is another factor in its geometry and may be rounded but irregular, ovate but irregular, linear but irregular, or markedly linear. These categories are arbitrary but serve as a guide.

It must be kept in mind when describing the external geometry of a basin that a clear distinction must be made between the present geometry and that which existed, or may have existed, prior to deformation or erosion and beveling of the basin. Palinspastic reconstruction of the basin, where the beds are gently folded or folded locally, may show little difference between the present and previous geometry. On the other hand, where large-scale thrusting along sole faults is involved, the difference may be appreciable.

The internal geometry of a sedimentary basin is determined by the solid geometry of individual stratigraphic units or sequences of units within the basin. For example, in the case of a basin that is asymmetrical, the thick portion may be composed largely of volcanoclastics, whereas the thin portion may be of shale and limestone with a local but considerable build-up of bioherms along a trend fringing the edge of the thin portion. In this case there are three units that may be considered separately from the point of view of solid geometry: the volcanoclastics, the limestone–shale beds, and the bioherm trend. Inherent in this division are considerations of the nature of the beds and of the cumulative thickness of beds, that is to say considerations of lithology. These, in turn, are related to considerations of depositional environments, so that the whole question of depositional environments and stratigraphic framework is inextricably interwoven but must nevertheless be separated by definitive and quantitative terms and definitions. To make a point, the thin portion of the sedimentary basin of shale and limestone referred to above may be essentially a carbonate platform in which the growth and facies are controlled by the bioherm development along its fringe; but in defining or describing the solid geometry of stratigraphic units, the term platform has no part as it carries a particular connotation with respect to depositional environment.

With reference to stratigraphic framework, however, the use of the term platform is valid. To recapitulate, stratigraphic framework refers not only to the solid geometry of the whole basin and of the stratigraphic units or sequences of units of which it is composed but also to the nature and consequently to the depositional environments of individual stratigraphic units.

An example illustrating stratigraphic framework is presented by Ohlen and McIntyre (1965) and shown in Fig. 2-12 of the Permian System in the

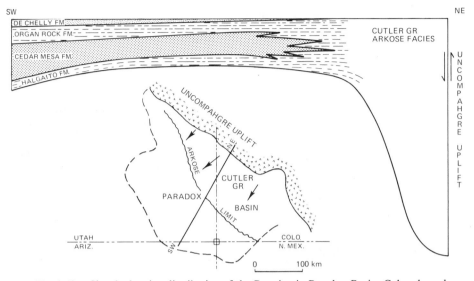

Fig. 2-12. Sketch showing distribution of the Permian in Paradox Basin, Colorado and Utah, and a generalized stratigraphic section of the Permian formations. Southwest section is less than 1000 m thick, northeast section is more than 5000 m thick. [After Ohlen and McIntyre (1965).]

Paradox Basin of Colorado and Utah. The figure shows a sketch of the outline of the present basin, the Uncompahgre Uplift, from which the Permian sediments were largely derived, and a SW–NE stratigraphic section across the basin. The present basin is merely a remnant of its former self, as major periods of erosion followed the deposition of Permian, Jurassic, and Cretaceous beds. Also it should be noted that the stratigraphic section is generalized but represents the asymmetrical stratigraphic framework of the Permian depositional basin.

The Cutler Group, which probably began its depositional history during the Pennsylvanian, consists of more than 5000 m of conglomerate, lithic sandstones (arkose), and red shale derived from the erosion of highlands situated to the present east in the area of the Uncompahgre Uplift. These sediments accumulated as fluvial deposits and alluvial fans along a fairly narrow frontal belt of the highlands. To the southeast of this frontal belt, less than 1000 m of sediments appear to have been deposited, comprising four formations. The oldest Halgaito Formation consists of red siltstone, sandstone, and shale; the Cedar Mesa Formation of sandstone; the Organ Rock Formation of red siltstone and shale, and the youngest De Chelly Formation of eolian sandstone. On the basis of stratigraphic framework the Cutler Group can be considered as a separate unit by virtue of its thickness and distribution as a narrow belt fronting the Uncompahgre Uplift and imparting a marked asymmetry to the solid geometry of the basin. The other formations can be considered as a whole, with the possible exception of the De Chelly eolian sandstones. These formations probably represent geographical extensions and outwash facies of the Cutler Group.

A more complex asymmetrical framework is shown in Fig. 2-13 which demonstrates the internal structure of major prograding stratigraphic units within an asymmetrical sedimentary basin forming today off the east coast of Nova Scotia. The oldest unit of consolidated sediments lies on the basement. Seismic surveys indicate that this unit has a compressional velocity of 3.7 km/sec. Berger *et al.* (1965) state that the consolidated sediment is not confined to the landward side of the main basement ridge, as shown by Officer and Ewing (1954), but appears to continue over the ridge. The ridge itself is a major feature of the sedimentary basin in that it divides two thick sections within the basin. One section consists largely of consolidated sediment and has a total thickness of approximately 4 km, as shown in Fig. 2-13; the other section comprises two separate prograding units of unconsolidated sediments and has a thickness in excess of 6 km. The older unit has a compressional velocity of 2.7 km/sec, the younger 1.7 km/sec. The section overlying the basement ridge separating these two thick sections has a thickness of approximately 2.5 km. The ridge appears to extend from the margin of the continental shelf east of Halifax, Nova Scotia, northeast to Sable Island. It is probable that at least one of the two thick sections continues with the ridge to form a thick linear belt within the basin.

Although the thickest section shown in Fig. 2-13 is of the same order of magnitude as the thick belt shown in Fig. 2-12, their relationships in the stratigraphic framework with reference to depositional environments and geographic situations are entirely different. Whereas the thick belt shown in Fig. 2-12 was adjacent to a rapidly eroding landmass, the thickest belt shown in Fig. 2-13 is being deposited farthest from land, at the edge of the continental shelf and on the continental slope.

Asymmetrical sedimentary basins with thick belts of sediments may be developed in either marine or nonmarine environments. In the former much of the fine sediment is carried out on to the continental shelf where, in the case of a large delta, it is deposited as a widespread sheet of prodelta mud. Some of the fine sediment is moved out to the continental shelf where periodic slumping of the accumulated sediment causes it to slide down the slope along canyons to form submarine fans. The bulk of fine sediment transported by rivers to the sea finds its way eventually to the continental outer shelf and slope where it builds up the thickest part of

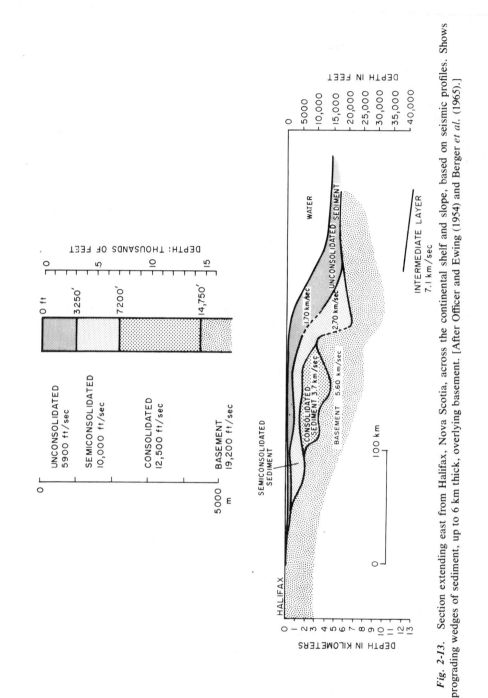

Fig. 2-13. Section extending east from Halifax, Nova Scotia, across the continental shelf and slope, based on seismic profiles. Shows prograding wedges of sediment, up to 6 km thick, overlying basement. [After Officer and Ewing (1954) and Berger *et al.* (1965).]

58

the sequence in the prograding pile of sediment. In such cases the thickest part of the sequence within the basin will be composed largely of lutites, although arenites transported by turbidity currents may also constitute a significant percentage. Limestones are more commonly deposited in the thinner parts of the sequence which developed in shallower water. Apart from lithostratigraphic features of the beds and the stratigraphic framework of basins formed on continental margins, the fossil content may indicate neritic or bathyal sedimentary environments.

In the case of nonmarine sedimentary basins with a marked asymmetrical stratigraphic framework, the thickest belt of sediments in the sequence lies close to the source of the sediments which are poorly sorted rudites and arenites, commonly conglomerate, coarse arkosic gritstone and sandstone, with minor siltstone and shale. Internal stratigraphic features and sedimentary structures indicate that these beds have been deposited rapidly, probably periodically in times of flood, and that they are largely of fluvial origin. Some beds may have formed by mass slumping of alluvium, and some may be lacustrine. In many such cases the thickest part of the sequence in the sedimentary basin was formed by the accumulation of coalescing fluvioalluvial fans underlain by grabens or flanked by an orogenic belt. The thinner parts of the sequence are of finer sediments, commonly sandstone, siltstone, and shale, with local beds of evaporites such as gypsum. These beds were formed from sediments washed outward from the fans on to flatter areas where lakes or playas were developed. Such sequences may interfinger with marine beds where the sea has transgressed a broad valley infilling with fluvioalluvial fans or where the fans flanking a mountain range sweep down to the coast. In such nonmarine sequences, regardless of whether they interfinger with marine beds, the presence of coal in the section is rare. On the other hand, in marine sections formed largely by a delta complex the paralic facies may be markedly coal.

Until the development of offshore seismic technology, little was known of the stratigraphic and structural framework of sedimentary basins forming today. The information that offshore seismic surveys are now yielding, studied in the context of recently developed concepts of crustal tectonics, has resulted in a rational interpretation of the relationships between crustal deformation and the accumulation of sediments, and in particular the role of plate tectonics in the formation of submarine trenches and sedimentary basins such as that being formed by the Bengal Deep-Sea Fan.

A longitudinal cross section of the Bengal Deep-Sea Fan, which constitutes a sedimentary basin, is shown in Figs. 2-14, 2-60. According to Curray and Moore (1971), the fan is 3000 km long, 1000 km wide, and

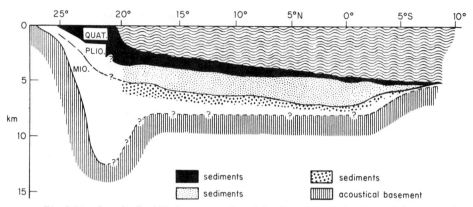

Fig. 2-14. Longitudinal N–S cross section of the Bengal Deep-Sea Fan off the coast of Bangladesh showing Miocene, Pliocene, and Quaternary sedimentary units, each having a different compressional velocity and each separated by an unconformity. [Modified after Curray and Moore (1971).]

more than 12 km thick. The sediments presently building up the fan are being transported by turbidity currents which periodically sweep down the continental slope along a submarine canyon that extends landward to the prodelta area of the Ganges–Brahmaputra Delta. The seaward extension of the canyon divides into a complex network of fan valleys. Older sediments were probably also moved by turbidity currents and probably also had their source in the ancestral delta which appears to have been in existence since the Miocene. The section illustrated in Fig. 2-14 is based on seismic profiles and shows the stratigraphic framework of this basin comprising three units of sediments, each with a different compressional velocity, and each separated by an unconformity. The unconformities are estimated by Curray and Moore (1971) to be Late Miocene and Early Pleistocene and to correspond to periods of orogeny in the Himalayas. It is also estimated from the present rate of sediment influx to the delta that the current regional rate of denudation of the Himalayas is more than 70 m in 1000 years. As the orogeny of the Himalayan belt of mountains is believed to be caused by the northward thrusting of the Indian crustal plate beneath the southern edge of the Asiatic plate, the development of the Bengal Deep-Sea Fan is directly the result of plate tectonics.

The structural nature of the basement can also determine the development and stratigraphic configuration of an individual sedimentary basin or of adjacent and contemporaneous basins. This has been illustrated in Fig. 2-13 of the sedimentary sequence off the east coast of Canada but is even more remarkable in the region of the East China Sea bounded on the west by the Peoples' Republic of China and on the north and east by South

Korea and Japan respectively. The sedimentary basin development and tectonic framework of this region have been described by Wageman *et al.* (1970). They point out that the development of basins and subbasins appears to have been controlled by a series of NE–SW trending basement ridges of folded sedimentary and igneous rocks. The ridges are believed to have served as dams separating large sediment-filled depressions. The sediments were derived from large rivers draining into the East China Sea, such as the present-day Yangtze Kiang River, and the Huang Ho River which flows into the Yellow Sea west of Korea. Wageman *et al.*, state that the large basins probably have a thickness of more than 5 km comprising mainly Miocene and Pliocene sediments but also including older Tertiary beds. Fig. 2-15 shows the distribution of basins and subbasins having thicknesses in excess of 2 km. Probably much of the sediment in these basins has been transported by turbidity currents.

Viewed on a larger scale covering the region of the East China Sea, South China Sea, Celebes Sea, Java Sea, Banda Sea, and the southwestern part of the Pacific Ocean, the distribution of sedimentary basins, troughs, and trenches is illustrated in Fig. 2-16. Parke *et al.* (1971) state that the Gulf of Thailand Basin, extending from Bangkok south to Singapore, is separated from the two basins to the east and northeast by a buried ridge of basement rock. These two basins, the Mekong Basin of Vietnam, and the Brunei–Saigon Basin which appears to comprise much of the Northwest Borneo Geosyncline, lie along a trend parallel to that of the Gulf of Thailand Basin. All three basins are more than 2 km thick and may attain thicknesses in excess of 4 km. Sedimentation in the Gulf of Thailand Basin probably began during the Late Cretaceous and subsequently in other basins of the region. Some of these basins, including the Gulf of Thailand Basin, the Mekong Basin, and the Brunei–Saigon Basin have been filled, according to Parke *et al.*, and now constitute part of the continental shelf or the landmass. The term filled, as used here, does not necessarily imply a topographic depression of the sea floor, although such may have existed, but rather a downsinking of the sedimentary pile through compaction of the sediments and subsidence of the underlying basement. The present apparent stability suggests that sedimentation is keeping pace with compaction and subsidence.

Tectonic Control and Structural Framework

Tectonic control is imposed on the development of sedimentary basins by major movements of the earth's crust underlying or adjacent to the basin. Until recent years these movements have not been well

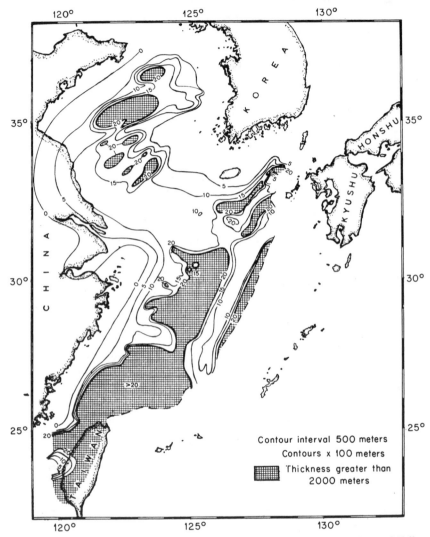

Fig. 2-15.   Distribution of Tertiary sedimentary basins in the East China Sea and Yellow Sea showing marked linearity resulting from separation by basement ridges. [After Wageman *et al.* (1970).]

documented, although the basic ideas behind the present-day concepts have been suspected and discussed for many decades. It was not until the interpretation of geophysical maps from offshore geomagnetic surveys led to the theory of sea-floor spreading that the concept of continental drift, proposed and supported by many geologists for several decades, was

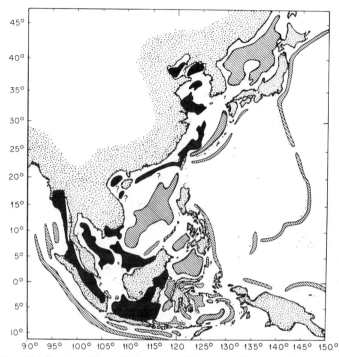

Fig. 2-16. Distribution of Late Cretaceous to Tertiary sedimentary basins in the southwest Pacific region showing filled basins (black) and partly filled basins (crosshatched). (After Parke *et al.* (1971).]

finally on a firm basis. The theory of sea-floor spreading and the old concept of continental drift were developed and presented in the form of a revised concept referred to as plate tectonics. This concept, which is supported by a wealth of evidence based on geophysical, lithologic, structural, stratigraphic, and paleontological investigations, implies not only that crustal plates are pulling apart causing continental masses to separate but also that in places crustal plates are impinging. The evidence of direction of drift of the crustal plates, as shown by geomagnetic surveys, and the distribution of arcs of crustal disturbance marked by volcanic chains, submarine trenches, and trends of earthquake epicenters strongly suggest that the latter are the result of deformation of crustal plate edges where one plate is being thrust under another. This underthrusting, termed subduction, is believed to result not only in the development of orogenic belts along or near the edge of the plate being rammed but also along or near the growth of linear tectonic depressions overlying or adjacent to the zone of subduction. These tectonic depressions may or may not be reflected in topographic or sea-floor depressions, depending upon the relative rates of subsidence and infilling of sediments, but they do provide a mechanism whereby thick piles of sedimentary and volcanic layers can be built up.

Looking back into the past history of sedimentary basins, one can find many examples where basin development can be interpreted in the light of present-day knowledge of current volcanism, earthquake activity, sea-floor topography, and sedimentation. In the case of older basins which have been metamorphosed and have become part of the basement for younger basins, it is more difficult to interpret the possible relationships between their development as basins and penecontemporaneous tectonism. The clues to such relationships may lie in the interpretation of geophysical data which indicate the geomagnetic or gravitational fabric of the basin. Indeed such data are essential in the early stages of study of any basin and are normally acquired by airborne surveys. In this regard it is of interest to note the magnetic framework of the basement beneath sedimentary basins in a large region of central and northern USSR, as shown in Fig. 2-17. The solid black areas are positive magnetic anomalies in the Late Precambrian to Early Paleozoic basement and indicate buried structural trends which coincide with structural trends in fold belts where the basement rocks are exposed. The linear northeasterly trend shown in the western sector of Fig. 2-17 marks the belt of the Ural Mountains where the basement rocks are exposed and which separate the Russian Platform on the west from the Siberian Platform on the east. Hamilton (1970) says that paleomagnetic orientations suggest a wide separation between these two platforms during the Early Paleozoic, a convergence during the Middle Paleozoic, and a collision of crustal plates during the Permian and Triassic, resulting in orogeny along the present belt of the Ural Mountains. Hamilton says, (p. 2553):

> The medial eugeosyncline of the Uralides consists largely of what may be oceanic material scraped off against the edges of the opposed subcontinents. Basalt-and-spilite belts may represent ocean-floor abyssal tholeiite, and the manganiferous cherts and other sediments upon them may be pelagic oozes. Andesite belts may have formed as island arcs within the ocean, swept subsequently against the continents.

Hamilton suggests further that during the Early Mesozoic the Uralides may have been continuous with the Ellesmerides of North Greenland and the Canadian Arctic Islands. This may be the case, as the mountain belt of northern Ellesmere Island consists in part of Permo–Pennsylvanian carbonates (Trettin, 1969).

Having outlined this example, where regional studies of surface expo-

Fig. 2-17. Geomagnetic map of central and northern USSR showing trends of positive
magnetic anomalies within the basement. Stippled areas are basins of Jurassic and younger
sediments; clear areas are basins of Paleozoic and Triassic sediments overlying basement.
Scale shows a distance of 1000 km. [After Demenitskaya (1967) and Hamilton (1970).]

sures and of geophysical data have led to inferences and conclusions
regarding the relationships between tectonism and sedimentary basin development throughout a period of time extending from the Late Precambrian to the Early Mesozoic, it is appropriate to reiterate some of the
statements made earlier regarding the terms sea-floor spreading, plate
tectonics, and subduction. It is not intended to discuss these subjects in
any depth, as they are extensively covered in numerous papers and
textbooks, but to briefly point out their relevance to the development of
sedimentary basins.

An example of a present-day zone of sea-floor spreading is given by
Luyendyk and MacDonald (1976) and shown in Fig. 2-18 which is a topo-

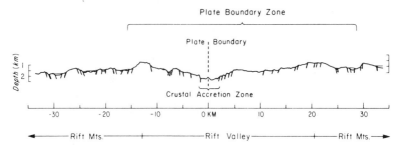

Fig. 2-18. Profile of sea floor across the Mid-Atlantic Ridge at latitude 37°N. showing the volcanic crustal accretion zone within the rift valley flanked by block-faulted ridges. No vertical exaggeration. [After Luyendyk and MacDonald (1976).]

graphic profile of the sea floor across the Mid-Atlantic Ridge at latitude 37°N. The profile shows a jagged topography with a central irregular rift valley having a width of about 30 km and a relief of up to 1.5 km. The crustal accretion zone in the center of this rift valley is formed by basaltic lavas which periodically well up along fractures and are squeezed out like toothpaste on the sea floor or erupted to form volcanic cones. This zone is the deepest part of the section across the ridge and is bounded on both sides to the east and west by faulted blocks of older basaltic rock. In the troughs formed by numerous grabens, clays and oozes of pelagic origin accumulate very slowly. Farther from the crustal accretion zone the depressions will have had a longer period of time in which to accumulate sediments; but it is unlikely that any accumulation would ever be sufficiently thick to be other than a very minor subbasin. In many locations subbasins would be cut off from any source of sediment on the continental shelf by intervening abyssal plains.

In other areas where crustal plates are drifting apart, there may be appreciable thicknesses of sediments deposited in the depressions formed by complex grabens. An example illustrated by Barberi *et al.* (1972) and Tazieff *et al.* (1972) is shown in Fig. 2-19 of the Gulf of Aden and the southern part of the Red Sea where the Arabian and Nubian crustal plates are splitting apart. New oceanic crust is forming in the medium rift along the Red Sea and Gulf of Aden, and a complex graben structure forms the structural framework of the sedimentary basin underlying the Red Sea. This basin has been formed from sediments, derived from the erosion of the Arabian and African landmasses, and from evaporite deposits formed along the coastal and adjacent areas. Sedimentation appears to have kept pace with basin subsidence. It is of interest to note that in the deeper parts of the central rift metallic sulfides, apparently derived from volcanic sources, are also being deposited. In the Danakil Depression,

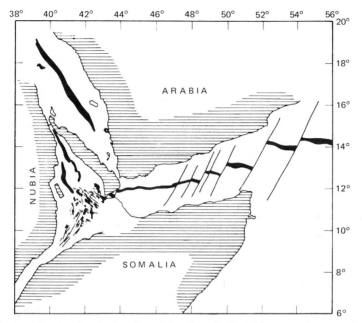

Fig. 2-19. Map showing the location of the triangular Danakil Depression at the junction of the Red Sea, Gulf of Aden, and East African rifts. Black areas are zones of generation of new oceanic crust. [After Barberi *et al.* (1972) and Tazieff *et al.* (1972).]

now a desert with salt pans in the topographic depressions, rifting began in the Early Miocene. The oldest sediments found in the depression are of detrital origin and crop out along margins of the depression.

On a much grander scale, discussed by Oxburgh (1974) and Fisher and Heezen (1971), sea-floor spreading in the Pacific region is shown in Fig. 2-20. In the area west of Central America, a crustal rift that was actively splitting apart prior to and during the Tertiary has resulted in the formation of two parallel bands of Miocene rock, one on each side of the rift. The eastern plate has moved eastward and been subducted under the crust of Central America (Fig. 2-10). To the west, sea-floor spreading has resulted in progressively older rocks underlying the sea floor. The contact between the Jurassic and Oligocene–Eocene in the western region of the Pacific corresponds to the present zone of crustal disturbance along the Marianas Trench which is believed to reflect an active zone of subduction.

An example of the control of sedimentary basin development imposed by such tensional crustal movements of the bathymetry of the sea floor in the region south of South Africa is shown in Fig. 2-21. Between Cape Town and Durban the continental slope is steep and in the areas of the

68   2. EVOLUTION AND DEVELOPMENT

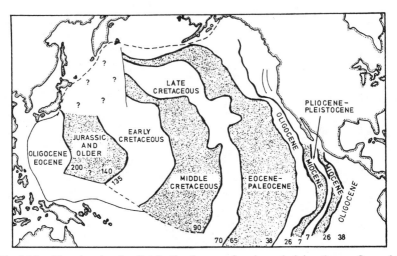

Fig. 2-20. Map showing the distribution by age of rocks underlying the sea floor of the Pacific Ocean. Numerals indicate ages in millions of years. The belt of Pliocene–Pleistocene rocks west of Central America forms the crest of the East Pacific Rise, a major rift in the earth's crust. The contact between Jurassic and Oligocene–Eocene rocks in the western Pacific corresponds to the present active zone of subduction. [After Oxburgh (1974) and Fisher and Heezen (1971).]

Transkei and Agulhas basins terminates abruptly on the floor of an abyssal plain. The Agulhas Plateau separates the Transkei and Agulhas basins and is itself separated from the continental slope by a trough more than 4700 m deep. Scrutton (1973) expressed the opinion, based in part on work by many others, that the steep, linear continental slope reflected a major shear fault associated with the Agulhas Fracture Zone and that the Agulhas Plateau is a feature of the oceanic crust, possibly an abandoned sea-floor spreading center.

Tectonic control of sedimentary basin development, particularly in the case of eugeosynclinal basins, is manifest in the areas adjacent to belts of crustal disturbance which are believed to correlate with belts of active subduction. Fig. 2-22 (Oxburgh and Turcotte, 1970) shows the distribution of belts of crustal disturbance in the western Pacific region. These belts are marked by earthquake epicenters and centers of volcanic activity. Lying parallel to them are submarine trenches in which thick stratigraphic sections can form by the deposition of turbidites and volcanic flows and debris. Belts and submarine trenches lie along the edges of crustal plates that are being rammed by other plates. In the zone of subduction faulting and volcanism are active, the edges of plates being subducted are scraped and deformed, and behind the edge the crust is buckled. This buckling is

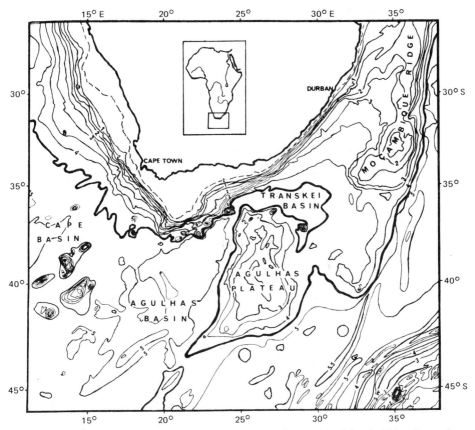

Fig. 2-21. Map showing the bathymetry of the sea floor south of South Africa. Parts of the Transkei and Agulhas basins are more than 4700 m deep. The steep, linear continental slope may overlie a major shear fault, and the Agulhas Plateau may be of oceanic crust formed by a now abandoned sea-floor spreading center. [After Scrutton (1973).]

reflected at the surface by the development of submarine trenches. It is evident from the distribution of trenches and zones of crustal disturbance shown in Fig. 2-22 that the crustal plate underlying the Pacific region is pushing in a general eastward direction, whereas the Australian crustal plate is pushing to the north.

The northward thrust of the Australian plate is illustrated in Fig. 2-23 which shows the inferred (Carter *et al.*, 1976) structure of the crust beneath the island of Timor and the Timor Trough which forms the eastern end of the Java Trench (Fig. 2-22). The Australian plate is shown to be subducted beneath the Indonesian island of Wetar and the Wetar Strait. The leading

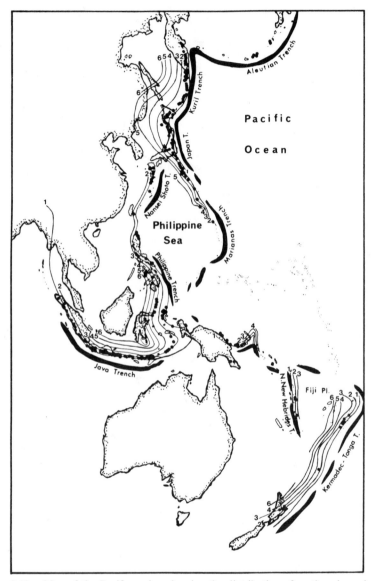

Fig. 2-22.   Map of the Pacific region showing the distribution of earthquake epicenters (black dots) and of submarine trenches (black arcs). [After Oxburgh and Turcotte (1970), based on data from Gutenberg and Richter (1954) and Hamilton and Gale (1968).].

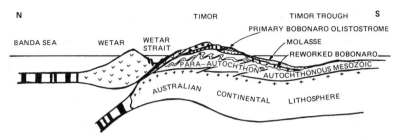

Fig. 2-23. Schematic structural section across the Indonesian islands of Wetar and Timor showing subduction of the Australian crustal plate beneath the Asian plate. Not to scale. [After Carter *et al.* (1976.]

edge of the Australian plate is internally deformed by thrust faulting and folding that occurred during the Miocene–Pliocene, contemporaneously with deposition of sediments on the flank of the Timor Trough.

The physical processes whereby the edges of the subducted and rammed crustal plates are deformed and the resulting structural configurations are tectonically related are open to various interpretations. There is general agreement among many geologists that the process involves a scraping of the strata on the subducted plate to form a sequence of thrust sheets, some along sole faults, that give rise to mountain ranges trending parallel to the subduction belt. An example of the general structural features involved is shown in Fig. 2-24 which is a schematic structural section of the Gulf of Alaska region. The section extends inland from the edge of the continental shelf to a massif of igneous and metamorphic rocks. Underlying the continental slope and shelf is a stratigraphic sequence formed by prograding wedges of sediments. Traced inland, the sequence becomes progressively more deformed as it is disrupted by thrust faults and distorted by folds within the thrust sheets. The thrust sheets are

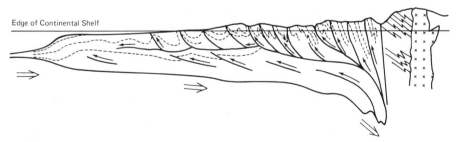

Fig. 2-24. Schematic structural section of the Gulf of Alaska region showing thrust sheets of Mesozoic–Cenozoic strata scraped off the surface of the Pacific crustal plane which is subducted beneath the Alaskan plate. Not to scale. [After Stoneley (1967).]

believed to have formed by northward movement of the Pacific crustal plate which is ramming the Alaskan plate. The stratigraphic sequence lying on the Pacific plate along its subducted edge has been scraped off as thrust sheets. These have become highlands which are eroding, the resulting sediments being carried to and deposited on the continental shelf. In this way the sediments are recycled. Stoneley (1967) states that the Mesozoic and Cenozoic sequence, which forms a structurally deformed sedimentary basin shown in Fig. 2-24, was deposited on a rapidly subsiding continental shelf. The sequence has been deformed by repeated oceanward thrusting and folding which increase in intensity toward a currently active strike-slip fault at the edge of the basin.

The same basic concept, but with variations, is expressed by Moore and Karig (1976) in Fig. 2-25. They point out that in arc systems where an oceanic plate with a thick cover of sediments is subducted, the sediments are scraped off and accreted to form ridges behind which younger sediments may be deposited as separate minor subbasins. Moore and Karig based their conclusions on studies carried out in the Indonesian region.

Deformation of stratigraphic sequences in thrust sheets and wedges believed to have been produced by subduction is commonly observed. Another feature which is also observed is a mishmash or mélange, consisting of broken blocks of sedimentary strata in a matrix of similar but finer material. Such mélanges commonly form irregular beds termed olistostromes. Some of these may have resulted from gravity sliding of incompetent stratigraphic layers; others are believed to have been formed by the crushing and mixing of strata scraped off the upper surface of a subducted crustal plate. The olistostrome so formed may be drawn down with the subducted plate to depths where it becomes metamorphosed, or it may in part be shoved onto the trench rise. Examples are given by Hall (1976) who described the origin of mélanges mapped in the region of the Bitlis Massif in eastern Turkey (Fig. 2-26). Hall's interpretation is illustrated in Fig. 2-27 which shows the Arabian crustal plate, overlain by a sequence of Late Cretaceous carbonates and shales, subducted beneath the present Bitlis Massif. Although the Bitlis Massif constitutes a complex structural and stratigraphic feature with an earlier history of tectonism, the subduction of Late Cretaceous strata indicates that its late history includes Alpine orogeny.

The Alpine orogenic belt of the Mediterranean region has been studied in great detail, and several occurrences of olistostromes and other mélange deposits have been mapped. This orogenic period took place during the Late Cretaceous and Early Tertiary. Orogeny occurred along a belt which Alvarez (1976) believes can be traced through Corsica to Sicily and southern Italy (Fig. 2-28). Later movements of crustal subplates dur-

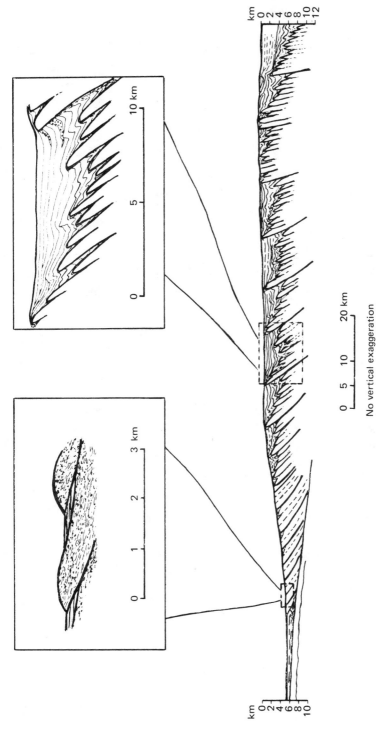

Fig. 2-25. Schematic model section, based on data from seismic reflection profiles, deep-sea drilling, and field work on Nias Island, Indonesia, showing thrust-faulted wedges of strata scraped off a subducted crustal plate. The wedges form ridges which have minor subbasins of younger sediments deposited between them. [After Moore and Karig (1976).]

73

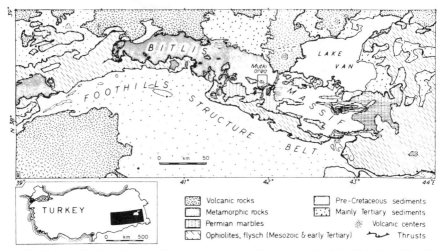

Fig. 2-26. Geological map of a region in eastern Turkey showing the Bitlis Massif and associated rocks which form the Taurus Mountains. [After Hall (1976).]

ing the Miocene and Pliocene dispersed parts of this belt and produced the Atlas orogenic belt of northern Africa and the Apennine belt of Italy.

An example of a mélange within the Apennine belt is described by Page (1963) who was of the opinion that it had been formed by gravity sliding. The occurrence is illustrated in Fig. 2-29 which is a section across an area south and east of Passo Della Cisa in the northern Apennines of Italy. The section shows a horst of autochthonous Oligocene sandstone flanked by an allochthonous mélange comprising blocks of Cretaceous, Eocene, and Oligocene limestone, sandstone, and sandy siltstone in a shaly matrix. Page was of the opinion that gravity is the only agent that could have moved the allochthon and that the relationships of blocks within the allochthon indicated tensional rather than compressional stresses. Other field workers may have different views.

It must be emphasized that although it is recognized that tectonic control is a vital factor in the location and development of sedimentary basins, there is still much to learn about the processes involved in crustal deformation and sedimentary basin development. Hypotheses abound, some theories are well established; but as knowledge increases, old theories are modified and new concepts are formulated.

Structural framework of sedimentary basins refers to the structural nature and configuration of the basement on which the basin has developed. The structural nature may, during the period of basin development, remain essentially the same; but where deformation of the basement is

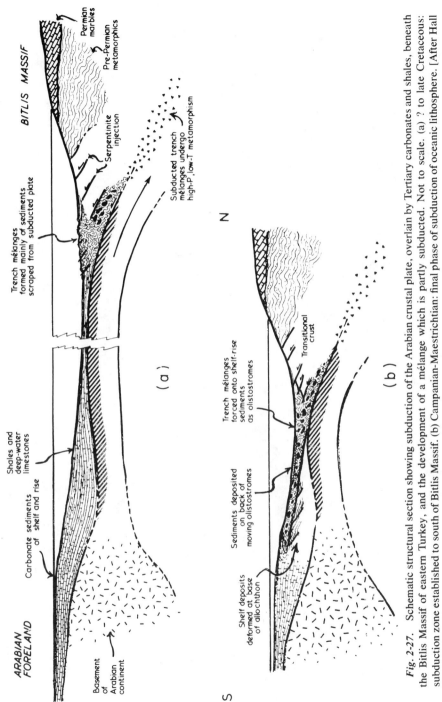

Fig. 2-27. Schematic structural section showing subduction of the Arabian crustal plate, overlain by Tertiary carbonates and shales, beneath the Bitlis Massif of eastern Turkey, and the development of a mélange which is partly subducted. Not to scale. (a) ? to late Cretaceous: subduction zone established to south of Bitlis Massif. (b) Campanian-Maestrichtian: final phase of subduction of oceanic lithosphere. [After Hall (1976).]

75

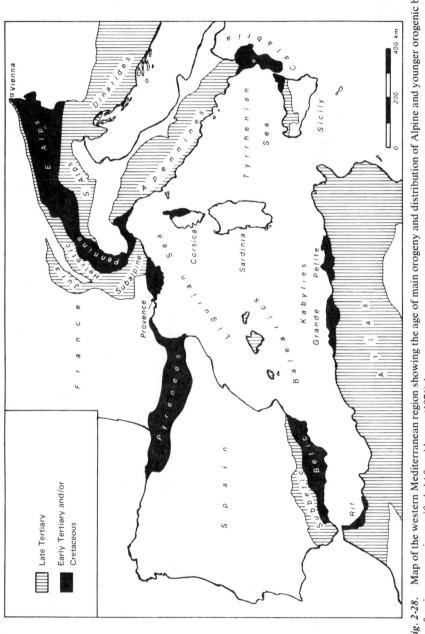

Fig. 2-28. Map of the western Mediterranean region showing the age of main orogeny and distribution of Alpine and younger orogenic belts. Age of main orogeny is specified. [After Alvarez (1976).]

76

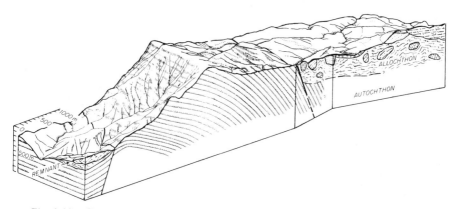

Fig. 2-29. Structural section of an area south and east of Passo Della Cisa, northern Apennines, Italy, showing a horst structure within an autochthon of Oligocene sandstone flanked by an allochthonous mélange of Cretaceous, Eocene, and Oligocene blocks of limestone and sandstone in a matrix of shale. A remnant of the allochthon is shown. Approximately to natural scale. [After Page (1963).]

penecontemporaneous with basin development, the configuration may change appreciably. This is commonly the situation during the development of many basins where the basement is broken by graben and horst structures which shift their relative positions during the growth of the basin and consequently exert some influence on the localization of sediment accumulations. Such basement movements may also influence the shift of depocenters which are the areas of thickest sediment accumulation during a particular interval of time; but in general the location of depocenters depends upon surficial factors, in particular the lateral shifting of sediment sources such as the mouths of large river systems.

It is important to keep in mind the actual scale of sedimentary basins. Commonly cross sections are drawn with greatly exaggerated vertical scales which accentuate features such as folds, diapiric structures, and bioherms and steepen the angles of faults. Such exaggerations may also result in the apparent vertical separation of strata to the extent that a fault is interpreted where none, in fact, exists. When looked at in natural scale, sedimentary basins appear to be remarkably shallow.

An interesting example is shown in Fig. 2-30 which illustrates sections across the Rhine Graben and the Viking Graben of the northern North Sea, between the Shetland Islands and Norway. Both sections, according to Ziegler (1975), are based on seismic refraction profiles and have a vertical exaggeration of approximately two. Taking the Viking Graben as an example, using the same vertical scale, and doubling the length of the section to show its true proportions in natural scale, it can readily be seen

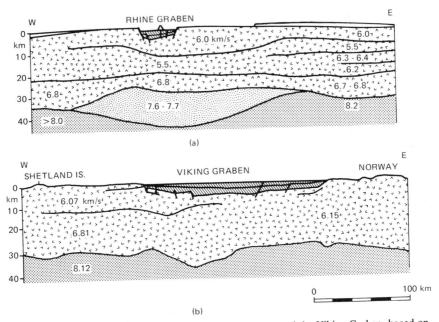

Fig. 2-30. Structural sections across the Rhine Graben and the Viking Graben, based on seismic refraction profiles. Compressional velocities of various layers shown in km/sec. Vertical exaggeration approximately 2×. (a) Refraction profile Rhine graben. (b) Refraction profile northern North Sea. [After Ziegler (1975).]

that although the Viking Graben sequence has a maximum thickness of approximately 5 km in the section shown, the shape of the Viking Graben basin is that of a very shallow saucer.

It must also be kept in mind that the structural framework of a sedimentary basin is inherently a part of and consequentially the result of control exerted by the regional tectonic framework. In many cases the two cannot readily be differentiated, and the establishment of categories for structural classification or nomenclature becomes somewhat arbitrary. A case in point has been illustrated by Bonnefous (1972). Figs. 2-31 and 2-32 show the location and main structural features of the Sirte Basin in northern Libya. Regional tectonic trends of the central Mediterranean region are shown in the major northwest-trending system of faults that traverse the Sirte Basin and control the trends and linearity of separate sedimentary troughs. These troughs, which measure up to 400 km in length and 150 km in width, are sufficiently large to qualify for the term basin. In fact, some basins form troughs, but not all troughs are large enough to be called basins. The term trough has the connotation of marked linearity commonly resulting from the structural control of elongate graben structures

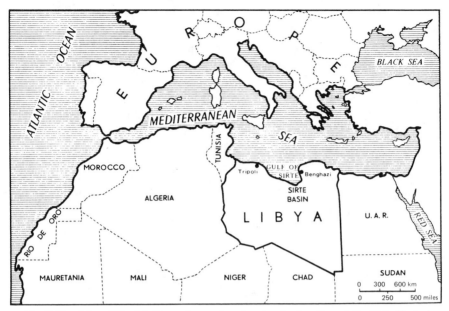

Fig. 2-31. Map of the Mediterranian region showing the location of the Sirte Basin in northern Libya. [After Bonnefous (1972).]

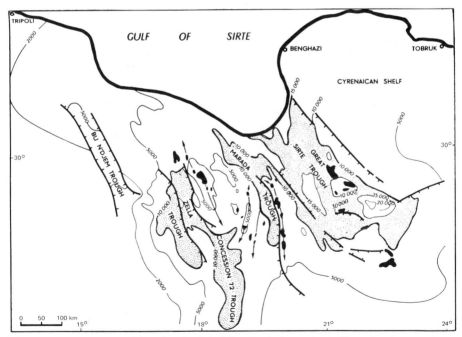

Fig. 2-32. Structural framework of the Sirte Basin in northern Libya showing a major system of northwest-trending faults which control the location of linear and parallel sedimentary troughs. Solid black areas are oil and gas fields. [After Bonnefous (1972).]

79

within a linear trend of major faults. Each of the troughs shown in Fig. 2-32 has its own minor structural framework of faults; and each trough framework is part of the major structural framework of the Sirte Basin.

In recent years one of the major contributions to the understanding of the structural framework of sedimentary basins has been made by Beck and Lehner (1974). Based on seismic profiles traversing continental shelves, slopes, and abyssal plains, these workers have constructed sections which elucidate the structural relationships of sedimentary basins to the continental and oceanic crust. These structural relationships, as seen particularly in the Atlantic area, in general can be grouped in four main categories, as shown in Fig. 2-33. All of these categories show the basaltic oceanic crust butting into the granitic continental crust. Not shown is a modification in which outlier blocks of continental crust have separated from the main mass and are surrounded by oceanic crust. All four catego-

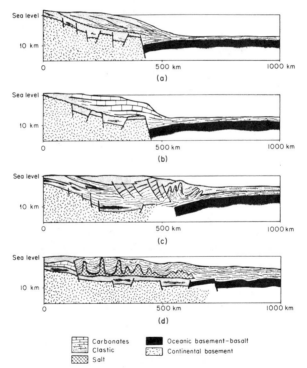

Fig. 2-33. Generalized structural framework of main types of sedimentary basins (a) marginal geosyncline, (b) carbonate platform, (c) major delta with gravity tectonics, and (d) salt tectonics, underlying continental shelves, slopes, and abyssal plains in the Atlantic area showing relationship of basaltic oceanic crust to granitic continental crust. Vertical exaggeration approximately 6×. [After Beck and Lehner (1974).]

ries are shown to have a basic structural framework of grabens and horsts. The sedimentary piles designated as forming a marginal geosyncline and a carbonate platform are shown to constitute prograding time-stratigraphic units, the maximum thickness of the piles overlying grabens situated beneath the continental shelf and slope. The category designated as a major delta with gravity tectonics would also form a sequence of strata that is essentially prograding but distorted by faults and folds formed contemporaneously with deposition by gravity slumping of the sedimentary pile. The category illustrating salt tectonics shows the relationship of salt diapirs to the overlying and intruded sedimentary strata which become folded and faulted by the salt domes thrusting upward. These diagrams must be regarded in perspective, and it should be kept in mind that they have a vertical exaggeration of more than six.

As mentioned in the opening remarks concerning structural framework of sedimentary basins, the configuration of a structural framework may change during the period of development of the basin where deformation of the basement proceeds contemporaneously with deposition of the basin. An example is given by Brink (1974) in Fig. 2-34 which shows seven stages of development, from Aptian to Recent, of the Gabon Basin in Gabon, west-central Africa. The basin has two main elements, the Atlantic subbasin and the eastern subbasin, each occupying a complex graben structure and separated by a complex horst structure. The diagrams have a vertical exaggeration of five. They show how the growth of the graben complexes has allowed the sedimentary basin to develop and also how the initially flat stratum of salt deposited in Stage 2 has been subsequently squeezed upward as diapiric structures that have deformed the overlying sedimentary strata. The reconstructions shown are interpretive, but they illustrate the principle that whereas the nature of the structural framework of a sedimentary basin may remain essentially the same, the configuration of the framework may change through the period of development of the basin.

This period of development is not always readily defined, even though the ages of the strata comprising the basinal accumulation have been determined. Some accumulations developed during long periods of time spanning several periods which may be included in two eras. Commonly, in such cases there is a major unconformity separating the strata in one Erathem from another. Such demarcations are not always evident but may be arbitrarily imposed. Within these accumulations spanning several systems there are commonly discontinuities of sedimentation marked by unconformities, disconformities, and paracontinuities. Also during the period of time in which the accumulation occurred, the basement has commonly been subjected to stresses which have caused a changing pattern,

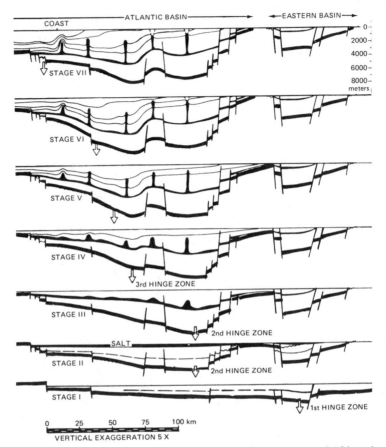

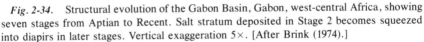

Fig. 2-34.   Structural evolution of the Gabon Basin, Gabon, west-central Africa, showing seven stages from Aptian to Recent. Salt stratum deposited in Stage 2 becomes squeezed into diapirs in later stages. Vertical exaggeration 5×. [After Brink (1974).]

or modifications to an existing pattern of the structural framework. As these changes proceed, older strata will be displaced and possibly uplifted and eroded. Younger strata may be deposited during a period of structural deformation in which they are penecontemporaneously deformed by folds and growth faults, or locally thinned over structural highs.

Thick sedimentary accumulations comprising strata of two Erathems may have developed within the same basic structural framework and exhibit the same basinal trends and similar limits to their depositional edges; or the sequence of strata within each Erathem may overlap by wide margins and show similar but divergent trends. In the latter case each sequence is defined as constituting or being part of a separate basin. In the former case the total pile of sediments, including both sequences, may be referred to as a single basin, particularly where the two sequences basically share a common structural framework. An example is shown in Figs. 2-35 and 2-36 of the Bonaparte Gulf Basin underlying Joseph Bonaparte Gulf in northern Australia. The basin has developed during a period of time spanning Carboniferous to Recent in a complex graben system within the Precambrian basement. The system has probably resulted from stresses set up by the movements of major basement blocks in juxtaposition, in particular the Kimberley, Sturt, and Darwin blocks. Some deformation of younger Mesozoic strata has also resulted from the intrusion of salt diapirs originating in Early Paleozoic deposits. These intrusions appear to be related to epeirogenic movements of the fault system. The salt beds from which they originated are believed by Edgerley and Crist (1974) to have formed on terraces during an early stage of development of the basin.

In the case of the Bonaparte Basin, the complex graben structure that constitutes the structural framework of the basin was probably caused by tensional stresses between adjacent Precambrian blocks of the continental crust. Where such blocks actually split away from the main mass of continental crust, they may be separated by basaltic oceanic crust which forms the basement of a sedimentary basin situated between the continental mass and the outlying block of continental crust. An example is shown in Fig. 2-37 which illustrates the general features of a sedimentary basin lying between the southern end of Mozambique and the southern part of Malagasy off the coast of southeast Africa. The basin has a maximum thickness in the section shown and comprises Mesozoic and Tertiary marine sediments. It is of interest to note that the section illustrates not only that Mesozoic sedimentation took place on a basaltic surface broken by faults which may in part have been penecontemporaneous with sedimentation but that it also shows later intrusion of basaltic magma into the Mesozoic and Tertiary beds. The section is an interpretation by Beck and Lehner (1974) based on seismic reflection profiles. A similar section, located about 500 km south, is illustrated by Dingle and Scrutton (1974) in Section 5 of Fig. 2-38. This section shows a sedimentary basin lying between the Mozambique Ridge (a continuation of the granitic continental crust underlying Malagasy) and the continental crust of the mainland. It also shows oceanic crust underlying sediment on the eastern flank of Mozambique Ridge but does not show the presence of oceanic crust west of the ridge, possibly because it was based on less evidence than that available to Beck and Lehner. The other sections, 1–4 inclusive, are across marginal basins off the coast of South Africa and are based on seismic refraction and reflection profiles, gravity surveys, and other data.

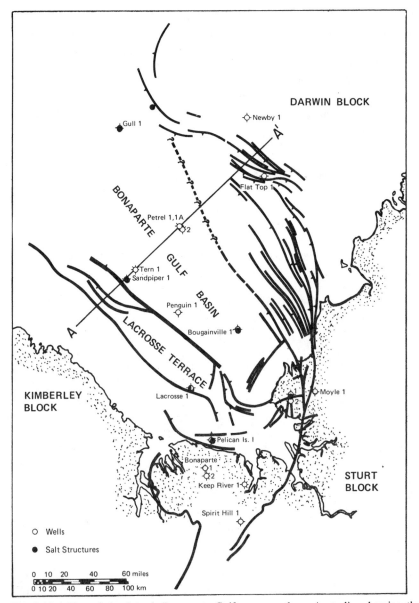

Fig. 2-35.   Map of the Joseph Bonaparte Gulf area, northern Australia, showing the trends of faults within a complex graben structure that forms the structural framework of the Bonaparte Basin. The basement is of Precambrian rock comprising parts of the juxtaposed Kimberley, Sturt, and Darwin blocks. [After Edgerley and Crist (1974).]

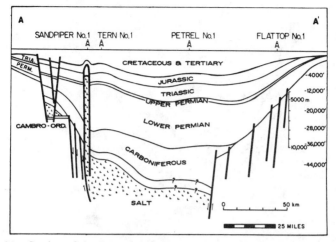

Fig. 2-36. Section of the Bonaparte Basin, northern Australia, showing the apparent conformity of stratigraphic sequences included in two Erathems. Contemporaneous faulting and deposition of the Carboniferous is suggested. [After Edgerley and Crist (1974).]

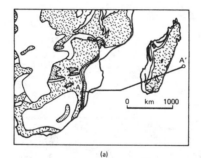

(a)

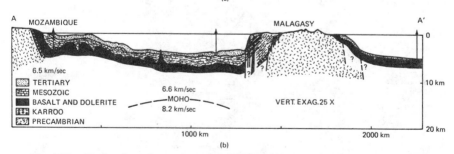

(b)

Fig. 2-37. Section from the southern end of Mozambique to Malagasy, based on seismic reflection profiles, showing Mesozoic and Tertiary sediments, up to 3 km thick, lying on a basaltic basement (black) broken by faults. The block of granitic continental crust underlying Malagasy may be separated from the main continental mass by oceanic crust. [After Beck and Lehner (1974).]

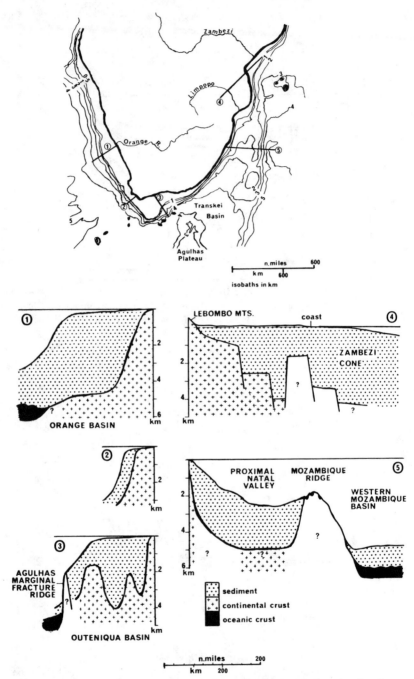

Fig. 2-38. Sections based on seismic profiles, gravity, and other data showing the structural configuration of sedimentary basins and basement off the coasts of South Africa and Mozambique. [After Dingle and Scruton (1974).]

Section 1 across the delta of the Orange River shows a typical continental shelf and slope profile underlain by a thick sedimentary section; Section 2 consists almost entirely of slope sediment; Sections 3 and 4 appear to have a similar structural framework of basement ridges and valleys, probably formed as horsts and grabens within the granitic basement. Section 4 also shows part of the Zambezi Cone which includes the delta of the Zambezi River. Of particular interest is the apparent disruption by tensional structures of the continental margin along the eastern and southern coasts of South Africa, suggesting that this trend was a zone of dislocation of the continental crust. The dislocation is probably related to stresses set up by rotational movements caused by the drifting apart of the African and South American continental masses. The apparent separation of the block of continental crust underlying Malagasy may also have taken place during this period of disruption.

Another example of the relationship between structural framework and stratigraphic configuration of the sedimentary pile is described by Sastri *et al.* (1973) and shown in Fig. 2-39 of the Cauvery Basin on the southeast coast of India off the northern tip of Sri Lanka. The Archean basement of this basin is broken by ridges formed by horsts trending generally northeast. The stratigraphic sequence, comprising Late Mesozoic and Tertiary beds, forms an easterly prograding accumulation within a basement depression that rises to the south and also to the north where the basement is at or close to the surface in the area of Pondicherry. The sections suggest that during the Late Mesozoic structural disruption of the basement and

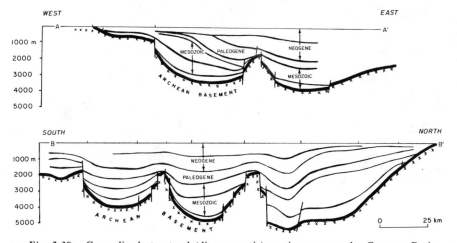

Fig. 2-39. Generalized structural (diagrammatic) sections across the Cauvery Basin, southeast India, showing northeasterly trending basement ridges and a stratigraphic sequence prograding from the west. [After Sastri *et al.* (1973) and Negi *et al.* (1968).]

deposition were penecontemporaneous but that during the Neogene (Late Tertiary) deformation of the younger beds was caused mainly by compaction over basement highs.

Similar relationships obtain in parts of the southern area of the Northwest Shelf off the coast of Western Australia, as described by Kaye *et al.* (1972) and illustrated in Fig. 2-40 which shows the general stratigraphic sequence and structural framework of Mesozoic and Tertiary beds northwest of the Dampier Archipelago. The Mesozoic beds, including Triassic, Jurassic, and Cretaceous, have been dislocated by faults, some of which extend through a Permian section into the Precambrian basement. Movement along the faults has been periodic, with consequent greater displacement of the older beds. This implies contemporaneity of later movement and the younger Mesozoic beds, with consequent local development of growth structures and depositional thinning over structural highs. The Tertiary beds have been deformed only to the extent of local warping by compaction over these structural highs. The basic features of the struc-

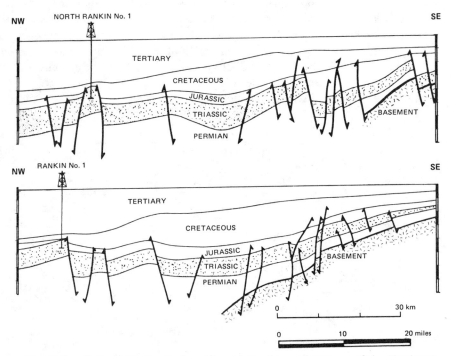

Fig. 2-40. Generalized structural sections across the southern area of the Northwest Shelf, northwest of the Dampier Archipelago, Western Australia, showing the Rankin structures which contain gas and condensate. [After Kaye *et al.*, (1972).]

tural framework in this region, as in the coastal area of South Africa, were probably initiated by tectonism during periods of crustal separation. It is of interest to note that some of the horsts and upthrust fault blocks, such as the Rankin structure shown in Fig. 2-40, form traps for hydrocarbons where permeable Triassic sandstone beds are truncated updip and sealed by younger and relatively impermeable beds. Also of note are the relative directions of thinning of the Triassic and Cretaceous sections, the former thinning landward and the latter forming a trough that thins both landward as it approaches the rising basement and seaward as it approaches a structural high within the Triassic. The general features of the structural framework of this Cretaceous basin were those of a broad and complex graben trending northeast. The Cretaceous deltaic sediments which filled this graben-type trough formed a sequence prograding generally to the northwest.

In the northern areas of the Northwest Shelf, a similar basic structural framework is recognized, the Mesozoic beds being disrupted by deep-seated faults and the Tertiary being affected only by compactional warping. The effect of structural highs in pre-Mesozoic beds on the deposition and structural configuration of Tertiary beds is illustrated by Powell (1976) in Fig. 2-41 which shows a structural cross section from the mainland area northeast of Broome through Scott Reef to the abyssal plain. Compaction of the Tertiary sediments over the flanks of a horst structure in pre-Mesozoic beds has resulted in a topographic high on which a Recent reef has grown. This is an interesting example of a present-day geomorphic and sedimentologic feature reflecting an ancient and deep-seated struc-

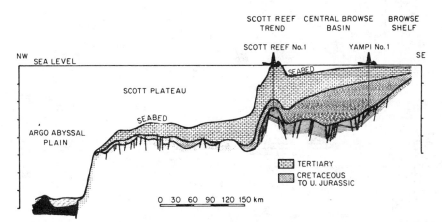

Fig. 2-41. Generalized structural section across the Northwest Shelf, northwest of Broome, Western Australia, showing the Recent Scott Reef developed on a deep-seated structural high. [After Powell (1976).]

ture. Also of interest is the interpretation showing oceanic basaltic crust underlying the abyssal plain.

The structural framework of a sedimentary basin is the product of regional tectonism which is commonly reflected in the lithologic characteristics of the sedimentary sequence. An interesting example is presented by Tjhin (1976) and illustrated in Figs. 2-42 and 2-43. The former figure shows a map of the Trobriand Basin area north of the eastern end of Papua–New Guinea and a generalized structural section between two wells drilled in

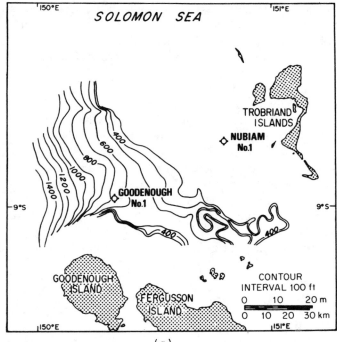

(a)

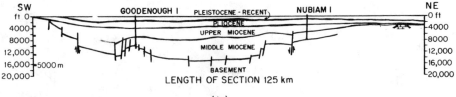

(b)

Fig. 2-42. Map of the area of the Trobriand Basin, eastern end of Papua–New Guinea, showing (a) location of wells and (b) a structural section between them. [After Tjhin (1976).]

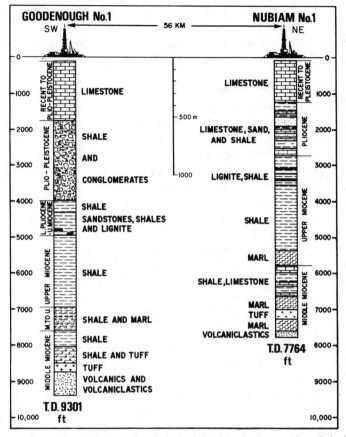

Fig. 2-43. Generalized lithologs of wells in Trobriand Basin, Papua–New Guinea, shown in Fig. 2-42. [After Tjhin (1976).]

the basin. The latter figure shows the general lithologic nature of the sections penetrated by the wells. Basement consists of Miocene and older volcanics and volcanoclastics and is broken by numerous faults which cut younger Miocene sediments. These faults form the structural framework of a complex graben trending east-west. The graben is believed to have been initiated by taphrogeny, the type of orogeny that forms major rifts. Tensional stresses produced a complex graben which rapidly filled with volcanic extrusions and debris. Later, Miocene marls, line muds, tuffs, and clays were deposited penecontemporaneously with continual faulting. The Late Miocene appears to have been a period of relative stability characterized by a shallow water to paralic environment, as evidenced by the presence of seams of lignite. Similar shallow water conditions with local uplift and erosion appear to have prevailed during the Pliocene; since then the area has seen shallow marine conditions in which organic reefs and limestones have formed. The total thickness of the Trobriand Basin is approximately 5000 m, a cumulation attained within a period of 20 million years.

The infilling of such graben-type troughs may be rapid, although the depositional conditions remain those of shallow marine to paralic environments. Volcanic flows, tuff, and volcanoclastics may be interbedded with clay muds from rivers or lime muds and organic reefs. Large rivers may form deltas which prograde into the sinking structural trough, depositing extensive sheets of prodelta muds and forming widespread beds of peaty coastal marsh. Topographically the area of such a sedimentary basin may be fairly flat, yet structurally the basin may be rapidly sinking. In other words, the sediment supply is adequate to keep pace with the rate of subsidence.

Rapid deposition of sediments on the block-faulted flank of a sinking basement depression is also a feature in the Beaufort Basin, Northwest Territories, Canada, where more than 9000 m of Tertiary sediments form a deltaic sequence prograding to the northeast. The general features of the structural framework underlying this sequence described by Hawkings *et al.* (1976) and shown in Fig. 2-44. Of economic interest is the fact that rapid accumulation of Tertiary sediments in the Beaufort Basin was accompanied by the development of growth faults and structures, some of which have formed traps for oil and gas.

Depositional, Eroded, and Displaced Margins of Basins

Geological maps of sedimentary basins show only the remnants of basin areas that previously were much more extensive. Regional uplift and erosion may have beveled the basin margins, removing them entirely where they are thin. Orogeny may also have been responsible for the removal of margins where fold belts have disrupted a basin. Major crustal movements may have split the basin apart so that the two portions are widely separated within a continent or form parts of two separate continental masses. Whatever the interpretation placed on the distribution of strata in sedimentary basins as seen and mapped today, it must be kept in mind that the map reflects only the results of younger events in a basin's history and that the subsurface geology has most of the clues to the past.

A sedimentary basin is a composite pile of strata comprising several subpiles, each of which has a separate maximum area of thickness or depocenter. As sources of sediment, such as the mouths of a river system,

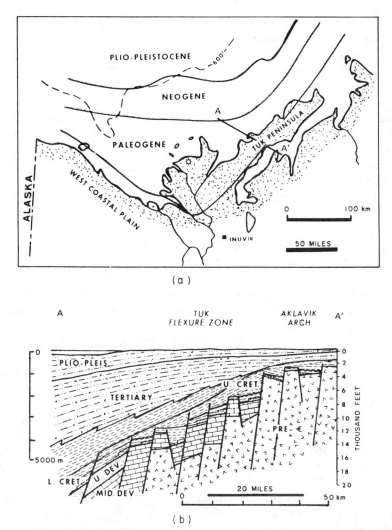

Fig. 2-44. Map of the Mackenzie River delta area of the Beaufort Basin, Northwest Territories, Canada, showing a generalized structural section of the south flank of the basin. [After Hawkings *et al*. (1976).]

periodically migrate laterally over a prograding sequence of strata, they build up a series of offlapping sediment lenses, each with its own depositional edges and depocenter. Most sedimentary basins also include unconformities, disconformities, and paracontinuities. Unconformities imply not only uplift but some degree of regional tilting or warping of the beds prior to or penecontemporaneous with erosion of the beds. Disconformities require only regional uplift so that the beds are beveled parallel to the bedding planes. Paracontinuities imply a hiatus of deposition, possible withdrawal of the sea, and weathering of the exposed beds but no significant erosion. It should be stated that regional uplift is a relative term but implies some degree of crustal movement. Exposure of beds to erosion may also be effected by the lowering of sea level. Unconformities and disconformities bevel the subpiles or lenticular accumulations of sediment surrounding the depocenters whereas a paracontinuity marks the upper surface of one or more lenticular subpiles.

The geological history of any basin has a beginning and ultimately an end, although its ghost may remain as a sort of gigantic schlieren long after the basin has been assimilated in the granitic continental crust. What is seen today in geological maps of the land surface, in paleogeological and other maps of the subsurface, and in bathymetric and sedimentological maps of the surfaces of basins forming beneath continental shelves and slopes is merely a passing glimpse.

Depositional margins of sedimentary basins may approximate the erosional edges in young basins still forming today. As the basin develops, progradation of the sediment layers causes the shoreline to migrate seaward, exposing the depositional edges of the older layers inland. Some erosion and beveling of these older layers may take place, but commonly the erosional edges are in close proximity to the preexisting depositional edges. Typical of situations where fair to good conformity exists between depositional and erosional edges are the Tertiary to Recent basins developing in large deltas prograding seaward from low-lying landmasses where erosion is minimal. In these situations the wedges of sediment which constitute sections of depocenters thin shoreward and extend under the land area of the delta where they form the landward margin of the basin.

An example of depositional thinning toward the margin of a basin, accentuated to some extent by erosional surfaces, is described by Whitten (1976) and illustrated in Fig. 2-45. This is a schematic section off the eastern coast of Cape Hatteras United States, showing the general features of the Mesozoic to Tertiary miogeosynclinal basin. The section shows a Triassic sequence enclosed within a graben of continental crust. This section is truncated by a major unconformity on which is deposited a sequence of lenses of Cretaceous to Recent carbonates, sandstones, and shales. The Triassic section has a thickness of up to 1500 m, the younger Mesozoic and Tertiary beds a thickness from 500 m to 4000 m.

In subsurface studies it is important to evaluate the significance of the zero thickness edge shown on the isopach map of a particular sedimentary

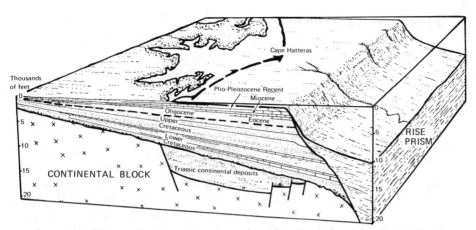

Fig. 2-45. Schematic section off the coast of Cape Hatteras, United States, showing the main stratigraphic features, including depositional thinning and erosional surfaces, within the miogeosynclinal basin. [After Whitten (1976) and Dietz and Holden (1966).]

unit. Is the edge an erosional edge or the limit of a depositional margin? These questions are not only of importance in understanding the basin's history but also may be of economic significance in the search for potential hydrocarbon traps. Determination of the nature of the edge may be assisted by lithofacies or biofacies maps of the stratigraphic unit concerned or by studies of the sedimentary structures and other internal features within the unit. Commonly, the paralic and fluvial facies are missing from a sequence, having been eroded during a brief period of uplift marked by a disconformity that may pass laterally toward the basin into a paracontinuity. Concomitant with an interpretation defining a basin's depositional margin is the concept that although a basin commonly comprises several time-stratigraphic units, some of which may have beveled edges on the landward side of the basin, the margin of the sedimentary basin as a whole may be considered as depositional.

Eroded margins of sedimentary basins are commonly observed in regions of tectonic disturbance where the strata are disrupted by block faulting or folded and thrust faulted along orogenic belts. These basins may be of the miogeosynclinal or eugeosynclinal type. Eroded margins are also observed in many intracratonic basins situated in comparatively stable tectonic environments but subject to regional tilting and consequent beveling of the upper strata over long periods of time. Where deposition has occurred in adjacent troughs separated by a low and broad ridge of basement, erosion may completely remove the strata overlying the ridge and separate the sedimentary sequences in the two troughs. Where these sequences are of a different lithologic nature, it might be assumed that they were deposited as parts of separate sedimentary basins, whereas they may be different facies of the same rock-time sequence.

Erosion of basin margins results in recycling of a portion of the sediments back into a fluvial system and eventually to a marine environment. This possibility must be kept in mind in evaluating the significance of fossil evidence, particularly in the case of microfauna and microflora which are readily reworked from older to younger sediments. Recycling is also observed in some sandstones which contain pebbles of sandstone having much the same composition of fragments as the sandstone in which they are enclosed. Repeated cycles result in a higher percentage of quartz grains, although many quartzose sands have been derived from the reworking of previous sand deposits rather than from sandstone. Recycled sedimentary and fossil material may be redeposited within the same sedimentary basin. When deposition has ceased and the growth of the basin has come to an end, subsequent erosion results in removal and transportation of sediments to new areas of deposition. These may be far removed from the basin being eroded, or they may be adjacent. In the latter case younger subbasins may overlie parts of older basins. Where the younger beds lie almost conformably on the older beds, it may be difficult to differentiate the two basins or to determine whether a bed or coal seam pinches out by deposition or is truncated by an unconformity.

A simple situation illustrating the erosion of a sedimentary basin margin has been outlined by Sloss and Scherer (1975) in Fig. 2-46. Downwarping of the basin with consequent accumulation of sediments is followed by uplift and erosion which results in complete removal of the youngest limestone strata. The basin finally becomes a remnant of its former self, showing only the beds deposited during the early part of its development. No evidence remains of its later history and possible relationships with adjacent basins, although indirectly it may contain some clues. For example, diagenesis resulting from burial metamorphism will leave its trace in the remaining beds. Calculations can be made on the basis of authigenic minerals and reflectivity of carbonized particles as to the temperature–pressure conditions that have obtained and consequently as to depth. Such calculations may be of economic significance, particularly in the case of beds in which the history of maturation of organic matter, kerogen, and hydrocarbon precursors is of particular interest with respect to the hydrocarbon-bearing potential of the beds.

Displaced margins of sedimentary basins may result from transcurrent faults, sole faults, and from folding along orogenic belts. The last two are commonly associated and recognized in the stratigraphic and structural

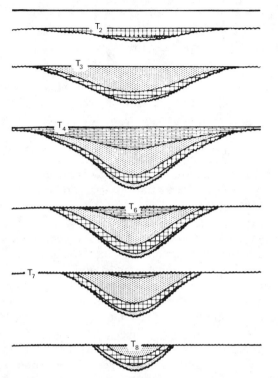

Fig. 2-46. Schematic sections across a sedimentary basin during various stages of subsidence with deposition and uplift with erosion. Width of basin in hundreds of kilometers; depth in kilometers. [After Sloss and Scherer (1975).]

relationships mapped today in mountain belts around the world. Not so evident are the possible basin margin disruptions that may have occurred as a result of major crustal movements within a segment of continental crust or as a result of separation of segments of continental crust. In the former category there is evidence to support the hypothesis that major movements of the continental crust have separated portions of Early Paleozoic basins in Australia (Crawford and Campbell, 1973). There is also evidence to suggest that basins in West Africa and in the region between Venezuela and northeastern Brazil were also once connected. Another example has been discussed by Cebull *et al.* (1976) who point out the apparent offset between strata of the Marathon Mountains in Texas with strata in the Ouachita Mountains of Arkansas, and between the latter and strata in the Appalachian Mountains of the eastern United States. They suggest that these apparent offsets are the result of Paleozoic transform faults which formed an *en echelon* pattern offsetting the Paleozoic basins developed along zones of subduction separating crustal plates, as illustrated in Fig. 2-47. Such hypotheses are stimulating but difficult to demonstrate. As Cebull *et al.* say (p. 111):

> The proposal that Paleozoic transform faults play an important role in the development of the Ouachita–Appalachian structural belt suggests that such faults might leave a significant tectonic–sedimentologic signature in the geological record. However, such a signature, if present, is difficult to confirm by direct observation of geologic relationships.

Could it be that the signature is there, but unrecognized?

Questions concerning intercontinental separation of parts of the same sedimentary basin are not without possible economic significance, as is evidenced by the occurrence of diamonds in conglomeratic beds in both West Africa and northeast South America. Problems concerning the displacement of sedimentary basin margins, or of the displacement of depositional margins of particular stratigraphic units, are much more easily solved and are commonly pertinent to the solution of other problems concerning exploration for hydrocarbons or coal. For example, the depositional margin of a stratigraphic unit may have been displaced horizontally by 100 km or more along low-angle thrust faults. Restoration of the original position of the depositional margin, with respect to the rest of the stratigraphic unit, depends on a knowledge of the structures involved in the disruption and displacement of the margin. By unfolding the folds and

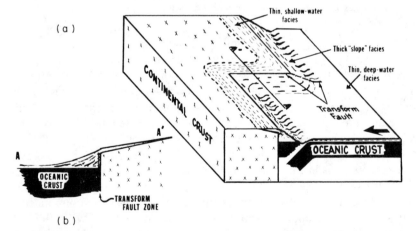

Fig. 2-47.   Schematic sections (a) and (b) illustrating possible growth of a transform fault cutting a zone of subduction and separating a developing sedimentary basin. [After Cebull *et al.* (1976).]

moving sections back along fault planes, the crumpled and disrupted stratigraphic unit can be unraveled to an approximation of its original dimensions. This process, known as palinspastic adjustment, is a very useful method in helping to determine the lithofacies distribution of particular stratigraphic units. It is of particular importance in reconstructing the paleogeographic setting, distribution, and trends of carbonate reefs or sandstone bodies that may be potential traps for oil or gas.

Growth and Dissolution Features

The growth of sedimentary basins depends not only on the rates of sedimentation of noncarbonate material such as lithic sands, silts, and clays derived from river systems but also on the rates of precipitation of carbonate and siliceous tests of pelagic organisms, on the quantity of air-fall volcanic ash that periodically settles on the sea floor, on the chemical precipitation of calcium carbonate mud on shallow but extensive carbonate banks, on the growth of organic bioherms and biostromes, and on the precipitation and subsequent migration of salt beds. These factors will operate qualitatively to varying degrees in different basins and quantitatively will vary from one location to another in the same sedimentary basin. Some basins will consist primarily of carbonates, others of shales and sandstones or of volcanics and siltstone. Most basins will comprise a number of rock types, some of which predominate within a particular vertical section of the basin. Traced laterally, such a section may occupy a different part of the basin sequence. For example, the middle part of a section across the center of a basin, traced laterally, may occupy the lower part of a section across the flank of a basin. Quantitatively also, the rock types within a vertical section will vary from one location to another within a basin, essentially because the sedimentary pile comprising the basin is a composite of several depocenters. Each depocenter or area of maximal sedimentation developed independently, and either successively or penecontemporaneously so that they overlap or merge along the margins. Rates of sedimentation vary not only from one depocenter to another but also within a depocenter from its center to its margin.

Concomitant with growth features formed of salts are features caused by dissolution of salts. These may be local collapse structures resulting in displacement and brecciation of the overlying beds, or broad basinal depressions measuring many kilometers in diameter. Such features may also result in part from the plastic deformation of salt beds which are squeezed laterally to form local diapirs. The thinning or removal of a salt bed by pressure results in subsidence of the overlying beds. By these processes contemporaneous growth and solution features develop within basins, and the growth and distribution of depocenters are influenced.

Another type of structure that probably occurs in some sedimentary basins, but which is difficult to demonstrate, is an astrobleme or fossil crater formed by a meteorite impact. Petrological evidence such as vitrification, strain features in quartz grains, development of minerals indicating conditions of high pressure and temperature, and the occurrence of shatter cones and brecciated rock have been taken as evidence to argue the case for both impact by a meteorite and a cryptoexplosion possibly caused by magma injecting a water-saturated zone. Such evidence is rarely recorded in sedimentary basins; but considering the pockmarked surfaces of other planetary and lunar bodies in our solar system, it would be unreasonable to think that the earth has not been struck in some of the areas now defined as sedimentary basins. Probably the best evidence indicating the dimensions and configurations of structures possibly caused by meteorite impacts or cryptoexplosions will be seen on seismic profiles. Such structures are themselves neither growth nor solution features but may initiate both.

Growth features of sedimentary basins are mainly those that originate as colonies of organisms, such as corals, algae, stromatoporoids, crinoids, and polyzoans, or those that are formed by diapiric structures of salt. An uncommon type of growth structure, but one that occurs in Tertiary beds of the Otway Basin of Victoria, Australia, is the volcanic cinder cone formed on the ocean floor. These structures seen on seismic profiles may be mistaken for small organic bioherms. Larger bioherms which commonly develop as reef trends along the outer and seaward margin of a carbonate shoal or platform are generally tabular in shape with relatively flat tops and steep sides. Sections are normally drawn with a vertical exaggeration sufficient to accentuate the bioherms and associated structures such as domes caused by draping of the overlying beds; but drawn to natural scale, many bioherms have the appearance of sections cut through a table top. A three-dimensional model could be made of many bioherms by drawing a plan view, measuring at least 10 by 20 cm, on a 1-cm-thick board and cutting out the outline with a fret saw. These structures grow in situ by the process of cementation of new skeletal framework on the old. Other colonial organisms with a skeletal framework, such as crinoids and polyzoans, may cover large areas of the sea floor where the environment is suitable. The debris resulting from the accumulation of skeletal remains is swept along by ocean currents and distributed as a layer or lens referred to as a biostrome. Strictly speaking, biostromes are not growth features, although in some cases they assume the proportions of a bioherm. The term reef, which is widely accepted, should be used only with reference to wave-resistant bioherms. In the case of corals the terms bioherm and reef are synonymous because the bioherm will not grow at depths below the zone of photosynthesis.

The development of bioherms in a sedimentary basin does not significantly affect the structural configuration of the basin but merely results in modifications, usually of a minor nature, such as draping of beds over bioherms. The main influence of bioherms on the development of a basin is seen in the relationships of reefs to carbonate sediments being deposited today in areas such as the Persian Gulf, Bahamas, and the Yucatan Peninsula of Mexico. Of much greater effect are the diapiric structures of salt which grow throughout the entire period of development of a basin, by plastic flowage and deformation of an original salt layer. These have a major influence on the stratigraphic and structural configuration of the basin.

The manner in which some of these diapiric structures grow, somewhat reminiscent of the early stage of growth of mushrooms, is illustrated in Fig. 2-48. This diagram shows the shapes and pattern of one type of structure in relation to the original thickness of the Permian salt bed from which they were formed in the North German Basin. Trusheim (1960) is of the opinion that these structures are the result of autonomous movement of salt under the influence of gravity and that movement has taken place over a long period of time, since the Triassic, at an average rate of flow of about 3 m in 10,000 years. The process by which this possibly takes place can be explained if one visualizes the effect of tilting a plate of soft toffee and leaving it for a while. The toffee tends to slump and the surface is heaved up in parallel wrinkles much the same as the linear trends of salt diapers shown in Fig. 2-49. Having a lower specific gravity than the rock in which they are enclosed, these salt structures tend to rise by geostatic adjustment. Thrusting upward individual apophyses of salt produce domes and faults in the overlying beds, some of which form traps for oil and gas. The linear trends from which these diapirs rise are commonly up to 50 km in length but can have a length in excess of 100 km as shown in Fig. 2-49.

The structural configuration of individual salt diapirs and of trends of diapirs is best revealed by means of seismic profiles such as shown in Fig. 2-50 and presented by Williams (1974). These profiles show the development of salt domes within Paleozoic beds offshore from Prince Edward Island in the Gulf of St. Lawrence, Canada. Two-way time in seconds, shown as the vertical scale, refers to the interval of time between the generation of a shock wave and its arrival (after being reflected from a rock layer) at a geophone. The interpretations shown are those of the

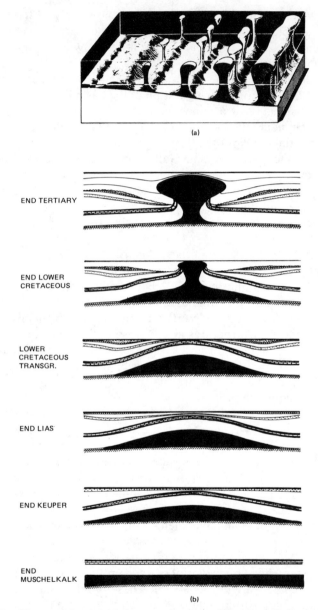

(a)

END TERTIARY

END LOWER
CRETACEOUS

LOWER
CRETACEOUS
TRANSGR.

END LIAS

END KEUPER

END
MUSCHELKALK

(b)

Fig. 2-48.   Diagrams showing (a) shapes and pattern of Permian salt diapirs in the North German Basin and (b) the manner in which these have formed since the Triassic. [After Trusheim (1960).]

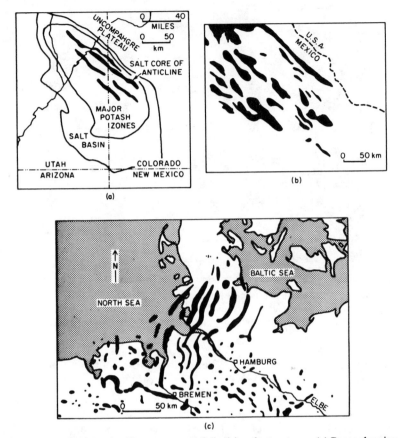

Fig. 2-49. Maps showing linear patterns of diapiric salt structures. (a) Pennsylvanian salt structures in the Paradox Basin, southwestern United States. [After Elston and Shoemaker (1963).] (b) Probably salt structures in northeastern Mexico. [After Murray (1966).] (c) Permian salt structures in the North German Basin. [After Trusheim (1960).]

Hudson's Bay Oil and Gas Company Limited. These reveal folding of the Riverdale beds into which the salt domes are intruded and piercement-type domes which have broken through the overlying beds.

Another example of piercement-salt domes is illustrated in Fig. 2-51 which shows a north–south seismic profile of the continental rise and slope off the coast of Nigeria. The diapirs are steep-sided and extend upward to close under the sea floor which shows slight elevations above the salt domes. The thickness of the sedimentary pile in this region of the continental rise is approximately 2.5 km. According to Mascle *et al.* (1973), the diapirs probably originated in a deep-seated Aptian-Albian salt

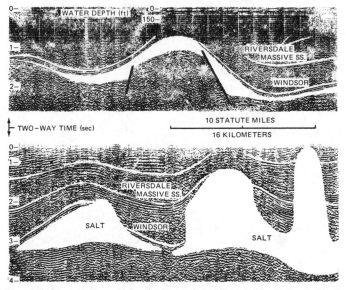

Fig. 2-50. Seismic profiles offshore Prince Edward Island, Gulf of St. Lawrence, Canada, showing Paleozoic salt diapirs and overlying dome structures. Interpretation by courtesy of Hudson's Bay Oil and Gas Company Limited. [After Williams (1974).]

bed which may be the same salt bed known to occur off the coast of Angola and Gabon. If so, this Early Cretaceous salt bed covers a very extensive area off the southwest coast of Africa.

Other intrusive structures which have the appearance of salt diapirs in seismic profiles may in fact be igneous plugs or volcanic cones. One example, in which the structure is interpreted by Bryant *et al.* (1969) as an igneous plug, is illustrated in Fig. 2-52 which shows a seismic profile across the Jordan Knoll in Florida Strait, eastern Gulf of Mexico. The intrusive structure is capped around the margin of its upper surface by an Aptian-Albian reef which is overlain by deep-water Late Cenomanian to Eocene limestone. The younger sediments are of Late Eocene to Pleistocene carbonate muds containing limestone clasts derived from older beds deposited on the knoll. The illustration has a vertical exaggeration of approximately 12×; and if shown to natural scale, the limestone beds capping the knoll would appear to have the shape of a mesa. When deposited, the oldest limestone bed was a carbonate bank rimmed by a chain of reefs which appear to have formed an atoll. Later subsidence drowned the reef and resulted in its burial by deep-water carbonaee mud. Subsequent uplift and partial erosion of the carbonate bank are believed to have occurred during the Middle Eocene.

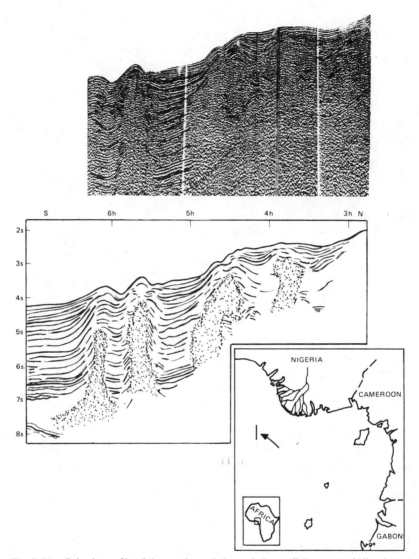

Fig. 2-51.   Seismic profile of the continental rise and slope off the coast of Nigeria (map inset) showing diapirs which probably arise from an Aptian–Albian salt bed piercing the strata and pushing up the sea floor. [After Mascle *et al.* (1973).]

Dissolution features in sedimentary basins are observed in both macroscopic and microscopic scales. The former are seen as breccia, distorted beds, caverns, cavities, and as disrupted or discontinuous strata. Some of these features can be seen in drill core, hand specimens, or in outcrops;

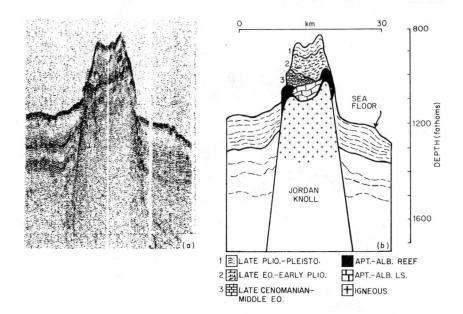

Fig. 2-52. Seismic profile across Jordan Knoll in Florida Strait showing (a, b) interpretation as igneous plug capped by Early Cretaceous reef that formed an atoll. Vertical exaggeration approximately 12×. [After Bryant *et al.* (1969).]

others can be seen only in stratigraphic sections or seismic profiles. Microscopic features include stylolites and dolomite formed by dolomitization during or after lithification. The presence of stylolites indicates dissolution, and an abundance of stylolites may indicate a considerable loss, possibly up to 10%, of the original thickness of the rock layer in which they occur. It is not possible to accurately quantify the probable loss of rock volume from data relating to stylolites but the cumulative loss resulting from such individual pressure–solution features may be significant. Dolomitization also results in a reduction of rock volume and is commonly selective within a carbonate sequence. In general, dolomitizing solutions move most readily through a bed of rock along paths of least resistance which, after lithification, are commonly parallel to the bedding planes. These paths, along which the flow of solutions is greater, are related to the original composition and texture of the sediment, to authigenic changes within the sediment, and to its history of compaction. The possible answers as to why certain zones within a carbonate bed are dolomitized can seldom be stated unequivocally, nor are the stratigraphic and genetic relationships of such zones always understood. Regardless of these uncertainties, pressure–solution and dolomitization must be considered as processes, albeit minor ones, responsible for loss of rock volume.

Dissolution on a major scale involves the removal of a layer, or selectively of part of a layer, resulting in collapse of the overlying beds. This collapse may be relatively rapid, causing brecciation of the overlying beds, or it may be sufficiently slow to allow distortion of the beds. Time is one factor, the competency of the collapsed beds and their relative thicknesses are others. Shale beds tend to become deformed and distorted, whereas limestone beds being more competent and brittle tend to become brecciated. These types of disruption are most evident in thinner beds, particularly in a sequence of interbedded shale and limestone. Thicker beds of shale will be deformed as monoclinal folds draped over the flanks of the subsiding area, and thicker beds of limestone tend to be disrupted in large blocks.

The dissolution features mentioned above are typical of those formed by removal of beds of salt, gypsum, and anhydrite. Solution features such as cavities and caverns formed in massive limestone sections do not always result in total collapse of the bed immediately overlying a cavern, as the limestone section may be sufficiently competent to compensate for stresses arising from these features. Occasionally, deep-drilling rigs have penetrated such caverns at depths of more than 2000 meters.

It should be stated that disruption and distortion of beds caused by collapse are commonly but not exclusively the result of dissolution of an underlying bed. They are certainly caused by removal of the underlying bed, but in the case of a salt bed removal may be effected by lateral flowage of the salt, particularly in areas where salt diapirs have developed. Lohmann (1972) states that seismic records of Permian and Triassic beds in the North Sea show collapse features measuring 1.5 km at an approximate depth of 1.5 km. These are apparently caused by salt removal; but to what extent the removal has resulted from dissolution or flowage is not know.

The process by which a small center of collapse may spread laterally to become a major stratigraphic feature is illustrated in Fig. 2-53 which shows a salt layer overlain by a progressively collapsing bed where it overlies a growing center of salt removal. In this model the salt is removed by solutions, and at no stage does a cavity exist; the space previously occupied by the removed salt is filled with brecciated rock from the overlying bed.

On a much larger scale collapse features may produce graben-type structures where the salt removal is effected by flowage and the development of salt diapirs. Such features can only occur where the initial salt bed

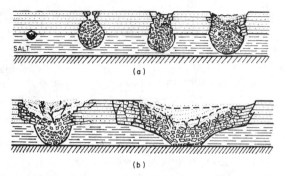

(a)

(b)

Fig. 2-53. Sketch showing (a) development and (b) lateral spread of a collapse feature where bed overlies a salt layer removed by solution. [After Lohmann (1972).]

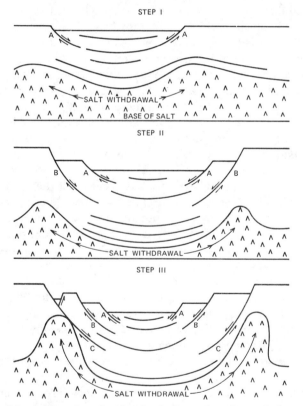

STEP I

STEP II

STEP III

Fig. 2-54. Schematic sections showing development of a collapse–fault structure caused by salt removal in Gulf Coast area, Louisiana. [After Seglund (1974).]

is thick enough to allow a graben to form over the area of total or partial salt removal. In such cases the original thickness of the salt layers is commonly in excess of 50 m. The manner in which such structures are formed is explained by Seglund (1974) and illustrated in Fig. 2-54 which shows the general cross-sectional features of collapse–fault structures in the Gulf Coast area of Louisiana. A plan view of one such structure underlying Calcasieu Lake shows the circular configuration of faults and the known and inferred positions of associated salt domes, as illustrated in Fig. 2-55. The circular collapse–fault structure has a diameter of approximately 25 km.

The interpretations of such structures which constitute major stratigraphic features depend largely on seismic surveys, the records from which can be assembled to form a seismic profile which illustrates the configuration of strata along the line of section. Major collapse features caused by salt removal are described by Lohmann (1972) and illustrated in Fig. 2-56 which shows a seismic profile of Permian and Triassic sedimentary and salt beds underlying the North Sea. The local removal of Triassic salt has resulted in two adjacent collapse features which together have a diameter of 4 km.

Of similar dimensions to the collapse–fault structure underlying Calcasieu Lake in Louisiana is the Rosetown collapse feature in Saskatchewan, shown in Fig. 2-57 and described by De Mille *et al.* (1964). This

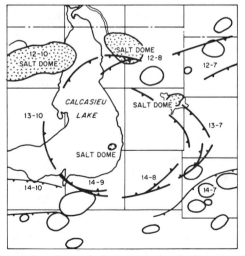

Fig. 2-55. Plan view of a collapse–fault structure showing known and inferred salt domes in Gulf Coast area, Louisiana. Grid system shows townships and ranges, each square comprising 36 square miles. [After Seglund (1974).]

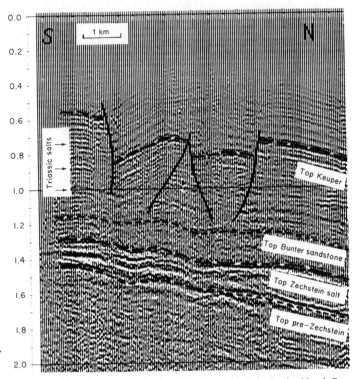

Fig. 2-56. Seismic profile in central part of English Zechstein Basin, North Sea, showing grabens formed by local solution of Triassic salt bed. [After Lohmann (1972).]

feature has been formed by the collapse of beds overlying an area of removal of the Prairie Formation salt which has a thickness of up to 200 m.

The Prairie Formation is part of the Middle Devonian Elk Point Group which includes evaporite beds of halite, potash, and anhydrite extending northward from North Dakota to the border area between Alberta and the Northwest Territories, as shown in Fig. 2-57. Salt diapirs are not recorded in the area, and the removal of salt in the Rosetown structural low was apparently caused by solutions. Wells drilled in the low have encountered only breccia within the stratigraphic interval previously occupied by salt.

It is of interest to note that the upper part of the Prairie Formation contains a commercially mined bed of potash and that the evaporite facies of the Elk Point Group show a zonation marked by an inner area of potash, a middle area of halite, and an outer area of anhydrite. This subject will be discussed later in the book.

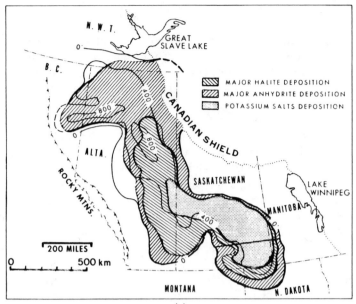

(a)

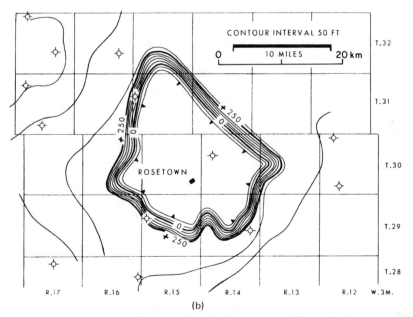

(b)

Fig. 2-57. (a) Evaporite facies of the Middle Devonian Elk Point Group, Western Canada. (b) Rosetown collapse feature, Saskatchewan, caused by solution of the Middle Devonian Prairie Formation. [After De Mille *et al.* (1964).]

Solid Geometry of Basins

The solid geometry of a sedimentary basin is not static but varies to some degree in configuration throughout the depositional history of the basin. Subsequently, when deposition has finished and the basin is folded, faulted, uplifted, and eroded, its configuration changes considerably. Most of the sedimentary basins known today are mere remnants of preexisting basins, yet the present configuration of these is commonly better understood than the configuration of most sedimentary basins still forming today. The reason is clearly that most of the ancient basins have been mapped, and many drilled and surveyed by geophysical methods, whereas the basins forming today are offshore, more difficult of access, and certainly much more expensive to explore. For these reasons descriptions of sedimentary basins based on their solid geometry or three-dimensional shape must clearly define the stage of sedimentary development or of orogenic and tectonic distortion, disruption, and bevelment that is inferred. This requirement imposes a condition that cannot always be fulfilled, particularly as it presupposes a fairly comprehensive understanding of the entire history of a basin. In most cases where the entire geological history of a sedimentary basin is under review, particularly with reference to its later history of deformation, the sequence of events and corresponding basin configuration are interpreted in part on the basis of inferences having varying degrees of validity. As with many geological investigations, the interpretations may be right, partly right, or wrong; but as long as they do not invalidate or disregard the available data, the interpretations can be accepted, or debated, but not ignored.

The stratigraphic and tectonic relationships of subbasins to major basins must be considered in working out the evolution and three-dimensional development of basins. Of particular interest is the spatial and apparent genetic relationship of adjacent basins that have evidently had different sedimentary and tectonic environments. A classic case showing the couple formed by a miogeosyncline and a eugeosyncline in the Appalachian region of the United States has been discussed by Kay (1951) and is illustrated in Fig. 2-58. The section crosses the present fold belt of the Appalachians in New York, Vermont, and New Hampshire and shows the Early Paleozoic basins as they may have been at the close of the Ordovician. The section has been the subject of various interpretations and modifications, all of which depend on palinspastic reconstruction of the fold belt. Of particular significance to the evolution of the three-dimensional shape of these basins is the relationship of each basin to the bathymetry and tectonic framework of the continental margin. The miogeosyncline is characterized by carbonates and other shelf-type sedi-

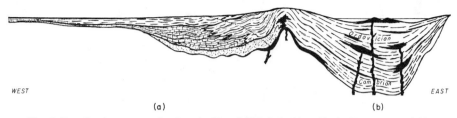

WEST   EAST

(a)   (b)

Fig. 2-58. Section across the Appalachian fold belt in New York, Vermont, and New Hampshire showing the Early Paleozoic (a) miogeosyncline and (b) eugeosyncline couple as it may have been at the close of the Ordovician. Section is 400 km long. Maximum thickness of eugeosyncline is 7500 m. [After Kay (1951).]

ments that were evidently deposited in neritic environments; the eugeosyncline is characterized by turbidites and volcanics that were deposited in deeper water. These differences in rock-type assemblage suggest that the miogeosyncline sediments were deposited on a continental shelf and the eugeosyncline sediments on a continental rise subject to volcanism but not necessarily separated from the continental shelf by a volcanic archipelago. Such couples, rather than being considered as separate geosynclines, are different facies of a major accumulation of strata that could be regarded as a megageosyncline. Subsequent uplift, as shown in Fig. 2-58, may result in structural separation.

Sectional views of sedimentary basins are effective representations of particular aspects of the solid geometry but must be regarded in the context of the whole basin. This can be illustrated in the form of an isopach map which shows the areal distribution of a basin, its linearity, and its symmetry as seen in sectional views. In general, sedimentary basins can be classified as linear or nonlinear and as symmetrical or asymmetrical. Many modifications and subcategories may be included in this general classification as suggested by the complexity of structures shown by Kamen-Kaye (1967) in Fig. 2-59. This figure illustrates the general features of cross sections seen in many basins that have been subjected to tectonic and orogenic activity. Cross-sectional views of the original basins may be considerably different, although basement faulting may be contemporaneous with deposition. The present-day solid geometry of a basin can be interpreted in different ways depending largely on the availability and quality of data; but there is only one correct solution. On the other hand, evolution of the present-day solid geometry is a matter of interpretation and cannot be stated unequivocally.

During their stages of sedimentary development basins may be oval to linear in general outline and lopsided in the area of thickest accumulation of sediments. That is to say, they may be asymmetrical in sections drawn

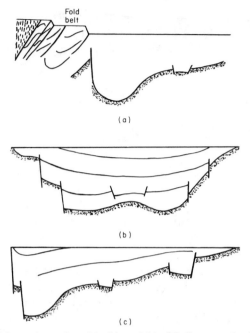

Fig. 2-59.   Schematic sections (a), (b), and (c) of sedimentary basins showing symmetrical (b) and asymmetrical (c) configurations and their relationships to basement faults (a). [After Kamen-Kaye (1967).]

both parallel and normal to the long axis of the basin. This relationship is particularly relevant where subbasins have developed. An example of asymmetry in a present-day sedimentary basin is illustrated in Fig. 2-60 which shows sections of the Bengal Fan based on Alam (1972) and offshore seismic surveys cited by Curray and Moore (1971). The Bengal Fan forms a sedimentary basin that is oval to linear, has a length of 3000 km and a width of 1000 km. It lies south of Bangladesh between the Andaman Islands and the east coast of India. The greatest thickness shown in this section, up to 12 km, lies beneath the continental shelf and slope off the coast of Bangladesh where the present-day delta of the Ganges–Brahmaputra river system is prograding. The sediments range in age from Miocene to Quaternary and form a wedge that thins to the south, away from the source of sediments. Asymmetry possibly results from basement subsidence along a trend extending northwest from the Java Trench.

A generalized section of the Bengal Fan, drawn normal to the linear section, is also shown in Fig. 2-60. This section extends from a stable shield area underlying shelf sediments on the west, through a thick section

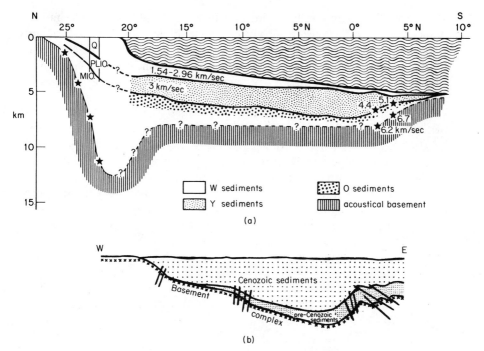

Fig. 2-60. (a) Section based on offshore seismic survey of the Bengal Fan south of Bangladesh, a sedimentary basin 3000 km long, 1000 km wide, and up to 12 km thick. [After Curray and Moore (1971).] (b) Generalized section of the Bengal Fan from India on the west to Burma on the east. Length of section is 500 km. Vertical exaggeration 9×. [After Alam (1972).]

of sediments, to a mobile belt on the east. It shows an asymmetrical section thickening toward the mobile belt, suggesting that the Bengal Basin is asymmetrical not only from north to south but also from east to west.

With few exceptions sedimentary basins are asymmetrical at the close of their depositional history and tend to become more so following orogenic disturbance. During the period of their growth sections normal to the general direction of sediment transport are commonly asymmetrical but may be symmetrical, whereas sections parallel to the direction of sediment transport, which in the case of a delta is also the direction of progradation, are always asymmetrical. The latter sections pinch out on the landward side to a zero isopach edge and thin out on the seaward side to an approximation of a zero edge. The section is always thickest underlying the continental slope and rise which tend to migrate seaward with the prograding stratigraphic units. In situations where the continental slope is underlain and probably localized by deep-seated and active faults, it may remain stationary. In such cases the sediments that would normally accrete on the slope and build it forward are swept down to the continental rise by turbidity currents triggered by periodic movements of the underlying faults. The result is a stationary continental slope and a prograding continental rise.

An important factor in the pattern of sedimentation and solid geometry of a sedimentary pile during the period of its growth is the configuration of basement. The influence this exerts on basin development is discussed by Scrutton and Dingle (1973) with reference to sedimentary accumulations on the western continental margin of southern Africa. Fig. 2-61 illustrates two schematic sections in this region, the upper section showing the continental margin off the coast of Cuanza Basin of Angola, and the lower section showing another prograding sequence of beds off the west coast of South Africa. As seen in the section off Angola, local thickening of the sedimentary pile may occur not only beneath the continental slope and rise but also where sediments are dammed by basement ridges.

Another factor in the sedimentary pattern and consequent configuration of the sediment pile is the growth and lateral shifting of local centers of deposition. The areas of maximum deposition in these centers are referred

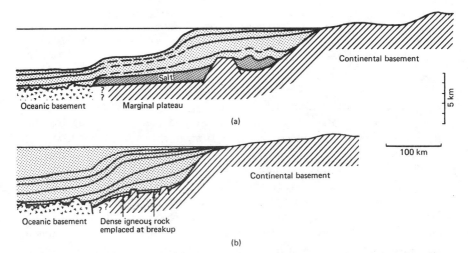

Fig. 2-61. (a) Schematic section of the continental margin off the Cuanza Basin, Angola. [After Scrutton and Dingle (1973), Reyre (1966), Brogan and Verrier (1966), Baumgartner and Van Andel (1971), and Leyden *et al.* (1972).] (b) Schematic section of the continental margin off the west coast of South Africa. [After Scrutton and Dingle (1974), Bryan and Simpson (1971), and Scrutton (1973).]

to as depocenters. These migrate laterally, depending on various factors such as compaction of the sedimentary pile and faulting, but primarily on shifts of the river system where deltas are concerned. These centers of deposition are underlain by lenticular sedimentary sequences which show an offlapping arrangement in sections parallel to the general direction of trend of the coastline and a prograding arrangement in sections normal to the coastline and parallel to the general direction of sediment transport. One such center of deposition of Late Pliocene and Pleistocene sediments off the coast of Louisiana is shown in Fig. 2-62. The sequence, deposited during a period of less than 5 million years, attained a maximum thickness of more than 3.5 km, indicating an average rate of sedimentation in the central area of the depocenter of 0.7 m in 1000 years. The solid geometry of this sedimentary sequence has been modified by the growth of salt

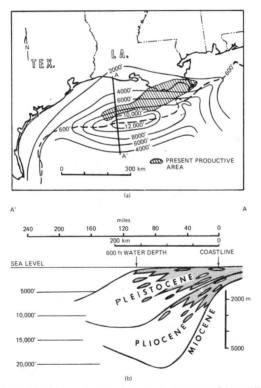

Fig. 2-62. (a) Generalized isopach map showing depocenter of Late Pliocene and Pleistocene sediments off the coast of Louisiana. Section (b) shows prograding arrangement of Pleistocene and Pliocene sedimentary sequences. [After Woodbury *et al.* (1973).]

diapirs formed as a result of the weight of sediments overlying more than a thousand meters of mobile salt beds. Woodbury *et al.* (1973) note that during the period of time in which this sequence was deposited, the depocenter shifted to the southwest by more than 300 km.

Of particular interest is the fact that hydrocarbon productive trends follow the shifting path of the depocenter of this Pliocene–Pleistocene sequence. The relationship of such trends to thicker sections, and consequently to the prospects for hydrocarbon accumulations in sections underlying the continental slope and rise, has been pointed out by Roberts and Caston (1975) who suggest that basins with less than 1 km of sediment are unlikely to be oil-bearing. Shallow basins do have a better chance of being gas-bearing. These comments refer to sedimentary basins that have not been subjected to appreciable erosion, as many present-day shallow basins have been deeply buried, uplifted, and substantially beveled. In many such cases the hydrocarbons have largely escaped. The prospects for hydrocarbon accumulations in thicker sections underlying the continental slope and rise are being seriously considered in view of technological achievements in drilling that are now possible. A generalized section showing the relative thickness of prograding sedimentary sequences overlying a tectonically passive continental margin of the type formed by rifting, is seen in Fig. 2-63. The solid geometry of many basins of this type shows some degree of asymmetry in sections parallel as well as normal to the trend of the continental margin. Sections approximately midway between the parallel and normal sections may show stratigraphic units lying with a parallel rather than a prograding arrangement, as seen in the section across the Grand Banks in Fig. 2-64. Section (b) of this figure shows a prograding arrangement of stratigraphic units and the effect on section thickness of a major basement fault underlying the continental slope. Austin (1973) states that the interfaces between seismic velocity layers 1 and 2 and between layers 2 and 3 are thought to be erosional and that layer 1 correlates with the Tertiary, layer 2 with the Upper Cretaceous, and layer 3 with Lower Cretaceous to Triassic.

The solid geometry of stratigraphic accumulations ranging in age throughout the Mesozoic is commonly complex, as shown in the offshore region of eastern Canada. As more information becomes available, it is possible to differentiate individual basins and subbasins within the total sedimentary pile. In other thick sedimentary piles of long-ranging age the stratigraphic units may be relatively conformable, although separated by disconformities or paracontinuities. An example is illustrated in Fig. 2-65. This figure shows a schematic section, from Saudi Arabia to Iran, of the stratigraphic sequence underlying the region of the Persian Gulf. This sequence, ranging in age from pre-Permian to Tertiary, comprises a gen-

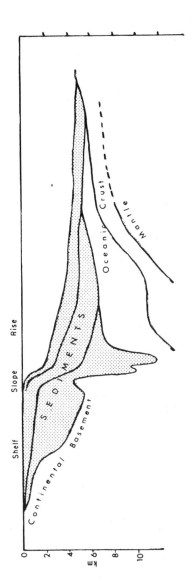

Fig. 2-63. Generalized section across a continental margin to show relative thickness of prograding sedimentary sequences and their spatial relationships to the continental shelf, slope, and rise and also to the underlying continental and oceanic crust. [After Roberts and Caston (1975).]

119

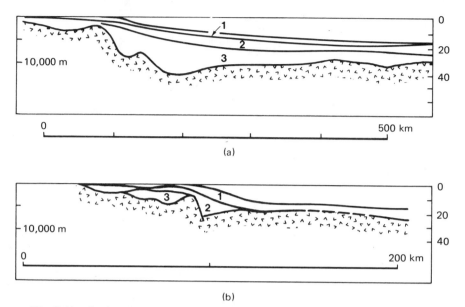

Fig. 2-64. Sections across (a) Grand Banks and (b) Scotian Shelf based on seismic refraction surveys off the east coast of Canada. Layer 1: Tertiary, Layer 2: Upper Cretaceous, Layer 3: Lower Cretaceous–Triassic. Vertical scale in thousands of feet. [After Austin (1973), Drake *et al.* (1959), and Officer and Ewing (1954).]

erally conformable arrangement of stratigraphic units. Commenting on this stratigraphic sequence Kamen-Kaye (1970, p. 2371) says:

> Continuous subsidence and sedimentation through Phanerozoic time resulted in the deposition of maximum thicknesses of 25,000 ft, and possibly more than 30,000 ft under the present mountains and foothills of Iran. The zone of maximum thickness may best be regarded as one relatively thick prism within an immense compound prism of sedimentary rocks. The sedimentary prism also may be regarded as having been the site of persistent epicontinental conditions in a persistent platform environment.

Some of the structural complexities encountered in stratigraphic accumalations within sedimentary basins forming today are illustrated in Figs. 2-66, 2-67, and 2-68. Fig. 2-66 illustrates a structural section off the coast of Maryland and New England from the continental shelf to the continental rise. Based on seismic reflection profiles, the section shows the configuration of major reflecting horizons, within the stratigraphic sequence, and the depth of the oceanic crust. Of particular interest is the interpretation showing a substantial fault within the sedimentary pile un-

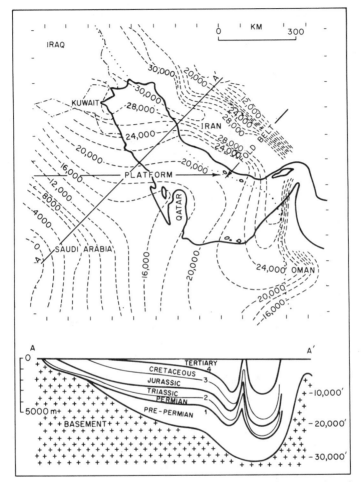

Fig. 2-65. Schematic section of the Persian Gulf region showing general conformity of stratigraphic units ranging in age from pre-Permian to Tertiary. [After Kamen-Kaye (1970).]

derlying the continental slope. This type of fault may be a growth fault active during the period of sediment accumulation and may reflect movement within an underlying deep-seated basement fault. Schlee *et al.* (1976) state that the Baltimore Canyon Trough, across which this line of section traverses, contains up to 14 km of presumed Mesozoic and younger sedimentary rocks. The adjacent Georges Bank Basin contains up to 8 km of sedimentary strata.

Fig. 2-67 illustrates the structural influence of salt diapirs on the configuration of sedimentary accumulations off the coast of Morocco. Also

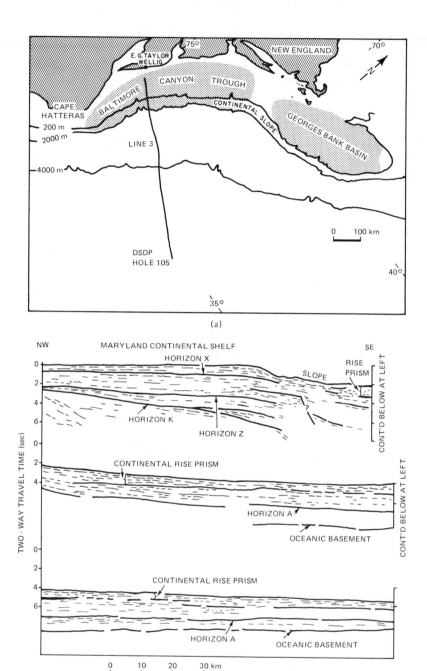

Fig. 2-66. Section across the continental margin east of Maryland, United States, showing major seismic reflection horizons, the depth of oceanic crust, and a probable fault beneath the continental slope. [After Schlee *et al.* (1976).]

122

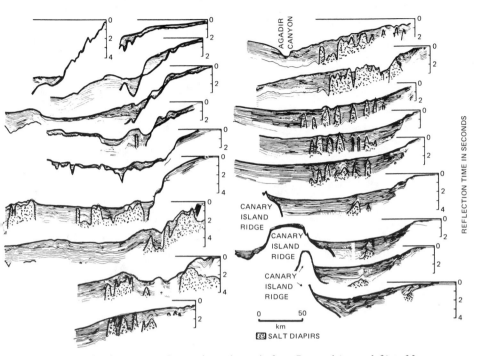

Fig. 2-67. Sections across the continental margin from Portugal (upper left) to Morocco and to the Canary Islands (lower right), showing the structural influence of salt diapirs on post-Triassic sedimentary accumulations, and the entrapment of a basin between the Canary Islands and Morocco. Sections based on seismic reflection, magnetic, and gravity surveys. Vertical exaggeration 13×. [After Uchupi *et al.* (1976).]

shown is the influence of the Canary Islands Ridge which forms a dam trapping sediments between itself and the mainland. According to Uchupi *et al.* (1976), the oldest layers overlying basement of the continental margin include evaporites and were deposited during the Late Triassic to Early Jurassic.

Fig. 2-68 is an interpretation by Barr (1974) and illustrates a section based on a seismic reflection profile across the continental margin off the southwest coast of Vancouver Island. It shows the influence of basement topography, controlled in part by faulting, on the accumulation of sediments. Also shown is the horizontality of sediments overlying the deep Cascadia Basin and the damming effect on sedimentation of ridges of basement rock.

Following the close of its depositional history, a basin is commonly uplifted and beveled but may also be dislocated or deformed by faulting and folding. Where uplift and beveling only are involved, the basin may

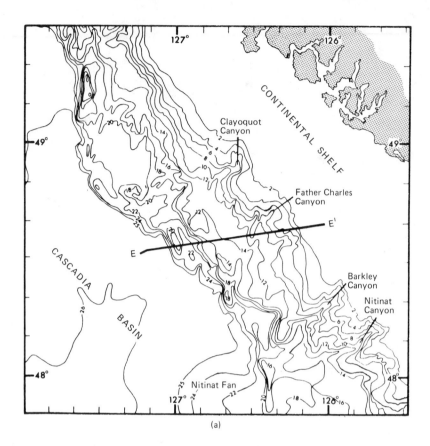

Fig. 2-68. Section across the continental margin off the southwest coast of Vancouver Island showing (a) effect of basement topography on sediment accumulation. Based on a seismic reflection profile (b). [After Barr (1974).]

124

preserve many of the characteristics it had during its period of deposition, although severe beveling will remove the depositional margins of the original basin. Examples illustrating the stratigraphic configuration and solid geometry of three uplifted and beveled Cretaceous to Tertiary basins in Australia are shown in Figs. 2-69, 2-70, and 2-71.

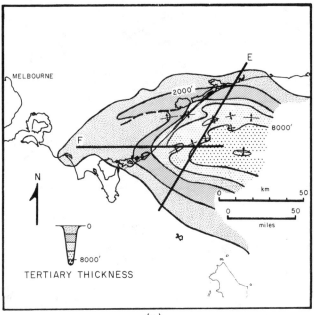

(a)

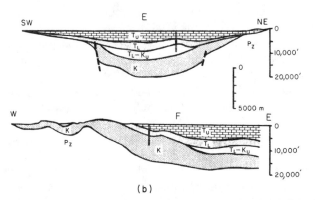

(b)

Fig. 2-69. (a) Isopach map showing sections and (b) Tertiary thickness of the Gippsland Basin, Australia. [After Weeks and Hopkins (1967).]

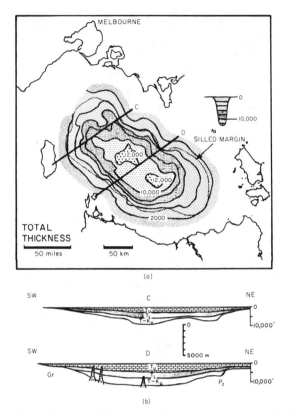

Fig. 2-70. (a) Isopach map and (b) sections of the Bass Basin, Australia. Total thickness is given in feet. [After Weeks and Hopkins (1967).]

Fig. 2-69 shows an isopach map and sections of the Gippsland Basin in Victoria. Weeks and Hopkins (1967) record that this basin, which lies mainly offshore, has an area of approximately 50,000 km² and a thickness of at least 3000 m of Late Jurassic to Cretaceous terrigenous sediments and 2500 m of Tertiary clastics and carbonates. The largest production of oil, condensate, and gas in Australia comes from Eocene sandstone beds in this basin. The sections and isopach map shown in Fig. 2-69 illustrate the general features of the solid geometry of the Gippsland Basin which is seen as being symmetrical in section E and asymmetrical in section F. In detail, the Late Jurassic to Cretaceous beds form a separate basin under-lying a Late Cretaceous to Tertiary basin. These basins have approxi-mately the same configuration and depositional trend. In particular, the Early Tertiary beds show a history of prograding deltas growing eastward and of periodic westward transgressions of the sea. During the Late Ter-

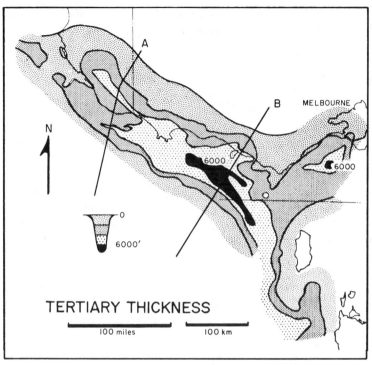

(a)

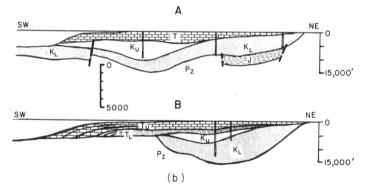

(b)

Fig. 2-71. Isopach map showing (a) sections and (b) Tertiary thickness of the Otway Basin, Australia. [After Weeks and Hopkins (1967).]

tiary there was widespread marine transgression accompanied by the development of extensive carbonate shelves. The northern flank of the Gippsland Basin is an erosional margin. The southwestern flank is separated from the Bass Basin by a basement ridge which must have had an influence in limiting the depositional margin. Seaward, the basin appears to be open-ended because of lack of information based on drilling and geophysical surveys. Contours on the isopach map shown have been extrapolated to the vicinity of the edge of the continental shelf, which interpretation suggests that the Gippsland Basin extends underneath the continental slope and continental rise.

The Bass Basin shown in Fig. 2-70, a deeply silled basin lying between the mainland and Tasmania, is bounded on both the west and the east by ridges of basement rock. The basin has an area of about 90,000 km^2 and a thickness of more than 3500 m of Late Cretaceous to Tertiary sediments. Of particular interest are the presence of Tertiary cinder cones within the sequence. On seismic records these can appear to be pinnacle reefs. The Bass Basin is probably one of the few examples of a sedimentary basin having an original basinlike shape that is preserved today. Little is known of the stratigraphy of this basin as it has been sparcely drilled. Structurally, the Bass Basin is an unusual example of a basin that in three dimensions is symmetrical.

The Otway Basin in Australia, shown in Fig. 2-71, has an area of approximately 115,000 km^2 and a thickness of 4500 m of Cretaceous and Tertiary sediments. This is an interesting example of an elongate basin comprising two basins having somewhat different depositional trends and structural configuration. The lower sequence of Cretaceous beds shows a northwest trend and is beveled and overlain by a Tertiary sequence which also has a northwest trend slightly offset from that of the Cretaceous. The Tertiary beds were deposited as a prograding sequence building seaward away from the present shoreline and in a direction approximately normal to the trend of the Cretaceous basin. The northwest margin of the Otway Basin is erosional; the southeast margin is probably in large part a depositional edge as it lies against a basement ridge which forms a separation from the Bass Basin.

Superposition of sedimentary basins within a major area of tectonic subsidence is a fairly common relationship. Although individual basin trends may be somewhat offset from the one below, the stratigraphic sequences composing each basin may be essentially conformable. Disconformities and paracontinuities representing periods of uplift, beveling, and nondeposition or erosion in situ may be recognized, but the total sedimentary pile in effect forms a single megabasin. An example is cited by Dickey (1972) in Fig. 2-72 which shows a section from west to east across the

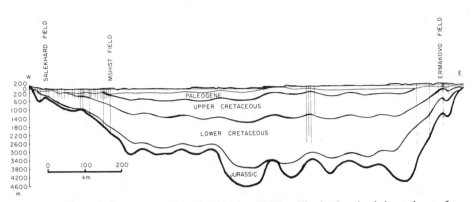

Fig. 2-72. Section across the basin underlying the West Siberian Lowland situated east of the Ural Mountains, USSR. Oil fields are shown. [After Dickey (1972) and Vasil'yev (1968).]

central part of the West Siberian Lowland situated east of the Ural Mountains, USSR. This megabasin is generally oval in shape, has a northerly trend and a width of more than 1000 km. It includes an area of about 1,750,000 km^2 and a thickness of up to 4500 m of Jurassic to Tertiary sediments. Several large oil fields and gas fields are producing from Jurassic and Cretaceous sandstone beds.

The solid geometry of sedimentary basins is commonly illustrated by two or more isopach maps, with appropriate stratigraphic sections, of selected stratigraphic sequences. The selection of these sequences depends on several factors such as the rock types within a sequence, the presence of unconformities and disconformities, and generally on the availability and quality of information.

CHAPTER

3

Stratigraphic and Structural Concepts

SEDIMENTATION

Rates of Physical Sedimentation

Physical sedimentation refers to processes whereby lithic and calcareous rudites, arenites, and lutites are transported and deposited. Some carbonate muds have formed by physicochemical precipitation and deposition of minute crystals of aragonite. Rates of physical sedimentation range from a few centimeters to several meters in a thousand years. The former rate may obtain in some abyssal depths not subject to turbidity currents where the only source of transported particulate sediment is the dust that falls into the ocean. At such depths there may be sediments such as clays and siliceous ooze which have been deposited by chemical precipitation. There may also be siliceous oozes formed by the accumulation of siliceous tests of dead plankton which rain down upon the sea floor. This is a biological process in a sense, and could be referred to as biophysical sedimentation. Such chemical and biophysical processes result in rates of sedimentation that may be less or more than the rate of sedimentation of settling dust. Commonly, where the accumulation of dust is the main factor in sedimentation, the rate is small, possibly of the order of 1 cm or less in 1000 years. But there are exceptions in oceanic areas where prevailing winds blow dust far out to sea from large deserts.

At the other end of the scale, the rates of sedimentation in depocenters of a delta complex, or on the upper slopes of a continental rise formed by a submarine fan, may be several meters in a thousand years. These generalizations are subject to modifications depending on the interrelation of a number of geographic and climatic factors. By way of quantifying the upper and lower limits of sedimentation within a period of 1000 years, defined arbitrarily as 1 cm in abyssal depths and 4 m in delta depocenters, the following rates can be calculated. During the interval of 5 million years since the beginning of the Pliocene, 50 m of sediment may be deposited in abyssal depths, whereas 20 km may theoretically be deposited on the sea floor in a delta complex. Even allowing for compaction of the sediments with burial, the latter figure is obviously too high. It can be explained by the fact that whereas the rate of sedimentation within a depocenter may be as much as 4 m in 1000 years, the positions of depocenters are continually changing, depending on fluctuations of the river course and other factors. In this regard it is relevant to refer to Fig. 2-60 which shows the maximum and average thickness of a wedge of sediment deposited since the beginning of the Pliocene in the Ganges–Brahmaputra Delta to be 4 km and 3 km respectively. In the Gulf of Mexico, as illustrated in Fig. 2-62, the maximum thickness of sediment deposited south of Louisiana during the same interval of time is about 6 km.

There is always a temptation to quantify rates of sedimentation and to relate these to thicknesses of sediment within a delta or to rock sequences within a sedimentary basin. In doing so, certain assumptions must be made regarding the rates of compaction within a sedimentary pile. Rates of sedimentation are commonly expressed in terms of thickness in surficial zones. A mud slurry deposited on the sea floor may have a porosity or water content of approximately 40% of its volume. At depth this porosity is reduced, more than half of the water being expelled during the first 500 m of burial (Fig. 3-16); but in sedimentary piles such as those of the Mississippi Delta the clay sediments can remain as stiff clays under sediment loads of 3000 m or more, at which depths the porosity has been reduced to nearly 15%. Lithification, with further loss of porosity requires processes of cementation and diagenetic alteration which may not be completed during the life of active sedimentation of the basin.

These comments serve to point out the many factors involved in relating rates of sedimentation to thicknesses of sediment and rock sequences. Descriptions of rock sequences commonly carry the implication that they were rapidly deposited, or slowly deposited in a sediment-starved basin. Such inferences may be valid but must be stated with caution. Furthermore, they may apply to part but not to the whole sequence of beds. For example, the accumulation of sediment within a depocenter in a delta

complex may be comparatively rapid. But as the depocenter shifts laterally, the area it previously occupied receives much less sediment. A further shift of the depocenter may again accelerate the rate of sedimentation. The problem of estimating rates of sedimentation in buried
sequences is consequently fraught with hazards.

Sediment-starved basins are commonly referred to and require some
qualifications. It is a relative term and implies a markedly lower rate of
sedimentation in one basin as compared with another, or in one part of a
basin with respect to another. In this sense abyssal depths, where rates of
sedimentation are of the order of a few centimeters in a thousand years,
may be regarded as sediment-starved. The term may also be applied to the
central parts of deep intracratonic basins filling in from the margins by
prograding deltas. The majority of intracratonic basins were probably
shallow throughout their entire depositional history and consequently are
likely to have been sediment-starved. In shallow basins sediments are
readily transported by currents and are swept out on the shelf far from
shore, thus distributing the load.

Apart from the obvious relationship of distance from a sediment source,
which may be a delta or an unstable accumulation of sediments on a
continental slope or rise, crustal movements may result in sediment starvation. For example, where the central part of a shallow intracratonic
basin is subjected to downwarping caused by block faulting, and provided
the rate of downwarping exceeds the rate of sedimentation, then the sea
floor in the central part of the basin becomes progressively deeper. The
result is that sediments which were previously swept over the marginal
shelves to the center of the basin are now deposited on the slopes of the
shelves where they form a prograding sequence. Some sediment may be
transported to deeper parts of the central basin by means of local turbidity
currents arising from the slumping of sediment on the slopes. But in
general the central part of the basin is effectively starved of sediment.

An example of a starved basin is described by Lineback (1969) and
illustrated in Fig. 3-1. The figure shows an isopach map and section of the
Mississippian Valmeyeran Series in the Illinois Basin. Limestone and
siltstone on the basin's margin, interpreted as shallow-water shelf sediments, commonly have a thickness of 100–200 m whereas shale and siliceous rock in the central part of the basin are less than 30 m thick. Of
interest is the rectangular outline of the part of the basin shown. In other
areas outlines such as this are thought to possibly reflect very large-scale
block faulting in the basement.

Fig. 3-1 was based on marker beds established from electric logs and
well samples. Lineback states (p. 112), "The tops and bottoms of marker   ·

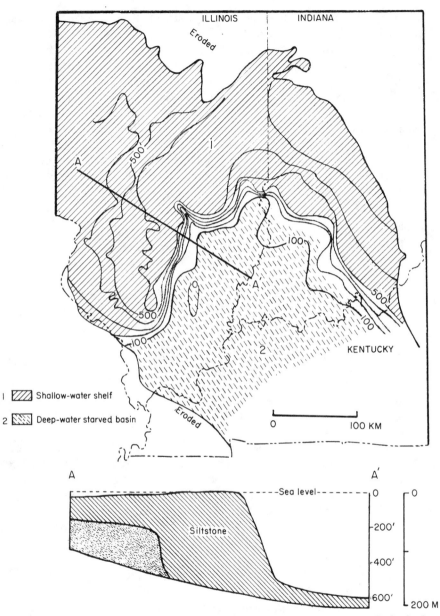

Fig. 3-1. Isopach map and section of part of the Mississippian Valmeyeran Series in the Illinois Basin. 1: Shallow-water shelf limestone and siltstone. 2: Deep-water shale and siliceous rock. Contours and depths in feet. [After Lineback (1969).]

beds are bedding planes that record the original sediment-water interface, and thus are time planes.'' He says further that the interpreted angle of slope of some bedding planes suggests that a sequence of several hundred feet of rock in one place may be stratigraphically equivalent to a few inches of rock in another.

Variable rates of sedimentation in which sediment starvation is at one extreme end of the range, combined with the subsequent effects of variable degrees of compaction in sediments of different composition, result in problems with subsurface correlations where there are no definitive electric log markers. In such cases there may be a multiple choice of correlations based on electric logs only, and careful analysis of well samples may be needed to resolve the problem.

Mass sedimentation at the other range of the scale is characterized by thick sedimentary deposits that have accumulated rapidly as a result of gravity sliding and transportation by turbidity currents. The latter are triggered off by masses of sediment that have slumped as gravity slides down unstable slopes, but which continue to flow down the sea-floor slopes as a homogeneous and immiscible fluid. Initial mass failure of sediment accumulations on depositional slopes may arise from overloading, earthquake shock, or hydraulic pressure from a combination of abnormal tides and waves.

The basic physical properties of noncarbonate sediments that determine their reactions to stress, and the mechanics of their failure, are lucidly outlined by Dott (1963, p. 104). He said:

> Clay-rich sediments, though commonly considered the least stable, possess greatest intrinsic bulk cohesion due to molecular and surface forces. Coarser materials derive stability from particle mass and physical packing; they are essentially cohesionless. Intermediate silt and fine sand actually tend to be least stable; they are but weakly cohesive and commonly accumulate with open, metastable packing. Hydrostatic pore pressure buoys saturated sediment masses slightly, reducing shear resistance. All fine sediments are susceptible to disturbance of their packing structures by forcible movement of fluids, thus slight shocks may deform metastable structure inducing gravity failure, even on half-degree slopes; such failures occur chiefly in areas of rapid deposition.
>
> Two fundamental dynamic boundaries distinguish elastic, plastic, and fluid behaviors. *Plastic flow* begins when the *yield limit* of cohesive sediments is exceeded; original stratification is preserved but contorted. *Viscous fluid flow* occurs if the *liquid limit* is exceeded; decreasing cohesion and churning destroy stratification and suspen-

sion forms. This continuum of movements can be arrested at any point as is illustrated by the sedimentary record (load flows, contorted strata, pebbly mudstones). But if *liquid limit* is reached, cohesion normally deteriorates very rapidly by spontaneous liquefaction, almost inevitably producing a turbid suspension. This apparently accounts in part for the much greater abundance of graded bedding compared with contorted strata and pebbly mudstones.

The main forms of mass sediment transport are illustrated in Fig. 3-2. These forms tend to merge and are subject to modification but constitute a

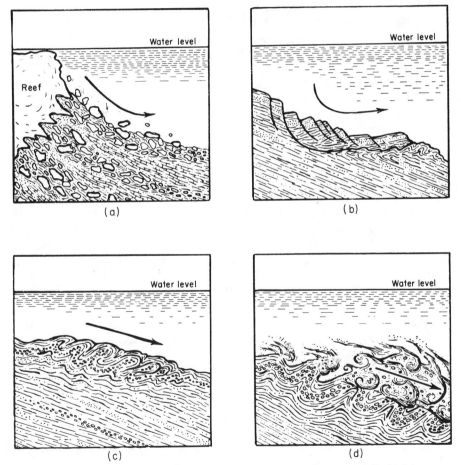

Fig. 3-2. Types of mass movement of sediment. (a) Subaqueous rock fall. (b) Subaqueous slump and sliding. (c) Subaqueous mass flow. (d) Subaqueous turbidity flow. [After Dott (1963).]

basic classification. Subaqueous rockfalls are talus deposits modified by current transportation. They include accumulations of carbonate detritus on the seaward or foreslope of reefs, volcanic debris on the flanks of submarine volcanoes, and sedimentary breccia formed by the breaking up of lithified carbonate beds overlying a slumping mass of sediment. They are essentially local features. Subaqueous slump and sliding occurs within a wide scale of dimensions ranging from a few to several thousand meters. Where slumping involves the upper few tens of meters of a section, the sediment slides down the slope in a sequence of steplike wedges slipping along a surface that is subparallel to the bedding planes. Where the slumping is on a megascale at depths of several hundred to thousands of meters, the slump surfaces form a complex pattern, commonly fan-shaped in section and branching in plan view, of growth faults as illustrated in Fig. 3-22. Subaqueous mass flow results from the downslope movement and plastic deformation of surficial beds of sediment that retain sufficient cohesion to preserve relict stratification. Such mass flow results from translation sliding of upper beds slipping over lower and forms convolute bedding. Commonly this type of sedimentary structure is overlain by turbidites, and it has been assumed that the convolutions resulted from the pulling up of sedimentary layers underlying the turbidity current. This may happen, particularly in the uppermost layer where flame structures are obviously formed by disruption of the surface over which the current flows. But in many examples seen in outcrop, the convolute structures are truncated and overlain by turbidites, indicating that they were formed prior to the turbidite deposition. In such cases it seems probable that surficial slumping of sheets of sediment produced the convolute structures and also resulted in upslope slide scars from which overriding turbidity currents originated. The turbidity currents arise where the sediment reaches the liquid limit and loses cohesion completely. This can occur very rapidly, particularly when the sediment is subjected to sudden shock such as caused by earthquake. This reaction can be demonstrated with thixotropic substances which on vibration or shock are suddenly converted from a gellike consistency to a liquid. Repeated flows of turbidity currents and deposition of sequences of turbidites can occur rapidly on the lower reaches of sediment fans at the mouths of submarine canyons. The lower reaches of these fans may be widespread over the floor of abyssal plains.

Flows of debris arising from slide scars on the continental slopes have been recognized on the Amazon cone, in the Gulf of Mexico, on the North American continental rise, in the Mediterranean, and off the northwest coast of Africa. According to Embley (1976), debris flows off the coast of the Spanish Sahara have traveled down slopes as low as 0.1° for distances of several hundred kilometers. These flows spread over the sea floor and cover an area of approximately 30,000 km². They are recognized by their acoustic character in sonic surveys, by their undulating topography and sharp contacts with underlying beds as seen in piston cores, and by their pebbly mudstone fabric. Fig. 3-3 illustrates an example of such a flow off the coast of the Spanish Sahara. Repeated mass flows of sediment may in some regions contribute substantially to the rapid accumulation of a sedimentary pile underlying the lower reaches of a continental rise.

Debris flows and turbidites on the floor of the Mediterranean Sea have been extensively studied. It is of interest to note that in the Mediterranean area other types of debris flows form gravity slide masses or mélanges within sequences of folded Late Mesozoic to Tertiary strata. These debris flows are described by Temple (1968). They are widely distributed, as illustrated in Fig. 3-4, and travel for long distances of 100 km or more. They are assumed to have resulted from orogenic movements, but the precise relationships of crustal uplift to gravity sliding are conjectural. As shown in Fig. 2-29, their stratigraphic and lithologic relationships are in general likely to preclude them from being interpreted as submarine flows of debris, particularly where large blocks of different rock types are present. But in some cases the origin of such bodies of debris is open to

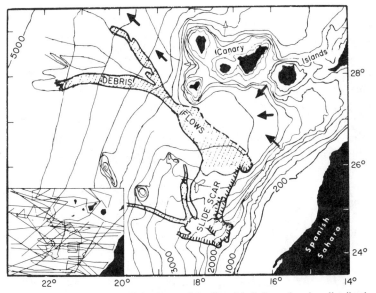

Fig. 3-3. Bathymetric map off the coast of the Spanish Sahara showing distribution of a slide scar and debris flow approximately 700 km in length. Arrows indicate paths of Pleistocene–Holocene turbidity currents. Contours in meters. Inset shows survey ship traverses on which this map is based. [After Embley (1976).]

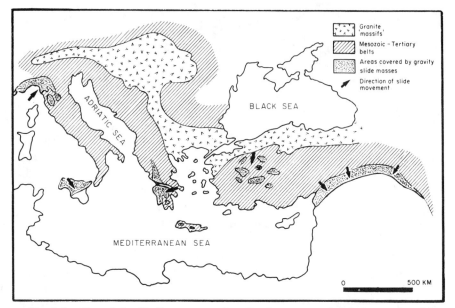

Fig. 3-4. Map of the Mediterranean region showing areas underlain by gravity slide masses of Late Mesozoic–Tertiary rock debris (areas of long distance gravity transport). [After Temple (1968).]

question. Examples may be found in carbonate bodies consisting largely of rubble from reefs. Where outcrops are not extensive, it may be difficult to tell whether they are parts of a mélange or of sedimentary talus flanking a reef trend. A third possibility is that the outcrops form part of a body of debris which originated as a submarine gravity slide from a carbonate shelf where reefs form solid masses and where cementation of particulate matter results in lithification at shallow depths of burial. In the latter case, as with all such submarine gravity slides, rates of sedimentation can be rapid over the period of deposition of successive debris flows.

Difficulties inherent in the recognition of the origin of rock masses called mélanges or olistostromes have been pointed out by DeJong (1974) with reference to such a mass in the area of Lago Titicaca in Peru. He says that although previous authors considered the chaotic mass to have been caused by folding and overthrusting, it was probably the result of subaerial sliding and is not of tectonic but of sedimentary origin. As such, DeJong considers the mass to be an olistostrome. He says further (p. 729):

> Large olistostromes, such as the Titicaca mélange, indicate sedimentary–tectonic mass transport and are transitional between structures formed by gravity tectonics (tectonic mass transport) and

clastic sediments (sedimentary mass transport). The close association of folds, a large olistostrome, and a thick pile of clastic sediments suggest that all three processes have been important in the area near Lago Titicaca.

Processes of Physicochemical Sedimentation

In recent years physicochemical processes have been increasingly recognized as important in the accumulation of some sedimentary sequences, particularly in the genesis and stratigraphic control of many syngenetic ore deposits. Various examples of these processes are found in both fresh and salt water bodies and at all depths. They consist essentially of matter derived by precipitation from solutions or colloidal suspensions and can be classified as carbonates, phosphates, silica gel, clays, metallic oxides and sulfides, and evaporites such as chlorides and sulphates.

The two most important of these processes from the point of view of bulk contribution to the mass of sediment are the precipitation of calcium carbonate on shallow carbonate banks and the precipitation of clay minerals from fresh-water colloidal suspensions emptying into the sea off the mouths of large delta distributaries. Calcium carbonate becomes dissolved in the sea from the addition of river water and from the dissolution of carbonate tests of dead plankton that form a continuous rain in many areas of the ocean. This rain of minute shells is particularly heavy in certain areas of the oceans where currents, temperature, availability of nutrients, and other conditions are ecologically favorable for rapid growth of the planktonic populations. Among these populations are plankton with siliceous tests which also form a rain of dead shells. Probably all of the siliceous tests reach the ocean floor at most abyssal depths, but it is thought that in the deeper abyssal zones the calcareous tests are dissolved and never reach the sea floor. Cold water can retain more carbon dioxide in solution as carbonic acid than warm water. Consequently, cold water per unit of volume can dissolve more calcium carbonate than warm water. Where cold ocean currents well up over shallow carbonate banks and shelves such as those of the Bahamas, which cover tens of thousands of square kilometers at depths of less than 10 m, the upwelling water is warmed. This results in the sudden precipitation of clouds of minute crystals of aragonite. The phenomenon, clearly seen from the air as a whitish patch in the sea below, was at one time thought to result from the stirring up of the sea floor by schools of fish. Once precipitated, the tiny crystals of aragonite are moved around by waves and currents. They form carbonate muds, and in some areas the particles accrete to form carbonate sands. Combined with the growth of corals, certain algae, crustaceans, gastropods, lamellibranchs, and other organisms which draw their supply of calcium carbonate from the seawater, the precipitation of aragonite can result in the accumulation of several thousand meters of limestone during a 10-million-year period.

Commonly associated with limestones and also thought to have been precipitated from upwelling cold ocean currents are certain deposits of phosphate rock. These do not constitute quantitatively a significant volume of rock in most basins in which they occur but are of special interest in view of their commercial value.

Of the noncarbonates precipitated clay minerals and salts form a large bulk, the latter being particularly evident in some sedimentary basins. Precipitation of clay minerals is brought about by the mingling of river water containing colloidal clays with seawater. This process normally occurs well out to sea from the river mouth, as the fresh water rides out over the salt water, forming a muddy fan that extends miles from the coast. Large volumes of clay ooze are deposited on prodelta areas by this process of physicochemical sedimentation.

Associated with the category of precipitation of silicates constituting clay minerals is the precipitation of silica as a gel. There is little substantive information on this subject, but it is thought that silica forms as nodules and thin lenses of gel in layers close beneath the ocean floor in some regions and that the growth of these nodules and lenses in part incorporates and in part replaces the particulate matter in which they grow. The source of the silica may in some regions be derived from the dissolution or partial dissolution of siliceous tests of plankton which settle to the sea floor. Some chert and flint nodules in limestone beds may have been derived in this way. The source in other regions may be emanations from submarine eruptions of lava, as suggested by the common association of basaltic lava, argillite, and beds of ribbon chert.

Well known are the iron and manganese-rich nodules commonly containing amounts of copper, cobalt, and nickel that are formed in profusion on some areas of the ocean floor. These become buried in accumulations of sedimentary strata and may later be incorporated in a sedimentary basin.

Other manganese deposits are obviously of fresh-water origin having probably been precipitated in bogs or shallow lakes. Some of the world's large iron deposits in Western Australia (Fig. 2-3) were also deposited in shallow water, presumably in fresh-water intracratonic basins filled by runoff from the surrounding hinterland of Precambrian rocks.

Of particular interest is the discovery in recent years of metallic sulfides precipitated in deep zones on certain areas of the sea floor in the Red Sea. These sulfide deposits appear to be associated with hot brine springs welling up along dislocations resulting from tension faults caused by sea-floor spreading. The discovery has added further evidence to support the syngenetic origin of some banded ores of copper, lead, and zinc such as those of the Mt. Isa mine in Queensland; although the problems related to the genesis of stratiform ore deposits are still subject to variations of interpretation. The precipitation of iron sulfide as marcasite and pyrite in muds deposited in euxinic environments and the association of disseminated sulfides of iron and copper in the Kupferschiefer shales of Germany are well-known examples of chemical precipitation of sulfides in restricted sedimentary environments. The Kupferschiefer is the basal unit of the Zechstein evaporite sequence, and was apparently deposited in a closed body of water having little or no circulation.

The precipitation of chlorides and sulfates, the commonest being halite and gypsum respectively, is a process that adds substantially to the accumulation of strata in some sedimentary basins. These deposits are commonly referred to as evaporites, the implication being that they were precipitated in closed basins of sea water such as lagoons or in sabkhas such as those along the coast of the Persian Gulf. Salt and gypsum are commonly formed in this way, and successive layers of evaporites may become buried by alternating layers of sediment in a cyclic sequence. Subsequent to burial, the salt layers tend to migrate by solution and plastic flow, and the gypsum is eventually converted to anhydrite which may remain as layers of laminae parallel to the bedding or migrate to fill fractures and pore space in the rock.

A classic example of halite and gypsum precipitation in a lagoon is described by Adams (1969) with reference to the Late Permian Capitan Reef of Texas and New Mexico. This reef forms an oval-shaped barrier enclosing a lagoon approximately 200 km long and up to 150 km wide. Within this lagoon up to 900 m of sediments of the Castile Formation were deposited. These consist mainly of alternating laminae of dolomitic limestone and anhydrite in a semicyclic sequence which appears to have resulted from periodic changes in the rate of inflow of water from the surrounding sea. Within the sedimentary sequence are several thick beds of halite which were deposited mainly within the northeastern half of the lagoon. No modern-day lagoons of this type are known. In this respect it must always be kept in mind that although physical processes of sedimentation and precipitation do not change, the conditions under which they operate, arising from environmental changes or catastrophic events, may modify or radically alter the degree to which they influence or control the nature of a sedimentary sequence.

It is a moot point whether some beds of salt hundreds of feet thick were formed as evaporites or were deposited as precipitates on the floor of a closed inland sea. They may also have been deposited within large, very shallow lakes where the water is so saturated that encrustations of salts form on any submerged object. An interesting example is seen in Lake Turkana (previously Lake Rudolf) area of Kenya where in certain bodies of water some nesting ibises have their legs so weighed down by encrusted salt that they cannot fly.

All the above-mentioned chemical precipitates whether deposited as discrete layers, or disseminated within the sediments, become part of the stratigraphic record of the history of a sedimentary basin. Interpretation of their genesis is commonly based on equivocal evidence, and precise relationships of the relevant physicochemical processes to sedimentation are not always entirely understood.

Superposition of Strata

It is axiomatic that within a sequence of strata that has not been over-turned the upper layers are younger than the lower. Interpretation of the geological history and sequence of depositional environments exhibited by core and drill cuttings samples from a well depend upon this principle. Where the stratigraphic sequences have been deformed by overturned folds and nappes, or dislocated by sole faults, the original sequence may be reversed. Such reversals are not always readily apparent in the field by examination of the rock types and sequence of strata in local outcrops. Geological mapping and fossil determination may be necessary to indicate a reversal of sequence. In areas of the subsurface drilled for hydrocarbon exploration, nappe structures rarely occur; although reversals, resulting from the thrusting of older beds over younger along faults subparallel to the bedding planes, commonly occur in the foothills and front range mountain belt of southern Alberta where gas fields in fractured limestone have been discovered.

In a marine sequence the occurrence of a sandstone section overlying a shale section indicates an influx of sediment deposited in a higher energy environment such as found in the shallow inner neritic zone of offshore sands or the deep-water abyssal zone of turbidites. In many sequences the overall direction of transport of the sand is away from the landmass, but in some situations a turbidity current may flow along a trough trending parallel to the coast.

Where a marine sandstone section overlies shale in a deltaic sequence, the relationship indicates regression of the sea which may arise from a fall in the absolute sea level, from tectonic uplift of the landmass, or as is commonly the case from rapid deposition of sediment at the mouth of a main distributary system and consequent local progradation of the delta.

The sandstone section forms a composite body deposited as distributary mouth bars, distributary channel sands, distal bars, interdistributary bars, and other offshore sand bodies. These sandstone bodies have planar bedding or low-angle cross bedding and may be sparcely fossiliferous. In the case of prograding sand bars, they also have grain gradation from finer below to coarser above. The sandstone section shows a fairly sharp contact with the underlying shale section which was deposited as a prodelta mud sheet and may be very fossiliferous with abundant bioturbation.

In interdeltaic areas regression of the sea is also indicated by a section of sandstone overlying shale, but in this case the sandstone section comprises a sequence of offlapping sandstone bodies deposited as offshore sand sheets and bars, some of which may have formed barrier islands. Internally the sandstone exhibits planar bedding or low-angle cross bedding and also has grain gradation from finer below to coarser above. The underlying shale section will be similar in lithology, but not necessarily in fossil content, to shales formed from prodelta muds.

Sandstone sections deposited as turbidites overlying shale sections deposited as abyssal clay and silt exhibit sequences of graded bedding, each of which may be truncated. The contact between the sandstone and underlying shale is sharp, and the shale section may have flame structures, pull-apart structures, convolute structures, or scour features in its uppermost part. Fossil assemblages within the shale will be those characteristic of a deep-water environment.

Sandstone and shale sections are also commonly interbedded with limestone or dolostone sections. Superposition of strata and variations in lithologic characteristics and fossil content are essential clues to the depositional environments in which beds were laid down. The association of limestone and sandstone, which is commonly quartzose, indicates deposition on the inner to middle neritic zones of a continental shelf. The sand, derived from beaches and bars adjacent to carbonate banks, is periodically swept over the area of carbonate deposition by currents which may form cross bedding within the sand. The association of limestone and shale may indicate deposition in somewhat deeper water, either in an outer neritic environment or as the result of tectonic subsidence and intermittent drowning of a carbonate reef and bank complex. It may, on the other hand, indicate intermittent influx of mud in an inner neritic environment from a nearby river system. The periodicity of influx may be of climatic origin, or result from lateral shifts of the river mouth. Clues to the possible origin of such interbedded sequences of limestone and shale hinge on lithologic features and fossil content.

Superposition of strata presupposes superposition of disconformities and paracontinuities. Paracontinuities represent hiatuses in sedimentation, without any appreciable removal of sediment, although they include surfaces that have been exposed to subaerial erosion in situ. Examples are commonly not recognized in older rock sequences but have been described in Pleistocene to Holocene carbonate sequences which have periodically been exposed and lithified by meteoric water. Examples also include fossil soils where the underlying beds have not been disconformably eroded. Disconformities, on the other hand, imply appreciable erosion of beds, the erosional surface being subparallel to the bedding. In consequence, adjacent beds which may be similar in rock type, or may appear to be part of the same depositional sequence, may in fact be separated by a considerable break in time. Commonly, there is a separation marked by a disconformity or unconformity between, for example, the Jurassic System and the Cretaceous System. But this is not always the case in all areas and sequences known in the Peace River Basin of British Columbia (Warren and Stelck, 1958) where sedimentation appears to have been continuous throughout the Portlandian and Berriasian Stages. It must also be kept in mind that thousands of meters of section represented by core and cuttings from a single borehole may have been deposited not only in a number of sedimentary environments but also in a variety of climatic conditions in different latitudes as a result of continental drift. The concept of superposition of strata is axiomatic but carries a number of implications which must be considered with care in the light of all possible field and subsurface evidence.

Transgressive and Regressive Cycles

A characteristic of shoreline facies in a noncarbonate environment, as seen in sectional views, is the irregular cyclic pattern of alternating beds of shale, siltstone, and sandstone. These beds were deposited in a paralic environment as shoreline or offshore muds and silts, as river mouth sand bars, or as interdistributary and interdeltaic offshore sand sheets. They commonly represent fluctuations in the position of the shoreline at the location of the stratigraphic section viewed. These fluctuations may arise from absolute changes in sea level, from tectonic uplift or subsidence of the landmass, or from local progradation of delta lobes of sediment. In general, sands have a distribution closer to shoreline than the muds, but this is not always the case. For example, off the coast of the Amazon Delta upwelling ocean currents moving shoreward carry muds from the Amazon River back toward the land where they are deposited along the coast. Also along parts of the Louisiana coast, sand beaches formed as cheniers during periods of normal sedimentation and winnowing have been drowned with mud and silt washed in by currents from nearby newly formed river mouths. In both cases the position of the shoreline has not materially changed other than by the normal accretion of sediment. Further, in areas of coastal vegetation such as salt marshes and mangrove swamps the muds and silts are trapped.

In part, the seaward distribution of sand, silt, and mud along a coastline depends on availability of the various grades of sediment. If there is no source of sand, then shoreline facies will consist entirely of fine sediment that will subsequently become siltstone and shale. Such lutites are likely to be carbonaceous, or even to include laminae of coaly matter. In part also, the nature of the sediment deposited along a coastline depends on the rate of sedimentation and on the degree of winnowing. Given the same degree of winnowing, but a rapid increase in the rate of sedimentation due to a shift in the river mouth, the result is a change from a sandy to a less sandy or muddy facies. Given the same rate of sedimentation but a sudden increase in the degree of winnowing due to a change of coastal currents or to heavier wave action caused by a period of storms, the result is a change from a sandy sediment to a sand, or from a poorly sorted to a well-sorted sand. There are many factors involved, and it cannot be assumed that in a sequence of sandstone and shale the former necessarily represents a nearer shore facies than the latter, although commonly this relationship does obtain.

Transgression and regression of the sea are commonly local phenomena that have nothing to do with absolute changes in sea level, or with tectonic movements of the landmass, but are related to progradation or subsidence of the coastal area. In an area of rapid sedimentation near the mouth of a delta distributary, the delta advances seaward and a marine to paralic regressive sequence is deposited. In areas where the river system has changed course, and where sedimentation is no longer active, the delta subsides by compaction of the underlying sediment. The sea consequently floods low lying areas of swamp, resulting in the deposition of a transgressive sequence. Such sequences can be rapidly reversed by subsequent changes in the river system.

There are also classic examples of transgression and regression, modified by minor reversals, on a regional scale. The sequence overlying the Ordovician St. Peter Sandstone (Fig. 3-5) of the Upper Mississippi Valley region is one example of a transgressive sequence. The basal unit of the St. Peter Sandstone is the Tonti Member, a fine-grained quartz sandstone deposited on an erosional surface as a transgressive sheet of sand. The overlying Starved Rock Member, of medium-grained quartz sandstone, was probably deposited (Fraser, 1976) as a barrier–island complex. The Glenwood Formation, of shale, mudstone, argillaceous sandstone, and

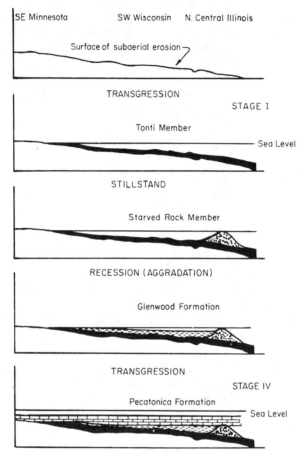

Fig. 3-5. Schematic diagrams illustrating development of the transgressive marine sequence beginning with the Tonti Member of the Ordovician St. Peter Sandstone in the Upper Mississippi Valley. [After Fraser (1976).]

dolomitic mudstone in the upper layers was deposited landward from the Glenwood as lagoonal sediments. The overlying Pecatonica Formation of arenaceous, brecciated, and laminated limestone was deposited in a tidal to subtidal environment as carbonate mud flats and carbonate banks.

Another example of a transgressive stratigraphic unit is the widespread Late Devonian Chattanooga Shale of central United States which can be traced laterally across the continent to the Early Mississippian Exshaw Shale of Western Canada. In this case the transgression occurred during a period of time linking the Devonian and Mississippian, resulting in the widespread deposition of a diachronous shaly unit. Many other examples of regional marine transgression can be cited, among which is the Early Cretaceous Birdrong Sandstone of northern Western Australia. This is a well-sorted permeable quartzose sandstone deposited in a beach environment. Such regional transgressions are the result of either rises in the absolute level of the sea, which may be caused by periodic melting of the polar ice caps, or tectonic subsidence of portions of the continental crust.

The basal unit of a transgressive sequence may be a sandstone, as in the case of the St. Peter Sandstone and the Birdrong Sandstone. In some areas a basal sandstone unit may be missing or only locally present as in the case of the Chattanooga Shale which in parts of Oklahoma is conformably underlain by the quartzose marine Misener Sandstone (Borden and Brant, 1941). The Misener unconformably overlies the Ordovician. Whether or not a sandstone unit forms the basal unit of a marine transgressive sequence depends on the availability of sand. Where, for example, the sea is transgressing a widespread flat area underlain by limestone, there may be no sand available. In such cases the basal transgressive unit may be calcareous and shaly.

Examples of regional regressive sequences are also numerous, and among these may be cited the Late Cretaceous Eagle Sandstone that forms the rimrock of the escarpment at Billings, Montana (Shelton, 1965). The Eagle Sandstone constitutes a sequence of offlapping, seaward-facing sandstone bodies. Similar stratigraphic relationships are found in the Early Cretaceous Viking Formation in Saskatchewan (Evans, 1970) and in the basal sandstone section of the Late Cretaceous Belly River Formation in Alberta (Conybeare, 1976). Such basal sandstone units in regressive sequences commonly show internal offlapping relationships, yet form widespread diachronous layers commonly referred to as sandstone sheets.

A typical section of a delta lobe is illustrated by MacKenzie (1972) in Fig. 3-6 which shows the distribution of sand and mud facies during a period of regression and subsequent transgression of the sea caused by progradation and subsequent cessation of deposition within a delta lobe. The lowermost part of the section shows shoreline sandstone bodies offlapping seaward. These sandstone bodies overlie marine shale and are overlain and locally scoured by delta distributary channel-fill sandstone bodies which are enclosed in nonmarine to paralic delta sediments, including coal layers formed from coastal swamp deposits. The stratigraphic relationships of facies shown in the lower part of Fig. 3-6 can be matched almost precisely in the basal section of the Belly River Formation in the eastern area of the Pembina oil field of Alberta (Conybeare, 1972).

Interpretation of the basal sandstone unit may depend on correlation of electric log characteristics which show the internal offlapping relationships. Externally the basal sandstone unit may appear to be a homogene-

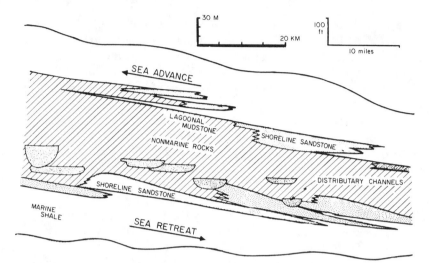

Fig. 3-6. Schematic diagram illustrating the internal facies relationships within a delta lobe during a period of regression and subsequent transgression of the sea. [After Mac-Kenzie (1972).]

ous sheet of sand rather than a composite bed of offlapping sandstone bodies. The contact with the underlying marine shale is sharp, whereas the upper contact may not be so well defined. Where distributary channels cut into the basal sandstone unit, the relationship may not be recognized unless the channel-fill sandstone body is drilled and the sand composition, texture, and grain gradation from coarser below to finer above are noted. The overlying paralic to nonmarine beds consist of shale and siltstone predominantly, with minor sandstone distributed as linear, anastomotic bodies deposited in river channels. The fine sediments with coal layers were deposited in coastal bays, salt marshes, backswamps, and on flood-plains. Following cessation of sedimentation on a delta lobe, the sea advances over a sinking wedge of sediment. Winnowing of the sediment forms sand bars behind which lagoonal muds accumulate. The sequence is repeated as the sea continues its advance, the lagoonal muds being over-lain by bars and other shoreline bodies of sand. Within several hundred to a thousand or more meters of section, quite a few of such delta lobes may interfinger with inner neritic marine facies.

An example of interfingering delta lobes is described by Masters (1967) and illustrated in Fig. 3-7 which shows a section of the Late Cretaceous Mesaverde Group in Colorado. The section is more than 1000 m thick and comprises three delta lobes of marine shoreline sand with associated la-goon and flood plain sediments. The vertical exaggeration shown in this

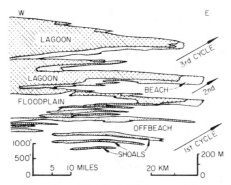

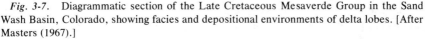

Fig. 3-7. Diagrammatic section of the Late Cretaceous Mesaverde Group in the Sand Wash Basin, Colorado, showing facies and depositional environments of delta lobes. [After Masters (1967).]

figure is more than 40×, so it is evident that the delta lobes themselves are sheetlike in geometry. Another example illustrating similar stratigraphic relationships to depositional environments within the Mesaverde Group is shown in Fig. 3-8. Three delta lobes interfingering with the marine Mancos Shale and Lewis Shale are shown. These constitute an irregular cyclic

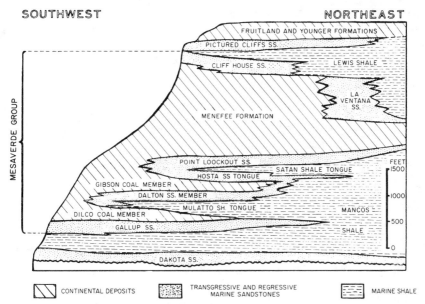

Fig. 3-8. Diagrammatic section of the Late Cretaceous Mesaverde Group in the San Juan Basin, New Mexico, showing facies relationships within a sequence of delta lobes. [After Sabins (1964).]

sequence resulting from periodic lateral shifts in the river system that formed the Mesaverde delta. It is of interest to note that Sabins (1964) records an upward increase of maximum grain size in those sandstone bodies deposited as prograding shoreline sands during periods of marine regression, a textural relationship that is consistent with grain gradation seen in similar coastal bodies of sand being deposited today.

In the central part of the United States several marine transgressions and regressions occurred during the Late Paleozoic. These are marked by widespread units of black fissile shale and limestone. Interstratified with these marine beds are clastic wedges of sandstone and shale with coal seams, formed as thin delta lobes. One such lobe, the Late Pennsylvanian Labette Formation, is illustrated in Fig. 3-9. It shows the intertonguing relationships of sandstone and shale units with beds of limestone. These widespread transgressions and regressions are recorded in separate basins and are believed to have been caused by eustatic shifts of sea level which may have resulted from changes in the glaciation of Gondwanaland or from global tectonism (Wanless, 1969).

Lithofacies

The concept of lithofacies implies a lateral change of rock type within a time-stratigraphic unit in response to a change in depositional environ-

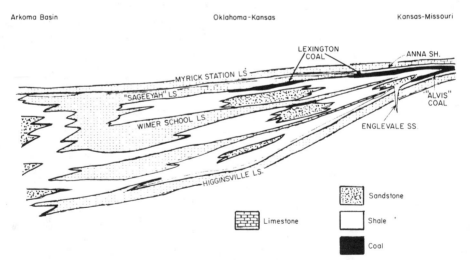

Fig. 3-9. Diagrammatic cross section illustrating relationships within the Labette Shale: the Late Pennsylvanian Labette Formation, central United States, showing a delta lobe of sandstone, shale, and coal within a marine sequence of limestone. [After Wanless (1969).]

ment. Such a change may involve depth of water, paleogeographic situation, climatic variations, fluctuations of ocean currents, seasonal tides,
and the effects of crustal disturbance and gravitational slumping on sediment masses. No lithostratigraphic unit maintains its characteristics when
traced laterally, although some diachronous units are remarkably widespread; it either pinches out or changes facies. The change in facies may
appear to be transitional, but on close examination it is commonly seen to
consist of an interfingering, in varying volumetric proportions, of the two
dominant rock types. For example, a time-stratigraphic unit that changes
facies from limestone to shale has a zone of interfingering tongues of
limestone and shale. Some lime mud is washed into the clay mud tongues,
and some clay mud into the lime mud tongues. If sufficient intermixing
takes place, the zone in which the facies change occurs may be incorrectly
interpreted as a transitional zone of argillaceous limestone or calcareous
shale. In such cases the interfingering may not be evident in outcrop but
will show up in electric logs. Similarly, a time-stratigraphic unit that
changes from sandstone to shale has a zone of interfingering tongues of
sandstone and shale. These tongues may be of silty or argillaceous
sandstone and silty shale, but the zone is not a transitional zone of
siltstone. Such facies changes are observed in time-stratigraphic units of
variable thickness, ranging from members and their correlatives a few
meters thick to formations hundreds of meters thick, and may involve
changes from one rock type to another or from one assemblage of rock
types to another. A change of lithofacies within a rock-time unit also
commonly involves a change of biofacies, particularly of benthonic
organisms.

Determination of the lateral extent of a time-stratigraphic unit is always
a problem, particularly in view of the effects of compaction and the common lack of reliable markers such as bentonitic ash falls, beds containing
pelagic fauna, spores, or pollens that have a narrowly restricted time
range. Many widespread units of marine sandstone appear in outcrop or in
the subsurface as blanket units of sandstone but are in fact a composite of
many offlapping sandstone bodies formed as shoreline sands along the
margin of a regressing sea. In these stratigraphic units, as also in basal
sandstone units such as that of the Ordovician St. Peter Sandstone of the
eastern United States, the rock type remains fairly consistent over a widespread area because the sediment was everywhere deposited by the same
laterally migrating depositional environment. Such units are recognized
with little difficulty as they are basal units commonly overlying a marine
shale in the case of a regressing sea, or overlying an unconformity or
paracontinuity in the case of a transgressing sea. With other time-
stratigraphic units it may be difficult to recognize their upper and lower

limits, yet such recognition may be essential to the construction of a lithofacies map, particularly where the map is quantitative.

Depth of water is obviously a limiting factor in the deposition of noncarbonate sediments which are moved by river systems to the continental shelf where they are in large part deposited. Commonly the sands are deposited nearer to the coastline than are the muds, but this is not always the case where ocean currents sweep shoreward, as has been previously mentioned with regard to the delta of the Amazon River system. Subsequent movement by gravitational slumping of sediments in submarine valleys on the continental shelf and slope may initiate turbidity currents which carry the sediment to abyssal depths. Carbonates, on the other hand, are indigenous to the sea, with the exception of some rather rare limestone beds deposited in fresh-water lakes. Commonly these marine limestones are of chemically precipitated carbonate mud, organic fragments, oolites, and aggregates of carbonate mud particles. Much of the particulate organic matter is so minute that it forms a carbonate ooze or mud. Carbonate precipitates and most particulate sediments are deposited in shallow areas of carbonate banks and shoals which may or may not form part of a continental shelf. Deeper water carbonate sediments, formed by the rain of coccoliths and other microscopic tests of dead organisms are deposited in neritic to bathyal zones. But dissolution of calcium carbonate at depth probably precludes the deposition of carbonate ooze in the deeper abyssal zones. On the other hand, siliceous oozes formed largely from the rain of radiolarian or diatom tests are deposited at great depths of up to 8000 m or more.

Lithofacies types and trends are also indicators of the paleogeography that existed at the time of deposition. Common examples are the progradation of deltaic facies over marine beds. In these sequences the basal unit of the deltaic facies is a sheetlike sandstone formed of seaward prograding lenses deposited as shoreline sands. This basal unit is overlain by delta sediments of sand, silt, and mud, deposited in distributaries, bays, lagoons, and river floodplains. Coal layers are commonly present. Study of the sequences in outcrop or in the subsurface indicates the general direction of growth of the delta and consequently the relative locations of sea and landmass.

With respect to sandstones, a particularly useful approach to the problem of provenance of sediments is the microscopic examination of thin sections to determine the mineralogical content and the nature of rock fragments in the sediment. Specific mineral types of garnet, zircon, tourmaline, etc., may indicate derivation from a particular area characterized by metamorphic rocks containing these particular varieties of minerals. Or they may indicate secondary reworking and derivation from preexisting sandstones. In either case, the direction of provenance may be established. Specific types of rock fragments may also be used as indicators of provenance where they can be matched to a possible source. For example, some of the sandstone beds of the Late Cretaceous Belly River Formation of Alberta contain fragments of dark gray cherty rock which under the microscope resemble ribbon cherts or siliceous argillites in the Permian Cache Creek Series of British Columbia. The basal deltaic sequence of the Belly River progrades eastward over marine beds in response to orogenic movements in British Columbia and withdrawal of the sea in Alberta, so it is not improbable that some of the sediment was transported eastward for 500 km or more.

Limestone sequences do not so readily reveal the paleogeography of their depositional environments. They may have developed on a shallow area of a continental shelf such as that fringing the Yucatan region of southern Mexico, on a broad and shallow platform similar to the Bahama Islands shelf which is separated from the mainland by Florida Strait, or within the shallow lagoon of an atoll far out in the sea. The presence of tongues of sandstone or shale indicates that the limestone sequence was deposited on a continental shelf, but their absence does not necessarily indicate a Bahamoid or atoll type of origin. Sandstone tongues are commonly quartzose and originated as well-washed and well-sorted banks of sand moved by currents into the area of carbonate deposition. Tongues of shale may indicate floods of mud resulting from lateral migration of a nearby river or from seasonal changes such as a period of heavy rainfall that brings a heavy influx of mud into the sea. In some paleogeographic situations, such as those prevailing in areas similar to the Red Sea region where arid mountain topography borders an inner neritic environment in which coral reefs and associated carbonate beds are developing, the limestone sequence may include tongues of arkosic sandstone. In other paleogeographic situations, where carbonate beds and reefs are developing off the coast of a desert region, windblown sand may accumulate in considerable quantity within the carbonate beds. Some limestones, probably formed in regions of desert, contain disseminated grains of quartz that are rounded and frosted, suggesting an eolian derivation. The frosting is not in itself a definitive criterion as it may have resulted from the replacement of quartz by calcite.

Mention should be made of fresh-water limestone beds. These are found in nonmarine sequences and consist of chalky or fine-grained limestone deposited in large part by the chemical precipitation of calcium carbonate in lakes. Where rivers drain limestone regions, the water may become saturated with calcium carbonate, so that on entering a lake and being warmed the calcium carbonate is precipitated. Such precipitates may be stirred by waves and currents to form particles of agglutinized lime mud, or ooliths. They also are commonly very fossiliferous, containing abundant fragments of calcareous algae, charophytes, ostracods, bivalves, and gastropods.

Lithofacies may also reflect climatic variations over periods of many hundreds or thousands of years. Red beds are examples of sequences that have been used as climatic indicators. Commonly these are nonmarine and consist of sandstone, siltstone, and shale, although reddish limestones are known. With regard to the latter, some reddish layers have originated as weathered zones and consequently indicate disconformities. Unfortunately, the evidence supplied by red bed sequences is not always sufficiently definitive to allow an unequivocal interpretation. Some sequences have been interpreted as having been derived from an arid landmass, particularly those that include a major volume of poorly sorted lithic or arkosic sandstone. The rationale suggests that these were rapidly deposited during infrequent periods of heavy rainfall that washed sediment from an arid terrain. Other sequences of reddish to brown shale, siltstone, and minor sandstone are interpreted as floodplain deposits derived from a red weathered regolith in a warm and humid climate.

Red muds are known to be widely distributed over abyssal plains. They consist of clay and very fine particles of dust which may be largely of volcanic origin. Because the rate of accumulation is very slow, probably of the order of a few millimeters in a thousand years, the sediment in places has a relative abundance of shark teeth which are periodically shed and replaced during the life of these fishes. Such red beds, where recognized in a rock sequence, have no climatic connotation except possibly as an indicator of worldwide volcanic activity but do indicate deposition within abyssal ranges of depth.

Fluctuations of ocean currents and seasonal tides may be reflected in lithofacies variations. Mention has already been made of facies changes within a seaward prograding sequence. Variations in facies may also occur as a result of landward-moving ocean currents which in some situations can sweep the fine sediments shoreward and distribute them along the coast. This situation, as previously discussed, is found off the coast of the Amazon Delta. Facies changes along a coastline result mainly from the distribution of sediments by ocean currents. The chain of barrier islands in the Gulf of Mexico has formed by the transportation of sand southwestward along the coast. Similar island chains off Long Island in New York, in the Netherlands, and fringing the Nile Delta of the United Arab Republic were formed by the coastal distribution of sand. Fluctuations of ocean currents and seasonal tides may have a profound effect on the distribution of biofacies and also on the occurrence of catastrophic events which are recorded in the geological record of rock sequences. For example, present-day occurrences of so-called "red tides" sweep colonies of microscopic red algae shoreward, killing thousands of fish which become strewn in masses on the beach. Such events in the past may account for some of the fish bone beds found in rock sequences. Others may have resulted from submarine volcanic eruptions, the shock of meteoric impacts, or pollution from natural causes.

Other catastrophic events, such as earthquakes attended by regional uplift or subsidence, may have a marked effect on the lithofacies of a time-stratigraphic unit deposited on a continental shelf. Such events may also trigger off mass slumping of sediment on the lower reaches of the continental shelf, on the continental slope, and within submarine canyons, resulting in turbidity currents. One of the most dramatic catastrophic events that may have happened within the latter part of the earth's history, and one which would be reflected in the sedimentary record, is the flooding of the Mediterranean basin by the Atlantic Ocean pouring through a gap south of Gibraltar. Interpretations of events prior to this possible catastrophy are equivocal; but it appears that during the Tertiary massive salt beds were deposited as evaporites within the Mediterranean Basin which is thought to have been an area of subsidence lying below sea level.

Biofacies

The subject of biofacies is not dealt with in this book for reasons previously stated. It is an extensive field in itself within the vast range of paleontological studies. Only insofar as biofacies relate to some of the factors controlling or influencing the distribution of lithofacies will the subject be discussed.

Stratigraphic zones characterized by a particular genus, species, or assemblage of genera or species can be designated as a specific biofacies. If the diagnostic fossils are of pelagic organisms that had a limited or restricted time range, then the biofacies is contained within a time-stratigraphic unit; and if the time range of the fossil is sufficiently restricted, the time-stratigraphic unit may serve as a time marker, regardless of lithofacies changes within it. Some time-stratigraphic markers defined by ammonites are widespread, others defined by pollens and spores are relatively local. On the other hand, if the diagnostic fossils are of benthonic organisms, they may well have a wide time range and be restricted to a particular ecological environment which is subsequently reflected in the lithofacies of a stratigraphic sequence. For example, a benthonic assemblage that favors development in prodelta mud will follow the lateral and vertical migration of this sediment facies as it fluctuates with transgression and regression of the sea and with lateral shifts of the river system in a delta. In the geological record such associations result in a close affinity between biofacies and lithofacies. Three factors that have been mentioned as significant in the distribution of sediments and lithofacies are fluctuations in ocean currents, paleogeographic situation, and depth of water. These factors are also of significance in the distribution of pelagic and benthonic fauna.

Fluctuations in ocean currents, with consequent changes in water temperature, have very marked effects on the number and variety of pelagic organisms living, dying, and being buried in the sediment underlying a particular area of ocean. Fig. 3-10 illustrates the predominant currents in the region of Taiwan. With reference to the distribution of planktonic forams obtained in bottom samples from this region, Huang (1972, p. 31) says:

> The distribution of planktonic foraminifers in the sediments of Taiwan Strait is controlled by a number of physical and biological variables that may affect the growth of the species. The introduction of large volumes of fresh water in the coastal areas of western Taiwan results in a lowering of surface salinity and the exclusion of planktonic foraminifers. On the other hand, large numbers of

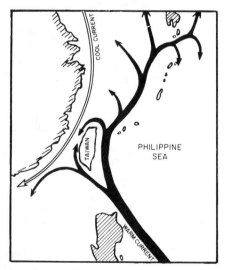

Fig. 3-10. Distribution of warm (black) and cold (white) ocean currents in the region of Taiwan. [After Huang (1972).]

planktonic foraminifers are found in the shallow basin in the middle of the strait.

Twenty-seven species of planktonic foraminifers are recorded. The presence of cool-water species in the assemblage is regarded as being due to the influence of the China Coastal Stream flowing from the north, while the warm-water species have been brought from the tropics by the Kuroshio Current.

Paleogeographic situation is an important factor in the distribution of benthonic fauna as it relates to a number of physical factors including type of sediment being deposited, depth of water, energy of wave or current action in the area, and water temperature. The distribution of biofacies resulting from the growth of benthonic colonies in various geographic situations may be on a regional or local scale. The distribution of biofacies on an intermediate scale is illustrated by Fig. 3-11 which shows the abundance and diversity of mollusks within an area of nearly 40,000 km² of the continental shelf in the Gulf of Batabano lying between the Peninsula de Zapata and the Isla de Pinos in Cuba. The isopleths indicate that both the quantity and diversity of species are much larger along a belt close to and parallel with the shoreline. Regarding these distributions, Hoskins (1964, p. 1680) says:

The distribution patterns of the biofacies are related to those of calcareous sediments, salinity, median grain size, sorting, and certain other ecological factors. Salinity, amount of carbonate mud, and water turbulence primarily control the distribution of the mollusks. The biofacies tend to occur in concentric patterns and in particular sequences relative to the land margins of the Gulf and the reef tract.

Depth of water as a factor in the distribution of Holocene brachiopods in the Antarctic region is illustrated in Fig. 3-12 which shows that the greatest diversity of species is found within a depth range of 400–1200 m and that tolerance to greater depths becomes restricted to fewer species. The distribution also suggests a decreasing tolerance to shallower depths, but as noted by Shabica and Boucot (1976), this may result from the destructive effects of ice. There is a good deal of controversy regarding the significance of certain fauna as indicators of depth. Organisms such as colonial corals are obviously of shallow-water origin, whereas some solitary corals can tolerate a wide range of depth. Where used in basin analysis to assist in interpretations relating to depth of sediment accumulation, biofacies data must be considered in context with the local and regional lithofacies.

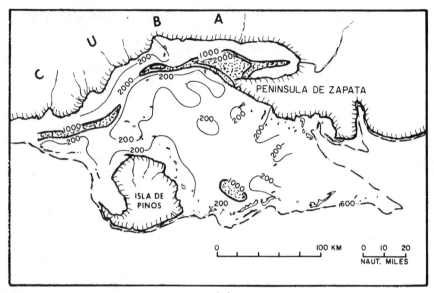

(a)

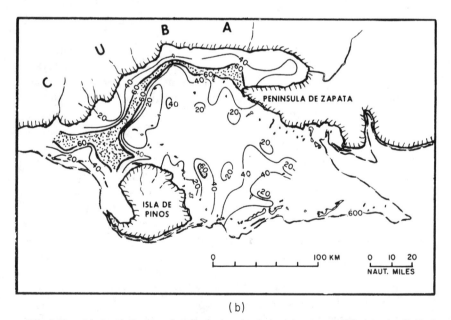

(b)

Fig. 3-11. (a): Isopleth map of mollusk abundance (specimens per 300 g) in the Gulf of Batabano, Cuba. (b): Isopleth map of mollusk diversity (species per 300 g). [After Hoskins (1964).]

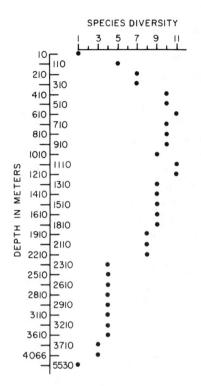

Fig. 3-12. Plot showing diversity of species of Holocene brachiopods found at ocean depth ranges in the Antarctic region. [After Shabica and Boucot (1976) and Foster (1974).]

Time-Stratigraphic Units and Diachronism

Time-stratigraphic units are stratigraphic units that include all the sediments deposited during a particular interval of time within the whole or part of a sedimentary basin. Where sedimentation was continuous during this interval of time, and where subsequent erosion has not removed part of the record, the time-stratigraphic unit is complete. Where sedimentary layers did not form during part of the interval of time, either because of a hiatus in sedimentation or because the area was uplifted and exposed, a paracontinuity results and the time-stratigraphic unit becomes incomplete. Time-stratigraphic units are not necessarily mappable units, although they may be defined in stratigraphic sections. In this respect they differ from rock-stratigraphic units or lithostratigraphic units which are rock units distinguished by a particular rock type or lithologic characteristic. Rock-stratigraphic units are commonly diachronous.

Time-stratigraphic units can seldom be defined with accuracy because the boundaries must not only be recognizable but must define events that occurred within an interval of time. Reliable marker beds that can be used to define boundaries are bentonitic layers that record widespread falls of volcanic ash, beds containing pelagic organisms having a limited stratigraphic range, such as certain ammonites, zones defined by particular assemblages of pollens and spores, and beds noted for an abundance of some organic remains, such as fish scales, or recognizable by a pronounced increase in radioactivity. Whatever the physical property or paleontological character that distinguishes the unit, it must be widespread and record some event that occurred during a comparatively short space of time. Some units may be evident in rock outcrops or even in drill core and cuttings; others are recognized only by some petrophysical character in electric logs.

The interpretation of lithofacies and the construction of lithofacies maps and sections depend on the recognition of time-stratigraphic units. Where the boundaries of such a unit are known only generally, the lack of information may be critical in deciding how to plot the distribution of lithofacies on a map. Recognition of time-stratigraphic units is consequently basic to reconstruction of the geologic history of any sedimentary basin. The shorter the interval of time recognized, the more accurate is the reconstruction of events that occurred during that interval. Where the time-stratigraphic interval covers too long a period of time, the geological history becomes somewhat blurred.

In the absence of suitable marker beds, the definition of time-stratigraphic units depends on the selection of recognizable stratigraphic boundaries such as disconformities, paracontinuities, the base of a deltaic sequence prograding over a marine sequence, transgressive or regressive beds characterized by a particular fossil assemblage, or other widely recognizable rock units. These are all diachronous, and regardless of their apparent lithologic continuity some may be diachronous to an extent that precludes their use as boundary features. The decision as to whether or not to use certain stratigraphic features as boundaries will depend on the knowledge and experience of individual geologists.

Time-stratigraphic units are more easily defined where sedimentological conditions and consequently rates of sedimentation are fairly constant. Such conditions obtain on vast carbonate banks covering thousands of square kilometers of the continental shelf and in bathyal to abyssal depths that are not subject to periodic influxes of turbidites. Stratigraphic sequences deposited in these environments exhibit a marked degree of widespread parallel bedding that in carbonate sequences particularly may show cyclic sedimentation. The apparent continuity, parallelism, and great thicknesses which may be more than a thousand meters give rise to wonder and contemplation on the part of geologists mapping the mountain ranges. How could these layers have formed? Are such sequences forming today, and if so where?

Time-stratigraphic units are much more difficult to define in nonmarine, deltaic, and submarine fan sequences in which facies changes occur within short distances. Fig. 3-13 illustrates the distribution of lithofacies within a Late Permian deltaic sequence of the Sydney Basin, Australia. The Permian in this region, according to Conolly and Ferm (1971), comprises up to 3600 m of fluvial, deltaic, and marine shelf sediments. The section shown is more than 250 km in length and has a maximum thickness of 300 m. Time-stratigraphic units are not known, but presumably the coal seams approximate time-rock units. These cut across the divisions showing prodelta, delta front, delta plain, and alluvial plain facies. The divisions, which are mappable units, are obviously diachronous. The offshore prodelta muds are overlain by a lower unit of offlapping shoreline sand sheets and bars with interfingering tongues of mud. This lower unit is overlain by units of distributary sands, bay muds, coastal marsh vegetation, and river floodplain deposits. Viewed in natural scale, it is apparent that the time-stratigraphic units and the rock-stratigraphic units designated as lithofacies are nearly parallel; so that the stratigraphic sequence appears to comprise a prodelta unit overlain by a delta front unit which, in turn, is overlain by delta plain and alluvial plain units. The sequence is one that indicates regression of the sea. Considered as a whole, the section illustrated in Fig. 3-13 is also a time-stratigraphic unit.

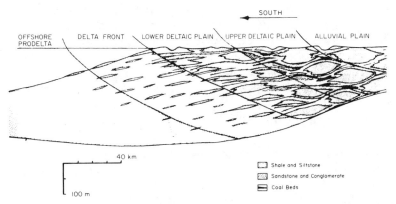

Fig. 3-13. Time-stratigraphic unit within a Late Permian deltaic sequence in the Sydney Basin, Australia, showing diachronous rock-stratigraphic units reflecting prodelta, delta front, and delta plain facies. Section shown is approximately 250 km long and 300 m thick. [After Conolly and Ferm (1971).]

Marker Beds and Selection of Datum

Marker beds are, as the name implies, stratigraphic units that can be recognized and correlated from one location to another. Preferably they are widespread, such as many bentonitic beds, but they may also be local, such as an individual coal seam. Where it is not possible to recognize individual seams, a coal-bearing zone may be the most easily recognized stratigraphic unit that can be employed. Marker beds may be time-stratigraphic units, such as a bentonitic bed, or rock-stratigraphic units such as a diachronous coal-bearing zone. Whatever the nature of the marker bed selected, the purpose is to enable correlations to be made of beds lying below the marker, in particular, but also above the marker where they are in fairly close proximity. These correlations are illustrated by means of stratigraphic sections which assume an approximate horizontality for the marker bed at the time of deposition, although in fact it may have been deposited on an undulating or sloping surface. In constructing a stratigraphic section, it is common practice to assume that the marker bed was originally a nearly horizontal surface and to use the bed as a datum from which the underlying section is hung.

The selection of a marker bed, or a surface of a marker bed, to be used as a datum may depend to some extent on the interpretation placed on the origin of the bed and the purpose for which it is to be employed. Where correlation of beds is the main purpose, the selection may not be critical. But where the purpose is to interpret the geomorphology of an underlying stratigraphic body, the attitude of the datum at the time of deposition and the probable degree of compaction of the section lying between the datum and the stratigraphic body are essential to interpretation. Many examples can be found in sandstone bodies where the selection of either an upper or lower surface, or of a marker bed close to these surfaces, as a datum will place a different emphasis on interpretation of the origin and geomorphology of the sandstone body. The selection of a datum above or at the top of a sandstone body may indicate by its shape that it was a channel sand, whereas selection of a datum below or at the base may indicate that it was a sand bar.

Reconstruction of the stratigraphic and structural configuration of a sedimentary basin during various stages of its geologic history depends on the use of marker beds for a sequence of datum surfaces. Each successively higher surface is assumed to be approximately horizontal so that underlying marker and other beds become progressively warped by compaction or otherwise disrupted by growth faults. The selection of a stratigraphic zone that can be used as a datum commonly depends on the recognition of a distinctive electric log feature. Preferably, this lies within a marine shale section which provides continuity of rock type over a wide area. Such an electric log marker may be recognizable in outcrop, or in drill core and cuttings, by some distinctive lithologic, mineralogic, or paleontological feature. For example, the Early Cretaceous Toolebuc Member of Queensland, Australia, a widespread dark, bituminous, shaly limestone several meters thick, is recognizable not only by its abundance of fish scales and bones but by its relatively high radioactivity which shows distinctly on logs measuring the natural gamma radiation of rocks. This fish scale zone is an excellent subsurface marker and datum for the purpose of regional correlation. Other types of marker beds may be used for local correlation. An interesting example is the use of pollens and forams in determining detailed correlations and consequently in working out the structural pattern of fault blocks within oil-bearing Tertiary sequences in Venezuela. Pollens have also proved to be particularly useful in regional correlation of Tertiary rocks in Venezuela.

Marker beds may also be indicated by seismic profiles which show traces indicating the depth and lateral extent along the line of traverse of reflecting surfaces between beds having a marked velocity contrast. Some of these traces disappear as the difference in density between adjacent beds decreases. Others persist over long distances but may locally disappear and reappear along the line of traverse, necessitating inferred correlations. In such cases experience and care are needed to ensure that the correlations are valid. For example, where reflections are from two coal seams that are not continuous or overlapping along the line of traverse, it may be difficult to determine whether they are the same or separate seams.

The construction of structural sections in which sea level or some horizontal plane is used as a datum also entails the need to determine a marker bed. Fig. 3-14 illustrates a structural section presented by Tixier and Forsythe (1951) to show the oil and gas-bearing Early Cretaceous Viking Formation sandstone in the area of the Excelsior Field, Alberta. A widespread electric log feature within a shale section above the Viking is easily recognized and serves as an excellent marker bed. It is of interest to note that the Viking Formation, deposited as shoreline sands, and the marine shale section between the Viking and marker bed maintain a remarkably constant thickness over a distance of approximately 35 km along the depositional trend of the Viking.

Correlation of Strata

Strata are correlated on the basis of individual beds or sequences of beds, the latter commonly having a wide range of thickness and a number

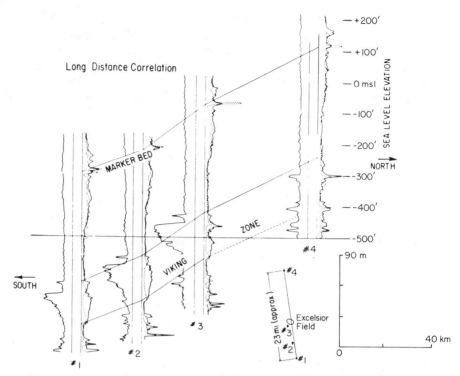

Fig. 3-14.   Structural section of the Early Cretaceous Viking Formation and adjacent shaly beds in the area of the Excelsior Field, Alberta. Section is along the depositional trend of the Viking shoreline sands which exhibit a typical funnel-shaped electric log characteristic. [After Tixier and Forsythe (1951).]

of beds within the limits defined by a formation. By definition, a formation is a mappable stratigraphic unit and therefore distinguishable from beds or sequences of beds above and below. Correlations may also be made between groups where the stratigraphy is not sufficiently known to correlate formations within the groups. But normally the geologist is concerned with correlation of members and formations.

Correlation of strata is based either on lithologic or time factors; in other words, a correlation is made between lithostratigraphic units or time-stratigraphic units. Lithostratigraphic units are commonly but not always diachronous. Exceptions such as marine deposits of volcanic ash and glauconitic layers on large Bahamoid-type carbonate banks may form excellent marker beds. Lithostratigraphic units may also exhibit facies changes within the basic rock type, such as changes from sandstone to silty or shaly sandstone, from clay shale to calcareous shale, and from calcarenite to calcilutite. Correlation of sequences in which the facies change from one rock type to another, such as from sandstone to shale, or from shale to limestone, particularly where the strata have variable ranges of thickness, depends on the recognition of time-stratigraphic boundaries. These may be defined by mineralogic zones such as a bentonitic bed rich in montmorillonite, or beds containing abundant glauconite or chamosite; or they may be defined on the basis of zones containing particular assemblages or species of pelagic macrofossils, or of microfossils such as forams, pollens, and spores. All such criteria defining the limits of a bed assumed to be a time-stratigraphic unit must be evaluated on sedimentologic and biologic bases to determine whether they represent an event that occurred during a limited period of time over a wide area or whether they reflect migrating sediment facies and consequent shifting of ecological environments.

Formations, being mappable units, are also lithostratigraphic units and may be diachronous. Correlations made from one side of a sedimentary basin to the other or from one basin to an adjacent basin may depend on the sequence of strata rather than on the presence of pelagic fauna. Sequences are determined by field mapping, drilling, and seismic surveys where these can be tied-in to the known subsurface section by means of a borehole velocity survey. The division of sequences into members, formations, and groups is not entirely unequivocal and may be controversial. Consequently, regional correlations are not always clearly defined nor generally accepted; which in some regions gives rise to a multiplicity of nomenclature in the naming of members and formations.

Correlation essentially involves matching sequences of stratigraphic layers from one location to another. These locations may be outcrop sections or boreholes in which the sequences exposed or drilled may not be complete. Where the distances between locations are considerable, particularly where facies changes are involved, it becomes difficult to correlate with any degree of certainty unless a marker bed can be established. In the absence of any such recognizable stratigraphic unit, correlation depends on matching the gross lithology and order of sequence of the beds. Sequences of sandstone and shale beds commonly change facies when traced laterally, becoming predominantly sequences of sandstone with minor shale or of shale with minor sandstone. In matching such sections, it must be remembered that compaction has altered the apparent relative stratigraphic position of beds and that the shale portion of the sequence must be expanded accordingly. The amount of expansion estimated will be approximate only, and where the cumulative thickness of shale is of the order of several hundreds of meters, the compaction factor will not be uniform throughout the sequence. Nevertheless, sections can be drawn to show the possible original thickness of a sequence, calculated to correspond to the known compaction of clay muds within the depth range that obtained when the upper surface of the sequence was the sea floor. When this is done, the correlation of sandstone units, which are relatively uncompactible, may become apparent. At best, this method depends, first, on certain assumptions regarding compaction and, second, on the best fit of the reconstructed sections.

Correlations within thick sections of limestone and dolomitic limestone, which are not readily compactible but susceptible to loss of volume by dissolution, depend largely on visible lithologic variations, mineralogic accessories, fossil content, and on petrophysical characteristics. Individual limestone beds within a shaly sequence are commonly conspicuous on electric logs and may serve as marker beds, particularly if they are further distinguished by fossil content. An example is the Cretaceous Ostracod Limestone of Alberta which comprises two or more thin ostracod-bearing limestone beds within a narrow but widespread stratigraphic interval.

Correlation of individual lithologic units within nonmarine sequences is commonly not possible over any appreciable distance as the unit either pinches out or cannot be distinguished from other units of similar rock type. The sequences themselves can be recognized as nonmarine by the absence of marine fossils, by the presence of fresh-water gastropods and bivalves, by sedimentary structures, and other criteria. They may also be correlated by their stratigraphic position with reference to sequences above and below or to an underlying unconformity. Such sequences may be widespread, and some present-day examples cover areas of thousands of square kilometers.

Similar difficulties are encountered in the correlation of individual units within a continental volcanic sequence, although the sequence itself may be readily correlated with other sequences having the same general stratigraphic position. These sequences comprise individual flows of lava, ignimbrites, and local falls of tephra, some of which may be washed into areas of sedimentation to form tuffaceous litharenites. Widespread falls of fine volcanic ash cannot be used as time-stratigraphic or lithostratigraphic units because, having fallen on irregular topography and being subject to erosion, they are not traceable when buried in the geological record of rock layers. Ash falling in a body of water may, on the other hand, form an excellent time and lithostratigraphic marker. Submarine volcanics also have a local distribution, although eruptions from a chain of volcanos fringing a sedimentary basin may occur during the same interval of time, forming a time-stratigraphic sequence within the basin. Of particular significance in this respect are volcanic ash beds deposited in bathyal and abyssal depths. Much of this ash has probably been airborne, but a significant volume may in some cases have been transported by ocean currents.

Sequences deposited in bathyal to abyssal environments on gentle slopes or abyssal plains commonly have a remarkable continuity of relatively thin layers, some of which may constitute marker beds. Radiolarian ooze may form marker beds at such depths, although known deposits of radiolarite are not confined to deep water. Some deposits, such as the Early Cretaceous Windalia Radiolarite of Western Australia, were evidently deposited in neritic depths on a continental shelf. In abyssal depths overlying oceanic crust sedimentation is normally very slow, and sedimentary sequences are commonly very thin by comparison with those overlying margins of the continental crust. Abyssal sequences are also commonly young, reflecting post-Cretaceous sea-floor spreading. Incontrovertible examples of this type of deep-sea sedimentation are probably rare, and lithologic or other characteristics assumed to indicate depth of sedimentation in such beds are commonly not definitive. Some ophiolite–wildflysch deposits in which the sedimentary rocks consist largely of thin-bedded ribbon cherts and siliceous argillites may fall in this category.

CONSOLIDATION OF SEDIMENTS

Compaction

Compaction is of primary consideration in the correlation of sandstone and shale sequences, particularly in those which have different sand–shale ratios and variable thicknesses. Whereas consolidated sand is relatively uncompactible, except at depths where crushing of the grains and diagenetic alterations occur, a consolidated fine-grained deposit consisting largely of clay is readily compacted, losing at least a third of its water content during the first 300 m of burial. The rate at which water is lost by expulsion at greater depths depends on the nature of the sediment, in particular the type and percentage content of clay minerals. As previously mentioned, compaction results in the juxtaposition of beds which originally had a considerable stratigraphic separation and may consequently lead to miscorrelations. Compaction also produces draping of beds over basement topography and resistant bodies such as sandstone lenses or limestone reefs. Such draping can be remarkably persistent, being reflected in beds several hundred meters high above the basement or reef.

Where correlations are required to construct a stratigraphic section or a lithofacies map, the problem of compaction commonly arises. Obviously, it would be incorrect to apply the same compaction factor to the total shale section, unless the interval is thin, as the deeper zones were compacted more than the shallower zones during any interval of time. Thus, in restoring a stratigraphic section depicting the vertical relationships of strata during the time of deposition of a particular bed, an estimate must be made of the possible progressive rate of compaction of the clay-layer component in the section. A good deal has been written on the subject of compaction with reference to correlation, including papers by Conybeare (1967) and O'Connor and Gretener (1974). Calculations of compaction based on theoretical considerations give approximations which in many cases are probably adequate for the purpose required. But in general, paucity of specific data on compaction rates of the various sedimentary layers being restored is a limiting factor that precludes both accuracy and lack of equivocation.

An example of the type of gross effects of compaction on a sequence which changes laterally from sandstone and shale in one area to shale in another is illustrated by Carver (1968) in Fig. 3-15. The upper diagrams

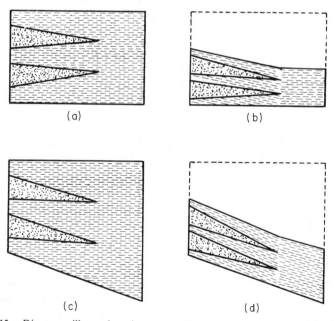

Fig. 3-15. Diagrams illustrating the gross effects of compaction on a sequence of sandstone and shale. (a) and (b) show deposition on a sloping surface. Blank areas in (b) and (d) show loss of volume by expulsion of water. Depth ranges in (a) and (c) are 1200–1800 m. Factor of compaction applied is constant. [After Carver (1968).]

assume deposition on a horizontal surface, the lower diagrams on a sloping surface. Both show the estimated total loss of volume by expulsion of water from the clay-rich layers within a depth range of 1200–1800 m. But in both diagrams a constant gross factor of expansion has been applied to the total thickness of shale in order to restore the sequence to what is assumed to be its original thickness. Obviously, this cannot be accurate because compaction of the clay-rich layers is taking place at a progressive rate during the entire period of deposition of the sequence. So the key problem in restoring a sequence of sandstone and shale to its original thickness, where the upper surface of the section was the sea floor, becomes one of estimating the degree of differential compaction throughout the entire sequence.

Rates of compaction of clays and claystones have been studied by many workers, and although specific porosity–depth curves for claystones of different compositions are not available, curves based on the average porosity of samples of variable composition have been drawn. These show a general agreement of configuration and indicate that 75% of the water in a firm clay is expelled within the first 500 m of burial and that further loss of porosity with depth of burial takes place at a much decreased rate to depths of about 3000 m, below which an increase of pressure produces only a slight decrease in porosity. The curves show some irregularities, and one of considerable significance, discussed by Powers (1967), is shown in Fig. 3-16 at depths of 6000–8000 feet (1800–2400 m). This figure illustrates a depth–porosity curve for samples of mudstone obtained from the Mesozoic in the USSR. Proshlyakov, 1960). The progressive decrease of porosity with increase in depth is reversed at a depth of about 1800 m and returns to its previous trend at a depth of about 2400 m. This aberration is apparently caused by diagenesis of the montmorillonite content of the claystones. The water bound within the montmorillonite is released at the depth range 1800–2400 m in which montmorillonite is transformed to illite. This free water initially occupies space created by reduction in the volume of solids resulting from this transformation. With expulsion of the water, porosity is reduced. The expulsion of water, at a depth and temperature range which is considered to be optimum for the generation of oil, may be an important factor in the primary migration of oil from source rocks that were deposited as organic-rich montmorillonite clays. Compaction of clays at high pressures not only results in the expulsion of water released during the diagenesis of montmorillonite but also, as recorded by Wijeyesekera and de Freitas (1976), in changes in the chemistry of the pore fluid and in the fabric of the clay.

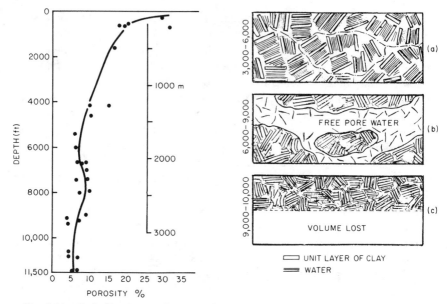

Fig. 3-16. Left: Depth–porosity curve for Mesozoic claystones in the USSR, showing aberration caused by diagenesis of montmorillonite to illite within depth range 6000–8000 ft (1800–2400 m). Right: Diagram illustrating release of water and consequent increase in porosity resulting from transformation (a) montmorillonite before diagenesis; (b) after diagenesis to illite; (c) after diagenesis and compaction. [After Powers (1967) and Proshlyakov (1960).]

Lithification and Diagenesis

Physicochemical processes involved in the lithification of sediments have attracted the attention of many workers who have presented various interpretations and different experimental results. In general, lithification is effected by the cementation of original particles in a permeable sediment, through the introduction of minerals such as calcite and quartz in solution, and by processes of diagenesis which result in mineralogical transformations of certain constituents within the sediment. Cementation does not necessarily require elevations of pressure and temperature but can take place in surficial zones. Examples are found in carbonate beach rock which has been lithified by cementation of carbonate sand soaked with meteoric water. Other examples are seen in pseudoconglomerate and sandstone formed along the banks of some lakes where incoming streams have no mouths but percolate through the beach gravel and sand. Diagenesis, on the other hand, commonly requires elevated ranges of pressure and temperature obtaining at depths of several hundreds or thousands of meters. Within these ranges, and concomitant with diagenesis, cementation also occurs.

Another factor which is significant where carbonate sediments are concerned is the cementing effect of colonial organisms which secrete calcium carbonate as their hard framework, or accrete carbonate particles and fragments of dead organisms. This cementation process is of particular stratigraphic significance with respect to the development and subsequent burial history of bioherms and biostromes. Not all colonial organisms cement their calcium carbonate framework to form a solid body, as seen in banks of sea lilies whose remains accumulate as a rubble. Outcrops of cross-bedded Mississippian encrinite in the Rocky Mountains of Alberta indicate that currents have moved these ancient deposits of crinoid remains. In general, lithification of carbonate sediments occurs at an early stage of burial, mainly by the mutual cementation of carbonate particles.

Noncarbonate sediments are not readily lithified in the absence of a cementing agency such as relatively fresh, moving water derived from a meteoric source. In some depositional situations where transgression of the sea has resulted in the formation of a basal sheet of permeable sand overlain by relatively impermeable deposits of mud, meteoric water seeping into the landward margins of the sand sheet moves down dip, eventually rising to the sea floor as fresh-water springs. In such situations, lithification by cementation of the sand sheet may result, the cement being commonly of carbonates and of oxides and hydroxides of iron.

Where cementation by moving fresh water is not a factor, and where the marine sediments remain buried with their content of connate salt water, lithification depends largely on subsequent diagenesis, with cementation resulting from chemical reactions that occur within particular ranges of pressure and temperature. The time factor in these reactions is undoubtedly of importance but difficult, if not impossible, to quantify. The ranges of depth within which these mineralogical transformations occur will depend on the temperature gradient of the sedimentary pile, the mineralogical constituents of the original sediment, and possibly on geological factors such as those producing situations of abnormally high pressure. Normally, the pressure at any given depth within a sedimentary pile is hydrostatic. Another probable factor is the rate of accumulation of the sediments which is, in effect, the time factor. In thick sedimentary piles that have accumulated rapidly, diagenetic alterations and accompanying lithification may take place within zones deeper than those in which diagenesis occurs in more slowly accumulating piles. In rapidly accumulating piles such as the Quaternary and Late Tertiary of the Mississippi Delta, clay sediments from some locations at a depth range of 3000–

3500 m are not lithified but remain as stiff clay, whereas sands are firm but very crumbly and friable.

Results of experiments aimed at reproducing the pressure or temperature conditions under which diagenetic alterations of clay sediments occur are not always consistent. A prime factor in diagenesis and accompanying lithification is the expulsion, resulting from compaction of the sediment, of connate water and chemically-bound water within clay minerals. Water expulsion resulting from alteration of montmorillonite to illite at a depth range of 1800–2400 m has already been discussed. The chemistry of expelled water undergoes changes with increases of pressure and temperature, attended by diagenetic alterations of the mineral constituents in the sediment. These changes have been studied by Hiltabrand *et al.* (1973) with respect to a temperature range of 100°–200°C which, at an average temperature gradient, is equivalent to a depth range of 4800–11,000 m. They have also been studies by Wijeyesekera and de Freitas (1976) with respect to a pressure range of 2000 lb/in² (140 kg/cm²)–6500 lb/in² (455 kg/cm²) which is equivalent at normal hydrostatic pressure to a depth range of 1800–6000 m.

Hiltabrand *et al.* used argillaceous sediment obtained from the Louisiana Gulf Coast and heated the samples in artificial sea water. As the sediment and fluid reacted at higher temperatures, the chemistry of the fluid was monitored. They noted an increase in the Na^+, K^+, and Ca^{++} concentrations and a decrease in the Mg^{++} and SO_4^- concentrations during diagenetic alterations which included the formation of chamosite and illite and the destruction of feldspars and kaolinite. Calcite, montmorillonite, and quartz remained, the soluble concentration SiO_2 stabilizing near the saturation value for quartz. Hiltabrand *et al.* concluded that much, if not most, of the mineral content of shales is the result of diagenesis rather than depositional environment.

Wijeyesekera and de Freitas used samples of kaolinitic clay and analyzed the chemistry of water expelled within a pressure range normally encountered in drilling deep wells. They found that the concentration of ions in the expelled water changed as compaction progressed. In particular, they noted a systematic depletion of the concentration of Na^+ and K^+ ions in the water as compaction progressed. They noted also that changes in the composition of expelled pore water are partly attributable to crystallization within the pores, that the development of preferred orientation of the clay particles is prevented by the presence of high pore–fluid pressures, and that the major change in the chemistry of expelled water is concurrent with a fall in pore–fluid pressure.

Diagenesis produces certain minerals that have been considered to be indicators of temperature–pressure ranges, and consequently of depth.

Without knowing the probable effect of time as a factor, caution must be observed in using such criteria. Even so, evidence on this basis, with evidence supplied by studies of the reflectivity of coal material in the sediments, will probably give an approximation of the depths to which a stratigraphic sequence has been subjected. Such an approximation may be of great importance to the search for oil in certain areas where the prospective sequence is within a depth range favorable for the generation and accumulation of hydrocarbons, but which has at some time been buried very much deeper. In this respect it must be said that age is not a factor in the lithification of sediments and that once the sediments have turned to rock, their subsequent metamorphosis depends largely on the temperature–pressure conditions to which they are subjected. An interesting case in point is seen in some of the Proterozoic and Early Paleozoic rocks of central Australia which in outcrop do not appear to be older than many of their Mesozoic or Tertiary counterparts in other parts of the world.

STRATIGRAPHIC FRAMEWORK OF BASINS

Lateral Migration and Progradation of Delta Lobes

Concepts relating to the nature of the stratigraphic framework of various types of sedimentary basins have been discussed in Chapter 2. Basins comprising a major volume of deltaic sediments have been formed by the progradation, lateral shifting, and coalescing of one or more large deltas, as pointed out by Weimer (1970) with reference to the Upper Cretaceous sequences of Wyoming and adjacent regions in the United States. These shifts, which may be in response to tectonic adjustments of the crust, are marked by the growth of depocenters of sedimentation which themselves may shift. The depocenters form prograding lobes that are lenticular in sections more or less normal to the direction of sediment transport. These lobes have an offlapping relationship, although adjacent lobes were not necessarily deposited during a continuous period of time. Sediments swept out to sea and deposited as a broad fan on a fairly flat continental shelf form a slightly elevated surface in the area of the depocenter. Subsequent sedimentation of a younger depocenter tends to be on the flanks of the older, in much the same relationship as that of offset channel deposits within the alluvium-filled course of a major river.

The rate of sedimentation within each depocenter lobe is variable; also, different lobes may have had different average rates. Correlation by marker beds within a particular lobe may be excellent; but correlation from one lobe to another may be indefinite. Another complicating factor in the correlation of stratigraphic units across the width of a major delta is the possibility that sedimentation was not continuous over the delta area during the entire period of sediment accumulation but that the sequence may be locally or regionally interrupted by unconformities, disconformities, and paracontinuities.

Some idea of the complexity of fabric and stratigraphic framework of a large delta may be obtained by a consideration of its age and of the probable periodicity of major lateral shifts of the river system. Deltas such as those of the Ganges–Brahmaputra and Mississippi have been actively growing since at least the beginning of the Pliocene, an interval of 5 million years. The patterns and processes of sedimentation and the forms of geomorphologic features observed today are recapitulations of those that existed during the Late Tertiary. If one assumes that major shifts in the main river course of a large delta are likely to occur within a period of 10,000 years, as is the case with the Mississippi Delta, then over a period of 5 million years at least 500 such shifts are possible. As mentioned earlier in this chapter, rates of sedimentation on a delta may be as much as 4 m in 1000 years, so that if the sediment fan remains in more or less the same position for 10,000 years, it will have a maximum of 40 m of sediment deposited at the surface. This will be compacted with burial to approximately half its original volume at depths in excess of 1500 m. These shifting fans, which may or may not be contiguous, constitute a larger unit of deposition, the central portion of which is referred to as a depocenter. The upper and lower limits of these units define them as time-stratigraphic units (Fig. 2-62). Within the framework of a deltaic sedimentary pile, there may be many time-stratigraphic units defined by depocenters, each unit consisting of a number of sediment fans. All of these units are lenticular in sections normal to the direction of sediment transport and exhibit the features of a prograding sequence in sections normal to the coastline.

Examples of these relationships have been described by Asquith (1974) with reference to the Upper Cretaceous sequence of Wyoming which includes numerous bentonitic time-stratigraphic markers. Asquith says that the Late Cretaceous shoreline prograded eastward as a series of pulses from constantly shifting centers of deposition and that transgression, regression, and stillstand probably occurred at the same time along this coastline in a manner similar to that of the coastline of Louisiana and Texas. Fig. 3-17 illustrates the prograding nature of lobes within the Late Cretaceous Pierre Shale of the Powder River Basin, Wyoming. The section is based on electric log correlations by means of bentonite and other markers and shows gentle depositional dips on a wide continental shelf

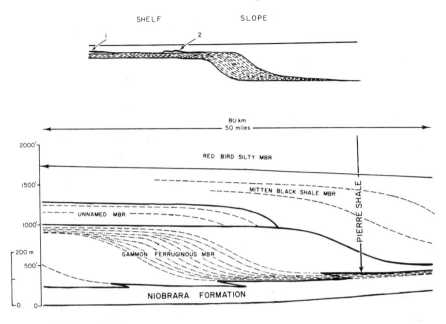

Fig. 3-17. Stratigraphic section, based on electric logs, of the lower part of the Late Cretaceous Pierre Shale in the Powder River Basin, Wyoming, showing lobes of sediment prograding seaward on a broad continental shelf. Slopes are those of the leading edges of the lobes. Upper diagram shows general configuration of a lobe, 1 and 2 being shoreline and bar sands respectively. [After Asquith (1974).]

that extends in the line of section for 80 km. The slopes shown are not those at the edge of the continental shelf but slopes of the leading edges of sediment lobes prograding seaward on the shelf. It should be noted that the vertical exaggeration is very considerable and that consequently the true dips of the leading edges are very slight. The upper diagram of Fig. 3-17 illustrates the general shape of a prograding lobe which is a time-stratigraphic unit. With reference to the apparent shifting pattern of deltaic sedimentation observed in the Upper Cretaceous sequence of the Powder River Basin, Asquith (1974, p. 2274) states, "If this pattern of shifting loci of deposition is correct, some stratigraphic concepts may require reexamination."

Contiguous lobes of sediment may locally form a continuous sequence with no hiatus in sedimentation. Elsewhere the same lobes may be separated by a paracontinuity, disconformity, or unconformity as illustrated in Fig. 3-18. All these features may be found within the stratigraphic framework of individual deltaic sequences and sedimentary basins comprising a

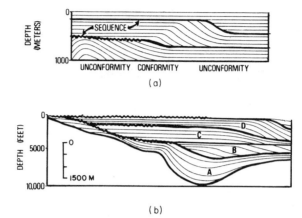

Fig. 3-18. Generalized sections of a sediment lobe sequence (a) and four lobes forming a deltaic sequence (b), showing possible stratigraphic relationships. [After Steele (1976) and Vail *et al.* (1975).]

number of such sequences. Interpretations of the stratigraphy and consequently of the stratigraphic framework of a sedimentary basin depend on recognition of these sediment lobes as time-stratigraphic units. Seismic data correlated with subsurface control can be a most useful tool in such interpretations, and has been commented on by Steele (1976, p. 67) as follows:

> One approach to the problem of extracting stratigraphic information from seismic data is through the concept of sequences which are defined as time-stratigraphic units that are bounded by unconformities or their correlative conformities. These sequences are distinct depositional units that may be thought of as the building blocks of sedimentary basins. Sequence boundaries can be recognized on good quality seismic data by offlapping, onlapping and truncated patterns of cycle terminations. These sequence boundaries can be mapped to determine the configuration and lateral extent of each sequence and to provide a time-stratigraphic framework for analysing a basin. In addition, the logical lateral variations of environment and lithofacies that may be expected within each sequence provide a basis for rational extrapolation of lithological trends beyond the well control.

The present-day surface of any large delta of a major river system reflects depositional patterns, environments, and geomorphologic features within the entire deltaic sequence, evidence that the present is a recapitulation of the past. A motion film, in which each frame is an aerial photograph taken at annual intervals, would show the dendritic growth of ever-shifting and ever-growing prograding lobes. It is essential to the interpretation of stratigraphic framework within a delta or sedimentary basin to visualize the internal features in three dimensions. Arcuate and bifurcating distributaries flowing outward on a lower delta plain are recorded by channel sand bodies that will ultimately in the subsurface form part of a three-dimensional system of anastomosing shoestring sands. Barrier island and other bar sands may form a sequence of offlapping sand bodies having an *en echelon* arrangement in sectional views. These concepts of lateral migration and progradation of delta lobes and of the consequent effects on the stratigraphic configuration of a deltaic sequence are essential to analysis of sedimentary basins.

Unconformities, Disconformities, and Paracontinuities

The recognition of breaks in a stratigraphic sequence, representing periods of nondeposition or removal of strata, is necessary in interpreting the geological history of any sedimentary basin. An unconformity is defined as a surface of erosion separating a younger from an older stratigraphic sequence. No time limit is implied, and the relationship between the two sequences may be angular or parallel. Where the relationship is angular, there is obviously a significant interval of time represented by the unconformity, during the development of which the underlying beds may have been tilted, folded, or faulted. But where the relationship is parallel or nearly so, the interval of time represented by the unconformity is commonly but not necessarily of shorter duration. Where the overlying and underlying sequences exhibit an angular relationship, the surface between has been referred to as an angular unconformity, but this term is not definitive as the recognition of very low angles is not always possible and therefore the distinction between angular unconformities and parallel unconformities termed disconformities is not always clearly defined. The angular nature of an unconformity may in some areas be determined by plane table mapping of outcrops, by the superposition of strata beneath the unconformity at different localities, and by correlations based on subsurface data. An angular unconformity may pass laterally into a disconformity which may, in turn, pass into a paracontinuity, as shown in Fig. 3-18.

Paracontinuities define surfaces, separating parallel sequences of beds, that are not surfaces of erosion but represent a hiatus in sedimentation. These may be more difficult to recognize than disconformities, although they may be numerous in some sequences of limestone. The term paracontinuity refers to small-scale physical and faunal breaks which represent time-stratigraphic breaks. Conkin and Conkin (1975, p. 2) describe the feature as follows:

> The term paracontinuity (Conkin and Conkin, 1973) is derived from *para* (nearly or almost) and continuity (a continuum); the implication is one of an apparently continuous sequence of sedimentation, which, however, is clearly seen (upon careful study) to be broken by a small-scale physical discontinuity (the surface of paracontinuity) and a slight though significant, coincident faunal discontinuity. The paracontinuity is, then, a kind of widespread diastem which is manifested by both a physical break and a faunal gap, and thus it offers a reliable means by which significant time-stratigraphic boundaries may be determined precisely and subsequently utilized for correlation.

Conkin and Conkin say further that essential to the concept of a paracontinuity is the presence of a basal transgressive detrital unit, commonly up to one meter thick, lying immediately above the physical surface of paracontinuity. This detrital zone is deposited near shore and may consist of quartzose sandstone, calcarenite such as encrinite, or quartzose siltstone with phosphatic nodules, glauconite grains, pebbles, oolites, fish teeth, and other remains.

Angular unconformities are normally recognized, particularly as there is commonly a significant difference in the ages of the underlying and overlying sequences. Disconformities are not so readily observed and their recognition depends, as it does with paracontinuities, on the interpretation of lithostratigraphic and biostratigraphic evidence. Fossil soils and subsoils may be recognized on the basis of texture, composition, and color. Where the weathered material was rich in alumina, local deposits of bauxite have been formed in some areas and subsequently buried by later sedimentary sequences. Such fossil deposits of bauxite have been mined in Europe. Other residual deposits of iron and manganese oxides and hydroxides have formed in this way and been buried. Disconformities, representing erosional zones, may also be noted by their reddish to brownish color in some sequences. But where there is no evidence of erosion in situ, of residual fossil soil or subsoil, recognition depends on the sequence of strata above and on the fossil content of sequences above and below the unconformity.

A disconformity may be overlain directly by a sequence of nonmarine sediments or by a transgressive sheet of marine sand such as the Tonti Member of the Ordovician St. Peter Sandstone (Fig. 3-5). Where nonmarine beds, originating as river channel sands and floodplain silts and clay muds, disconformably overlie marine beds, the demarcation is recognized primarily by the nature of stratification and sedimentary structures in the overlying beds and secondarily by the absence of invertebrate fauna, or change of faunal assemblage in the upper sequence.

Marine beds may also overlie older marine beds, the disconformity marking a considerable interval of time. These relationships occur where neritic deposits such as shoreline sand bodies and prodelta deposits of silt and clay muds are raised above sea level by regional uplift. This may be a gradual and regional event caused by crustal tilting, or a rapid and local event caused by earthquake as has happened on the northeast coast of New Zealand during this century. Where there is no appreciable erosion and the period of exposure is short, the depositional hiatus is marked by a paracontinuity. But where a disconformity results, the period of exposure may be considerable and difficult to recognize. An example is seen in the Triassic Wagina Sandstone, a quartzose sandstone that in places overlies Permian quartzose sandstone in the northern part of the Perth Basin of Western Australia. Both sandstones originated as marine shoreline sands. In such cases recognition of the disconformable relationship may depend on the fossil content of the sandstones.

Marine beds commonly overlie nonmarine beds, particularly in a deltaic sequence where transgression and regression of the sea are periodic events (Fig. 3-6). Marine beds may also overlie nonmarine beds where the sea invades a flat cratonic landmass of ancient rock covered with a thin veneer of nonmarine sediments. This type of situation obtained during the Early Cretaceous in northern Australia, where widespread outcrops of nonmarine beds overlain by marine beds are 50–100 m thick. The basal beds of a marine sequence are commonly shoreline sand bodies which form a sheet of sand, irregular in plan view and discontinuous in some sections, overlying a paracontinuity or discontinuity. In such stratigraphic relationships a disconformity may pass into a paracontinuity in a seaward direction. Similarly, a disconformity may pass into an angular unconformity in a landward direction.

Nonmarine beds are also known to disconformably overlie nonmarine beds, examples being seen in the western part of the Surat Basin of Queensland, Australia, where Jurassic fluvial and lacustrine deposits overlie Triassic deposits having a similar origin. Identification of the age difference in the two sequences is based on the palynological evidence of pollens and spores. Precise boundaries cannot always be determined in such stratigraphic situations, and the best zone of demarcation may be the base of an alluvial sequence as determined by outcrops, electric log characteristics, or drill core.

Where nonmarine beds disconformably overlie nonmarine beds there may be no significant differences in the fossil assemblages of the two sequences, both of which may contain the same types of bivalves, gastropods, and ostracods. Differentiation may be possible on the basis of pollen and spore zonations. Where marine beds disconformably overlie marine beds, there is commonly a difference in faunal assemblages which reflect ecological differences in the depositional environments of the transgressive marine shoreline sands and the neritic marine deposits over which they transgress.

Discontinuities and paracontinuities are a common feature in carbonate sequences, reflecting cyclic sedimentation. Such sequences comprise successive stratigraphic units, of variable thickness but similar pattern of layering, that were deposited during a series of depositional cycles resulting from rhythmic fluctuations in the relative levels of sea and land.

Basin Hingeline

A basin hingeline is defined as a line along which the rates of thickening and of dip of the bedding within a sedimentary basin begin to increase. This line is somewhat arbitrary but is determined within recognized limits. The increase in rate of dip is not a structural feature in the sense of tectonic deformation but a parameter of the stratigraphic framework resulting from the increase in thickness of the section and consequent greater degree of compaction of the layers in the lower part. In many cases the hingeline appears to roughly correspond to the depositional limit of shelf deposits of sand and silt and to mark the edge of the shelf where thick deposits of silt and clay muds accumulate on steeper depositional slopes. Compaction of the sediments accentuates these stratigraphic relationships.

In many areas of the world accumulations of hydrocarbons tend to be concentrated in sandstone bodies situated along the trend of a hingeline. It has been suggested that these hydrocarbons or their precursors were generated in organic-rich clay sediments deposited on the steeper slopes and that as the sediments were compacted, the hydrocarbons were expelled and migrated up-dip into bodies of partly lithified sand. Migration may have been effected in large part by the solubility of hydrocarbons in the expelled water (Conybeare, 1970) which would have moved laterally and upward, following the more permeable stratigraphic zones of least resistance. On reaching the upper part of the slope, migrating oil and gas would be trapped within permeable bodies of sandstone. Hydrocarbon-saturated solutions would, on reaching higher zones at lower pressures, release hydrocarbons which would also become trapped. Primary migration of hydrocarbons is related to the stratigraphic framework of a sedimentary basin and results in accumulation of oil and gas in stratigraphic traps. Secondary migration may form accumulations in stratigraphic, structural-stratigraphic, or structural traps in both the stratigraphic and structural framework of a sedimentary basin. Determination of hingeline trends is consequently an important aspect of sedimentary basin analysis from an economic viewpoint.

A sedimentary basin may have one or more hingelines which may be straight or arcuate depending on the configuration of the basin. Linear, asymmetrical basins tend to have a single fairly straight hingeline along the more gently dipping flank. Symmetrical basins that are linear may have a hingeline along both sides. These hingelines become arcuate where they trend around the ends of a basin or around irregularities in the configuration of the flanks. The change in rate of dip marking the limits within which the hingeline is determined may be so slight that it would be apparent only in sections with considerable vertical exaggeration. Furthermore, determination depends on the presence of a recognizable marker and on sufficiently close control to delineate the limits within which to place the hingeline. This is not to imply that a hingeline is necessarily present in all sedimentary basins. The development of a stratigraphic hingeline requires that sedimentation take place on a shelf and on the slope of the leading edge of the shelf. In noncarbonate environments sediments are transported seaward as prograding sheets or lobes, the coarser sediment being largely deposited on the shelf, whereas much of the finer is carried out to the proximity of the continental slope. In environments where carbonate shoals of calcilutite, calcarenite, and patch reefs have formed on a shelf, the leading slope of the shelf will be underlain by fine-grained sediments which may in large part be calcareous clay muds. In both cases the slopes may delineate the trend of the stratigraphic hingeline as the sediment pile accumulates and compaction proceeds. Examples are seen in prograding deltaic sequences in particular; whereas in shallow-water carbonate environments the relationship between a shelf edge and stratigraphic hingeline may be less well defined.

With reference to the approximate coincidence of depositional slope and hingeline, the slopes mentioned are not those on the leading edges of individual prograding lobes or sheets but rather those of larger time-stratigraphic units, the seaward slopes of which coincide with or are in the vicinity of the continental slope. As a sedimentary basin or portion of a basin develops by the progradation of such time-stratigraphic units, the continental rise and slope move seaward, resulting in migration of the stratigraphic hingeline.

It is essential to distinguish between a stratigraphic and structural hingeline. The former is determined by the construction of stratigraphic sections in which marker beds indicate a steepening of dip resulting in large part from compaction but also in part from the initial steeper angle of the depositional slope. Commonly this zone of steepening dip also coincides with a zone of increasing thickening of the section. A structural hingeline, on the other hand, is a flexure caused by tectonic warping of the margin of a sedimentary basin. Such flexures are caused by domes, monoclinal folds, or basement faults and can be illustrated by structural sections. Coincidence of a stratigraphic hingeline and underlying structural flexures may also occur. Other structural–stratigraphic complications arise where a hingeline is situated above growth faults which terminate upward in monoclinal folds.

As shown on a map, a hingeline is merely a directional trend and is a surficial parameter only with reference to the solid geometry of a sedimentary basin. A hingeline is, in effect, the surficial trace of a plane intersecting the sedimentary pile through stratigraphic intervals where the rate of dip begins to increase. As the hingeline migrates basinward, with progradation and thickening of the sedimentary pile, this imaginary plane becomes undulating and sloping, dipping landward. In essence, a hingeline is conceptual.

Superposition of Basins

A sedimentary basin constitutes a pile of strata, normally of sediments but commonly including volcanics, that are lenticular in any section and which fill a depression of the underlying rocks. The underlying rocks may be a basement of igneous and metamorphic rocks or an older sequence of strata belonging to a separate or preexisting basin. Within the basin itself there may be numerous paracontinuities and disconformities but, in general, no pronounced angular unconformities. An angular unconformity that can be seen in outcrop indicates not only a cessation of sedimentation but tectonic uplift, tilting or folding, and erosion with consequent destruction of the preexisting depositional environment. Subsequent deposition in the same region results from the development of a younger sedimentary basin which may overlap the older. Where overlap occurs, the two sequences separated by an angular unconformity may, for reasons of practicality or insufficient data, be referred to as constituting a single sedimentary basin; but by the definition of a basin as stated above, the two sequences should be regarded as parts of separate basins. An example can be made with reference to the Permian Bowen Basin and the Triassic–Jurassic Surat Basin of Queensland, Australia. The main development of the Bowen Basin lies to the north of the Surat Basin but extends beneath the latter in the area west of Brisbane. The angle of unconformity is slight in sections having no vertical exaggeration, and there is no significant difference in the metamorphism of the two sequences. The Mesozoic comprises fluvial, lacustrine, and paralic sandstone, siltstone, and claystone, whereas the Permian sequence is largely of marine and paralic beds including sandstone, siltstone, claystone, coal, and limestone. The combined sequences overlie a basement of igneous and metamorphic rocks in their area of overlap, but each sequence is clearly defined as belonging to a separate sedimentary basin.

The strata within a sedimentary pile constituting a basin may comprise two or more sequences belonging to different Systems or Series but separated by disconformities or paracontinuities. A case in point could be made with reference to the boundary between the Jurassic and Cretaceous Systems in Alberta and British Columbia where locally there appears to have been continuous sedimentation during the Late Portlandian and Early Berriasian, as defined by an *Aucella* fauna (Warren and Stelck, 1958).

Where minor and local angular unconformities occur, particularly where they pass into disconformities, the overlying and underlying sequences may for convenience be placed within the same sedimentary basin, keeping in mind that erosion may have removed the greater part of the overlying sequence which may in fact have formed a separate but adjacent basin. It is perhaps worth reiterating at this juncture that sedimentary basins seldom reveal their depositional edges but only a truncated portion of the thicker sections of the sedimentary pile. One must unfold the sequences and estimate by lithostratigraphic and biostratigraphic analyses the original extent and sedimentary environments of the beds.

Where major unconformities occur within a System, such as that separating the Early Permian from the Late Permian in the Cooper Basin of South Australia, or the Early Triassic from the Late Triassic in the Surat Basin of Queensland, it is preferable to treat these separate sequences as belonging to separate sedimentary basins, even though the depositional area of the upper basin may have occupied much of the lower. At best, defining the parameters and terminology pertaining to the genesis, development, lithologic constitution, and solid geometry of sedimentary basins is a hazardous exercise frought with reservations, equivocation, and differences of concept or interpretation.

STRUCTURAL FRAMEWORK OF BASINS

Basement Faults

The previous section on superposition of basins stated that sedimentary basins fill depressions in the underlying rocks which may be igneous and metamorphic, or an older sequence of strata. The base of a sedimentary basin is always an unconformity but not necessarily defined as basement which refers to granitic and associated metamorphic rocks. In consequence, a sedimentary basin may unconformably overlie a portion of another, which in turn may overlie part of another which is separated by a major unconformity from the basement. The term major unconformity can be applied not only to the surface separating an igneous–metamorphic complex from an overlying sequence of strata but also to unconformities separating sequences of strata that show a distinct angular contact in outcrop. The term also implies a considerable interval of time, marked by uplift and erosion, separating the two sequences.

It is obvious that the early development of a sedimentary basin, its internal structural configuration during growth, and its subsequent solid geometry following its depositional history and period of uplift and erosion, will be strongly influenced by movements of the basement floor underlying the basin. The initial growth of a sedimentary basin within a depression of the basement may have resulted from downward movement of fault blocks within the basement, followed by isostatic adjustment of the blocks under the weight of the growing sedimentary pile. Once established, basement faults become zones of weakness along which crustal movements may occur over long periods of time. The structural framework of a sedimentary basin thus depends primarily on the structural pattern of the basement.

Downward movements of basement grabens create depressions in which sediments may accumulate and form basins, provided that subsidence and deposition continue long enough. Upward movements of basement horsts create sea-floor ridges or rises which trap sediments within troughs that are commonly linear (Fig. 2-15). The sedimentary piles within these troughs may be discrete, forming separate sedimentary basins, or they may overlap to form separate subbasins in the lower part of the section and a single basin in the upper. In many such cases subsequent uplift and erosion would remove much if not all of the upper section, leaving the lower portions of the section as separate subbasins. Although the sections in each such subbasin may have been deposited during the same interval of time, they may differ considerably in the sequence and nature of their strata and also in their faunal assemblages.

The initial basement of a sedimentary basin, as seen in the basal beds of the sedimentary pile, may have been beveled to a fairly even or gently undulating surface traversed by the traces of fault planes. Subsequent movements along these fault planes may have occurred during the period of growth of all or part of the sedimentary pile. The initial basement may, on the other hand, have had a marked topographic relief during the period of deposition of the basal beds, and this relief may have been accentuated by further tectonic adjustments during the growth of the basin. Both situations can apply to predominantly nonmarine or marine sequences. In this regard it is of interest to note the relationships between basement structure and depositional environments in the San Diego region of southern California where basin and range topography on land extends off the coast to the edge of the continental mass. As there is practically no continental shelf, the offshore basement blocks form rises and subbasins. Sediment accumulation on the rises is comparatively slow, the deposits consisting of clay mud and pelagic detritus. Accumulation in the subbasins is comparatively rapid, the sediments consisting of the normal thin deposits of clay mud and pelagic detritus interbedded with relatively thick deposits of turbidites. On land the same type of basement topography results in ranges of high hills or mountains separated by intermontane subbasins filled with fluvial, lacustrine, and alluvial fan deposits that accumulate to thicknesses of a few thousand meters.

Periodic adjustments of basement blocks may occur throughout much of the period of growth of a sedimentary basin, then cease during the later stages of development. Sequences of beds that are buried at a time of basement adjustment will be faulted, whereas sequences being deposited may be disrupted in their lower beds and monoclinally folded in their upper. Also the upper part of the sequence commonly reflects the basement movement by depositional thinning in beds overlying the higher basement block. An early history of basement faulting during the development of a sedimentary basin indicates a period of crustal instability that may be associated with regional orogeny or major adjustments of the continental crust such as those resulting from the separation or impinging of continental masses. As shown in Fig. 2-40, illustrating a section underlying the northwest continental shelf of Western Australia, faults originating in the Precambrian basement of igneous–metamorphic rock can be traced throughout the stratigraphic sequence to the Tertiary beds which were deposited after the period of crustal instability believed to have been caused by continental drift. In the Gippsland Basin of Victoria, Australia, basement movements displaced the Paleogene beds to a greater extent than the Neogene beds, the Pliocene being apparently uninterrupted, as shown in Fig. 3-19 (Franklin and Clifton, 1971).

Basement movements may have profound effects on depositional environments and sedimentation where the depth to the sea floor is shifted significantly, as may be the case where there is rapid displacement of the basement underlying a continental shelf. A possible example is cited by von der Borch (1967) who postulates that the Beachport Plateau shown in Section G–H of Figs. 2-7 and 2-8 may be a block which was downfaulted

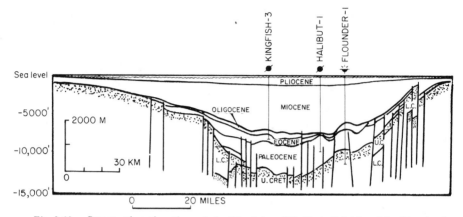

Fig. 3-19. Structural section, through the Kingfish and Halibut oil fields, of the Gippsland Basin in Victoria, Australia. Vertical exaggeration 13×. [After Franklin and Clifton (1971).]

from the shelf in pre-Cenozoic times. The plateau in the Great Australian Bight traversed by Sections C–D and E–F of Fig. 2-7 may also be a downfaulted block of the continental shelf. As stated by von der Borch (1967, p. 309), "In both cases the plateaus have acted as localised sediment traps. They have prevented the prograding shelf sediments of the Tertiary from slumping downslope to build up notable abyssal fans at the base of the continental slope." Depositional thinning is also a feature of sedimentary sequences deposited on higher topographic features of the sea floor. Such features commonly overlie basement highs which may be mobile. The effects of an elevated sea floor on sedimentation patterns, and on compaction of the sediments, can be reflected throughout hundreds of meters of a stratigraphic section. In limestone sequences basement highs may be reflected in reef development, as illustrated in Fig. 2-52. There may also be a relationship between some reef trends and underlying fault trends in the basement, the relationship being indirect in that the trend of basement faulting may determine the edge of a carbonate shelf along which reefs will develop. This concept has been considered with reference to Devonian reef development in the vicinity of a known major fault trend in the Precambrian basement extending southwest from Great Slave Lake in Canada. Whether such a relationship obtains in this case is conjectural.

All basin studies involving lithostratigraphic analysis of the entire basin logically begin at the base of the sequence and trace the development of the basin to the final records of its history. The structural evolution of the basement during this interval of time is an essential component of the basin's history, and a controlling factor in its structural framework.

Growth Faults and Associated Structures

Growth faults within a sedimentary pile arise from the weight of sediments and the release of stress along a fault plane that commonly transects the beds in the upper part of the pile and trends parallel to the beds in the lower part. Such faults are in effect gravity faults caused by slumping and slippage on a grand scale of great wedges of unconsolidated sediment. The process of faulting is probably related to such factors as compaction, angle of repose, and instability of particular clay beds which may act as a lubricating layer along which an overlying sequence may slip. The triggering mechanism may in some cases be crustal movements within the basement underlying the sedimentary pile. Associated with these primary growth faults are secondary faults resulting from tension in the upper part of the slumped wedge adjacent to the primary fault. These secondary faults give rise to graben-type structures comprising downfaulted slices of the sedimentary section, producing structural complexity (Fig. 3-22) in the subsurface configuration of many rapidly accumulating sedimentary piles such as those of large deltas. Contemporaneity of deposition and movements along fault planes results in the interrelationship of the stratigraphic and structural framework of a sedimentary pile as adjustments of the wedges of sediment may be reflected in the depositional pattern of the sea floor. The extent to which this may happen depends largely on the rate of sedimentation as opposed to the rate of subsidence of the downfaulted wedge.

Continual adjustments of sediment wedges and slices give rise not only to the growth of complex fault patterns but also to the growth of domes and monoclinal folds within the sedimentary pile. This may be of considerable significance to the accumulation of oil and gas during the early period of migration of these hydrocarbons. Associated with structures arising from the movements of sediment wedges are others possibly caused by the weight of a predominantly sandy sequence overlying a thick sequence of clay beds. Lobes of the upper sequence may sink into the lower, being displaced along growing faults. The underlying clay is itself displaced and in some cases is actually forced to the surface where it appears offshore in a delta as an island of mud. All of these structures are associated with the growth and stratigraphic framework of a sedimentary pile comprising a deltaic complex. They are subsequently preserved in the geological record and become part of the structural framework of a sedimentary basin.

Movement along growth faults that extend upward to the vicinity of the sea floor commonly trigger the slumping of surficial sediment and the possible initiation of turbidity currents. The features associated with surficial slumping are illustrated in Fig. 3-20 which shows the type of movement that produced the slide scar and debris flows illustrated in Fig. 3-3. The sediments lying between the head and toe of the slump have been moved predominantly by subaqueous slumping and sliding, as illustrated in Fig. 3-2, whereas those at the toe may exhibit the characteristics of a mass flow merging into those of a turbidity flow. One such example from the upper continental slope off the east coast of North Island, New Zealand, is cited by Lewis (1971). He states that continuous seismic profiles show that surface sediment 10–50 m thick has slumped down bedding planes sloping at 1°–4°. One such slump covers an area of approximately 250 km². Lewis postulates that the slumps observed were probably caused by earthquake shocks, a reasonable assumption in view of the tectonic instability of that region. Lewis further says that in many outcrops of Tertiary strata in New Zealand, there is evidence of ancient marine slumps involving layers up to 200 m thick.

In the deeper zones of a sedimentary pile, the development of growth faults and associated structures may result not only from gravitational sliding and the effects of sediment blocks sinking into and displacing masses of clay but also from the mobility of salt masses caused by dissolution and growth of salt domes. Growth structures resulting from the movement of thick salt layers are illustrated in Figs. 2-33, 2-48, 2-50, 2-51, and 2-67. These masses of salt and associated structures are a prominent feature off the northwest coast of Africa and are encountered elsewhere in the world underlying the outer fringe of the continental shelf and the continental slope. Fig. 3-21 illustrates the possible structural relationship of an underlying salt mass to faulted blocks of strata, as described by Shelton (1968). The formation of reverse drag or downfaulted slices may

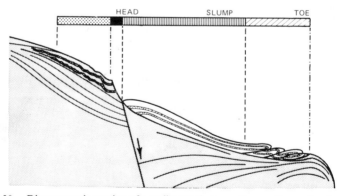

Fig. 3-20.   Diagrammatic section of a sediment slump on the sea floor, showing the head marked by a slide scar overlying a growth fault and the toe marked by compressional folding and thrusting. [After Lewis (1971).]

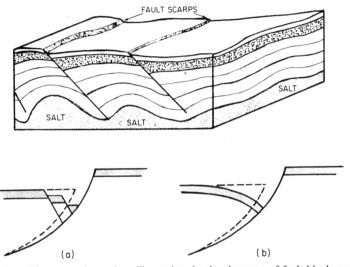

Fig. 3-21.   Diagrammatic sections illustrating the development of fault blocks resulting from movement of an underlying mass of salt and the mechanism by which antithetic faults (a) or reverse drag (b) are produced. [Upper after Shelton (1968); lower after Hamblin (1965).]

depend largely on the depth beneath sea level at which dislocation occurs and on the competency of the beds involved. Less competent beds at shallower depths may develop reverse drag. Hamblin (1965) says that in the western area of the Colorado Plateau reverse drag may pass both vertically and laterally into antithetic faults. The development of reverse drag is of considerable significance with respect to the early migration and accumulation of hydrocarbons in a sedimentary pile.

The effects of reverse drag and antithetic faults in the development of domal structures during the growth of a sedimentary pile are illustrated in the seismic profile shown in Fig. 3-22. The profile shows growth faults flattening at depth to become bedding plane faults. These faults range to a depth of 6000 m but are most prolific within the range to 3500 m. The combined effect of growth faults and compaction on the structural framework of the sedimentary pile is illustrated in the lower right diagram of Fig. 3-22. This shows the burial and progressive warping of stratigraphic units 1–4 to form local domes. As noted previously, the structural configuration of such complex growth fault systems is partly determined by the relationships between rates of deposition of the sediments and rates of subsidence of the sedimentary pile. Possible variations in the structural pattern of a growth fault system are illustrated in the upper diagram of

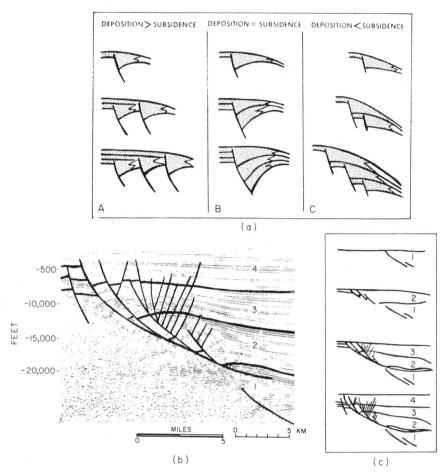

Fig. 3-22. (a) Diagrams illustrating structural patterns of growth faults developed by various rates of subsidence and deposition. (b) Seismic profile showing complex pattern of growth faults and domal structures in a sedimentary pile 6000 m thick. (c) Diagrams illustrating stages of growth of the fault pattern and domal structures shown in the seismic profile. [After Bruce (1973).]

Fig. 3-22. Bruce (1973) says that the three types shown are recognized in the Tertiary of southern Texas. With reference to Type A he says that the faults were formed during a period of regressive deposition in an area where the amount of sediment exceeded the space available for deposition, resulting in the development of a prograding sequence and successive shifts in centers of maximum deposition. Type B is believed to have formed in areas where stillstand depositional conditions prevailed and the rate of downward faulting was sufficient to accommodate the rate of sedimentation. Type C is considered to have resulted during transgressive phases of deposition in areas where the rate of subsidence exceeds the rate of sedimentation. Types A and B are thought to have developed beneath fairly flat sea floors, whereas Type C is thought to be associated with depositional areas where the sea floor is sloping.

The development of growth faults and associated structures within a sedimentary pile is believed to continue during various stages of compaction in which clay is converted to undercompacted shale. The release of water from pressured clay sediments may in certain stratigraphic situations be impeded by less permeable barriers above, with the consequence that the clay sediment becomes less dense with a higher water content than is normal for the depth range in which it is buried. The possible structural relationships of such low-density, high-pressure clay sediments to overlying growth fault patterns are illustrated in Fig. 3-23. The upper diagram illustrates the general features of these relationships within a sedimentary pile deposited as a large delta complex such as that of the Mississippi and its ancestral river systems. The lower diagram illustrates the possible stages of burial of a large mass of shale and the development of an overlying complex pattern of growth faults. These relationships are believed to obtain within the Tertiary of southern Texas, with reference to which Bruce (1973, p. 878) says:

> Regional contemporaneous faults of the Texas coastal area are formed on the seaward flanks of deeply buried linear shale masses characterized by low bulk density and high fluid pressure. From seismic data, these masses, commonly tens of miles in length, have been observed to range in size up to 25 mi in width and 10,000 ft vertically. These features, aligned subparallel with the coast, represent residual masses of undercompacted sediment between sandstone–shale depoaxes in which greater compaction has occurred. Most regional contemporaneous fault systems in the Texas coastal area consist of comparatively simple down-to-basin faults that formed during times of shoreline regression, when periods of fault development were relatively short. In cross-sectional view, faults in these systems flatten and converge at a depth to planes related to fluid pressure and form the seaward flanks of underlying shale masses. Data indicate that faults formed during regressive phases of deposition were developed primarily as the result of differential compaction of adjacent sedimentary masses. These faults die out at depth near the depoaxes of the sandstone–shale sections.

Previous mention has been made of the significance to petroleum exploration of antithetic faults and reverse drag within a deltaic sequence. A

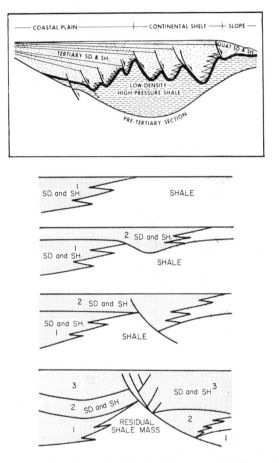

Fig. 3-23. Diagrammatic sections of a deltaic sedimentary pile, illustrating development of a growth–fault pattern resulting from the depression and displacement of an undercompacted mass of clay sediment by an overlying sequence of compacted sand and clay sediment [After Bruce (1973).]

specific example is shown in Fig. 3-24 which illustrates a normal growth fault and reverse drag in part of the Tertiary section of the Niger Delta in Nigeria. The Benin Formation in the upper part of this section is predominantly sandy, and the Agbada Formation which includes a biostratigraphic marker in its uppermost part is predominantly of sandstone and shale. Short and Stäuble (1967) say that rollover anticlinal structures formed by reverse drag form the main objectives of oil exploration, the hydrocarbons being found in sandstone reservoirs of the Agbada Formation.

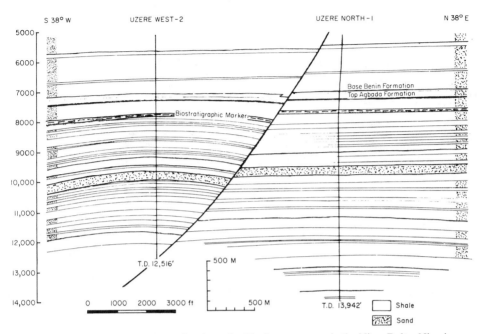

Fig. 3-24. Diagrammatic section through a Tertiary sequence in the Niger Delta, Nigeria, showing a rollover anticlinal structure formed by reverse drag against a normal growth fault. Scales in feet. [After Short and Stäuble (1967).]

Further examples of various types of hydrocarbon traps in the Niger Delta are described by Weber and Daukoru (1975). With reference to the subsurface geology they say, p. 209,

A sequence of under-compacted marine clays, overlain by paralic deposits, in turn covered by continental sands, is present throughout, built-up by the imbricated superposition of numerous offlap cycles. Basement faulting affected delta development and thus sediment thickness distribution.

The four types of structures shown in Fig. 3-25 are within a Tertiary section comprising the undercompacted Akata Formation of marine clay sediments, and an overlying paralic sequence of sandstone and shale within which the entrapment of hydrocarbons occurs. The types of structures in which the oil reservoirs are found include simple rollover structures or domes formed by reverse drag against a normal growth fault, and other structures formed by multiple growth faults, antithetic faults, and collapsed crestal structures. These structures result in the development of multiple reservoirs within the same field area.

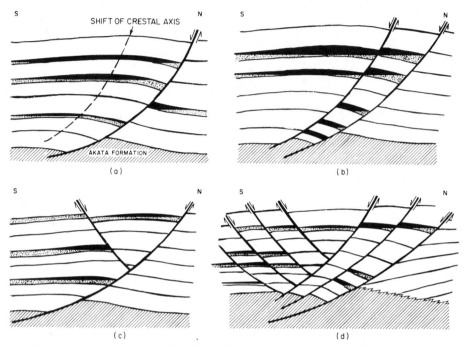

Fig. 3-25. Diagrammatic sections illustrating hydrocarbon traps formed in various types of growth structures in the Tertiary section of the Niger Delta, Nigeria: (a) Simple rollover structure. (b) Structure with multiple growth faults. (c) Structure with antithetic fault. (d) Collapsed crest structure. The underlying Akata Formation is of undercompacted clay sediments; the overlying sequence is of sandstone and shale. [After Weber and Daukoru (1975).]

The structural-stratigraphic relationships described above are commonly found in other parts of the world. Busch (1975, p. 217) says:

Tertiary sediments of the Burgos basin of northeastern Mexico are broken abundantly by faults which are of two types, namely, growth and postdepositional. The growth faults have a sinuous, north-south trend, and are many kilometers in length. Collectively they make up a series of sub-parallel blocks of sediments which are faulted in a down-to-basin direction. Although these sediments thicken basinward, they also commonly exhibit rollover (reverse drag) on the downthrown side. In addition, they commonly thicken locally in the opposite direction in close proximity to the growth faults. The structural closures on the downthrown side of such faults are asymmetric and increase in amplitude with depth. Vertical profiles of these faults describe hyperbolic curves which tend to flatten with depth.

He further says (p. 217):

> Postdepositional faults are more numerous than the growth faults, more closely spaced, and generally have less throw. Their distinguishing characteristic is that the sedimentary section on either side of a postdepositional fault is of the same thickness, whereas the sediments on the downthrown side of a growth fault are always thicker than their counterpart on the upthrown side.

The structural pattern described by Busch is illustrated in Fig. 3-26 which shows the development of growth and postdepositional faults in the Oligocene Frio Formation of the Burgos Basin. The Frio comprises a sequence of shoreline sand bodies that form an arcuate and linear trend

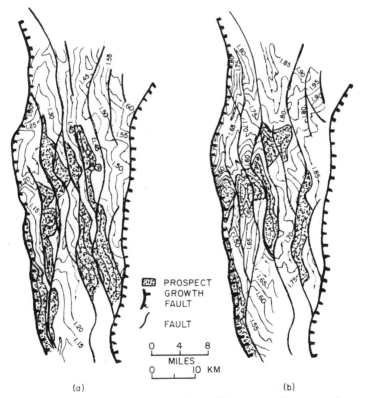

(a)   (b)

Fig. 3-26.   Interpretation, based on seismic data, of the structural patterns of growth and postdepositional faults within the Oligocene Frio Formation in the McAllen–Reynosa block, Burgos Basin, Mexico. (a) Upper Frio. (b) Middle Frio. Stippled areas show prospective areas for oil entrapment. [After Busch (1975).]

from Texas to Mexico. Sandstone bodies within the formation are important reservoirs for oil.

Growth faults, penecontemporaneous folding, and postdepositional faults all contribute to form a complex structural framework in many thick sedimentary piles such as those of deltas which have accumulated rapidly. The term postdepositional refers only to the deposition of the faulted beds, and not to the period of deposition of thicker sequences within the total sedimentary pile. In this sense, postdepositional faults are penecontemporaneous with the growth of some part of the sedimentary pile. Contemporaneity of structural and stratigraphic development is reflected in the depositional history, as seen for example in the thicker sequence on the downthrown side of a growth fault, and emphasizes the relationship between the structural and stratigraphic framework of sequences within a sedimentary basin.

CHAPTER

4

Depositional Environments

GEOMORPHIC AND SEDIMENTARY ASSOCIATIONS

The depositional history of a sedimentary basin is reflected primarily in the lithologic sequences and distribution of lithofacies and biofacies within the pile of strata. It is reflected secondarily in the stratigraphic relationships of geomorphic features that are preserved. The identification of such features depends on the interpretation of sedimentary structures, grain gradation, and other internal characteristics of a rock or sediment body, and also on the external relationships of the body with respect to the enclosing strata. Recognition may be based on rock sections exposed in outcrops, or on subsurface data and drill core from one or more wells. Delineation of the distribution of such geomorphic features depends on adequate surface or subsurface control and commonly is based on some degree of extrapolation. One of the main problems in the recognition of geomorphic features, or of the probable presence of such features within a stratigraphic sequence, is that they have limited areal distributions and consequently are not uniformly exposed in outcrop or consistently encountered in the subsurface. Examples of such features include barrier islands and river distributary channels.

Lithologic sequences as observed in a series of outcrops or borehole can be correlated, and individual layers of rock can be traced laterally and delineated with a greater degree of certainty than geomorphic features such as barrier islands. Added to the problem of demonstrating the internal features, stratigraphic relationships, and distribution of geomorphic sedimentary bodies is the equivocal nature of the interpretations placed on the data. Geologists are not always in agreement on the origin of certain sedimentary bodies. These disagreements may stem from a time gap, during which new ideas have been published, from differences in the availability of data, from variations in experience and outlook, and from bias in points of view. Many of these differences have appeared in the geological literature of the past, and many will no doubt appear in the future. The criteria on which these interpretations hinge are continually being updated, increased in number, and sharpened both qualitatively and quantitatively; but the evaluation of such criteria depends on individual professional judgment.

Stratigraphic sequences are commonly referred to as marine or non-marine; but these designations, generalized as they may be, do not satisfactorily categorize many sequences within a deltaic complex where the beds have been deposited in a paralic environment. Such sequences may include carbonaceous shales with thin coal seams overlain or underlain by shales containing arenaceous forams and a marine to brackish-water fauna. The former were deposited as mud and rotting vegetation in coastal marshes and swamps, the latter were deposited in open bays of the marshy terrain. Local and possibly minor transgressions and regressions of the sea account for alternations of shale in the sequence.

Fig. 4-1 illustrates environments of deposition with respect to depth of water. The littoral or shoreline zone falls within the paralic category and extends from high to low tide. The neritic zone extends seaward to the edge of the continental shelf which is commonly defined as the 200 m bathymetric contour. The neritic zone thus covers the entire continental shelf. Division of this zone into inner, middle, and outer neritic subzones is somewhat arbitrary but useful where the continental shelf is broad. The continental slope and upper part of the continental rise are designated by the oceanic zone. Its lower limit, bordering on the abyssal zone, is arbitrarily considered to be at a depth of 2000 m. The remains of pelagic fauna are found in sediments at all these depths, whereas the distribution of particular assemblages of benthonic (i.e., benthic) fauna is restricted to certain depth ranges characterized by sediment types and environmental conditions that combine to make a favorable ecosystem. In this respect it should be kept in mind that many organisms show an amazing tolerance for pressure and lack of light, provided there are nutrients to sustain a chain of life. A fascinating example is described by Corliss and Ballard (1977) who discovered fish, giant bivalves, crustaceans, brittle stars, and

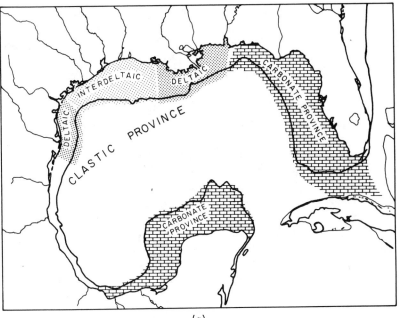

(a)

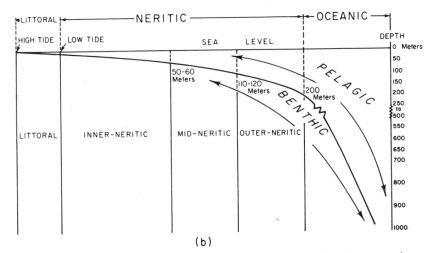

(b)

Fig. 4-1. (a) Map of the Gulf of Mexico showing areas of clastic and carbonate provinces. (b) Diagram illustrating depositional zones on a continental shelf and slope.

other organisms in the pitch-black depths of 2.5 km in the Galapagos Rift. At this depth several ecosystems were found, each centered around a submarine hot spring which gave off hydrogen sulfide and raised the temperature of the surrounding water to 17°C. The hydrogen sulfide is metabolized by certain bacteria which form the basic link in the food chain.

Fig. 4-1 also illustrates the two types of sediment province found on continental shelves, the clastic and the carbonate. The former depends on a terrestrial provenance and results from the transportation of clastic material and clay minerals by river systems to the sea where deltas are commonly built out on the continental shelf. The latter depends on sea water that is suitably warm to sustain colonial corals, on shallow sunlit water necessary for the growth of calcareous algae, and on the lack of muddy water such as that which spreads out in a great fan at the mouths of many large rivers. Interdeltaic areas are fed by smaller rivers and by longshore currents that move sediment from the deltas. The continental slopes fronting the deltas in the Gulf of Mexico merge with continental rises which consist of sediment washed down canyons on the slopes. Carbonate provinces are commonly widespread and very shallow but likely to have steep slopes off their margins.

The components of a clastic province are illustrated in Fig. 4-2 which schematically shows the spatial relationships of a delta spreading out on a continental shelf to the vicinity of submarine canyons that debouch on deep-sea fans extending to the abyssal plain. It is emphasized that the term delta is used in the sense of Moore and Asquith (1971) who consider a delta to comprise a subaerial and submerged contiguous mass of sediment.

Both the subaerial and submerged configurations of deltas depend on several factors including the number of rivers flowing into a delta complex, the number and pattern of distributaries, the sediment types and load being carried by these rivers and distributaries to the delta, the width and slope of the continental shelf, and the effects of waves, currents, and tides. The classic example of a bird-foot delta is seen in the extremity of the Mississippi Delta. Other large deltas throughout the world exhibit different subaerial configurations. Among these, three are shown in Figs. 4-3, 4-4 and 4-5. Fig. 4-3 illustrates the configuration of the Po Delta in Italy and shows a coastline to which sediment is accreted from a number of smaller rivers. The Adige River adds sediment to the Po Delta which is flanked by extensive areas of bays and marsh. The shelf on which the Po Delta is prograding slopes gradually beneath the Adriatic Sea such that within a range of 30–70 km from shore the water depth is less than 35 m. Fig. 4-4 illustrates two larger deltas, the Doce Delta and the Paraiba Delta

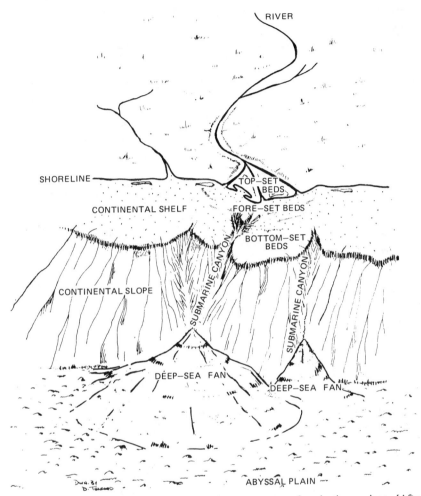

Fig. 4-2. Schematic diagram illustrating the components of a clastic province. [After Moore and Asquith (1971).]

of Brazil. These deltas are separated by a distance of 200 km and are fed principally by two large rivers. The subaerial configurations of these deltas present smooth coastlines which face the vigorous actions of the South Atlantic Ocean. Fig. 4-5 shows the delta of the Sao Francisco River in Brazil. The subaerial configuration of this delta is subdued, barely showing as a minor feature of a comparatively smooth coastline. Other minor rivers along the coast empty into the South Atlantic Ocean as estuaries.

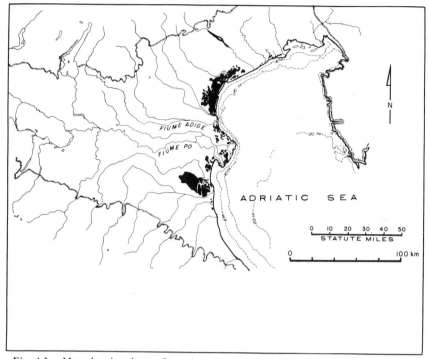

Fig. 4-3.   Map showing the configuration of the Po Delta and adjacent shoreline, Italy.

Sediments moved along a coastline by longshore currents sweeping from a delta, and those carried into the sea by minor rivers entering an interdeltaic coastline, commonly exhibit a general gradation seaward from coarser to finer. This distribution is modified by the topography of the sea floor and by the local conditions of wave and current action. As previously mentioned, the sea floor off the coast of the Amazon River Delta shows a reverse gradation resulting from upwelling, shoreward-moving currents that sweep the muds back toward the land. But in general, and with modifications, the distribution of sediments shows the coarser to be concentrated adjacent to the shoreline, as shown in Fig. 4-6. This figure illustrates a sediment pattern off the coast of Oregon. It shows a concentration of sand near the shoreline, muds farther seaward, and glauconitic sediments distributed in deeper water. Kulm *et al.* (1975) say that in this region, during the winter, long-period surface waves stir the bottom to depths of 200 m and that bottom currents transport the stirred sediment, including very fine sand, for up to 45 km across the shelf off the mouth of the Columbia River.

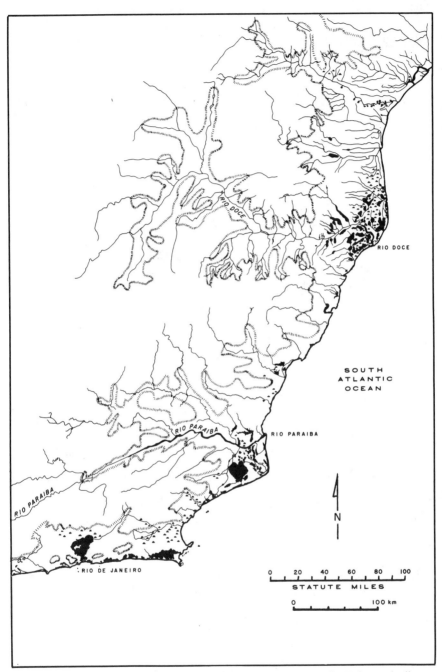

Fig. 4-4. Map showing the configuration of the Doce Delta, Paraiba Delta, and adjacent shoreline, Brazil.

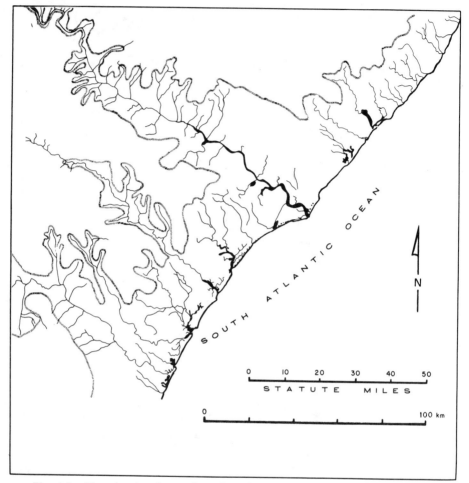

Fig. 4-5. Map showing the configuration of the Sao Francisco Delta and adjacent shoreline, Brazil.

With reference to sedimentary patterns associated with physiographic features in outer neritic, bathyal, and abyssal environments, the term geomorphic as used in this book refers not only to the evolution and configuration of landforms but also to features on the sea floor.

Terrestrial Environments

Lakes and swamps on low-lying landmasses and coastal plains have left significant records in geologic history. Sequences believed to be of glacial

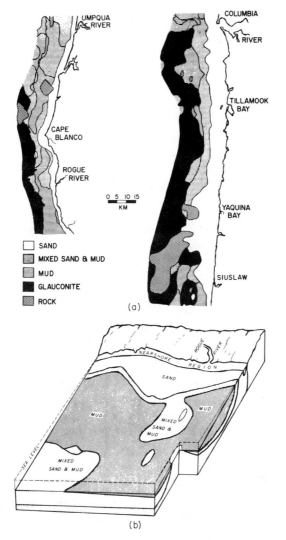

Fig. 4-6. (a) Map showing the distribution of sediments off the coast of an area in Oregon, United States. Glauconite pattern indicates glauconite-rich sediments. (b) Block diagram showing detail of sediment pattern off the mouth of the Rogue River. [Reprinted from Kulm *et al.*, *J. Geol* **83**(145–175). Copyright 1975 by the University of Chicago Press.]

and lacustrine origin have been described in the Proterozoic; and the extensive swamplands of the Carboniferous and Permian have contributed some of the large coal deposits known today. Many of the large bodies of fresh to brackish water which existed in the past have been referred to as vast lakes or inland seas. Seas have been defined as large

bodies of salt water, and the nomenclature commonly hinges on this crite-
rion. For example, the Dead Sea of the Middle East and the Salton Sea of
California are tiny compared to the size of the Great Lakes of North
America. The Great Salt Lake of Utah is also larger than either the Dead
Sea or the Salton Sea. The Caspian Sea of central Asia is described by
Dickey (1968) as a huge lake of brackish water. In this book it is accepted
that a lake is a standing body of inland water, and very large bodies
regarded as inland seas are consequently included in the category of lakes,
regardless of their degree of salinity.

The distribution and nature of sediments deposited on a lake bottom
and the variations in rate of sedimentation depend on a number of factors.
These include the areal extent and bathymetry of the lake, the input and
output of flowing water, the sediment load of rivers and streams emptying
into the lake, the rates of evaporation, and seasonal factors resulting in
changes of wave and current intensity or in the formation and drifting of
ice. An interesting study by Thomas *et al.* (1973) of the sedimentation
patterns in Lake Huron shows the bathymetry of the lake to be an important factor. Fig. 4-7 illustrates the distributions of quartz and clay minerals

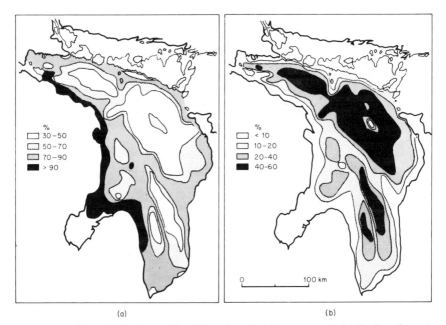

(a)   (b)

Fig. 4-7.   Map of Lake Huron, North America, showing patterns of distribution of quartz
(a) and clay minerals (b) in surficial sediments of the lake bottom. Bathymetry of the lake
shows the quartz (sand and silt) to be distributed mainly in shallow coastal areas and the
clays (mud) to be distributed in deeper areas. [After Thomas *et al.* (1973).]

in surficial sediments of the lake bottom and shows the quartz to be concentrated in the shallow offshore area adjacent to the western shoreline and the clays to be concentrated in the deeper areas of the lake. In terms of sediment type and grade, the study indicates that sand and silt are deposited mainly near shore in shallower water and that muds are deposited mainly farther from shore in deeper water.

The distribution of sediments in the Caspian Sea of central Asia has been described by Dickey (1968) on the basis of work done by Soviet geologists. This huge inland sea, which covers an area of more than 400,000 km^2, has no outlet. It is fed mainly by the Volga River in the north but also by the Emba and Kura rivers. The delta of the Volga River is more than 150 km wide and is the main source of sediment deposited in the northern part of the Caspian Sea. The bottom sediment distribution pattern has been described by Dickey (1968) and is illustrated in Fig. 4-8 which shows a concentration of sand in the northern area of the Caspian Sea and a concentration of mud in the central and southern areas. The predominance of sand in the north results from the proximity of the Volga Delta which progrades onto a broad and comparatively shallow shelf. The central and southern areas are much deeper, the depths ranging over 300 m. The bottom sediments of the eastern fringe of the Caspian Sea include sand and calcareous muds. An eastern embayment, up to 200 km in length and more than 100 km in width, is flanked by desert to the east and has restricted circulation to the Caspian Sea. The bottom sediment of this embayment contains evaporites. It is apparent that the sediment distribution is related to the bathymetry of the Caspian Sea, the sand being deposited primarily in shallower water, particularly on the shelf adjacent to the Volga Delta, and the muds being deposited in deeper water.

Sequences of clay sediments deposited in lakes commonly show a cyclic stratification which reflects both seasonal and long-term fluctuations in the weather pattern, as exemplified in varved clays. Pollen and spore assemblages within the various layers of sequences deposited in lakes can also be indicative of changing climatic conditions. These may be borne by the wind, falling directly into a lake, or carried from the mouth of a river by a fan of muddy water spreading far out over a body of salt or brackish water. Falling pollen can be so prolific that locally it forms a scum on the surface of swampy ponds and inlets draining into the distributaries of a delta. Thick accumulations buried under anaerobic conditions at the bottom of such ponds have in the past formed deposits of cannel coal.

Swamps are well recorded in geologic history, and, apart from differences in flora, the swamps of the Amazon Delta, the Florida Everglades, and the upper reaches of the Nile River have been cited as analogs for swamps that existed many millions of years ago. Swamps are commonly

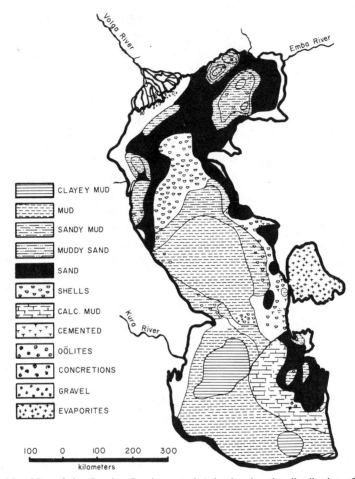

CLAYEY MUD

MUD

SANDY MUD

MUDDY SAND

SAND

SHELLS

CALC. MUD

CEMENTED

OÖLITES

CONCRETIONS

GRAVEL

EVAPORITES

100   0   100   200   300

kilometers

Fig. 4-8. Map of the Caspian Sea in central Asia showing the distribution of bottom sediments. Sands are deposited on shallow shelves, muds in deeper water. [After Dickey (1968) and Klenova *et al.* (1956).]

but by no means exclusively found in tropical climates. Vast areas within the Arctic Circle are underlain by muskeg which is similar to the peat bogs of northern Europe. Climatic conditions determine the types and rates of growth of flora; but the main criteria for the development of swamp and marshland are low-lying topography, poor drainage, and commonly but not necessarily an abundant influx of water such as may be supplied by river distributaries branching out on to a delta.

Stratigraphic sequences deposited in such environments may be entirely nonmarine. If they have been deposited in a paralic environment, they may include intercalations of sediments containing a marine fauna. Nonmarine sequences consist of sandy, silty, and shaly sediments interstratified with lignitic or coal beds; but some Quaternary sequences may rarely include thin beds of freshwater limestone or of diatomaceous earth. Limestones formed in such environments are chalky and commonly contain abundant ostracods, charophytes, gastropods, and remains of calcareous algae. Diatomaceous earth consists of an earthy to chalky deposit composed largely of diatom tests.

Of the many examples of swamps that could be cited, the following four are presented in Figs. 4-9, 4-10, 4-11 and 4-12. Of these, one is in an area of cold temperate climate, one in a warm temperate climate, and two in tropical climates. Fig. 4-9 is a map of the Amur Delta situated in southern Siberia adjacent to the northern end of the island of Sakhalin, north of Japan. The meandering river system within the delta flows through an area of swamps and lakes for a distance of more than 300 km before entering an estuary. Fig. 4-10 shows the delta of the Danube River prograding on to the western shelf of the Black Sea. The delta, which covers an area of more than 3500 km^2, includes large areas of swamp similar to that of coastal areas adjacent to the main distributaries of the Mississippi Delta. Water fowl in tens of thousands flock into these swamps of the Danube and Mississippi which, unless polluted, will remain preserved for the future. Of particular interest is the appearance of drowned topography south of the present delta, large bays being essentially cut off from the sea by the coastal buildup of barrier bars and associated marshy spits. Fig. 4-11 shows deltas developed in the tropical climate of southern Mexico by three rivers—the San Juan, Grijalva, and Papalopan. The latter two have built a delta complex that covers an area of more than 12,000 km^2, much of it consisting of marsh and lakes. Fig. 4-12 shows the area of the Magdalena Delta on the coast of the Caribbean Sea in northern Colombia. The delta drains an inland region of tropical swamp and lakes that cover an area also in excess of 12,000 km^2. It is from such large areas of marshy terrain that extensive coal beds have been formed.

Rivers and floodplains have left an impressive record in geologic history. Many of the stratigraphic sequences deposited by river systems are found in ancient deltas, but some have been preserved on erosional surfaces. From an economic point of view, sandstone bodies that originated as river point bar sands and other channel deposits have formed important reservoirs for oil and gas as summarized by Busch (1974) and Conybeare (1976). River floodplain deposits, consisting of varicolored layers of sandstone, siltstone, claystone, shale, and coal bands form spectacular

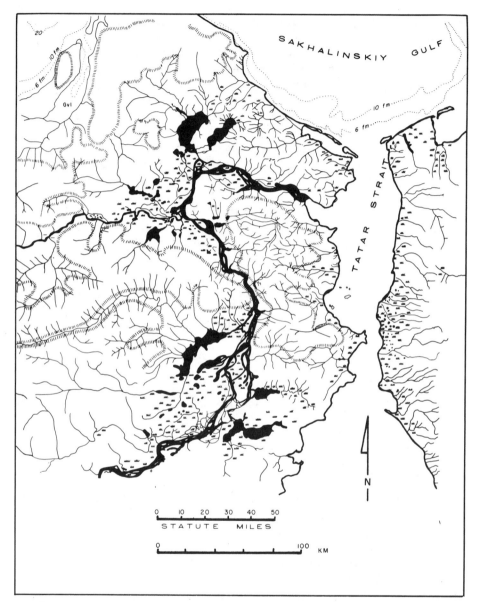

Fig. 4-9. Map of the Amur Delta, southern Siberia, adjacent to the northern end of Sakhalin, showing the river system meandering through an area of swamps and lakes.

fine sandstone and siltstone deposited as broad levees, thin-bedded to laminated siltstone, claystone and shale deposited on floodplains, and lignitic to coal beds of claystone and shale deposited in backswamps. Sequences including these rock associations commonly exhibit several periods of cut and fill by the river which migrates back and forth across the valley. The normal sequence deposited by a laterally migrating river comprises coarser sediments and cross-bedded sandstone in the lower part, with horizontally bedded sediments, commonly lignitic to coal, in the upper part. Subsequent changes of the river course may result in the partial or total removal of the upper part of the sequence in some localities. For this reason it may be difficult to correlate such beds for more than short distances. These sequences are characteristic of valley-fill deposits in the lowlands above a delta. Deltaic sequences, particularly those of river distributaries in the lower reaches of a delta, have somewhat different characteristics and will be discussed in the section dealing with paralic environments.

Examples of two different types of river systems are illustrated in Figs. 4-13 and 4-14. The former shows the region of the Colorado and Negro Rivers south of Bahia Blanca, Argentina. Both these rivers follow remarkably straight courses, flowing across coastal and inland plains that have a low relief and an elevation of less than 150 m. The very much larger Amazon River, shown in Fig. 4-14, follows a tortuous course, being fed by a maze of tributaries and eventually losing itself among clusters of islands within the estuary that forms the river mouth. The configuration of the Amazon and adjacent coastal region lying between Belem and Amapa displays the characteristics of a drowned topography. It is conceivable that this region of Brazil is one of the best analogs of the great regions of swamp that formed some of the Pennsylvanian and Permian coal deposits. Ancient floodplain deposits may include some reddish beds, particularly where the river was draining a reddish regolith developed in a warm and moist climate. Oxidation of the soil and subsoil above a shallow water table in a warm climate commonly results in the formation of a reddish-weathering regolith which in the wet season provides a supply of reddish mud to the river system draining the area. This mud is in large part deposited on the lower reaches of the river's floodplain, or in the swampy bays, bayous, and lagoons into which the river may debouch. Cores of mud and silt from such areas may show layers of reddish clay that record previous periods of floods that drained areas of reddish-weathered soil. Of the many examples of ancient river systems that can be cited, one notable example is described by Kranzler (1966) and illustrated in Fig. 4-15 which shows the regional trend of sandstone members in the lower part of the Early Pennsylvanian Tyler Formation in Montana. These

Fig. 4-10.   Map of the Danube Delta, (Bulgaria–Romania–USSR), showing the adjacent coasts in Bulgaria (south) and USSR (north).

badland topography in various parts of the world. The Late Cretaceous Edmonton Formation that is exposed in the badlands of Drumheller Valley, Alberta, was deposited on the floodplain of rivers meandering on a delta prograding eastward in response to initial uplift and erosion of the belt now forming the Rocky Mountains. Of particular interest are the dinosaur skeletons in the Edmonton Formation, discovered in 1884 by Dr.

Fig. 4-11. Map of a portion of southern Mexico adjacent to Guatemala showing the delta of the San Juan River and the delta complex of the Grijalva and Papalopan rivers.

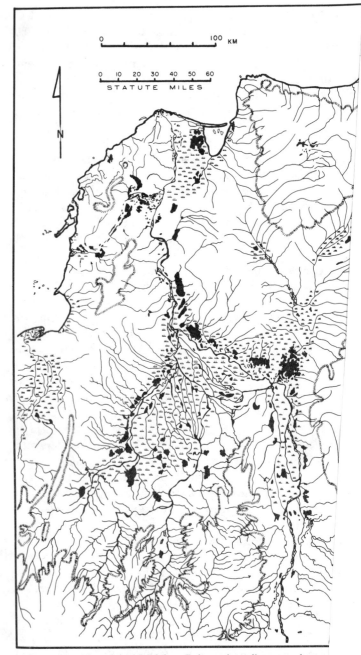

Fig. 4-12. Map of the Magdalena Delta region adjacent to the coa in northern Colombia showing a large area of inland swamp and lak

J. B. Tyrrell of the Geological Survey of Canada. Subsequent excavations carried out in particular by Dr. Charles M. Sternberg for the National Museum of Canada brought wide attention to these ancient floodplain deposits.

Large rivers migrating laterally are capable of cutting very extensive and broad valleys underlain by sequences of fluvial sediments that may be preserved in the geological record. The valley of the Mississippi River, for example, has a length of approximately 1000 km and a width of up to 150 km. It covers an area of more than 50,000 km² underlain by up to 100 m of fluvial sediments. Within these fluvial sequences preserved are lenticular bodies of conglomerate deposited as gravel in the deeper parts of river channels, lenticular bodies of sandstone deposited as point bars, layers of

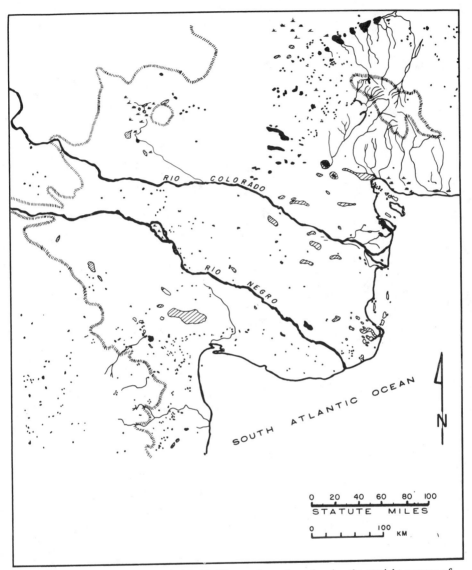

Fig. 4-13.   Map of the coast of Bahia Blanca, Argentina, showing the straight courses of the Colorado and Negro river deltas across lowlands.

sandstone members, which fill channels in a broad valley trending east–west through central Montana, are oil producing. The valley is cut into Mississippian limestone and shale. Another example of a Pennsylvanian river system is seen in the Red Fork Sandstone of the Anadarko Basin in

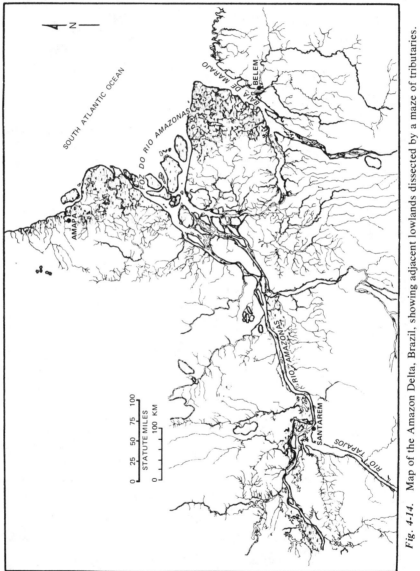

Fig. 4-14. Map of the Amazon Delta, Brazil, showing adjacent lowlands dissected by a maze of tributaries.

216

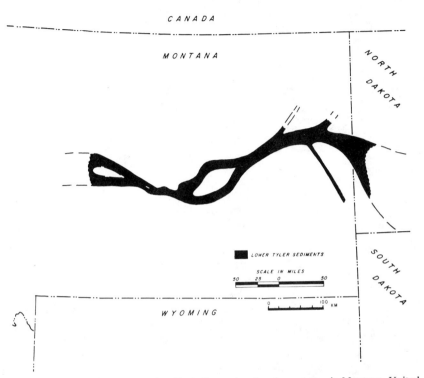

Fig. 4-15. Meandering trend of an Early Pennsylvanian river system in Montana, United States. Valleys within the system are filled with sandstone members which comprise the lower unit of the Tyler Formation (lower Tyler distribution). [After Kranzler (1966).]

Oklahoma. Some of the sandstone bodies of the Red Fork were deposited as river channel sands, others as bar sands. The Red Fork Zone constitutes a complex of sandstone bodies and associated sediments, including coal. Differentiation of the various bodies of sandstone, and determination of their origin, depends largely on the interpretation of core and electric logs. Fig. 4-16 illustrates the stratigraphic context of the Red Fork (Withrow, 1968) and also the electric log characteristics of sandstone bodies within the zone. Of particular interest is the apparent, but not actual, contemporaneity of a channel sand and a bar–complex sand, the channel having been cut down to the level of the bar. For general purposes of correlation as both a time-stratigraphic and lithostratigraphic unit, the various sandstone bodies that form this complex of channel, distributary, and bar sands are included within a single stratigraphic interval referred to as the Red Fork Zone.

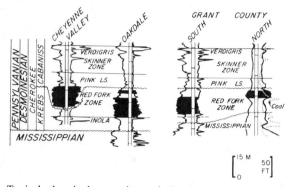

Fig. 4-16. Typical electric log sections of the Pennsylvania Red Fork Sandstone, Anadarko Basin, northwestern Oklahoma, showing a river sand filling a channel cut into a delta and offshore bar sequence. [After Withrow (1968).]

Glacial outwash plains are underlain extensively by fluvioglacial and glaciomarine sediments deposited by a continental ice sheet. Covering a region of coastal plain and adjacent lowlands, an ice sheet may reach to the sea where its leading edge breaks up into drifting blocks that carry debris which is eventually dumped on the shallow seabed. The dropping of erratics and other clastic material enclosed in the melting ice may be gradual, in which case the load is scattered, or it may be precipitous where an ice block is grounded. In the latter case an unsorted pile of debris will be dumped on the sea floor. Numerous examples can be seen in Permian glaciomarine beds exposed on the coast of New South Wales south of Sydney, Australia (Gostin and Herbert, 1973). As an ice sheet melts and its leading edge retreats inland, it dumps debris which form moraines. Meltwater from the ice washes gravel, sand, and silt from these moraines, forming sheets of fluvial sediments over which a system of braided streams develops. Finer sands and silts are carried farther from the melting edge of ice, being deposited largely on the outer fringes of the fluvial fan or carried into small lakes or ponds filling local topographic depressions. As the ice retreats, the coastal region rises with respect to sea level in response to isostatic adjustments, and a series of arcuate gravel deposits, each formed as a beach during a period of stillstand, may be preserved. Such raised beaches, formed during the Pleistocene and now more than 50 m above sea level, have been described in the Canadian Arctic (Christie, 1967). As uplift of the land proceeds, the previous shallow sea floor underlain by glaciomarine sediments is exposed. The whole coastal and adjacent lowland region becomes poorly drained and marshy in the lower areas, covered by numerous small lakes and ponds, many of which eventually become filled with peat to become water-soaked muskeg. Present-day glacial deposits can be applied as analogs to the interpretation of some Permian and Carboniferous glacial deposits; but other glacial deposits, such as those of the Ordovician in North Africa or the Proterozoic in Australia, were formed prior to the evolution of vegetation.

Preservation in the geological record of glacial deposits and features depends in large part on their elevation above sea level and distance from the coast. Alpine glaciation is obviously not going to be recorded, as the mountains will be eroded. Deposits of continental glaciation several hundreds of kilometers inland from a sea coast may be preserved provided they are at a low elevation in an area not being actively eroded and provided also that they are subsequently buried by marine, lacustrine, or fluvial deposits. Low-lying coastal regions are favorable for the preservation of glacial deposits that at one time were overlain by glacial outwash plains. Two examples in frigid climates, one in Alaska and one in northeastern Siberia, are illustrated in Figs. 4-17 and 4-18 respectively. Fig. 4-17 is a map of the region of the Yukon Delta which has a width of approximately 80 km and forms a large fan of sediment built out on the shallow continental shelf for 40 km. The lowlands which extend inland for a few hundred kilometers are covered by a myriad of little lakes and ponds left as relics of impounded blocks of ice. The whole region of lowlands constitutes an abandoned glacial outwash plain formed during the Pleistocene. Fig. 4-18 illustrates the delta of the Kolyma River in northeastern Siberia and shows an extensive coastal area covered by marsh and numerous small lakes. This area, which formed a glacial outwash plain, extends underneath the East Siberian Sea to form a very shallow continental shelf. As shown, at a distance of more than 150 km from shore, the sea floor is less than 20 m deep. Examples such as these, where extensive and shallow continental shelves rose above sea level to form a swampy lowland terrain in a region previously covered by an ice sheet, may well serve to depict Permian glacial and interglacial periods in Australia and elsewhere.

In searching for analogs to compare with Paleozoic and Early Mesozoic sequences, it must be borne in mind that the probable range of latitudes and relative position with respect to present-day landmasses were far different from what they are now. Australia, for example, was joined to Antarctica as part of the vast continent of Gondwanaland and during the Permian occupied a range of latitude with respect to the South Pole similar to the present range of the Arctic Islands and Northwest Territories of Canada with respect to the North Pole.

Alluvial fans consist of eroded material washed off the slopes of uplands and deposited as a fan-shaped topographic feature at the base of valleys

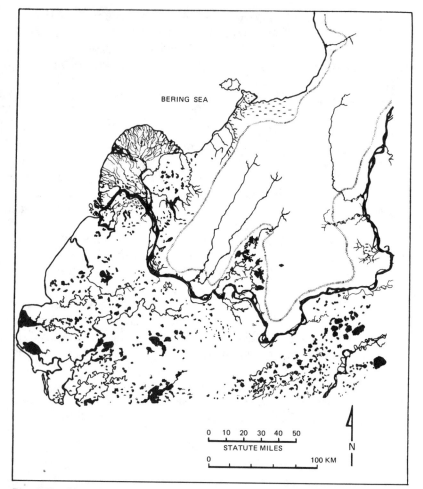

Fig. 4-17. Map of the Yukon Delta and adjacent coastal plain and lowlands in Alaska showing a myriad of small lakes left as relics of impounded blocks of ice on an ancient glacial outwash plain.

draining the higher areas. Some of the material is water-laid, some is dumped as a slurry by flows of debris. The water-laid sediment may have been deposited in the channel and levees of a braided stream system flowing over the fan, as sheetflood deposits during periods of torrential runoff. The debri–flows include those precipitated by the slumping of material on water-soaked slopes, and those produced by the slow down-hill creep of material. The grade of these sediments ranges from gravel to fine sand and silt but may also include coarse cobbles and boulders; the

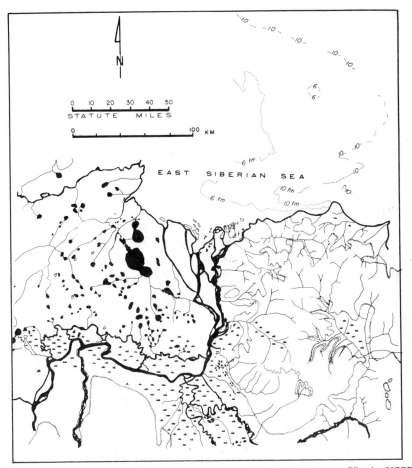

Fig. 4-18. Map of the delta region of the Kolyma River in northeastern Siberia, USSR, showing the adjacent swampy lowlands and very shallow continental shelf that formed a Pleistocene glacial outwash plain.

texture commonly ranges from poorly sorted to unsorted. In section alluvial fans show a sequence of lenticular bodies of gravel, sand, and unsorted debris.

The active area of deposition on the surface of an alluvial fan shifts laterally in much the same way as the lateral migration of a delta, so that an alluvial fan sequence comprises a number of separate lobes. Not only is each fan a composite accumulation of several lobes but many sequences of alluvial fan deposits are the result of the coalescence of separate interfingering fans, some of which attain thickness of several thousand meters.

The appearance of the stratigraphic framework of an alluvial fan sequence varies with the section exposed. Sections across the depositional slope show a marked lenticular arrangement to the beds whereas those along the slope show much less lenticularity. The internal features of alluvial fan deposits have been well described by Leopold *et al.* (1964), Selley (1970), and Bull (1972).

Examples of ancient alluvial fans exposed in rock outcrops have been cited in the literature. Selley (1965) describes Late Proterozoic alluvial fan deposits in northwest Scotland. He says that the clastics derived from the Lewisian gneiss are coarse and poorly sorted. Radiating depositional slopes are also mentioned as additional evidence of their origin. Numerous younger examples could be cited, many of which have not been called alluvial fans. Some sections of the Late Paleozoic to Mesozoic sequences in the Middle East, generally referred to as the Nubia Sandstone, were probably deposited as alluvial fans. In particular, some of the sections of conglomerate and sandstone that occur along the margins of the Saharan and Arabian Precambrian Shields are probably of this origin. Referring to such beds, Conant and Goudarzi (1967, p. 721) say, "Widespread over southern Libya is a thick sequence of continental beds long known as the Nubian Sandstone, which is also present in much of northeastern Africa and the Arabian Peninsula, though locally known by different names." Selley (1970) describes a Cambro–Ordovician sequence in Jordan (Fig. 4-19) that was deposited as a series of alluvial fans. This sequence comprises more than 700 m of cross-bedded, conglomeratic sandstone thought to have been deposited by braided stream systems flowing over a piedmont plain formed by coalescing alluvial fans. The original fans were derived from erosion of the Precambrian Shield; subsequent fans were in part formed of clastics reworked from older fans as the result of repeated uplift and pedimentation.

Deserts of vast proportions must have existed in the past before the development of land vegetation during the Late Devonian to Mississippian. In subsequent geological periods desert conditions developed from time to time in various parts of the world, the result of climatic changes taking place within as short a space of time as a few tens of thousands of years. These desert conditions may have been in hot or cold climates which would have had an effect on the degree of oxidation of surficial material, producing a reddish regolith in warmer regions. The previous existence of deserts is recorded in stratigraphic sections by unconformities overlain by detritus, poorly sorted conglomeratic sandstone commonly referred to as granite wash, and by cross-bedded sandstone that originated as dunes. Deserts are largely areas of deflation where the wind

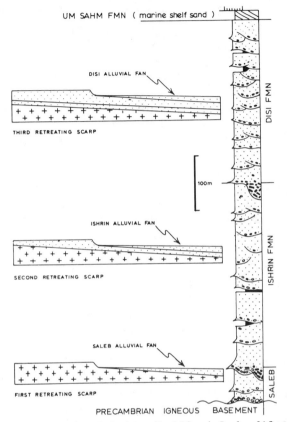

UM SAHM FMN (marine shelf sand)

DISI ALLUVIAL FAN

THIRD RETREATING SCARP

DISI FMN

100m

ISHRIN ALLUVIAL FAN

SECOND RETREATING SCARP

ISHRIN FMN

SALEB ALLUVIAL FAN

FIRST RETREATING SCARP

SALEB

PRECAMBRIAN IGNEOUS BASEMENT

Fig. 4-19. Sequence of Cambro–Ordovician alluvial fans in Jordan. [After Selley (1970).]

removes the fine sand, silt, and dust, leaving a lag deposit of debris. Periodic rainfall washes gravel and sands into the wadis and other topographic depressions. Elsewhere sand dunes may develop by climbing one on the other like sand ripples in a point bar deposit, accumulating to thicknesses commonly in excess of 100 m but ranging to 300 m. The morphology of modern and fossil deposits of detritus, poorly washed clastics, and sand dunes has been well described by Bagnold (1941), De Almeida (1953), Sanford and Lange (1960), Stokes (1961), Tanner (1965), McKee (1966, 1969), Conybeare and Crook (1968), Glennie (1970), Pettijohn *et al.* (1972), Reineck and Singh (1973), and others. Notable among examples of fossil sand dunes are the Jurassic Navajo Sandstone of Colorado, the Jurassic Entrada Formation of New Mexico, the Triassic Botucatú Formation near Sáo Paulo, Brazil, and the Pennsylvanian to

Permian Caspar Formation of Wyoming. Possibly the largest known of the ancient desert deposits is the Botucatú which is commonly more than 200 m thick and covers an area of 1,300,000 km². The Botucatú consists of reddish sandstone, the grains of which are well rounded and pitted. The sandstone, which lies on an unconformity and is unconformably overlain by Triassic volcanics, exhibits the typical sweeping cross beds of sand dunes. It is of interest to note that the Botucatú locally contains residual deposits of asphalt. Other well-known hydrocarbon-bearing fossil sand dunes are in the Early Permian Rotliegendes Formation of the North Sea region. These are associated with a red bed–evaporite facies. Glennie (1972) states that erosion of the Carboniferous coal measures caused inland regression of an escarpment cut by wadis which fed alluvial fans in a desert. Sand dunes, sheet sands, and surficial debris extended from the fans to sabkhas fringing the sea. Subsequent marine transgression resulted in reworking of the uppermost desert deposits and the deposition of cupriferous shales and the Zechstein evaporites. Hydrologic studies of desert regions indicate that although the surficial material is dry, water may be slowly flowing along channels at depths of several meters. It is interesting to speculate that copper-bearing water draining a desert region may percolate through buried stream channels to deposits of organic-rich mud where the copper content of the water may be retained. Associations of desert and swamps are not uncommon, and three examples are presented in Figs. 4-20, 4-21, and 4-22.

Figure 4-20 is a map of the region lying between the Gulf of California and the Salton Sea. The region borders the Gila Desert of Mexico and lies in a broad valley which forms an extension of the depression occupied by the Gulf. An extensive area surrounding the Salton Sea is below sea level, and the Colorado River drains into the valley through a fan-shaped system of distributaries, some of which enter into the Salton Sea and some into a marshy area at the head of the Gulf. Fig. 4-21 shows the delta of the Euphrates and Tigris rivers in Iran and Iraq and the adjacent desert region of Kuwait. The proximity of desert to the marshy delta illustrates the possible stratigraphic relationships that may result where a transgression of the sea causes marshy coastal areas to migrate inland over a desert terrain. Fig. 4-22 is a map of the Amu Darya Delta and adjacent area in the USSR north of Iran and Afghanistan. Near the delta on the southern coast of the Aral Sea a fan-shaped cluster of distributaries splays out on to the desert. During flood periods some of the runoff water in these distributaries drains into marshy ground bordered by desert, and some drains into a basin area lying below sea level. The marshy delta and adjacent desert region also illustrate the possible juxtaposition in the geological record of beds deposited in these distinctly different environments.

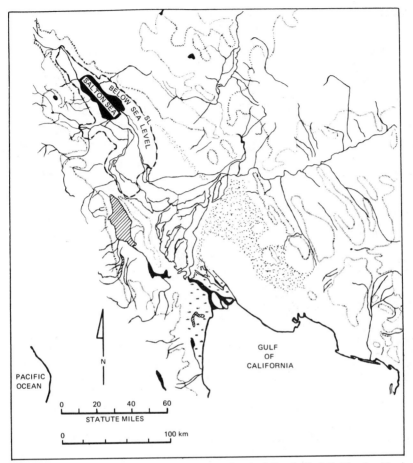

Fig. 4-20.   Map of the Colorado Delta at the head of the Gulf of California, Mexico, showing the depressed area of the Salton Sea in California.

Volcanic flow sheets can be erupted over widespread regions during particular intervals of time, building up thicknesses of many hundreds of meters of basic lava and volcanic debris such as the sequences of the Deccan basalts of southern India and the Columbia River basalts of British Columbia and Washington. Associated with the basic lavas of andesite and basalt are agglomerates, tuffs, and tuffaceous sediments. The stratigraphic interval in which a volcanic flow sheet occurs is thus a composite section of volcanics and sediments laid down during a sequence of events. As such, it is a time-stratigraphic interval in which the sections at different localities may show considerable variations in thickness and

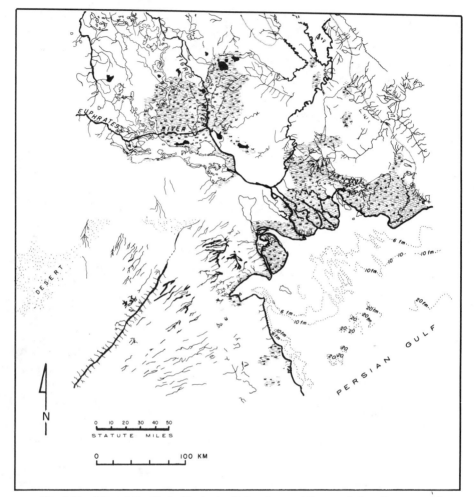

Fig. 4-21. Map of the Tigris–Euphrates Delta, Iran and Iraq, and the adjacent area in Kuwait, showing the proximity of desert and swamp.

composition. For this reason it may be difficult or impossible to correlate individual beds of volcanics or sediments from one locality to another within the major stratigraphic interval containing the volcanic flow sheets.

Volcanic flows and debris fill topographic depressions and are therefore commonly intercalated with fluvial gravel and sand. Finer volcanic matter such as ash may fall directly on an area or be washed in by streams. Falls of ash on a lake bottom are commonly preserved whereas those on a land surface may subsequently be removed by erosion. Volcanic ash falls are

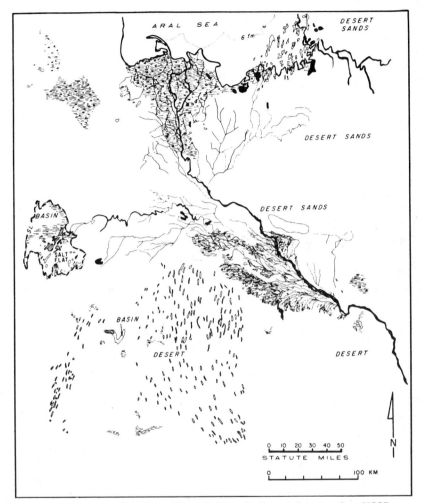

Fig. 4-22.   Map of the Amu Darya Delta and adjacent desert region, USSR.

recorded in the geologic history of an area by thin, widespread layers of bentonite, a soft rock composed largely of the clay mineral montmorillonite. The distribution of such layers is commonly elongate in the direction of source, indicating the approximate location of volcanoes from which the ash was erupted.

Much of the record of terrestrial volcanism in the past has been destroyed by erosion, and for the most part only those volcanic beds deposited on lowlands and coastal plains have been preserved. Numerous examples of Late Tertiary volcanism may be cited to the contrary but most of these will ultimately not be preserved. Many volcanic deposits on coastal plains extend offshore and will be discussed in the following section on paralic environments.

Paralic Environments

Delta distributaries and their systems have various shapes and patterns depending on a number of factors which include the size of the river and its volume of water, the number of distributaries, the shallowness of the continental shelf over which the delta is prograding, and the energy of wave and tidal action. Distributaries may be straight, such as those of the present bird-foot portion of the Mississippi Delta, or meandering, such as those of the Catatumbo Delta in Venezuela. The main distributary passes of the Mississippi River are building seaward on their own levees in very shallow water. Tidal variations are slight, and, except during periods of storm, the wave action has a comparatively low level of energy. In the Mississippi River system the greatest volume of water is transported by three main distributaries whereas in the Catatumbo River system, which carries a much smaller volume of water, the dispersal across the delta is through several meandering distributaries. Fig. 4-23 illustrates the delta of the Catatumbo and shows the drainage pattern from highlands to the south and west of Lake Maracaibo. The river systems drain into a low, swampy area that forms a southern extension of the topographic depression in which Lake Maracaibo is situated. The distributaries meander across this swampy area to the present delta that is building out across a width of more than 50 km. The distributaries of the Godavari and Kistna rivers in southeastern India are illustrated in Fig. 4-24 which shows the three distributaries of the Kistna Delta to the south to be fairly straight, whereas the several distributaries of the larger Godavari Delta to the north are somewhat meandering. Both deltas are prograding over a narrow continental shelf where the wave and tidal actions are vigorous, which accounts for their smooth, truncated outlines.

Sediments transported by the distributaries of a river system within a delta will in large part be carried to the sea and dispersed by the action of currents. This applies to the finer sediments and particularly to the silt and colloidal clay which may be carried out to sea for several kilometers before being precipitated on the sea floor as prodelta sediments. Coarser sediments moved along a distributary are in part deposited within the channel between the distributary levees and in part transported to the mouth of the distributary. Where several distributaries meander over a coastal plain, the load of sand they carry will be dispersed along the shore

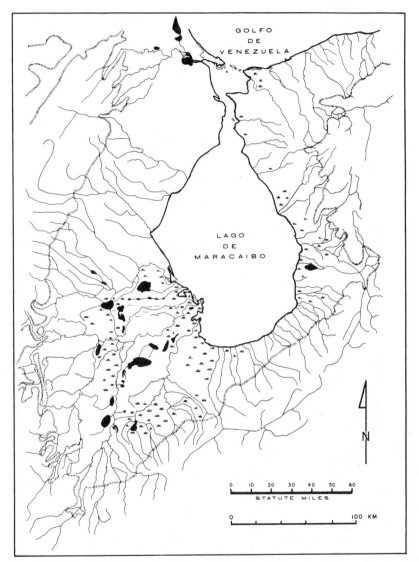

Fig. 4-23. Map of the Catatumbo Delta and Lake Maracaibo region, Venezuela, showing distributaries meandering across the swampy delta.

to form a sheet of sand as the delta advances. Sand bodies of this type are referred to as delta-front sand sheets. On the other hand, where a few distributaries build themselves out to sea on levees, as do the main distributaries of the Mississippi River, then the load of sand transported by a

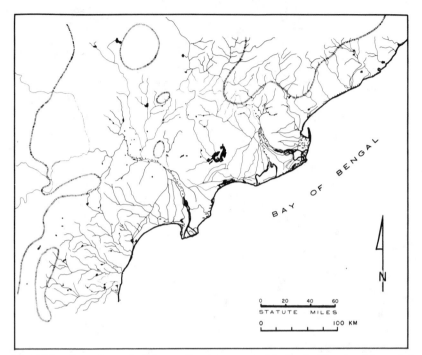

Fig. 4-24.   Map of a part of the southeastern coast of India showing the deltas of the Kistna River (south) and the Godavari River (north).

distributary is deposited mainly as a fan-shaped bar at the mouth of the distributary. As the distributary advances, it continually overrides this bar to form a linear body of sand referred to as a bar-finger. In a three-dimensional view of a delta, bar-fingers form an anastomotic pattern of linear sand bodies referred to as shoestring sands. Where structural closure exists, both sand sheets and bar-fingers can form good reservoirs for oil and gas, and numerous reservoirs of these types have been found in the Mississippi and Niger deltas.

Figure 4-25 illustrates the geomorphology and solid geometry of sand bodies deposited by river distributaries in the Mississippi Delta, as described by Reineck (1974) and Fisk *et al.* (1954). The classic bird-foot delta of the Mississippi delta complex is actively growing today. The main distributaries and their subsidiaries grow seaward like a huge vascular system. Building outward on their levees, the distributaries enclose ever-growing areas of bays and coastal marsh. In less active areas of sedimentation the sand is transported by numerous changing and meandering channels that disperse their loads more evenly in a manner similar to that

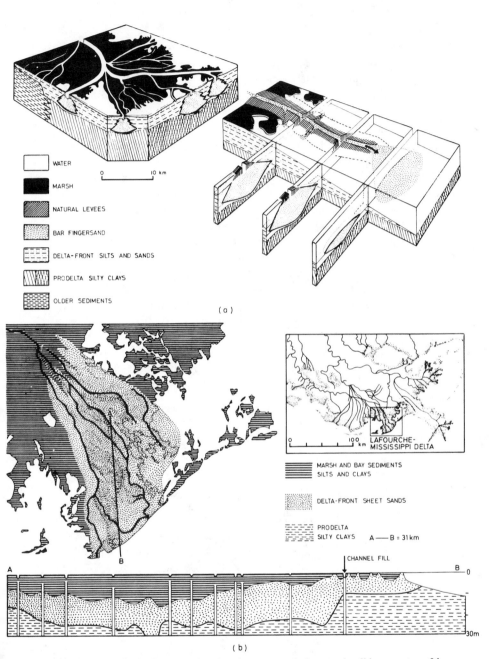

WATER

MARSH

NATURAL LEVEES

BAR FINGERSAND

DELTA-FRONT SILTS AND SANDS

PRODELTA SILTY CLAYS

OLDER SEDIMENTS

0   10 km

(a)

LAFOURCHE-
MISSISSIPPI DELTA

0   100
            km

MARSH AND BAY SEDIMENTS
SILTS AND CLAYS

DELTA-FRONT SHEET SANDS

PRODELTA
SILTY CLAYS   A——B = 31km

CHANNEL FILL

A   B ─ 0

                                                        ─30m

(b)

Fig. 4-25. (a) Block diagrams showing the geomorphology and solid geometry of bar-finger sand bodies formed in the bird-foot delta of the Mississippi River. (b) Plan and sectional views of a delta-front sand sheet formed in the Mississippi Delta. [After Reineck (1974) and Fisk *et al.* (1954).]

of a meandering stream system in a broad valley. Waves and currents further spread the sands along the coast, moving the silt and mud seaward, or to areas of lower energy where it may be deposited to form tidal mud flats.

Recognition of delta distributary sand deposits in the subsurface, including bar-finger sand bodies and delta-front sand sheets, depends on both external and internal features. The nature of underlying and overlying beds based on stratigraphic sequence, rock types, sedimentary structures, and faunal evidence indicates the depositional environment and possible paleogeographic setting in which the sand body was laid down. Where control is sufficiently close, the general shape and trend of the sand body will further indicate its origin. Internal features may be observed directly in core, or indirectly by means of electric logs, which may show upward coarsening grain gradation typical of bar-fingers, sand sheets, and other prograding bodies of sand.

One of the classic examples of a system of ancient delta distributaries is illustrated in Fig. 4-26 which shows the distribution of the Early Pennsylvanian Booch Sandstone in eastern Oklahoma. The Booch Sandstone forms a branching system of shoestring sands which locally form reservoirs for oil. The distribution, as illustrated by Busch (1971), is a composite picture of two stratigraphic zones. The main channels of this system contain sandstone bodies more than 60 m thick. The sandstone is commonly lithic and fine-grained but locally is medium- to coarse-grained.

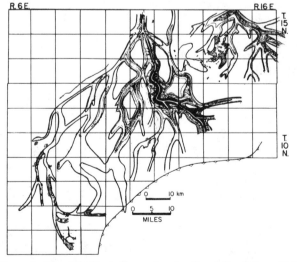

Fig. 4-26.   Isopach map of the Early Pennsylvanian Booch Sandstone, Oklahoma, showing the branching pattern characteristic of delta distributaries. [After Busch (1971).]

Delta distributaries are continually being formed and abandoned. As the sequence of beds within a delta complex grows, the buried channels of former distributaries form an anastomotic three-dimensional pattern which in section appears as numerous lenticular bodies of sand. Distributaries in a prograding delta system are commonly abandoned due to shifts in the river system brought about by the growth of new distributaries breaking through the levees. The abandoned channels of the previous distributaries become buried by swamp and bay deposits, the origin of which may be indicated by thin coal beds adjacent to shaly beds containing brackish-water fauna including arenaceous forams.

Bays and coastal swamps are formed extensively within the outer fringe of many deltas. Areas of salt marsh invade the bays and other bodies of water, cutting them off from openings to the sea, so that they become brackish and eventually filled with fine sediment and vegetation. As the delta front continues to prograde, these salt marsh deposits are buried by levee and backswamp deposits of the main river system. Where the delta front migrates in response to a major shift of the river system, the coastal fringe of swamp and bays may be cut off from a source of sediment and will consequently subside by compaction. Local transgression of the sea over an area of subsidence is recorded in the geological record by a wedge of marine sediment overlying shaly and silty beds with thin seams of coal. Subsequent growth of river distributaries over the area will again result in the deposition of interdistributary bay and swamp sediment, with accumulations of rotting vegetation, so that a sequence of interfingering marine, bay, and swamp deposits may ultimately be seen in the stratigraphic record. Such interfingering does not indicate fluctuations of sea level but rather fluctuations of the coastal land level as the area alternately subsides beneath or rises above the sea in response to lower or higher rates of sedimentation.

Areas of interdistributary bays and coastal swamps may extend inland to low-lying areas of grasslands or backswamp, depending on topography and climate. In the past, very extensive areas of backswamp within delta regions must have existed for long periods of time as seen, for example, in the coal deposits of the Early Eocene Latrobe Group of Victoria, Australia. The Latrobe was deposited as a deltaic complex in which paralic deposits of mud and vegetation formed coal seams up to 6 m thick and in which the accumulation of inland swamp vegetation formed coal seams up to 90 m thick. Of particular interest is the association offshore of interbedded oil- and gas-bearing sandstone bodies and carbonaceous mudstone with thin coal seams. These hydrocarbon-bearing sandstones contain the reservoirs for oil and gas fields in Gippsland Basin.

Bays and coastal swamps are also formed on the flanks of deltas and on the coastal regions between deltas. Fig. 4-27 is a map of the coast on either side of the Zambeze Delta in Mozambique, showing extensive areas of swamp and marsh into which drain numerous streams. Estuaries and small bays have also developed along this coastal lowland. Fig. 4-28 shows the region of the Ogooue Delta in Gabon where large dendritic bays are developed along the coast. The shape of these bays indicates a drowned topography, probably resulting from subsidence of the coastal region. Coastal currents have transported and deposited sediment, effectively blocking off the seaward entrances to the two northern bays. The southern bay retains an entrance to the sea and probably for this reason is referred to not as a lake but as Lagune N'Dogo. The differentiation of coastal bays and lagoons is not clearcut but depends in general on the

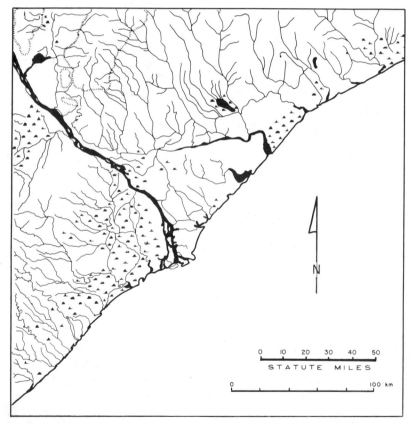

Fig. 4-27.   Map of the Zambezi Delta region in Mozambique showing extensive areas of coastal marshland.

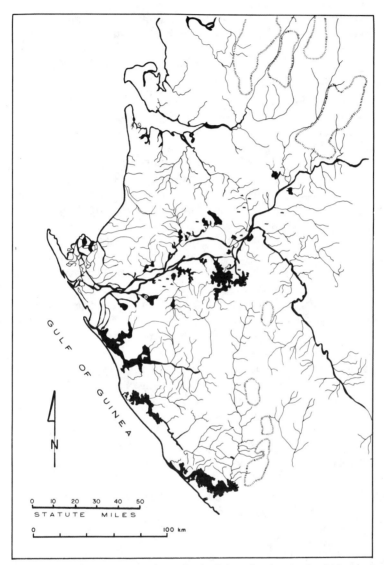

Fig. 4-28. Map of the Ogooue Delta region in Gabon showing the dendritic development of large coastal bays.

nature of the entrance to the sea. When the entrance is restricted, the name lagoon is commonly used and where the entrance is open the enclosed body of water is called a bay. Many lagoons are also elongate parallel to the shore, being cut off from the sea by sand bars.

Lagoons have access to the sea and consequently contain salt water, the salinity of which is variable depending on the influx of fresh water from streams and of salt water from tidal inlets between bars and barrier islands. Locally, the salinity can be sufficiently high for crystals of gypsum to grow within the mud of the lagoon floor, as evidenced by dredging of parts of the ship channel in southern Texas. Lagoons commonly range in width up to several kilometers but may be less than one or more than twenty.

Smaller lagoons in general have a more restricted circulation and higher content of organic matter. Many are completely cut off from the sea during periods of low tide and consequently become bodies of stagnant water. The mud on the bottom of such restricted lagoons is black and fetid with macerated rotting vegetation. The shallower parts of open lagoons in tropical areas are normally covered by thick growths of mangroves which harbor a prolific fauna of crustaceans and other life where tidal fluctuations allow the intake of fresh seawater. The lagoon sediments in such locations are considerably disturbed by bioturbation. As the beds are subsequently buried, preserved will be not only crustacean, bivalve, and marine worm burrows but also the outlines of penetrations made by mangrove and other roots. These sedimentary structures, taken in context with the sedimentary sequence and other internal features, may prove to be indicative of the depositional environment.

Lagoons associated with coral reefs are regularly flushed with fresh seawater as tides rise over the outer reef fringes and are characterized by carbonate muds and sand and by patch reefs. Lagoon sediments of this type are associated with atolls and reefs fringing coastal areas and islands. Depositionally the sediment types and sequences are those seen in large areas of carbonate banks and shoals such as form part of the shallow sea floor off the Bahamas.

Lagoons associated with the deposition of noncarbonate sediments are illustrated in Fig. 4-29 which shows the coastal region south of the Rio Grande Delta in Mexico. The Rio Grande does not flow throughout the entire year but carries large amounts of clay and silt to the delta. During the dry season this fine sediment is windblown, forming dunes of silt. Along the outer fringe of the delta, sand is carried by longshore currents and deposited as barrier bars and islands which form an effective barricade to the open sea. Fig. 4-29 shows a chain of barrier islands flanking the outer fringe of the Rio Grande Delta and extending north and south to enclose large lagoons on either side of the delta. These lagoons are collectively known as Laguna Madre. South of Tampico a former lagoon has been completely cut off from the sea by the coastal drift and deposition of shoreline sand.

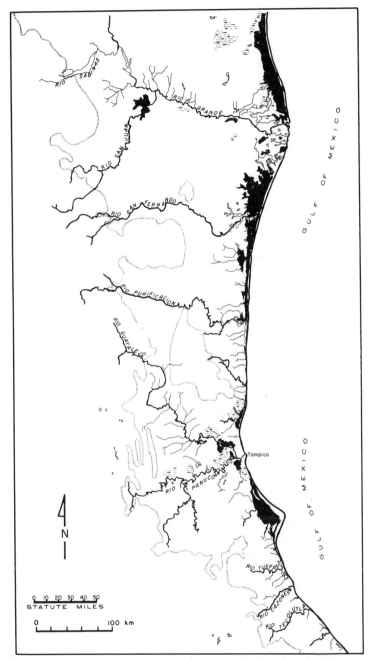

Fig. 4-29. Map of the coastal region adjacent to the Rio Grande Delta, Mexico, showing the Laguna Madre (Rios Panuco, Guayalejo, Purificacion, San Fernando, and Grande deltas) and enclosing chain of barrier islands on either side of the delta.

Lagoon sediments are predominantly muddy as wave and current action is much less vigorous than it is along the coast open to the sea. In large lagoons the sediments are locally subjected to winnowing, and banks of fine sand can be formed. Along the shores of many lagoons, particularly in areas where tidal variations bring an influx of fresh water and nutrients, colonies of oysters grow profusely among the mangrove roots. Once established by adhering to rocks and other objects, colonies grow upon themselves and commonly form oyster beds up to one meter thick. Many examples of such oyster beds are known in the Cretaceous and Tertiary in various parts of the world and are particularly well recorded in the literature referring to the stratigraphy of western and southern North America.

There is no way of distinguishing lagoon from bay sediments in outcrop or in the subsurface other than by the sequence of beds where, for example, a sandstone bed is recognized as an ancient barrier bar. As the barrier bar builds seaward, it is followed behind by the lagoon which in turn is encroached upon by coastal marsh on the mainland. A typical sequence of beds deposited in such an environment comprises a basal bed of upward coarsening sandstone deposited as a bar sand, a middle shaly to silty bed of lagoon sediment with bioturbation structures and abundant arenaceous forams, and an upper coastal marsh bed of carbonaceous and shaly siltstone with thin seams of coal. This sequence may in turn be overlain by fluvial and floodplain deposits.

Tidal mud flats can form extensive sheets of sediment along a coast. On the seaward side they shelve into very shallow water; on the landward side they are encroached upon by marsh or tussocks of coarse grass. Intertidal sheets of mud are commonly deposited where wave and current action is not strong and where the load of fine sediment being transported along the coast is fairly heavy. Mud sheets are transitory and at any locality may develop, be washed away, and reappear again. Such alternations may reflect lateral migrations of a nearby river distributary which debouches a heavy load of mud into the sea. It may also reflect climatic variations and periodic floods during which exceptionally large volumes of mud are carried to the sea, transported by longshore currents, and deposited along the coast. Where the action of waves and currents is not sufficiently vigorous, the sediment is poorly winnowed and the sand content remains dispersed in a much larger volume of silt and clay mud. In this situation the relatively low hydrodynamic energy of the coast is unable to cope with the heavy load of sediment, and a coastal mud sheet develops. If, for some reason, the sediment load transported along the coast decreases, possibly due to cutting off or migration of a nearby distributary, the wave and current action may be sufficiently energetic to carry away the mud, leaving a residual beach deposit of shelly sand. Such deposits form beach ridges. Subsequently, a mud flat may accrete to the beach ridge, and where this alternation continues, a series of arcuate beach ridges known as cheniers is formed. In outcrop these appear as a disconnected string of sandstone lenses within the same narrow stratigraphic interval measuring several meters in thickness. Along the section of outcrop the horizontal distance between sandstone lenses varies according to the angle at which the section of outcrop intersects the depositional trend of the cheniers. Cheniers are commonly up to 5 m thick and 1500 m wide in the subsurface and may be several kilometers long.

Extensive mud sheets, which on the landward side form intertidal mud flats, can rapidly accrete to a coastline. Mud flats can build seaward from a chenier at rates of a few meters a year, and the rate of progradation of the seaward fringe of a mud sheet may be even more rapid, particularly where the mud sheet is spreading over a very flat and shallow continental shelf such as that off the Irrawaddy Delta in Burma (Fig. 4-37). The association of large areas of subtidal and intertidal mud flats and salt marsh is shown in Fig. 4-30 of the Netherlands coast which includes the Rhine Delta. Fringing the coast are areas of beach and sand dunes which continue in a northeasterly trending arc as a chain of barrier islands. Behind these areas of sand formed by wave, current, and wind action are very much larger areas that, prior to cultivation of the land, were inlets, lagoons, bays, intertidal mud flats, salt marsh, and swamp. The recent sedimentologic history of this region has been outlined by Hageman (1969). The geographical distribution of sedimentary facies seen today is merely the surface expression of a prograding sequence of strata constituting the delta complex of the Rhine River system. As mud sheets build seaward and the coastline advances, intertidal mud flats are encroached upon by salt marsh vegetation which traps more sediment and builds up a swampy terrain. Provided the rate of deposition is sufficient to compensate for subsidence of the coastal region, the delta complex will continue to prograde and mud sheets to extend seaward. But if deposition does not keep pace with subsidence, then the coastal region sinks beneath sea level and preexisting mud flats and salt marsh are inundated by bays and lagoons. The interaction of sedimentation and subsidence during the past history of such a delta can be interpreted from the stratigraphic distribution in the subsurface of coastal sand bodies and beds of silty and clayey carbonaceous sediments.

Carbonate–salt mud sheets are widespread deposits of carbonate mud interbedded with layers of evaporites. In bodies of water open to the sea, the depositional environment of such sheets may be in part subtidal and in

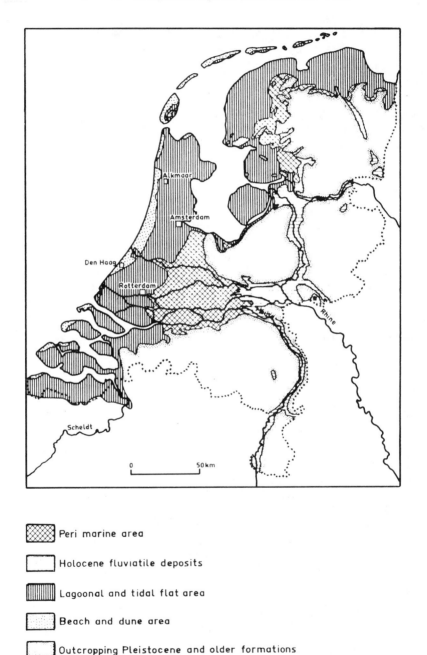

Peri marine area

Holocene fluviatile deposits

Lagoonal and tidal flat area

Beach and dune area

Outcropping Pleistocene and older formations

Fig. 4-30.   Map of the Netherlands showing the distribution of sedimentary areas during the Holocene. [After Hageman (1969).]

part intertidal; in closed bodies they are deposited mainly underwater. The carbonate muds are commonly dolomitic, may be clayey or silty, and contain scattered to abundant crystals of gypsum. These muds are deposited in large part by the chemical precipitation of aragonite, and aragonite mud being converted to dolomitic mud in hypersaline environments such as found in the intertidal zones associated with sabkhas. The evaporite layers consist either of halite or gypsum which on burial are rapidly converted to anhydrite. The sequence of layers of evaporites and carbonates is commonly cyclic on both large and small scales. Major layers which show a cyclic sequence may themselves be composed of cyclic varves. These reflect periodic rising and falling of sea level with consequent fluctuations of salinity within inner neritic to paralic environments; periodic climatic changes producing fluctuations in the flow of streams entering closed bodies of salt water; subsidence by compaction of the sedimentary pile and, as in the case of the Dead Sea sedimentary sequence (Neev and Emery, 1967), by downward faulting of a basin containing a saline body of water.

In large, very shallow inland bodies of water such as Lake Eyre which covers about 15,000 km^2 in South Australia, drainage from the surrounding hinterland during times of flood fills the lake and brings in dissolved salts. During times of drought the lake level falls rapidly, and large expanses of flat lake bottom encrusted with deposits of evaporites are exposed. Windblown silt, and mud from subsequent rapid floods, may cover these evaporite deposits. In some situations dolomitic mud is also deposited. Further drought completes the cycle, and in this way laminated evaporite deposits can accumulate. Similar deposits have formed in the flat areas of the Danakil Desert lying beneath sea level in the region of Eritrea in Ethiopia. They also occur on a much smaller scale in Death Valley, California. Where such bodies of saline water overlie grabens, as in the case of the Dead Sea lying between Israel and Jordan, a considerable thickness of interbedded evaporites and associated sediments can accumulate. Fig. 4-31 illustrates the sequence underlying the south basin of the Dead Sea. It is a schematic section from the El Lisan Peninsula in the north to the area of Safi in the south, a distance of some 40 km. The section, approximately 1500 m thick, comprises six main bodies of halite interbedded with marl and sandy marl. The southern part of the section that is not underlying the Dead Sea consists of alluvial fan deposits of gravel and sand.

The manner in which salt bodies such as those beneath the Dead Sea are formed is not well understood. In supersaline solutions seeding of tiny salt crystals results in growth of larger crystals. This method has been used by local inhabitants in regions of Africa where small bodies of supersaline

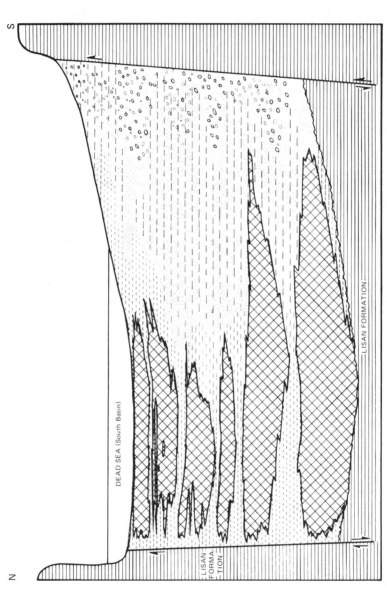

Fig. 4-31. Schematic section of the sedimentary sequence underlying the Dead Sea of Israel and Jordan showing bodies of halite (cross-hatched) interbedded with marl and sandy marl. The southern part of the section consists of alluvial fan deposits of gravel and sand. Section approximately 40 km long and 1500 m thick. [After Neev and Emery (1967).]

242

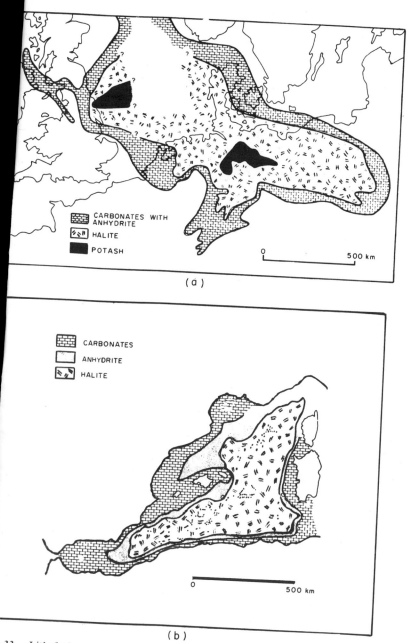

CARBONATES WITH ANHYDRITE

HALITE

POTASH

0   500 km

(a)

CARBONATES

ANHYDRITE

HALITE

0   500 km

(b)

Fig. 4-33. Lithofacies map of the Permian Zechstein sequence (a) in northern Europe and the Late Miocene Balearic Basin evaporitic sequence (b) in the Mediterranean region. [After Hsu (1972).]

water are found. Branches of bushes are dipped in the water and waved gently until dry. Then the salt crystals are shaken into the water. Subsequently, a salt mush is formed on the floor of the body of water and can be scooped up. Precipitation and growth of salt deposits on the floor of larger bodies of very salt water, possibly resulting from long periods of drought and consequent increase in salinity, may be the processes by which large bodies of salt have accumulated. A hypothetical sequence of dolomite, gypsum, and salt deposited in a closed saline basin is discussed by Hsu (1972) and illustrated in Fig. 4-32 which shows initial deposition of dolomite, followed by gypsum and finally by salt. Modifications to this sequence and stratigraphic configuration result from geographical factors such as the extent to which a body of water is cut off from an influx of seawater and the volume of inflow of fresh water from streams.

Climatic factors such as periodic droughts or excessive rainfall are also significant. The origin of giant salt deposits has many puzzling aspects. Some individual salt beds are hundreds of meters thick and cover areas of thousands of square kilometers. They are known to exist in many parts of the world and in recent years have been discovered by seismic surveys to underlie extensive regions of the outer continental shelf and continental slope of northwest Africa and eastern North America (Beck and Lehner, 1974; Scrutton and Dingle, 1974; Mascle *et al.*, 1973; and Williams, 1974). Although evaporite deposits are certainly formed in desert basins, the latter cannot account for most of the very large salt deposits that are

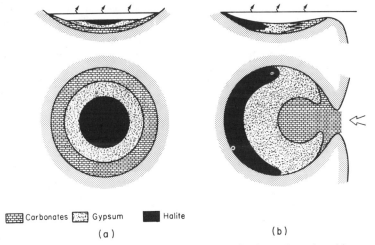

Carbonates   Gypsum   Halite

(a)   (b)

Fig. 4-32. Schematic diagrams showing possible distribution and stratigraphic sequence of evapoite deposits formed in (a) closed and (b) partially open bodies of salt water. [After Hsu (1972).]

known. Such salt deposits must have formed over long periods of time by the precipitation of salt in bodies of water subject to the influx of water containing salts in solution and to the loss of water largely or entirely by evaporation. A further requirement is subsidence of the sedimentary pile, by compaction and faulting, sufficient to accommodate the volume of the salt body. These conditions could obtain in geographical, climatic, and structural situations, such as those of the Dead Sea region, or in vast subsiding areas of salt lakes, playas, and salt flats.

Hsu (1972) has suggested such an origin for the Late Miocene salt deposits underlying the Mediterranean. He postulates that the Mediterranean Basin may have been depressed well below sea level during the Late Miocene. During this period the basin would have received a flow of meteoric water from the surrounding landmass but may also have had an influx of seawater from the present area of the Strait of Gibraltar. It is interesting to speculate that at some period prior to a possible catastrophic breakthrough of the Atlantic Ocean, there may have existed between Gibraltar and Tangier a salt waterfall many times larger in height and volume of water than the Niagara Falls.

In looking around the world today, one is hard pressed to find possible models of depositional environment to match in magnitude those that apparently existed in the past to form giant deposits of salt. On the other hand, the association in carbonate provinces of outer reef facies, middle limestone to dolomite facies, and inner evaporite facies is well documented in the literature, particularly with reference to the Devonian of North America. Fuller and Porter (1969) illustrate a regional facies map of the Late Devonian Stettler Formation in western Canada, extending for 1500 km from western Alberta to North Dakota, showing a transition over this distance from limestone to dolomite, anhydrite, salt, and red beds. In this type of distribution organic reefs are commonly found along the boundary between the limestone and dolomite facies. Detailed studies of sedimentation on the coast of the Persian Gulf (Illing, *et al.*, 1965; Kendall and Skipwith, 1969) have shown the relationships within an inner neritic to paralic environment of patch reefs growing offshore, lagoonal carbonate sands and muds, algal mats growing on flats between high and low tide levels, evaporite deposits formed in the dolomitic muds of sabkhas, and the outwash alluvial fans from outcrops of tertiary strata. Similar relationships have been observed in the distribution of stromatolitic, reef, and associated facies within sequences of ancient carbonate rocks, indicating that in some respects the present-day regions of carbonate deposition can be used as models for depositional environments in the past. The stratigraphic sequence of recently deposited sediments along the coast of the Persian Gulf can also be used as a model to explain the relationship of

ancient carbonates and noncarbonate clastics. ? Houlik (1973, p. 498) with reference to the Late ? Wyoming. He says:

> The repeated interbedding of carbonate tidal-? with detrital silicate sabkha and dune deposi? Amsden and Tensleep Formations resulted fr? transgressions and regressions, during which? removed from the coastal–continental sabkh?

One of the classic carbonate–salt facies assoc? eolian sediments is the Permian Rotliegendes and? north-central Europe and the North Sea descr? Sokolowski (1970), Glennie (1972), and others. Tl? unconformably overlies the Carboniferous coal n? variations from a desert environment in the south? evaporitic tidal flats in the north. Subsequent tr? resulted in deposition of the Zechstein sequence? deposits. These deposits underlie an area approxim? 500 km wide and have a thickness ranging up to at l? (Hsu, 1972) illustrates the distribution of the Zechs? ers a larger area but shows a similar pattern to the ? the Late Miocene Balearic Basin which occupied tl? Mediterranean Sea. Another interesting sequence? beds constitutes the Late Permian Castile Formation? Formation of the Delaware Basin in Texas and New? members have not been removed by solution, the Ca? approximately the same thickness of 500–600 m. ? mainly of salt, whereas the Castile comprises tens of? of carbonate, anhydrite, and salt laminae which are l? seasonal variations of sedimentation. These layers pe? distances and within the Castile can be traced for? according to Anderson *et al.* (1972). Of particular sign? rence in the Castile and Salado of stratigraphic units? breccia. Where these occur, the section is much thinn? sitional thinning is a factor, the comparative thickness? tion of salt has locally removed up to 500 m of sectic?

A possible example of salt deposition within a closec? Late Silurian Syracuse Formation of the Cayugan Se? the Great Lakes of North America. An isopach ma? presented by Alling and Briggs (1961) and shown in Fi? remarkable circular feature enclosing a section up to? Syracuse, which includes several salt beds commonly?

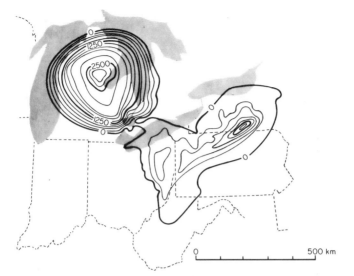

Fig. 4-34. Isopach map of the salt-bearing Late Silurian Syracuse Formation in the Cayugan Series of the Great Lakes region in North America. Thickness in feet. [After Alling and Briggs (1961).]

in the range of 10–20 m, comprises beds of dolomite, anhydritic and argillaceous dolomite, and dolomitic calcilutite.

Numerous questions remain concerning the genesis of large salt deposits, many of which have been mobilized to form irregular intrusive bodies and salt domes. Salt flowage undoubtedly is a major factor in the deformation of bodies of salt; dissolution and recrystallization may also have important roles in the growth of many piercement-type salt bodies.

Delta-front sand sheets are layers of sand deposited along certain parts of the leading edge of a delta. They may be shoreline sands that extend seaward from beaches, or bodies of sand deposited well out on a continental shelf in a neritic environment. The latter situation is unusual but has been described by Gibbs (1976) with reference to the Amazon Delta of Brazil. Normally, sand sheets are deposited in paralic to inner neritic environments by coastal currents that distribute the sediments carried out to the delta front by distributaries. Large distributaries form bar-finger sand bodies where they are able to prograde over a shallow continental shelf having a low-energy environments. These bar-fingers have previously been described as branching shoestring sands that in three dimensions constitute an anastomotic sequence of linear sand bodies trending generally normal to the coastline. Some sand from the fan-shaped front of these bar-fingers is swept away by currents and deposited elsewhere along the coast where it may form underwater bars or sand sheets. Where the sand is transported to areas flanking a delta, it may also contribute to the growth of barrier islands trending along the coastline. Sand sheets also exhibit some degree of linearity along the coastline and may be transgressive or regressive. A generalized diagram illustrating the distribution of a bird-foot-type delta front, described by Shannon and Dahl (1971), is shown in Fig. 4-35. The normal sediments deposited in prodelta areas are muds; those in delta-front areas are muds, silts, and sands, the latter forming bar-fingers or delta-front sand sheets.

In regions such as that of the Niger Delta in Nigeria, where strong current, wave, and tidal action preclude the development of a bird-foot-type delta, sand carried to the coast is distributed as shoreline sand sheets which are also referred to as coastal barrier sands. Along the shoreward margin these sands are thrown up by wave action to form barrier bars and beach ridges behind which lie lagoons and mangrove swamps. To what extent the coastal barrier sands present in the uppermost part of the Niger Delta sequence will be preserved in the geological record is open to question as they will probably be reworked and partly replaced by tidal–channel sands.

The anomalous depositional pattern off the coast of the Amazon Delta must be kept in mind with respect to interpretations of paleogeography based on lithofacies relationships. Gibbs (1976, p. 45) has described the nature of the depositional pattern as follows:

> The physical circulation of water shows Amazon River water thrusting out across the continental shelf and over the sea water with some

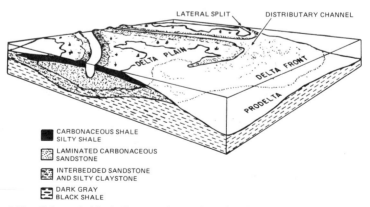

Fig. 4-35. Schematic block diagram of a portion of a bird-foot-type delta front showing plan and sectional views of prodelta, delta-front, and delta-plain areas. [After Shannon and Dahl (1971).]

entrainment and mixing with the sea water. The brackish plume from the Amazon is then turned northwestward along the outer shelf by the Guiana current and longshore currents to parallel the coastline for 500 to 700 km. Sea water upwells and flows landward under this plume.

The sediments of the Amazon River are thrust out onto the outer shelf, where a gradual depositional process occurs: the sand is deposited first, followed by the silt, and most of the mud is carried shoreward by the landward-moving bottom waters. This combination of processes results in the existing depositional facies pattern; modern mud deposits along the shoreline grade outward into silt deposits and finally into modern sand. The depositional process occurring is contrary to the classic model with sand along the shoreline.

The sand lithofacies forming off the Amazon Delta is obviously neither paralic nor inner neritic but rather outer neritic. It is included here by way of contrast as an interesting example of marine sand distribution that does not follow the normal pattern.

Recognition of delta-front sheet sands in the subsurface or in outcrop is based on the stratigraphic sequence of beds, faunal content, and internal structures. These sand bodies commonly prograde seaward, although they may move landward where the sea is locally transgressing some part of a delta. In the former case they are deposited over prodelta muds and silts and are themselves in turn overlain by carbonaceous beds deposited in coastal bays, lagoons, and marshes. Prodelta sediments are normally thin-bedded to laminated, dark brown to gray, and rich in organic matter. Commonly they contain an abundant benthonic fauna that churns the sediment to such an extent that bioturbation may completely destroy the original bedding. Prograding delta-front sand bodies overlying prodelta sediments are commonly quartzose and fine-grained but may show some degree of grain gradation. Where present, grain gradation is upward coarsening and may be observed in the funnel shape of the self-potential curve of electric logs. The overlying beds may contain thin coal seams and an abundance of arenaceous forams.

Barrier islands and bars are formed along coastlines where the action of waves and currents is sufficient to winnow the influx of sediments from river distributaries situated farther along the coast. Longshore currents carry the sand along the coast, and wave action piles the sand into bars which may rise above sea level to become barrier islands. Wind action further contributes to the building of barrier islands by sweeping sand off the beaches and depositing it as sand dunes between the beach and inner lagoon. Longshore currents not only transport the sand along the shore where it accretes to the seaward flanks of bars and barrier islands but also erode the upcurrent ends of the barrier islands and deposit sand as curved spits on the downcurrent ends of the islands. Consequently, chains of barrier islands, such as the classic example in the Gulf of Mexico, are continuously changing shape, each island being eroded or accreted at varying rates. Currents within the tidal channels separating the barrier islands normally prevent them from joining above sea level, but it must be kept in mind that the surficial expression of the islands is not of geological importance but merely reflects humps on a linear sand body trending along the coast. Such composite sand bodies are commonly up to 10 km in width and may be a hundred or more kilometers in length. These ribbon-shaped sand bodies are comparatively thin, commonly having a thickness of considerably less than 50 m.

Recognition of this type of sand body, the exposed portions of which are barrier islands, depends on determination of the geometry and internal features of the body, including a characteristic upward coarsening in grain gradation. The geometry can seldom be determined from outcrops only but, where sufficient control is available, can be readily demonstrated in the subsurface by means of isopach maps of the sand body. The self-potential curves of electric logs may also indicate, by their funnel-shaped configuration, the upward coarsening of grains. This type of grain grada-tion reflects the distribution of mean grain size with depth and, conse-quently, with energy level on the seaward slope of the prograding sand body. Progradation may also be reflected in the offlap sequence of bedding which may be visible in outcrop (Shelton, 1965) or inferred from electric log characteristics (Evans, 1970; Conybeare, 1976).

Sandstone bodies that originated as offshore sands, the exposed parts of which formed barrier islands, have been prolific producers of oil in North America, notably in the Gulf Coast region. The reservoirs may be purely stratigraphic but commonly result from a combination of structural and stratigraphic factors. Delineation of the trends of such sandstone bodies is consequently of prime importance in the lithostratigraphic analysis of sedimentary basins.

It has commonly been assumed that barrier islands are composed of lithic sands and that the stratigraphic associations with these geomor-phologic features are noncarbonate. Evans (1970) has pointed out, as illustrated in Fig. 4-36, that barrier islands can also form in environments of carbonate deposition. His studies of the sedimentation patterns and geomorphologic features in the Netherlands and southeastern region of the Persian Gulf lead him to the following conclusion, as stated on p. 493:

> The paper stresses that comparable patterns of environmental condi-tions have produced similar geomorphological forms and analogous

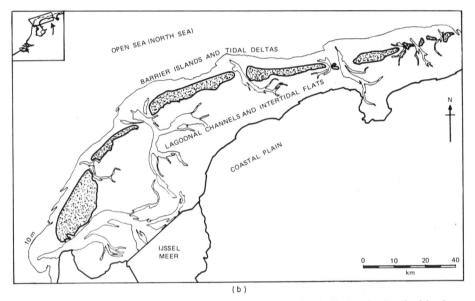

Fig. 4-36. (a) Map of an area of the Trucial Coast, Persian Gulf, showing barrier islands formed in a carbonate–evaporite depositional environment. (b) Map of part of the coast of the Netherlands showing barrier islands formed in a noncarbonate depositional environment. [After Evans (1970).]

patterns of sediments which do, however, reflect the differences in sediment source and climate in the two areas. If the sequences are ultimately preserved, the Trucial Coast will produce a limestone–evaporite sequence and the Dutch Wadden Sea a sandstone–shale–coal sequence.

The barrier islands described in the Persian Gulf are of carbonate sands, whereas those off the coast of the Netherlands are quartzose to lithic sands. Interpretation of carbonate sand bodies and the possible determination of ancient carbonate barrier islands in outcrop or in the subsurface are made difficult by the lack of lithologic contrast with other beds in the sequence. Furthermore, cementation by calcite or dolomite and recrystallization of the carbonate have commonly masked or destroyed the original textures.

Ignimbrite and lava flow sheets are not usually considered as a feature of deposition in paralic environments, although their preservation may depend on burial within a sequence of water-laid beds. Volcanics accumulated under subaerial conditions are commonly not preserved over a long range of geologic time, although numerous examples are known in North America of Miocene and younger basaltic flows preserved in Tertiary valleys. Most of these examples will probably disappear by erosion in time, and few will ever be preserved within sedimentary basins. On the other hand, volcanics laid down in shallow or deep parts of the ocean will be preserved and incorporated in the pile of strata constituting part of a sedimentary basin. In the ocean depths outpourings of basic lava are characteristic of regions of sea-floor spreading, whereas zones of subduction are marked by volcanic arcs, the so-called rings of fire. As a generalization this statement is valid, although zones of sea-floor spreading also include volcanoes such as those of the Iceland region. In shallow depths, particularly the paralic and inner neritic zones adjacent to a landmass formed largely of extinct and active volcanoes, the outpourings of lava and incandescent material from volcanic explosions are commonly blasted over the coastal areas into the sea. The continental shelf on which these volcanics are deposited is likely to be narrow and form part of a eugeosynclinal margin. The depositional sequence on such a margin may not be characteristic of the stratigraphic sequence within the thicker parts of a eugeosyncline which may consist largely of volcanics and turbidites, whereas the margins may include volcanics, lithic sediments, reef and other carbonate rocks. Paralic environments are consequently a repository for volcanics in regions of proximity to volcanism and cannot be considered anomalous where they underlie or overlie coastal deposits. In this respect it is of interest to visualize the conditions under which such volcanics may have been laid down in shallow coastal waters. An eyewitness account of such an event, the explosion of Mont Pelée on Martinique at 7:45 A.M. May 8, 1902, is given by the master of the ship Roddam, Captain E. W. Freeman. Roddam, at the time, was anchored in the harbor of St. Pierre. Freeman's vivid account is recorded in *Pearson's Magazine* (1902, p. 316) as follows:

There came a sudden roar, that shook the earth and the sea. The mountain uplifted, blew out, was rent in twain from top to bottom. From the vast chasm there belched up high into the sky a column of flashing flame, and a great black pillar of cloud.

That was all—just the one big roar of the shattering explosion, one flare and then the cloud, shooting out from the rent, rushing down the mountain-side onto the doomed city. It came down like a tornado, destroying everything as it passed, spreading out fan-wise as it neared the bottom—clearing the lower hills, it sprang upon St. Pierre, where the people were hurrying to mass, enveloping every street in darkness and in dust, in an instant—and then sweeping, leaping on down upon the shipping in the harbour, it rushed straight for the Roddam—a devouring tornado of fire from the bowels of the mountain.

There was no flame, save for that one great flash when the mountain blew up—it was simply a cloud of fine dust—powdered pumice-stone, I think, heated to an unknown temperature, that struck the city. In one minute and a half it raced from the heights of Pelée down to St. Pierre's harbour five miles distant below—searing the mountain-side and the hills, destroying in its frenzy every house in St. Pierre, killing every single one of the 40,000 persons, and every living thing that it touched, even to the insects in the air. As it struck the sea, it lashed the water into great waves that swamped the little sailing fleet, sending the Roraima on to her beam ends, the Roddam careening over so that she was half-buried under water, and the Grappler, as I think, straight to the bottom. The roar of it filled the air, and the blackness of it shut out the light.

The heat of this mass of hot gas and dust scorched the rigging and wood of the ships and fatally seared many of the seamen. The ships referred to are the steamships Roddam and Roraima and the cable repairship Grappler. This unique account adds a piquant touch of human interest to what otherwise might be an impersonal description of a *nuée ardente*. These incandescent clouds of gas, dust, and fine debris are collectively referred to as pyroclastic flows which, when consolidated and cooled to form a rock, are termed ignimbrites. An excellent description of the stratigraphy and rock types comprising the Mont Pelée volcanic pile is given by Roobol and Smith (1976).

Preservation of ignimbrites and associated volcanics laid down in a paralic environment is probably uncommon but is noted here as an example of a stratigraphic association diagnostic of a paleogeographic situation.

Neritic Environments

Prodelta mud sheets are thin-bedded to laminated deposits of mud laid down in front of a prograding delta. The mud consists of clay and silt transported by a freshwater plume riding out to sea from the mouths of river distributaries. Carried seaward and swept parallel to the coast by longshore currents, the plume of fresh, muddy water becomes mixed with seawater, and its load of colloidal clays is precipitated. Commonly these sediments are rich in organic matter and dark brown to grey. They are also the habitat for an abundant benthonic fauna which churn the sediment with living and feeding burrows. This bioturbation may in some localities completely destroy the original bedding.

As a delta advances seaward, the prodelta mud sheet progrades over deeper water sediments of middle to outer neritic environments and constitutes the lower unit of the delta stratigraphic sequence, as shown in Fig. 4-35. The beds over which this mud sheet rides may be sediments deposited in situ in the same depth of water, or reworked sediments originally deposited in much shallower water. The occurrence of modern sand far out on continental shelves in a middle neritic environment is, in some localities, the result of a rise in sea level during the Late Pleistocene. Much of this sand was originally deposited as shoreline sand bodies. Where a prodelta mud sheet overlies sand, or interbedded sand, silt, and mud, the depositional sequence may appear to be anomalous unless the possible fluctuations of absolute sea level, or of relative sea level, with respect to earth movements are taken into consideration. These possibilities must also be weighed in the interpretation of ancient strata. A further anomalous distribution of a mud sheet within a deltaic complex has been noted previously with regard to the Amazon Delta. Gibbs (1976) notes that the modern muds carried by the Amazon River are deposited along the shoreline and that they grade seaward into silts and finally into modern sand. This distribution is the reverse of the usual pattern but results from the upwelling of shoreward-moving seawater that carries the mud landward.

Prodelta mud sheets are best developed where deltas are building out on a broad, shallow continental shelf such as that underlying the Andaman Sea south of Burma, illustrated in Fig. 4-37. In this region the continental shelf slopes gradually seaward at an average rate of approximately 1 m in 2 km. On such broad and gently sloping sea floors, progradation of delta-front mud sheets can be rapid over a widespread area.

Recognition of prodelta sediments in outcrop or in the subsurface is based primarily on the stratigraphic sequence of units characteristic of deltaic sedimentation. This sequence normally consists of a basal unit of

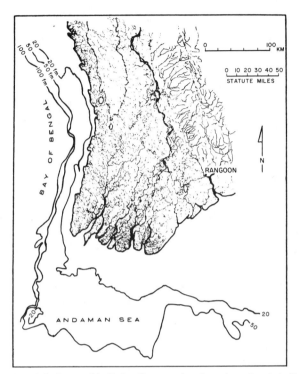

Fig. 4-37. Map of the Irrawaddy Delta, Burma, showing a broad and shallow continental shelf to the south. Bathymetric contours in fathoms.

dark brown to dark gray prodelta shale overlain by a delta-front unit of interbedded siltstone, shale, and sandstone deposited as shoreline sand bodies, and an upper delta plain unit of interbedded siltstone, sandstone, carbonaceous shale, and thin layers of coal formed from the accumulation of vegetation debris in coastal marshes and river backswamps. Recognition may also be assisted by the thin-bedded to laminated nature of the shale and by the varying degrees to which the laminae have been disturbed by bioturbation. These features and the sequence of units can be recognized in cores as well as in outcrops. Where cores are not available, interpretation of the nature and sequence of stratigraphic units may be greatly assisted by electric logs which indicate the lithology of various beds and the grain gradation within sandstone bodies.

Tidal current ridges are sea-floor features developed on continental shelves where strong tidal currents flow with velocities of up to 5 km/h. They consist of very fine, well-sorted sand and silt forming parallel ridges up to 50 km long, 5 km wide, and 50 m thick. Internally they have inclined bedding of a low-angle planar type similar to that of sand dunes, as indicated by sparker surveys. Tidal current ridges have been noted in many parts of the world and are described by Off (1963), Stride (1963), Swift and McMullen (1968), Houbolt (1968), Pettijohn *et al.* (1972), Blatt *et al.* (1972), and Boggs (1974).

Tidal current ridges form discrete bodies on a sandy to silty sea floor. They have not been recognized in outcrop or in the subsurface but have probably been preserved in the geological record. If so, they would be described as lenticular and linear sandstone bodies having no appreciable grain gradation and resembling barrier bar sand bodies in both shape and dimensions. Possible clues to their origin as tidal current ridges rather than a barrier bar complex may be the close parallel arrangement of the lenticular sandstone bodies and the absence of grain gradation and offlapping relationship. Off (1963, p. 335) suggested that the Late Mississippian Palestine Sandstone in the region of Kentucky may have been deposited as tidal current ridges and says:

> In a recent study of the Chester sands of the Illinois Basin (Potter *et al.*, 1958), attention was called to the alignment of sand bodies parallel with the current directions indicated by cross-bedding, and perpendicular to regional strike. The relative positions of some of these sand bodies (as shown by Potter's Figure 13 for example) and their unique dimensional scale, both as to individual thickness and distance of separation, combined with the presence of cross-bedding and suggestion of a marine environment, strongly suggest that they are tidal current ridges.

This interpretation differs from that of Potter *et al.* who considered the Palestine Sandstone to have been deposited in channels.

Examples of modern tidal current ridges, described by Blatt *et al.* (1972), and Houbolt (1968), are illustrated in Figs. 4-38, and 4-39. The former shows the distribution of ridges in the southern bight of the North Sea, off the coast of Norfolk, where they form discrete bodies up to 50 km long, 3 km wide, and 25 m high, trending in parallel arcs. The latter shows the distribution of ridges on the continental shelf of the Gulf of Korea situated in the northern part of the Yellow Sea. The ridges, which have about the same range of length and width as those of the North Sea, are situated within a depth of 10–20 m at distances of 20–50 km from the shore. Off (1963, p. 327) refers to these tidal current ridges as follows:

> Here, the ridges are spaced with an average of 3 miles between crests and rise an average of 65 feet from their base. They are oriented approximately parallel with the west coast of Korea and perpendicular to the north end of the Yellow Sea.

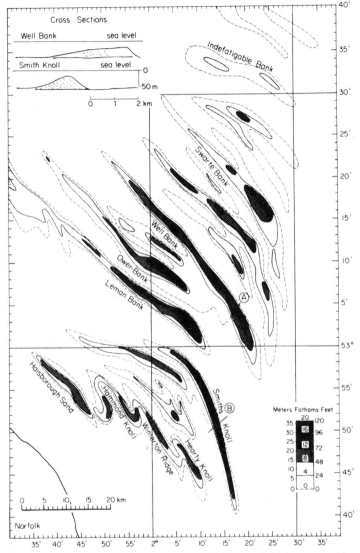

Fig. 4-38. Map showing the distribution and thickness of tidal current ridges in the southern bight of the North Sea off the coast of Norfolk. [After Blatt *et al.* (1972) and Houbolt (1968).]

According to this account, the height of ridges in the Gulf of Korea is of the same order of magnitude as the height of ridges in the North Sea; and, consequently, in size and shape the ridges in both regions are remarkably similar.

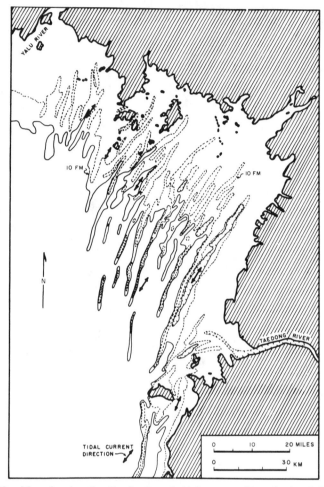

Fig. 4-39. Map showing the distribution of tidal current ridges in the Gulf of Korea. [After Off (1963). Redrawn from U.S. Navy Hydrographic Office Charts.]

Carbonate banks and reefs consist of extensive deposits of carbonate muds and sands and of the skeletal remains of colonial organic accumulations growing as barrier reefs around the periphery of the bank and as patch reefs within the bank. The muds are formed initially from the chemical precipitation of aragonite; the sands of various textures from the accretion of mud particles and fragmentation of the calcium carbonate remains of organisms. Muds accumulate in the quieter water of lagoons; the sands are formed in areas of shallow shoals and beaches where current or wave action is more vigorous. These carbonate banks lie in very shallow water, commonly in depths of less than 10 m, and may extend over areas of 100,000 km² or more. They form on broad continental shelves, such as that off the coast of Yucatan, Mexico, or develop on a rise separated from the continental shelf, such as the Bahama Bank on which Andros Island is situated, separated from the mainland by Florida Strait.

Of particular significance to the interpretation of the geologic history of carbonate sedimentary basins is the evidence of rate of growth of deposits in the Bahama Bank and elsewhere. Paleontological and other evidence based on drilling indicates that many thousands of feet of carbonates have accumulated during the Quaternary and Late Tertiary in these regions apparently as the result of carbonate sedimentation in a very shallow-water to subaerial environment keeping pace with tectonic subsidence of the crustal block underlying the region of the carbonate bank. Viewed in this light, the origin and history of stratification of such magnificent exposures of limestone as those of the Rocky Mountains in Alberta and British Columbia become more comprehensible. These great ramparts present to geologists an astonishing sequence of strata which provoke wonder and speculation as to their origin and depositional environment. Detailed stratigraphic and petrologic analyses of many reef and other sections in western Canada and other parts of the world have shown remarkable similarities to features of sedimentation and reef development seen in modern carbonate banks. It is of interest to reflect that in the Rocky Mountains and the subsurface of many regions in the world, the limestone sequences were formed from calcium carbonate derived from the sea, either by direct precipitation as aragonite mud or assimilation by organisms and certain types of algae. Slow as these processes may seem, given enough time, they can result in the accumulation of hundreds or thousands of meters of limestone.

The literature on carbonate petrology and sedimentation (including Wilson, 1975) is prolific. Illustrated material relating to the stratigraphy and sedimentology of the Bahama carbonate bank is well known, and examples from classic regions of the Persian Gulf are presented elsewhere in this book. Of the many examples that could be cited of modern environments of carbonate sedimentation, two are given here: Fig. 4-40, of an area off the west coast of Florida, and Fig. 4-41, of an area of the Great Barrier Reef in Australia.

Fig. 4-40 illustrates the geomorphology of the Ten Thousand Islands area in the vicinity of Everglades in southern Florida and shows a myriad of islands, channels, bays, and lagoons underlain by beds of vermetid reef rock, peat, accumulations of oyster shells, and quartz sand deposited during the past 5000 years in a period of fairly continuous marine transgression, according to Shier (1969). This sequence, which lies unconform-

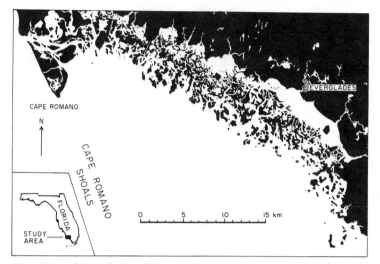

Fig. 4-40. Map of Ten Thousand Islands area, southwest coast of Florida; a drowned topography underlain by vermetid reef rock, oyster beds, quartz sand, and peat. [After Shier (1969).]

ably on Miocene limestone, has a maximum thickness of 8 m. It is of interest to note that given the same rate of deposition, a thickness of more than 2500 m could accumulate during the Quaternary. During the period of marine transgression of southwestern Florida, the coastal marsh was overlain by lagoon deposits of peat, oyster beds, and clayey sand which in turn were overlain by barrier reef deposits of vermetid rock formed from the cemented tubes of the gastropod *Vermetus*.

Fig. 4-41 is a map of a small area of the northern part of the Great Barrier Reef. The reef extends along the coast of Queensland, Australia, for nearly 2000 km, and the outer barrier reefs, which enclose an area containing numerous patch reefs, are at distances of 50–250 km from the mainland. The area illustrated in Fig. 4-41 shows a chain of barrier reefs extending for about 200 km in an arc trending parallel to and approximately 50 km from the coastline. Within this barrier reef trend are numerous patch reefs and the granitic Lizard Island which is surrounded by reef. Under the area of the Great Barrier Reef, the reef carbonate section is comparatively thin, consisting of a few hundred feet of limestone. This limestone section overlies several hundred feet of Tertiary noncarbonate rocks which are locally, and possibly regionally, largely of lithic sandstone and shale. Unconformably below the Tertiary the section is either of granite and metamorphosed Paleozoic rocks, or locally of Mesozoic rocks. Very little is known about the stratigraphy of the Great

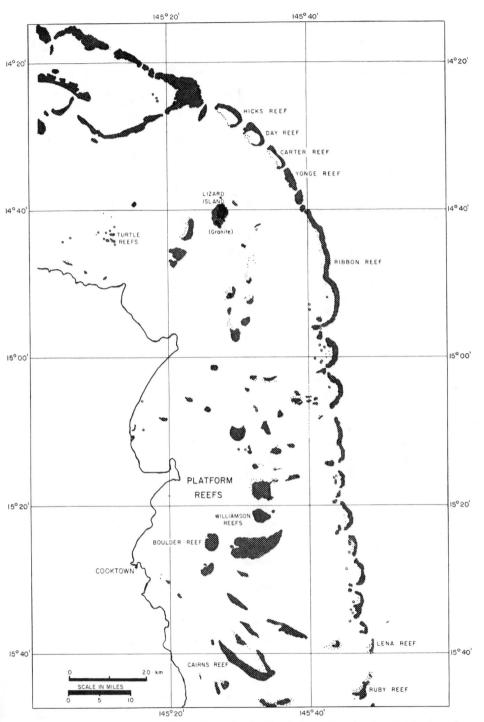

Fig. 4-41. Map of part of the Great Barrier Reef, Australia, showing distribution of barrier and patch reefs. [After Admiralty hydrographic charts.]

Barrier Reef region because, with the exception of a hole drilled on Wreck Island in the Capricorn Group to the south and three deeper holes drilled in the nearby offshore extension of a Mesozoic Basin southeast of Mackay, the area has been closed to oil exploration activities for reasons of conservation. The small portion of the Great Barrier Reef illustrated in Fig. 4-41 shows the geographical distribution, pattern, and trends of barrier and patch reefs plotted by hydrographic surveys. It is an interesting model for the geometry and trends of ancient reefs, and in this respect may be compared, as to size and areal distribution, with Mississippian reefs shown in Fig. 4-42. The reefs of the Great Barrier Reef are largely of colonial corals, with significant contributions from the calcareous algae *Halimeda;* the Late Mississippian reefs of the Williston Basin in North Dakota and Montana are algal pelletoid reefs developed in the Ratcliffe zone of the Charles Formation. With regard to size and distribution the Mississippian reefs compare favorably with patch reefs in the Great Barrier Reef. Furthermore, they appear to have had a similar genesis for, as stated by Hansen (1966, p. 2260):

> The reef deposits appear to have formed in slightly shallower parts of the shallow Ratcliffe sea.

The stratigraphy, lithology, and sedimentological features of carbonate bank sequences have been classified by Wilson (1974), Armstrong (1974), and others. These are variously referred to as carbonate platforms or shelves; but in this book they have been called banks which are defined (AGI, 1976) as elevations on the sea floor of large area, surrounded by deeper water, or as submerged plateaus. The inter-relationship of factors determining the stratigraphic sequence and lithologic features of carbonate bank sections are complex and not always well understood. With reference to carbonate sequences formed on shelf or platform margins Wilson (1974, p. 810) says:

> The development of such carbonate-shelf margins is a natural consequence of the normal processes of prolific carbonate production, trapping, baffling, encrustation, organic frame construction, and inorganic and organic cementation. These processes operate in clear warm shallow water on slopes at some distance out from low-lying landmasses. Hydrography, sea-level fluctuation, organic evolution, and tectonics all operate to influence the balance of these factors and to contruct particular types of shelf margins.

Cyclical sequences of many hundreds of meters of limestone with minor shale and sandstone deposited as sediment on carbonate banks have been observed in the Mississippian of the Cordilleran region in the western

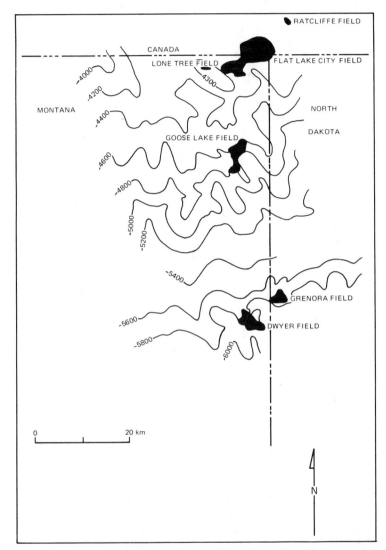

Fig. 4-42. Structure contour showing the Late Mississippian Ratcliffe Zone and distribution of oil-producing algal pelletoid reefs in the zone. Contours in feet below sea level. [After Hansen (1966).]

United States, Canada, and elsewhere. The origin of such sequences has been discussed by Rose (1976) whose explanation is illustrated in Fig. 4-43. He postulated a situation where the changes in sea level are minimal and where terrigenous sediments are transported across the carbonate

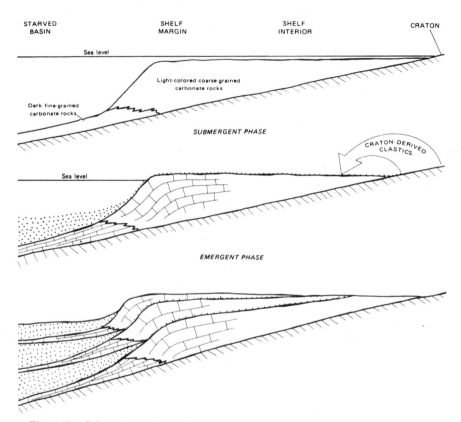

Fig. 4-43.   Schematic sections of a carbonate bank showing stratigraphic result of repeated submergent and emergent cycles of deposition. [After Rose (1976).]

bank during periods of emergence. Cyclical submergence and emergence of shallow areas on carbonate banks evidently take place, as seen in exposures of Quaternary limestone, notably in the Bahamas. Here there is no source of noncarbonate sediments, but the cementing and dissolution effects of meteoric water on the exposed carbonates produces beach rock. The fluctuations of sea level that cause emergent and submergent phases probably result from glacioeustatic variations. Such cyclical sequences of limestone, and of predominantly limestone beds, are deposited alternatively in neritic and paralic environments The emergent phase may be characterized by lagoons and mangrove swamps which trap not only carbonate mud but silt and clay washed in by streams. Where the carbonate bank has developed off a barren and sandy coast, emergence will result in the deposition of eolian and shoreline sand. The former may be evident as rounded and frosted sand and silt-sized grains of quartz disseminated in the limestone. The latter may form mobile shoals of cross-bedded sand.

Descriptions and classifications of the internal petrological features and facies relationships of carbonate banks are numerous and mentioned here only in general terms. Fig. 4-44 illustrates the typical sequence of facies within a carbonate bank and of the reef margin of a bank, as described by Motts (1968), and Namy (1974). The shelf evaporite facies is a paralic facies characterized by sabkhas; the shelf carbonate facies comprises dolomitic sediments deposited in a paralic environment and calcitic sediments deposited on a carbonate bank in a neritic environment; the reef zone facies includes bioherms developed as barrier and patch reefs along

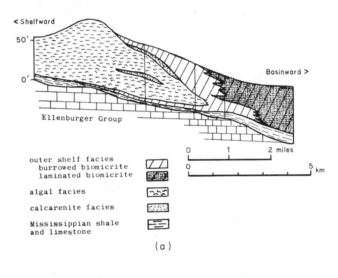

(a)

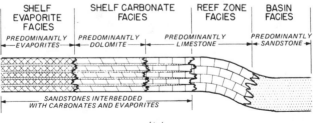

(b)

Fig. 4-44. (a) Cross section of an algal reef in the Early Pennsylvanian Marble Falls Group, Texas, showing general relationships of lithofacies. Vertical scale in feets. [After Namy (1974).] (b) Schematic cross section of the Permian Guadalupe reef complex, New Mexico, showing general relationships of lithofacies within a carbonate bank. [After Motts (1968).]

the outer margin of a carbonate bank where it terminates against the basin facies of noncarbonate sediments. Cyclical emergence and submergence of the shelf results in the development of an interbedded sequence of limestone, dolomite, and evaporites, the latter being commonly gypsum as anhydrite. Such cyclical sequences may have features of particular interest such as the dinosaur tracks which are found only in the dolomitic beds which were apparently deposited as shoreline mud during periods of shallowing when restricted bays or lagoons were formed.

The lithofacies of barrier reefs throughout the world are similar in that the inner reef or lagoonal facies consist of calcarenite and calcilutite; the core of the reef commonly includes lenticular beds of calcarenite formed as sand bordering the inner margin of the reef, and the outer margin of the reef consists of a zone of rubble which merges into a basin facies. The biofacies of barrier reefs can be stated in general terms only insofar as one is referring to a particular type of reef developed within a defined age range. Such reefs may be composed largely of the framework of colonial corals, stromatoporoids, calcareous algae, tube-building vermetid gastropods, or serpulid worms. Other reeflike mounds of the remains of crinoids or polyzoans may be formed within the area of patch reefs on a carbonate bank. Not all of these were formed in situ as indicated, for example, by planar type cross bedding in some Mississippian encrinite beds cropping out in the mountains of Alberta.

The general relationships of lithofacies within a reef complex are illustrated in Figs. 4-44 and 4-45. The former shows a cross section of an algal reef in the Early Pennsylvanian Marble Falls Group in Texas, the latter is a cross section of an oil-producing algal reef in Early Permian carbonate beds in New Mexico, described by Malek-Aslani (1970). In detail they differ, but in general relationship of lithofacies they are similar. It is essential to recognize that the spatial relationships of the lithofacies, as shown in the cross sections, have been very much distorted by vertical exaggeration and that the rock bodies of different lithofacies form comparatively thin layers. A natural scale cross section of the Early Permian reef is shown for comparison in Fig. 4-45.

An example of a stromatoporoid and algal reef in the Late Devonian Swan Hills Formation of Alberta is described by Leavitt (1968) and illustrated in Figs. 4-46, and 4-47. This oil-producing reef, referred to as the Carson Creek North reef complex, covers an area of 60 km² and has a maximum thickness of 100 m. The geometry of the reef suggests that it had a development similar to that of an atoll. Within this reef complex brachiopods, corals, and other fossils are common, as shown in Fig. 4-47, but the reef-building organisms were stromatoporoids and algae. Of particular interest is the restored structure map of the reef, showin in Fig.

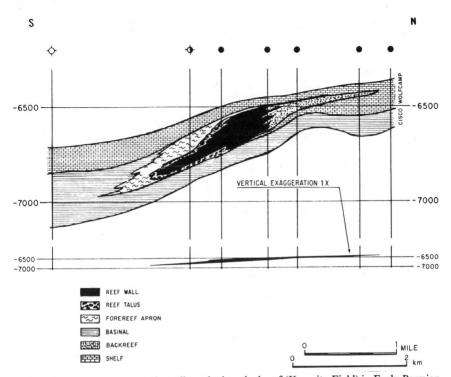

REEF WALL
REEF TALUS
FOREREEF APRON
BASINAL
BACKREEF
SHELF

Fig. 4-45. Cross section of an oil-producing algal reef (Kemnitz Field) in Early Permian carbonate beds in New Mexico showing general relationships of lithofacies. Vertical scale in feet. Note the vertical exaggeration. [After Malek-Aslani (1970).]

4-46. This map shows the structure of the top of the reef complex prior to regional tilting and indicates the presence of reentrants which resemble tidal channels of the Great Barrier Reef illustrated in Fig. 4-41.

Another oil-producing stromatoporoid reef within the Swan Hills Formation is the Goose River Field shown in Fig. 4-48. This reef has been described by Jenik and Lerbekmo (1968) who considered it to have developed as an atoll with an outer slope, reef rim, reef crest, and a central lagoon. They defined three biofacies. The forereef facies was characterized by large stromatoporoids, brachiopods, and crinoids, the organic and backreef facies by large stromatoporoids and *Stachyodes,* and the lagoon facies by *Amphipora* with minor occurrences of ostracods, forams, and calcispheres. Of interest is the recording by Jenik and Lerbekmo of abundant stylolites which, they say, indicate a reduction in reef thickness of 9%. This factor of volume reduction resulting from dissolution of carbonate on stylolite surfaces may be of significance with reference to stratigraphic analyses of carbonate sections in some sedimentary basins.

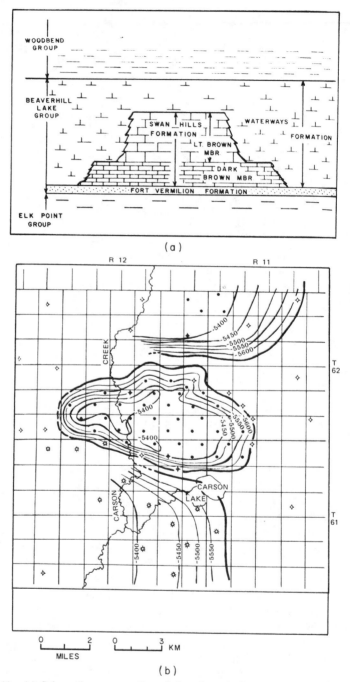

Fig. 4-46. (a) Schematic cross section of the oil-producing Carson Creek North reef complex in the Late Devonian Swan Hills Formation, Alberta. (b) Restored structure map (structure contours on restored top of Swan Hills Formation) showing possible tidal channels on both sides of the atoll like Carson Creek North reef complex. Elevations in feet. Contour interval 50 ft. [After Leavitt, (1968).]

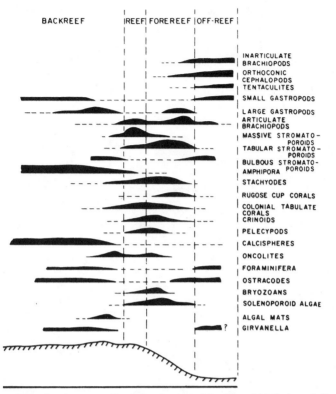

Fig. 4-47. Map showing generalized distribution of fossils and biofacies relationships in the Carson Creek North reef complex shown in Fig. 4-46. Not to scale. [After Leavitt (1968).]

The distribution of the Goose River, Carson Creek North, and other reefs in the Swan Hills Formation with respect to the contemporaneous landmass, carbonate bank, and basin facies is shown in Fig. 4-48. The Peace River area formed a landmass surrounded by a reef carbonate bank of dark brown limestone. To the east of this bank, which had a width of 100–200 km, lay a deeper marine environment in which noncalcareous muds were deposited. Reefs developed on the outer fringe of the carbonate bank, possibly as a cluster of atolls. Other descriptions of Late Devonian massive stromatoporoid reefs in western and northern Canada are given by Embry and Klovan (1971) and Klovan (1974). Embry and Klovan describe stromatoporoid reefs within the Weatherall Formation on Banks Island in the Artic Islands of Canada. These grew in shallow water on a carbonate bank but ceased their growth when the bank was subsequently covered by terrigenous sediments as a result of withdrawal of the sea.

Fig. 4-48. Paleographic map showing the distribution in Alberta of Late Devonian stromatoporoid reefs (black) within the Swan Hills Formation. Landmass (crosses), carbonate bank (horizontal lines), and deeper sea basin facies are shown. [After Jenik and Lerbekmo (1968).]

An example of a very large and ancient lagoon that developed on an extensive carbonate bank is given by Fisher and Rodda (1969) with reference to rudist reefs in the Early Cretaceous Edwards Formation of Texas. As illustrated in Fig. 4-49, the lagoon had a length of about 275 km and a width of more than 100 km. The lithofacies within the area of the lagoon is of dolomitic and anhydritic carbonate rocks with beds of anhydrite. This facies is surrounded by cherty dolomitic rocks which in turn are ringed by cherty limestone, the whole complex lying within a limestone bank. Rudist reefs have grown around the periphery of this complex, mainly within the cherty facies, but also within the surrounding limestone bank. Within the dolomitic carbonate sequence original depositional features are evident in the thin-bedded (less than 0.5 m) stratification, mud cracks, and

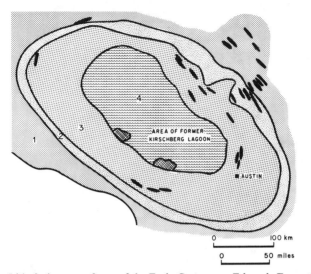

Fig. 4-49. Lithofacies map of part of the Early Cretaceous Edwards Formation in Texas showing the distribution of rudistid reefs (black) around the periphery of a lagoon facies: (1) limestone, (2) cherty limestone, (3) cherty dolomite, (4) dolomite and anhydrite, (5) anhydrite. [After Fisher and Rodda (1969).]

stromatolite structures which indicate sedimentation in a very shallow water to subaerial environment. Fisher and Rodda state that within this laminated to thin-bedded sequence, dolomitization resulted from pre-lithification replacement of calcium carbonate muds by magnesium-enriched brines in an environment of supratidal and intertidal mud flats along the margin of the lagoon.

Another example of rudist reef development forming an oval pattern is described by Coogan *et al.* (1972) with reference to the late Early Cretaceous El Abra, Tamaulipas, and Tamabra Limestones in eastern Mexico. The El Abra is in part the same age as the Edwards Formation in Texas. The pattern of reefs, as shown in Fig. 4-50, forms an oval trend enclosing an area approximately 160 km in length and 80 km in width. On the western side of this pattern, two parallel trends are known, an inner Golden Lane of rudist reefs and an outer Poza Rica trend of reefs comprising both corals and rudistids, as shown in Fig. 4-51. The rudist reefs of the Golden Lane trend are thought to have developed on a shallow-water carbonate shelf or lagoon; the coral–rudistid reefs of the Poza Rica trend grew in a more open marine environment along the margin of a deeper water basin. The whole carbonate complex probably formed a bank lying under very shallow water and surrounded by deeper water. Oil production is obtained from the reefs and reef debris.

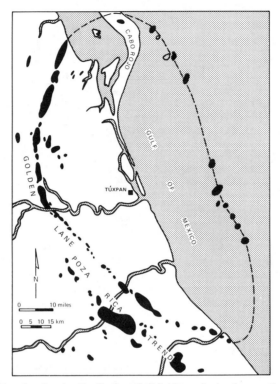

Fig. 4-50.   Map showing the distribution of oil fields producing from rudistid reef debris in the Early Cretaceous El Abra, Tamaulipas, and Tamabra Limestones of eastern Mexico. [After Coogan *et al.* (1972).]

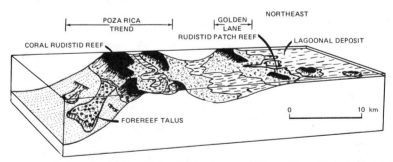

Fig. 4-51.   Schematic section of the Golden Lane and Poza Rica oil-producing trends showing the depositional environments of inner rudistid and outer coral–rudistid reefs in the Early Cretaceous El Abra, Tamaulipas, and Tamabra Limestones of eastern Mexico. [After Coogan *et al.* (1972).]

The study of reef growth and sedimentation on modern carbonate banks has made considerable and very significant contributions to our understanding of ancient reef complexes and their relationships to the carbonate sequences in which they occur. Detailed studies of lithofacies and biofacies within a reef and its surrounding beds depend on close control which is only available by drilling. Consequently, the best-known reefs are those that produce oil or gas. Interpretation of the regional geomorphic features, depositional environment, and facies distribution of stratigraphic correlatives requires adequate control, but not so close as needed in the area of the reef complex. The degree of control and proximity of control points on a map depends on the variability in thickness and lithology of the stratigraphic sequence being studied. In general, thicker sequences can be interpreted by painting a facies map with a broader brush, which may be a helpful method in understanding the continentwide distribution of ancient landmasses and seas. An example of this generalized treatment of a time-stratigraphic unit spanning the whole of the Middle Devonian series in North America is illustrated in Fig. 4-52 which shows the distribution of sequences consisting predominantly of clastics, carbonates, and evaporites, as interpreted by Hriskevich (1970). The extrapola-

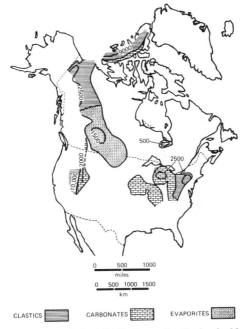

Fig. 4-52. Lithofacies map of the Middle Devonian Series in North America. [After Hriskevich (1970).]

tion involved in delineating the distribution of particular facies and in connecting discrete areas of the same facies is largely a matter of personal interpretation. On a detailed basis, the extent to which modern reefs can be used as analogs is also a matter of personal interpretation. In this respect Klovan (1974, p. 799) has injected a note of caution by saying:

> The ultimate form, geometry, and facies distribution of fossil reefs are dependent on a complex history of development as well as oceanographic shaping factors. Recent reefs can serve as models of development but should not be used as explicit analogues.

The association of beds of quartzose sandstone and limestone is not uncommon and has been observed in many sequences including the Triassic of northeastern British Columbia and the Pennsylvanian and Permian of Texas. Some sequences show a cyclic arrangement of carbonate and noncarbonate beds, one such example being illustrated in Fig. 4-53 which shows an alternation of sandstone and shale sections with limestone and black shale sections in the Late Pennsylvanian Cisco Series and Early Permian Pueblo Formation of west-central Texas. This sequence is described by Jackson (1964) as comprising alternating clastic and limestone-black shale units, the thickest part of each clastic unit lying basinward of the thickest part of the preceding clastic unit. The sequence

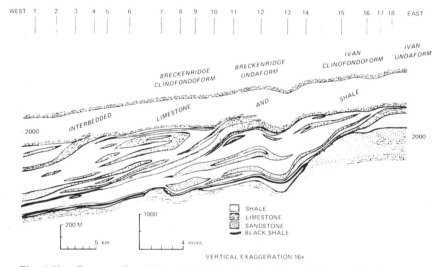

Fig. 4-53. Cross section of the Late Pennsylvanian Cisco Series and Early Permian Pueblo Formation in west-central Texas (Nolan and Taylor counties) showing the offlapping relationship of alternating units of clastics and limestone–black shale. The sequence is prograding and diachronous. Numbers 1–18 show well control. [After Jackson (1964).]

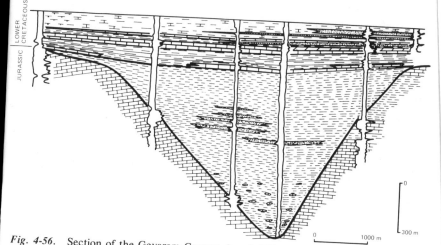

Fig. 4-56. Section of the Gevaram Canyon, Isarel. This submarine channel is cut into Late Jurassic limestone and filled with Early Cretaceous marine shale and sandstone. [After Cohen (1976).]

Helez Formation. The canyon, which has been traced by drilling for a length of at least 16 km, has a width of up to 7 km and depths exceeding 1000 m. The lower part of the canyon contains rubbly shale, including large blocks of limestone.

Examples of probable submarine channel deposits in the Tertiary of the French Maritime Alps have been studied by Stanley (1967). He noted that in this region the submarine channel deposits are generally of coarse clastics deposited by normal currents, whereas the associated flysch deposits laid down as submarine fans on a continental slope are turbidites. Stanley also noted similarities in the geometry, stratification, texture, and assemblage of sedimentary structures in ancient and modern submarine channel deposits but observed that the latter remain logistically difficult to study. Technological advances have enabled oceanographers to gain a better understanding of sea-floor topography and sedimentation; but information on the nature of channel deposits beneath the sea floor depends on geophysical methods or drilling.

Abyssal plain deposits have either been transported to their site of deposition by currents moving over the sea floor, or they have accumulated by the fall of dust, volcanic ash, and the remains of pelagic organisms. In all deep areas of the ocean there is a continuous but variable rain of extremely fine clastic and organic debris onto the sea floor. In some deep areas, notably those flanking continental rises subject to tectonic disturbances, periodic surges of turbidite flows spread widely over the abyssal

---

consequently shows an offlap relationship indicating progradation of a depositional shelf and infilling of part of a depositional basin. Periodic influxes of clastic sediments, possibly resulting from changes in a sea level, were deposited as wedges on the leading edge and slope of a carbonate bank. Developing in this way it is possible, in situation where the basin is shallow, for a limestone–clastic sequence to extend for considerable distances and to form a thick section over a wide region. Such sequences are obviously diachronous. In Fig. 4-53 the limestone bed shown in the western part of the section is the Early Permian Flippen Limestone, whereas the limestone shown in the eastern part if the Late Pennsylvanian Gunsight Limestone. Recognition of such diachronous relationships is essential to the interpretation of stratigraphic units and to sedimentary basin analysis.

Bathyal and Abyssal Environments

Submarine fans develop at the mouths of submarine canyons along which sediments are transported from the continental shelf by strong currents. Commonly these fans are developed on the lower regions of a continental slope where they spread out to form a complex of channel and levee deposits. Lateral migration and coalescing of numerous fans build a submarine delta or cone which prograles seaward and becomes a major element in the stratigraphic framework of a sedimentary basin. This element has been referred to by Galloway and Brown (1973) as a slope wedge, the general features of which are illustrated in Fig. 4-54. Near the

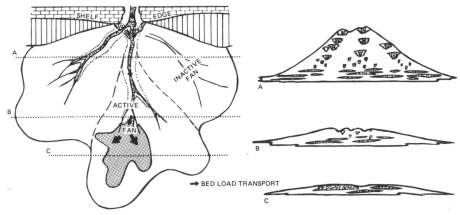

Fig. 4-54. Sketch (model) showing plan and section views of a submarine slope wedge comprising an active fan flanked by developing and inactive fans. Not to scale. [After Galloway and Brown (1973).]

mouth of a submarine canyon the sands are deposited mainly in channels, with spillover deposits of sand and silt laid down on the levees. Near the margins of the slope wedge, at the distal ends of the channels the sand spreads out as a sheet deposit. Progradation results in a sequence grading upward from sheet-type sand deposits to channel sand deposits. Periodic surges of sediment down a submarine canyon result in a cyclic or semicyclic layering of coarser and finer clastic material which in individual beds may exhibit graded bedding. Commonly, but by no means invariably, the beds comprising a slope wedge are turbidites interbedded with finer clastics deposited by normal currents, particularly in areas of tectonic instability.

Recognition of ancient submarine fan deposits depends on interpretations that may be regarded as equivocal, but examples can be cited. Galloway and Brown (1973, p. 1185), in referring to the Late Pennsylvanian sequence of the Midland Basin in north-central Texas, say:

> The Sweetwater slope system is composed of several slope wedges, or fans, each of which includes shelf-margin, slope-trough, and distal-slope sandstone facies, as well as slope mudstone facies. Terrigenous sediments were transported across the shelf by prograding fluvial–deltaic channels that locally extended through the Sylvester shelf-edge bank system and onto the slope where deposition in the deeper basin constructed submarine fans.

Submarine channel deposits have been extensively studied in recent years by means of sea-floor sampling, coring, and geophysical methods employed in oceanographic surveys. The channels are obviously pathways for transportation of sediment from the continental shelf to the slope, but the physical processes by which they have developed are not well understood. Some submarine channels may have originated as subaerial valleys which were subsequently inundated by the sea; others, including those off the southeast coast of Sicily, illustrated in Fig. 4-55, appear to have had a submarine origin, possibly as the result of erosion by turbidity currents. Cores from within these submarine channels in the Ionian Sea contain graded beds of sand and silt, whereas those from the sea floor between channels contain layers of tephra and mud with abundant pelagic fauna, according to Ryan and Heezen (1965). The channels in the Messina Cone, a large submarine slope wedge, are 3–8 km wide and commonly more than 180 m deep. In the lower parts of the continental slope the channel floors widen and become flat as they approach the abyssal plain.

Ancient submarine channels have been recorded in North America, Australia, Israel, and Europe. Examples in North America include Ter-

Fig. 4-55.   Bathymetric map of the Messina Cone off the southeast coast of S submarine channels cutting the cone and extending to an abyssal plain. [Af Heezen (1965).]

tiary channel deposits described by Hoyt (1959), Dickas and Pa Martin (1963), and Halbouty (1959). In the Gippsland Basin of of Victoria, Australia, the Marlin Channel, described by G Hodgson (1971), has been excavated to depths of up to 450 Eocene Latrobe deltaic sequence of sandstones, shales, and and has been filled with the Oligocene Lakes Entrance marine and siltstones. The origin of the Marlin Channel does not ap entirely submarine, and it may have been initiated by subaeri An interesting example of a submarine channel in Israel has scribed by Cohen (1976) and is illustrated in Fig. 4-56. The cl ferred to as the Gevaram Canyon, is cut into Late Jurassic lim filled with dark grey marine shale and minor sandstone of the I taceous Gevaram Formation which is, in turn, overlain by an in sequence of Early Cretaceous limestone and calcareous shale

plain, scouring and burying normal deposits of pelagic debris and clay. Repetition of these events produces a cyclic sequence of thin beds of sand and silt showing graded bedding, and of laminated pelagic clay which may have an abundant fauna. Turbidites exhibit not only an upward decrease of grain size within a bed representing a single turbidity current surge, as described by Bond (1973) and shown in Fig. 4-57, but also a decrease laterally away from the source. Consequently, although turbidites always show graded bedding, they may consist of medium to fine sand on the lower slope of a submarine fan and of extremely fine sand to silt at the distal end of a turbidite flow on an abyssal plain.

Studies of the Aleutian Abyssal Plain are described by Chase *et al.* (1971), Hamilton (1973), and Stewart (1976). This sea-floor feature is one of the three main abyssal plains, lying within the Gulf of Alaska and the northeast part of the Pacific Ocean, and extends for more than 2000 km south of the Aleutian Trench, as illustrated in Fig. 4-58. Stewart describes the plain as being underlain by a southward-thinning turbidite apron which ceased to grow about 30 million years ago during the Oligocene. Since then the apron has been covered with a thick blanket of pelagic deposits. Near the Aleutian Trench the apron is up to 800 m thick, whereas to the south, where it terminates near the Surveyor Fracture Zone, the apron is about 200 m thick. At one location in the northern area of the Aleutian Abyssal Plain, the apron of turbidites overlies a 5-m section of nanofossil limestone and varicolored clay which rests on basalt, presumably oceanic basement. The turbidite section at this location is about 250 m thick and consists of olive-gray pelagic clay interbedded with thin beds of graded

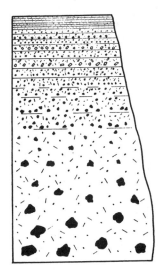

Fig. 4-57. Sketch illustrating graded bedding within a turbidite bed deposited by a single turbidity current surge. [Reprinted from Bond, *J. Geol.* **81**(557–575). Copyright 1973 by the University of Chicago Press.]

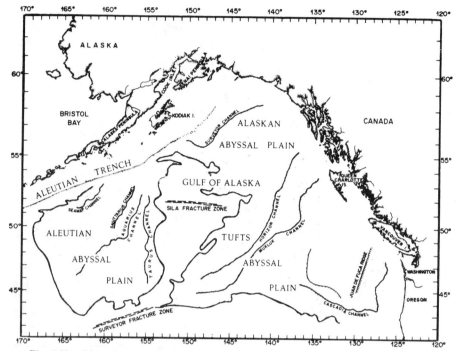

Fig. 4-58. Map of the northeastern area of the Pacific Ocean showing the distribution of abyssal plains and other main topographic and tectonic features of the ocean floor. [After Stewart (1976), Hamilton (1973), and Chase *et al.* (1971).]

silt and silty sand. Overlying the turbidite section is approximately 250 m of pelagic clay and diatomaceous debris.

In deep areas of the ocean not subject to turbidity currents or volcanism, the only source of sediment is the continuous downward drift of meteoric dust, volcanic ash, and the debris from pelagic organisms. The rate of fall of dust and ash is variable and depends on location, climatic variations, and periodic volcanism. The continuous rain of organic debris is also variable and depends on the abundance of pelagic organisms, particularly of planktonic forms with siliceous tests or framework. The abundance of these forms, in turn, depends on factors such as water temperature and the chemical composition of the seawater in various areas of the ocean. The role of ocean currents in the distribution of heat and dissolved substances is of prime importance to the deposition of pelagic sediment. In this respect it is of interest to note the correlation between the distribution of areas of maximum dissolved phosphate at a 100-m depth and of most abundant siliceous organisms in sea-floor sediments in the Pacific Ocean, as shown in Fig. 4-59. The areas of distribution show a remarkable agreement probably because the abundance of diatoms and radiolarians is

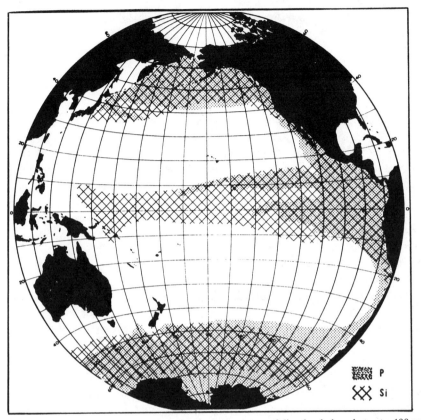

Fig. 4-59.   Pacific Ocean, showing distribution patterns of dissolved phosphate at a 100-m depth and of abundance of siliceous organisms in sea-floor sediment. [After Berger, (1970).]

related to the nutrient content of the seawater. The relationship appears to be qualitative but cannot be regarded as quantitative, particularly in view of the uncertainty of such factors as the solubility of silica with depth, as discussed by Berger (1970).

Locally associated with abyssal plain sediments are basaltic volcanics extruded through fractures in the oceanic crust or transported by ocean currents sweeping the flanks of submarine volcanic cones. Basaltic extrusions are a well-known feature of sea-floor spreading in the Atlantic; and in the Pacific, Fisher and Engel (1969) record the recovery by dredging of ultramafic and basaltic rocks at depths of up to 9400 m in the Tonga Trench. The association of banded argillite and chert with chloritic greenstone beds formed from the alteration of basic volcanics has in some regions been attributed to deposition in a deep-water environment. The problem of assigning a genesis to such sequences lies in the fact that metamorphism may not only have destroyed the original depositional features but resulted in the introduction or migration of silica.

LITHOFACIES AND STRATIGRAPHIC ASSOCIATIONS

Stratigraphic sequences possess particular lithologic, sedimentologic, mineralogic, chemical, and other characteristics that set them apart and define the depositional environments and tectonic setting in which they were formed; but some stratigraphic sequences do not enable precise definitions to be made, and the evidence supplied by their various characteristics may be equivocal. In general, most sequences can be placed in broad lithologic categories, depending on whether they are predominantly or significantly of carbonates, noncarbonate clastics, claystones, or volcanics. Difficulties in assigning a category may arise where qualitative and quantitative variations occur in the association of different rock types within the same sequence. Difficulties may also arise in terminology and definition with respect to individual rock layers and sequences of layers. The term graywacke, for example, has been variously used and to some geologists connotes a particular depositional environment or lithofacies. Where misunderstandings of substance or inference may arise through the use of such terms, it is preferable to employ words that are purely descriptive and preferably quantitative as well as qualitative. In the case of the term graywacke, it is preferable to substitute it by the general term litharenite which may then be suitably qualified by further objective description.

Some inferential rock-type terms with genetic connotations are so well entrenched in the literature and so clearly associated with particular sedimentary processes or depositional environments that they can clearly stand on their own merits. Such terms include turbidite and ignimbrite. Similarly, with reference to stratigraphic sequences, the terms molasse, flysch, carbonate, euxinic, and coal facies connote not only the lithologic character of the contained beds but also their depositional environments. These terms must be used with some caution, and the depositional environments and lithologic nature of each sequence so described must be clearly defined.

Molasse Sequence

Molasse sequences consist essentially of clastic sediments, mainly sandstone and siltstone, with lesser amounts of shale. Conglomerate, conglomeratic sandstone, and coal are commonly present whereas limestone, where present, is a very subordinate constituent. Total thickness may range up to several thousand meters. Within this thick sequence are both marine and nonmarine sediments. The former are predominantly inner neritic to paralic including shoreline and deltaic sediments; the latter are fluvial, lacustrine, and swamp sediments which accumulated as alluvial fans and as river, lake, delta, and floodplain deposits. The sandstone beds commonly exhibit either the planar type high-angle cross bedding formed in a fluvial environment or the low-angle type formed in offshore sand bodies. The latter may also exhibit planar cross bedding where tidal currents deposited the original sand. Both sandstone and shale deposited in paralic to neritic environments are commonly very fossiliferous, although rapid accumulation of sediments may locally preclude the abundance of fossils.

The entire molasse sequence consists of an interstratified assemblage of marine and nonmarine beds which individually range in thickness to 30 m or more and display an interfingering and lenticular stratification. The sequence is marked by local unconformities, disconformities, and paracontinuities in places associated with varicolored or red beds. Coal beds and carbonaceous shales are common in the nonmarine sections. The paralic sections may contain abundant ripple marks in the sandstone and siltstone beds.

A molasse sequence consists of sediments derived from the erosion of a landmass being uplifted by orogeny. It consequently forms a lenticular wedge that thins both in the direction of the orogenic belt and in the direction of the sea, having its maximum thickness underlying the area of deltaic sedimentation. Thick sequences of molasse or flysch sediments are commonly referred to as facies; but the term as applied to these sequences, which comprise sediments deposited in a number of different depositional environments, is not a satisfactory designation. These sequences include distinct sedimentary units which are also referred to as particular facies, depending on their geomorphic, lithologic, and other characteristics or on the paleogeography of their areas of deposition. The term facies is properly reserved to designate sediments having a particular lithologic or biologic reference to the conditions or environment of deposition.

Flysch Sequence

Flysch sequences consist mainly of argillite or shale and silty shale, with lesser amounts of lithic arenites. The sequence may also include a variable content of tuff, basic volcanic lavas, limestone, and cherty beds. The total thickness of a flysch sequence may range up to 8000 m or more. A general characteristic is even layering and rhythmic alternation of thin beds of lithic arenites and shales, a feature which, in the early days of geological mapping in the United Kingdom, evoked the name flagstone grits. The thin-bedded to laminated nature of the strata and its local fissility have given rise to the name flysch. Flysch sequences have also been referred to as a graywacke suite, a term relating to the content of lithic arenites. Beds are commonly less than one meter thick, and unlike beds in a molasse sequence, they maintain their thicknesses for considerable distances, showing no evidence of lensing or cut-and-fill. They differ also from molasse sequences in that they do not show cross bedding of the planar type, although very small scale cross bedding associated with current ripple marks may be abundant in the silty beds. Graded bedding is common and may be associated with slump structures such as load casts and convolute bedding or with strong current structures such as flow casts, striations, flame structures, and intraformational breccia.

Flysch sequences are deposited in the bathyal depths of continental rises and on abyssal plains where sedimentation can be sustained at a fairly rapid rate by sediment-loaded currents moving down submarine valleys and sweeping out over fans at the base of the continental slopes. Submarine trenches, in places flanked by volcanic chains, may be filled with thousands of meters of sediments and volcanics transported down the slopes and along the trench floor. Where this is the situation, directional current features such as striations may be formed parallel to the existing shoreline. Certain negative features may also be diagnostic of a flysch facies. For example, whereas molasse sequences may contain very fossiliferous beds, flysch sequences generally do not, apart from microfauna. Radiolarian cherts are commonly well preserved, but benthonic fauna may locally be unable to adapt to conditions of frequent disruption by turbidity currents.

Flysch sequences may form a lateral extension of molasse sequences or be overlain by paralic and neritic sediments. The latter situation is illustrated by Fig. 4-60 which shows a Late Cretaceous flysch sequence, in the Pyrenees region of southwest France, overlain by sediments deposited in continental and neritic environments. The sequence, described by Ricateau and Villemin (1973) and by Winnock et al. (1973), forms part of the Aquitaine Basin and has a thickness of up to 4000 m. It is believed to have been deposited in a deep trough, up to 1000 km long, which tended to migrate northward as a result of Pyrenean orogenesis.

Carbonate–Quartz Sand Facies

The carbonate–quartz sand facies consists mainly of limestone or dolomite with subordinate amounts of quartzose sandstone and minor

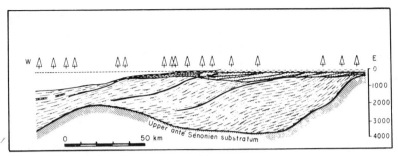

(a)

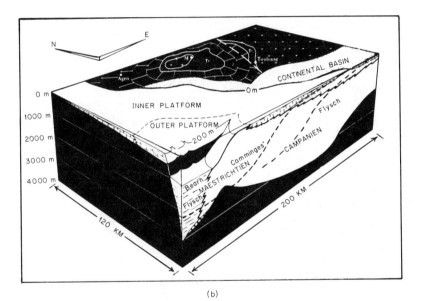

(b)

Fig. 4-60. (a) Section of a Late Cretaceous flysch sequence overlain by continental and neritic sediments in the Aquitaine Basin of southwest France. [After Ricateau and Villemin (1973).] (b) Block diagram showing the same flysch sequence. [After Winnock *et al.* (1973).]

amounts of shale. The carbonates are well bedded and comprise calcirudites, calcarenites, calcilutites, biostromes, and bioherms. The quartzose sandstones may vary in composition and texture from protoquartzites to orthoquartzites or friable quartz sandstones and generally consist of well-sorted and rounded grains. The beds commonly show cross bedding of both the high-angle planar type formed by fairly strong currents, such as those of tidal channels, and the low-angle sweeping type formed on the shifting and undulating surfaces of offshore sand bodies, including seaward extension of beaches. The sandstones are generally whitish or pale in color, the shales being gray to greenish-gray and calcareous.

Sections comprising carbonates, quartzose sandstones, and shales commonly range in thickness up to 1000 m or more. They differ from Bahama-type sections in two main respects. First, the sections consisting of carbonate beds deposited on extensive shallow banks cut off from the mainland, such as those of the Bahamas, may be very much thicker than those consisting of beds deposited on a shallow continental shelf. The Bahamas section, for example, is more than ten thousand meters thick. Second, the Bahama-type sections are devoid of sandstone as the banks on which they developed were obviously inaccessible to sand derived from a landmass. Such isolated banks may, in certain geographic and climatic situations, receive windblown silt which becomes dispersed in layers of sediments to form silty carbonates. Floating sand grains of frosted quartz dispersed in a carbonate bed are likely derived as windblown sand from an adjacent desert contiguous with the carbonate bank. In these respects the carbonate–quartz sand facies differs from the carbonate facies developed in isolation from a landmass. The latter has already been discussed in a previous section of this chapter on carbonate banks and reefs.

The carbonate–quartz sand facies is essentially a shallow and warm-water marine facies developed in the inner neritic environment of a carbonate bank adjacent to a coastline fringed by beaches and offshore bodies of quartzose sand. These sand bodies periodically migrate as bars or sand banks over the areas of carbonate deposition, forming lenticular beds of cross-bedded sand which eventually become buried by carbonates. Some sand may become dispersed in the sediment to form sandy carbonate beds. Other sand may be blown in from the land where sand dunes and desert areas are situated on the adjacent landmass.

Studies of the regional variations of lithofacies within a sedimentary basin are essential to understanding the history and paleogeographic setting of the basin. An example is given by Wilson (1963) with reference to the carbonate–quartz sand facies of a sedimentary basin in the Andean region of central Peru, illustrated in Fig. 4-61. This figure shows the lithofacies distribution of the Early Cretaceous Goyllarisquisga Formation. The western area of the region underlain by the formation is characterized by a carbonate–quartz sand facies, the adjacent eastern area by a quartz sand facies, and the southern area adjoining the eastern area by a quartzose to lithic sand facies deposited closer to the contemporaneous landmass. The section including the carbonate–quartz sand facies is up to 3000 m thick and is believed to have formed on a sinking shelf that flanked a tectonic trough lying west of the present Andes mountains. The clastic

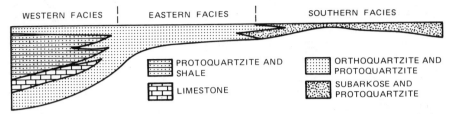

Fig. 4-61. Generalized section showing the distribution of lithofacies within the Early Cretaceous Goyllarisquisga Formation of the Andean region, central Peru. [After Wilson (1963).]

material came from the east and southeast, the latter being the direction in which a landmass is thought to have been situated.

Euxinic Facies

The euxinic facies consists predominantly of fissile black shale interbedded with laminae of dark gray silty shale and minor gray siltstone. Beds of dark gray siliceous limestone, chert, and shale containing an abundance of siderite, pyrite, or phosphatic minerals may also be present. Fossils of benthonic fauna are commonly absent, with the possible exception of phosphatic brachiopods; but organic matter that fell into the sediments, such as fish scales and shark teeth, are well preserved. Chert laminae may also contain abundant, though poorly preserved, remains of radiolarians. Total thickness of sections comprising beds of the euxinic facies is commonly less than 100 m.

The euxinic environment requires a lack of oxygen and consequent reducing conditions that permit the preservation of organic matter, the precipitation of ferrous carbonate and sulfide, and the inhibition of many forms of bottom-dwelling life. Such conditions exist in some bodies of water having a restricted circulation such as closed lagoons fringed with extensive mangrove swamps. The muds in these lagoons are black and fetid. Similar reducing conditions exist at depth in parts of large isolated bodies of water such as the Black Sea. Under such conditions the rates of sedimentation are very slow, and the euxinic facies has been regarded by some geologists as evidence of a starved sedimentary basin. This inference must be regarded with some reservation as slow sedimentation and reducing conditions may exist locally in a depositional basin whereas elsewhere a considerable thickness of deltaic or shelf sediments may be accumulating. Where preserved within a sedimentary basin, the thin section deposited as a deep-water euxinic facies becomes the time-stratigraphic equivalent of a much thicker section of shallow-water sediments. Sediments deposited in a euxinic environment may also be interbedded with turbidity-type sediments which became mobilized by slumping and rolling in a mass of slurry down the lower slopes of a depositional basin.

Many questions remain regarding the depositional environments of sediments referred to the euxinic facies and also of sediments containing a fauna preserved in sea-floor muds that appear to have suddenly become poisonous. An example of the latter is seen in the late Early Cambrian Stephen Shale near Field, British Columbia, where C. D. Walcott of the Smithsonian Institute discovered a remarkably well-preserved fauna, including holothurians, in which the visceral outlines are clearly seen. Scavengers do not appear to have damaged the greater part of the fauna which also contains crustaceans, and it may be concluded that some sudden event, possibly an influx of hydrogen sulfide, may have locally killed most life on the sea floor. The question remains a mystery.

With reference to the presence of hydrogen sulfide in deep areas of the ocean, it is of interest to note that, at a depth of 2.5 km in the Galapagos Rift west of Ecuador, Corliss and Ballard (1977) record oases of prolific fauna surrounding submarine springs emitting hot, hydrogen sulfide-bearing water. The hydrogen sulfide is apparently metabolized by certain anaerobic bacteria which form the bottom ring of a chain of life. This situation is biologically anomalous in that abundant life exists in what would otherwise be a barren environment.

Coal Facies

The coal facies consists mainly of interbedded lithic sandstone, siltstone, shale, and coal. Sandstones may locally be conglomeratic but commonly are fine-grained and cross-bedded; the shales may be varicolored, buff to brown, or dark-gray. All the clastic beds associated with coal may be carbonaceous, and the coal seams can range in grade from lignite, through subbituminous to bituminous in quality. Particular varieties such as cannel coal are derived from the accumulation of waxy and resinous organic matter such as pollen and spores. Anthracite coal occurs only where the coal-bearing beds have been subjected to low grade metamorphism, commonly during a phase of folding that accompanies orogenesis. The total thickness of a section of beds comprising the coal facies is as variable as the thickness of the coal beds themselves. The coal measures of the Eocene Latrobe Formation in the Gippsland Basin of Victoria, Australia, contains beds of soft, brown subbituminous coal up to

160 m thick (Brown *et al.*, 1968). The same formation extends southward under the sea where at a distance of 50 km it includes oil- and gas-bearing beds interfingering with coal seams up to 5 m thick.

Some of the sandstone units interbedded with the thin coal seams off shore contain a marine fauna, indicating a coastal marsh environment for the accumulation of plant remains from which the coal was derived. The thick coal seams on shore were obviously not formed from vegetation that accumulated in a paralic environment but rather from vegetation that grew in vast inland swamps that may have been the backswamp region of a river system.

The depositional environments in which vegetation accumulated to ultimately form beds of coal can be duplicated in modern geographic settings, although possibly not on as grand a scale as in the past. Regions such as the delta of the Amazon River and the Everglades of Florida may approximate the environments in which thick coal seams containing fossil roots and trunks originated. Thinner coal beds associated with lacustrine and floodplain deposits may have formed in situ from the accumulation of peat in extensive areas such as found in northern North America and Europe. They may also have formed from the accumulation of macerated vegetation washed into bodies of water.

Rates of accumulation of vegetation depend directly on local rates of growth, surface water movements that may deposit or remove dead vegetation, and conditions of oxidation within the water or mud in which the vegetation remains are buried. These depend on climatic conditions, geomorphic features, and paleographic settings. The accumulated vegetation is converted to peat or similar matter under conditions of reduction and in part by anaerobic bacteria. Tollund Man, discovered in 1950 near Silkeborg, Denmark, was buried at a depth of 3 m in peat. Carbon dating of body tissue and knowledge of the periods of cultivation of types of grain found in his stomach indicate the age of the corpse to be 1900 years. This suggests that the peat has accumulated at an average rate of approximately 1.5 m in 1000 years. Subsequent metamorphism of peat to coal requires physicochemical changes brought about by heat and pressure during burial of the strata. The ratio of peat to coal involved in this conversion is difficult to quantify but may be in the order of ten meters of peat to form one meter of subbituminous coal. Given the amount of vegetation and the period of time required to form one meter of peat, as indicated by the presence of medieval artifacts in the peat deposits of Europe, it is evident that many coal seams were formed from accumulations of vegetation in extensive swamps or marshes that existed for a period of many tens of thousands of years and that some must have existed for periods of several hundred thousand years.

In the context of the earth's history of sedimentation, and the development of plant life, it is significant that whereas coal beds are abundant in strata of all ages since the Mississippian, none occur within strata of the Devonian or older periods.

Criteria for Recognition of Depositional Environments

The following individual criteria and characteristics are not necessarily definitive with respect to the interpretation of depositional environments; but considered in the context of their stratigraphic sequence and lithologic relationships, they serve as a guide to interpretations. The outline consists of a summary of some of the main features that characterize depositional environments.

Continental

1. Limestone. Commonly absent but may include chalky or oolitic carbonates and accumulations of the remains of ostracods, freshwater bivalves, gastropods, and calcareous algae deposited in a lacustrine environment.
2. Sandstone. Commonly abundant, conglomeratic, cross-bedded, lithic, coarse- to fine-grained with upward fining of grains, deposited in fluvial to lacustrine environments.
3. Shale and Claystone. Commonly abundant, varicolored, including red beds; may be carbonaceous, interbedded with sandstone and locally with coal beds deposited in floodplain, backswamp, and lacustrine environments.
4. Fossils. Scarce but may be locally abundant, including freshwater bivalves, crustaceans, gastropods, plant remains, including pollen and spores, and vertebrates.

Paralic

1. Limestone. Commonly absent.
2. Sandstone. Commonly abundant, quartzose, well sorted, medium- to fine-grained with upward coarsening of grains, fossiliferous, deposited as offshore sand bodies, including bars and barrier islands (indicated by isopach maps showing net sandstone linear trends parallel to mean direction of coastline).
3. Shale and Siltstone. Not abundant, commonly dark-colored and carbonaceous, fossiliferous, interbedded with sandstone, deposited in bays and lagoons.
4. Fossils. Commonly abundant, including forams in the lutites.

Deltaic

1. Limestone. Absent except for coquina beds such as those formed as oyster banks.
2. Sandstone. Fairly abundant, locally cross-bedded, medium- to fine-grained, lithic to quartzose, interbedded with siltstone, claystone, and shale; forms lenticular and linear bodies (indicated by isopach maps showing net sandstone trends normal to mean direction of coastline) deposited in river distributary channels.
3. Shale. Commonly abundant, silty, gray to brown, kaolinitic; may be carbonaceous with interbeds of coal, deposited on floodplains, in oxbow lakes, and in backswamps.
4. Fossils. Locally abundant, including fresh- to brackish-water bivalves, gastropods, crustaceans, vertebrates, and plant remains; with abundant forams, pollen, and spores in the lutites.

Tidal Flat

1. Limestone only. Calcilutite predominant, dolomitic, locally anhydritic; stromatolites and mud cracks common, rare fossils, deposited in a sabkha environment.
2. Silty shale only. Silty shale, varicolored, thin-bedded to laminated with wavy layers formed by algal mats, abundant fossils, local mud cracks formed on upper levels of mud flat.
3. Fossils. Commonly abundant in noncarbonate tidal flat deposits, particularly heavy-shelled bivalves, and gastropods.

Prodelta

1. Limestone. Absent.
2. Sandstone. Uncommon but may be present as laminae or thin layers, fine- to very fine-grained, ripple marks, deposited as sand sheets washed seaward by currents.
3. Shale and Siltstone. Laminated shale and siltstone predominant, dark gray, micaceous, commonly calcareous, marked bioturbation, very fossiliferous, deposited seaward from a delta in shallow water commonly less than 15 m deep.
4. Fossils. Benthonic and pelagic macrofossils and microfossils locally abundant, including bivalves, gastropods, sea urchins, sponge spicules, shark teeth, fish scales, cephalopods, and abundant trace fossils on the bedding surfaces.

Neritic

1. Limestone. Predominantly calcarenite and calcilutite but including calcirudite and bioherms, deposited on carbonate banks in depths of less

than 10 m, with the exception of chalky limestone, composed mainly of coccoliths and forams that may have been deposited in deeper water of the outer neritic environment.

2. Sandstone. Commonly more abundant in the inner than in the outer neritic environment, mainly quartzose; may be glauconitic, medium- to fine-grained, well sorted, subrounded grains, commonly with calcareous matrix, fossiliferous, deposited as elongate sand sheets commonly trending along the coastline (as indicated by smooth, parallel contours of a net–sand isopach map).

3. Shale. Commonly more abundant in the outer than in the inner neritic environment, mainly dark gray to dark greenish gray, less well bedded or laminated than prodelta shale, commonly calcareous; may have abundant illite, fossiliferous with abundant pelagic debris, deposited relatively slowly by the settling of clay carried far out to sea by fans of muddy water debouching from rivers.

4. Fossils. Pelagic and benthonic forms abundant in limestone; may be abundant in shale; pollen and spores present in shale but commonly not abundant.

Bathyal

1. Limestone. Uncommon, may be interbedded with shale, fine-grained, compact to chalky; contains abundant coccoliths, forams, sea urchin and sponge spicules, macerated organic debris; may have abundant pelagic macrofauna, rare sedimentary breccia, including edgewise conglomerate, deposited on gentle continental slopes and on submarine plateaus by accumulation of pelagic debris.

2. Sandstone. Commonly more lithic than quartzose, gray, poorly sorted, thin-bedded with cross bedding and current ripple marks; some beds show graded bedding, sparcely fossiliferous, interbedded with dark gray fossiliferous shales, deposited on continental slopes, in submarine valleys, and on submarine fans.

3. Shale. Commonly the predominant rock type, dark gray, silty; may be micaceous, calcareous, or carbonaceous; commonly fossiliferous with pelagic fauna more abundant than benthonic, thin-bedded to laminated, interbedded with thin beds of sandstone, deposited on continental slopes, in submarine valleys, and on submarine fans and plateaus.

4. Fossils. Pelagic fauna may be more abundant than benthonic, including shark teeth, fish scales, cephalopods, and numerous planktonic forms.

Abyssal

1. Limestone. Absent.

2. Sandstone. Lithic, gray, medium- to very fine-grained, poorly

sorted, thin-bedded with graded bedding, nonfossiliferous to sparcely fossiliferous, interbedded with thin-bedded shale and siltstone containing sparce pelagic fauna and trace fossils; current grooves and striations on surfaces underlying individual sandstone beds may be interbedded with pyroclastics or basic volcanic flows, deposited as turbidites on the bottom slopes of continental rises and on the floors of abyssal plains flanking the rises; sand absent on abyssal plains remote from continental rises.

3. Shale and Claystone. Dark gray, reddish, laminated, argillaceous and commonly nonsilty; may be siliceous with chert bands; may be interbedded with thin beds of volcanic ash, commonly unfossiliferous to sparcely fossiliferous with the exception of trace fossils, radiolarians, and shark teeth, deposited very slowly by the rain of clay particles, dust, and organic debris in very deep parts of the ocean.

4. Fossils. With the exception of radiolarians and shark teeth, few are preserved; but trails and other imprints made by holothurians, worms, gastropods, crustaceans, and fish suggest that life may locally have been abundant.

Analysis of Outcrops, Drill Core, and Cuttings

MACROSTRATIGRAPHY

Sedimentary basin analysis depends primarily on the physical examination of exposed rock sections within the basin. This approach entails geological mapping and the analysis of rock samples. Initial determinations of lithology are essentially physical and qualitative; subsequent descriptions may be modified by quantitative parameters. These modifications, based on microscopic or chemical examination, may necessitate a review of the original field data or spot checks in the field. The analysis of a sedimentary basin depends secondarily on the availability of subsurface data based on geophysical surveys or drilling. Stratigraphic correlations based on such data in turn depend largely on the distribution of control points, the quality of data, and the accuracy of interpretations placed on the data.

These remarks apply equally to correlations based on the mapping of rock outcrops. For example, the distribution of outcrops between two sections being compared may be nearly continuous or scattered. They may also consist of various parts of the sections being compared. Field notes recording observations of these outcrops by different geologists over periods of years are likely to show variations in length, style, emphasis, and to some extent in lithologic descriptions. This latter point is of particular significance to the accuracy of subsequent interpretations of correlation, stratigraphic succession, or depositional environment. Hurried examination of outcrops toward the end of the day when there is still far to go on the traverse is not conducive to accuracy of observation and may result in tedious reexaminations to check questionable evidence or in unfortunate gaps of information where it is no longer practicable to revisit the area. In this regard it is essential that the field geologist not only qualifies and quantifies the rock types he maps but that he calibrates his standards with those of other geologists working in the same or adjacent area. In practice this procedure is usually followed, but its importance cannot be too strongly stressed. Differences of opinion with respect to lithology and nomenclature can be sorted out by discussion based on more leisurely and detailed examination and testing of rock specimens.

Surface Mapping

There are several good textbooks on geological survey methods and procedures, including the interpretation of air photos and remote sensing data. The purpose of this chapter is not to infringe on these matters but to outline a philosophy of approach to the problem of sedimentary basin analysis. Determination of the lithology and stratigraphic succession of exposed rock layers within a sedimentary basin is essential to analysis and is normally preliminary to further subsurface investigations. Geological mapping of the surface may be a comparatively easy operation, or extremely difficult, depending on the nature and accessibility of the terrain and on the structural complexity of strata within the basin. The latter condition is illustrated by Harris (1976) in Fig. 5-1 which shows the sequential development of a complex structure of folds and thrust faults within a stratigraphic sequence of Paleozoic rocks overlying basement in the Valley and Ridge province of Tennessee. The nature of the structural complexity of the strata and the structural framework of the basin in which the strata are contained must be broadly understood as a prerequisite to the analysis of any sedimentary basin. These features must be viewed in their regional contexts. Essential elements of the structural and stratigraphic frameworks may be highlighted on separate maps to illustrate physical and historical features of significance. The main point of sedimentary basin analysis is to understand and illustrate the development of a basin, usually with particular reference to particular geological or economic aspects; and in so doing, the study is open to any imaginative innovation or approach that clarifies the presentation.

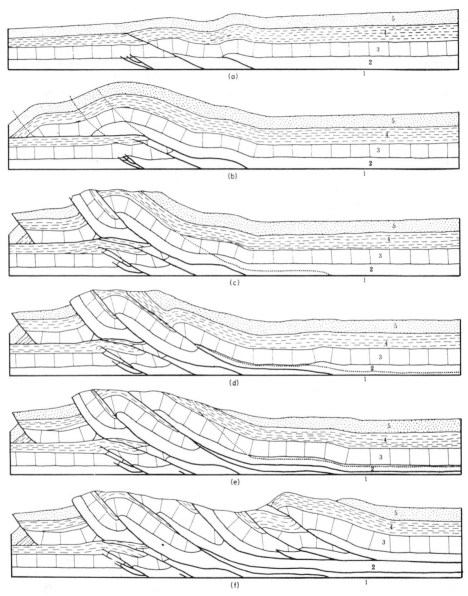

Fig. 5-1. Sequential development (A–F) of a complex structure within a Cambrian (2) to Pennsylvanian (5) sequence overlying basement (1) in the Valley and Ridge province of Tennessee, United States. [After Harris (1976).]

The scale in which a map is presented varies according to the view and in this sense can be regarded as analogous to a zoom lens that shows a range of views from panoramic to restricted. On the broadest possible scale, the worldwide continental distribution of sedimentary basins with respect to the areas of exposed Precambrian rocks is illustrated by White (1972) in Fig. 5-2. This type of overview may be helpful in reconstructing the paleogeographic setting of a basin with respect to an adjacent Precambrian shield or to other nearby basins. A much closer but nevertheless very broad view of certain geological features within a particular continental area may be obtained by selectively highlighting areas in which the exposed rocks have some common tectonic or sedimentologic significance. Such a map is illustrated by Fig. 5-3 which shows the distribution of areas of exposed granitic rocks in California and Nevada, as described by Crowder *et al.* (1973). The information supplied by such a map is of particular significance in evaluating possible relationships between the structural and stratigraphic frameworks of adjacent or contiguous basins.

Focusing on a position of the area shown in Fig. 5-3, and altering the definition of the map, Fig. 5-4 illustrates the distribution of granitic and associated metamorphic crystalline rocks, ultramafic rocks, and the Jurassic Franciscan Formation mapped by Burch (1968). The Franciscan comprises well over 3500 m of shale, radiolarian cherts, siltstone, and sandstone. The map shows a structural wedge, bounded by the San Andreas Fault and the Nacimiento Fault, that includes only the granitic and associated rocks. On either side of this wedge are areas characterized by exposures of Franciscan Formation into which bodies of ultramafic rock

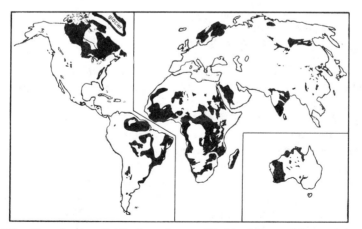

Fig. 5-2. Map showing distribution of areas (black) of exposed Precambrian rocks throughout the world. [After White (1972).]

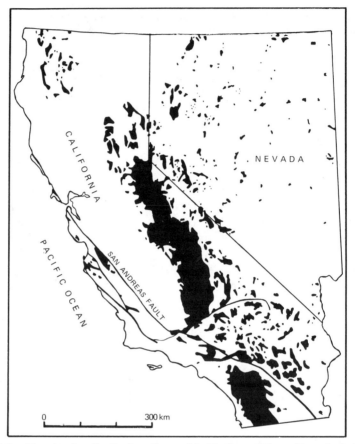

Fig. 5-3.   Map showing distribution of areas (black) of exposed granitic rocks in California and Nevada. [After Crowder *et al.* (1973).]

have been emplaced. This simplified picture, as shown in Fig. 5-4, serves as an aid to outline the main sequence of tectonic and stratigraphic events in the geologic history of the coastal region to the north and south of San Francisco. In a similar way, the distribution of areas of Precambrian rocks and of sedimentary basins in Wyoming, as illustrated in Fig. 5-5, serves to assist in understanding the main structural and stratigraphic relationships of basement rock and adjacent contiguous basins. The lower stratigraphic sequences between such basins may be separated by basement rock that is not exposed and may or may not be correlative, whereas the uppermost sequence may be continuous from one basin to another.

Sorting out these basic relationships is essential to working out the geologic history of adjacent and nearby basins. Interpretation of the

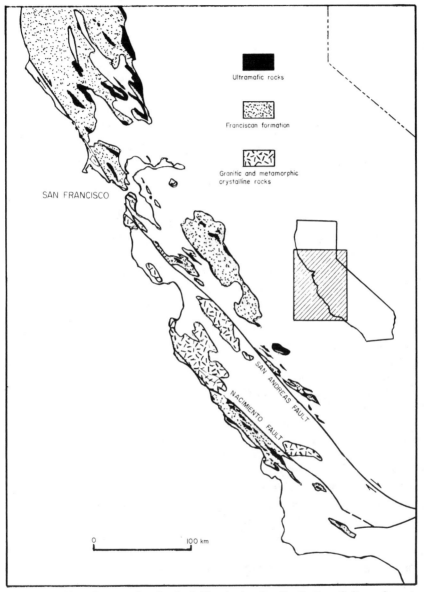

Fig. 5-4.   Map of a coastal region in California showing distribution of ultramafic rocks, granitic rocks, and the Franciscan Formation which represent separate geological events. [After Burch (1968).]

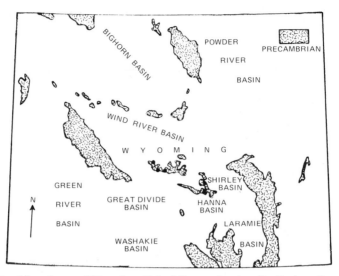

Fig. 5-5.   Map of part of Wyoming, United States, showing distribution of areas of exposed Precambrian rocks and of sedimentary basins. [After Rosholt and Bartel (1969).]

paleogeography and distribution of exposed basement rocks during various periods of time in the history of such sedimentary basins may also lead to hypothesis concerning the origin and emplacent of sedimentary ore deposits, such as those of uranium which may have been derived by drainage from areas of weathered basement rock (Rosholt and Bartel, 1969).

In the above discussion reference is made to areas of exposed rocks. Depending on the scale used, but obligatory where regional relationships are shown, a degree of latitude is allowed in extrapolating one outcrop to another and consequently in showing continuity of rock types exposed or buried at shallow depths. The latitude used depends largely on the experience of the geologist and his knowledge of the geology of the area mapped. Where outcrops are scarce with respect to the size of the area, a map based solely on rock exposures may not be meaningful unless a considerable degree of extrapolation is introduced. Where outcrops are numerous, it may be a waste of time to plot the distribution of individual exposures. For example, Fig. 5-6 shows the distribution of outcrops of bedrock in an area of southeastern Sweden, as plotted by Agrell (1974). The distribution shown has the advantage of indicating a fabric and structural pattern within the basement rock, but on a regional map the area could be shown as one in which the basement rock is wholly exposed.

Apart from the selective mapping of main lithologic, stratigraphic, and structural features, each of which may highlight a particular event in the

Fig. 5-6. Map of the Sommen–Åsunden region, southeastern Sweden, showing distribution of patches of exposed basement rock. [After Agrell (1974).]

geologic history of a sedimentary basin, the mapping of minor features may be of particular significance. Such minor features may suggest or indicate a depositional environment or physical event during a period of time within the geologic history of a sedimentary basin. To what extent the evidence may be used and to what limits of interpretation a hypothesis may be stretched are open to question. As a rule, an interpretation may be accepted as a possibility provided it takes into consideration all available information and does not invalidate the evidence. A case in point may be cited with reference to Dott (1974) who suggests that boulders of quartzite in Cambrian strata of Wisconsin, United States, may have been subjected to storm waves in a tropical latitude. The evidence presented suggests that Dott's interpretation may be right, although others may argue differently. More tangible evidence, with respect to Ordovician glaciation in North Africa, is presented by Spjeldnoes (1973) in Fig. 5-7 which illustrates the directions of movement of ice, based on glacial striations. Both interpretations are valid, but the degree of certainty of the latter exceeds that of the former. In this respect it must be kept in mind that the acceptability of particular criteria may vary with individuals and with time and that interpretations placed on sedimentary basin analysis should be based on the cumulative results derived from a diversity of approaches. Individual analyses of particular stratigraphic, lithologic, or structural elements of a basin may be definitive or otherwise, and where they bear each other out, their validity is enhanced provided they do not depend on a common set of premises.

Lithology and Quantification of Parameters

Determination of the gross lithology of an outcropping stratigraphic sequence requires examination of rock samples by physical and optical methods and estimation of the qualitative and quantitative relationships of

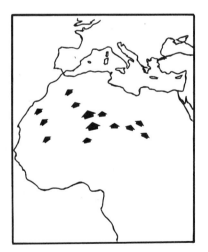

Fig. 5-7. Map of North Africa showing the directions of movement at various localities of Ordovician glaciers. [After Spjeldnoes (1973).]

the interbedded strata. This may be a fairly simple procedure in the case of a sequence comprising massive beds of sandstone or limestone. It may also be difficult to generalize, and description may require a considerable degree of judgment and consistency. This last point is of particular importance in that field notes are a significant factor in working out the correlation of stratigraphic sequences at various localities. Consistency is required not only in the field notes of each geologist but also in the field notes of different geologists working in the same area.

Where individuals disagree on matters of stratigraphic nomenclature, it is essential that they understand the nature of the sequence described in order to determine the validity of possible correlations, regardless of differences of opinion with respect to lithology. Many such differences may be resolved by physical tests, or by optical examination, either in the field or laboratory. For example, a mudstone may be calcareous or noncalcareous, a property easily demonstrated in the field by a simple test of the powdered rock with dilute HCl. In the case of a carbonate the vigor with which the powder reacts with the acid is a rough estimate of the ratio of calcite to dolomite and, consequently, an indication of whether to call the rock a limestone, dolomitic limestone, or a dolostone. Quantitative determinations of clay and other noncarbonate constituents require laboratory procedure, the carbonate rock sample being crushed and weighed, dissolved in acid, and the residue washed, dried, and weighed. This procedure may be of value in detailed studies of stratigraphic sequences including beds ranging in lithology from calcareous mudstones to argillaceous limestones. The silt content of a carbonate rock may also be quickly determined by dissolving a few grains of the rock in acid placed in a watch glass and stirring the residue with the rounded end of a glass rod. The scratchy feel of silty grit is readily detected. These simple field or field–laboratory tests are useful in quantifying the descriptions of stratigraphic sequences mapped in the field. They can be further added to by determinations of other parameters, using methods employing more sophisticated equipment in established laboratories.

Quantification of other lithologic parameters such as grade and texture is also essential, not only from a descriptive point of view but also as an aid to the evaluation of potential economic characteristics. This latter factor has been discussed by Robinson (1966) with reference to the classification of potential oil and gas reservoir rocks by their texture in fresh rock samples obtained from outcrops. Robinson based his conclusions on the examination of approximately 2000 samples and said (p. 547):

The rocks were first examined in polished section with the binocular microscope and classified according to surface texture. They then were analysed for reservoir properties of porosity, permeability, and pore-size distribution. An empirical correlation was found between these rock textures and reservoir properties.

This approach has also been studied by Beard and Weyl (1973) who examined the influence of texture on porosity and permeability of unconsolidated sand. Although this approach disregards the commonly significant effect on these parameters of intergranular matrix and cementation in sandstones, the classifications employed are useful in quantifying the nature of arenites mapped in outcrops. Beard and Weyl present a visual comparator, illustrated in Fig. 5-8 and based on studies by Rittenhouse (1943), for estimating the two-dimensional sphericity of sand grains. They also present charts, based on ranges of grade and sorting, of average porosity for various mixtures. A useful comparator of this sort can be devised by fixing measured quantities of sand sizes into a graduated sequence of square cubicles, as shown in Fig. 5-9. Such a device can be most useful in quantifying the texture of a sandstone specimen. Another

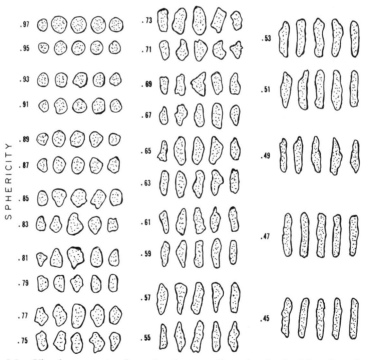

Fig. 5-8. Visual comparator for estimating two-dimensional sphericity of sand grains. [After Beard and Weyl (1973) and Rittenhouse (1943).]

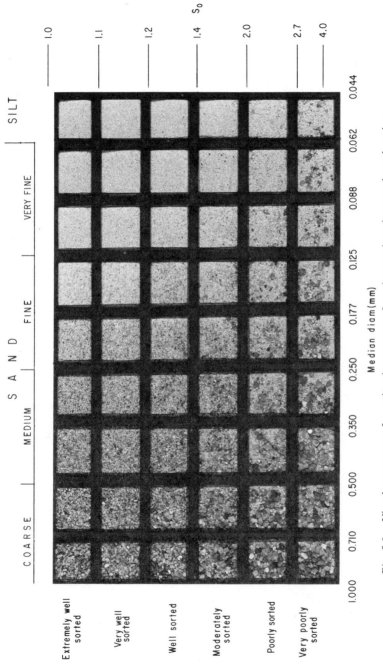

Fig. 5-9. Visual comparator for estimating texture of sandstones based on grade and sorting.

indispensable tool for calibrating the grade of sandstones is a pocket sand gauge in which various standard samples of grade and sorting have been classified. This provides a ready reference for rock chips which can be matched with the samples.

The points discussed above may appear to be self-evident, yet they are frequently ignored by field workers who do not adequately qualify their descriptions of the rocks mapped and so make difficult the later problems of interpretation and correlation carried out by themselves and others. The scale of mapping and the constraints of time are limiting factors in the degree of detail recorded, but every effort should be made to record significant details. For example, where cross bedding is recorded, there is commonly no mention of type or scale. Yet these dimensions may be definitive with reference to the depositional environment of sandstone. A limestone may be recorded without reference to whether it is a calcilutite or calcarenite, yet these distinctions can usually be made with a hand lens.

One aspect of the qualification of outcrops that is subject to personal view is the description of color. Distinct shades of red are commonly noted by all field workers, yet shades of gray and brown are variously described. Also it is not always clear from field notes whether these hues represent the fresh rock or the weathered, the overall appearance of the outcrop as viewed some way off, or the color of a hand specimen.

Such details may be of help in correlating surface sections with the subsurface. It must be kept in mind that whereas the shade and intensity of color shown by outcropping strata viewed from a distance may vary to the individual with time of day and conditions of weather, the color of a hand specimen of rock remains unchanged. Accurate description of the color requires comparison with a color chart designed for this purpose. The use of such charts is not always practicable in the field, but they should be available in the base camp for frequent reference in order to calibrate the descriptions written in the field from day to day. A further condition to be observed is that dry samples be described, as wet samples appear much darker. Consistency in this respect is essential.

Where it is not possible to sample a section of strata, or where only a part can be sampled, the color of the outcropping rock as a whole is recorded with an appropriate notation as to distance and light conditions. Such observations should always be made regardless of the accessibility of the section and availability of hand specimens. Seen from a distance, many stratigraphic relationships such as the nature of interbedding, cyclic sedimentation, lenticularity, and local unconformities are evident. Viewed from close at hand, they may not be readily apparent or may be taken out of context with the total stratigraphic sequence. Color variations seen only from a distance may mark such features.

A case in point, certainly one of the best examples to be found, is a view of the Grand Canyon of the Colorado River where some 1500 m of Paleozoic and Mesozoic strata are laid bare to the Precambrian basement. The color variations, as well as the physical appearance of the individual beds, accentuate the stratigraphic relationships of the entire sequence. In other parts of the world, including the Rocky Mountains of Canada and areas of Iran and Iraq, sequences of limestone, shale, and sandstone are magnificently exposed in sections many hundreds of meters thick. Where these sections are viewed from a distance, the variations in color or intensity accentuate the stratigraphic features.

Sedimentary and Biogenic Structures

Observation and recording of the nature and scale of sedimentary structures produced by physical processes, and of biogenic structures produced by organisms, are essential to interpretation of the depositional environment of strata exposed in outcrops or observed in cores from the subsurface. A number of books refer to the interpretation of such structures, including Reineck and Singh (1973), Pettijohn et al. (1972), Selley (1970), Allen (1970), Conybeare and Crook (1968), and Potter and Pettijohn (1963). It is not intended to reiterate any part of these books but to emphasize the importance of observation of sedimentary features with reference to the interpretation of outcrops and cores as a facet of sedimentary basin analysis. With this object in mind, a few types of sedimentary structures and plots of field data are presented as examples of the application of these structures to interpretation of the regional depositional history and paleogeographic setting of particular sedimentary units. For broader coverage and detailed treatment the reader is referred to numerous papers comprising the literature on this subject.

Interpretations based on sedimentary structures may be invalidated if they are taken out of context with the sequence of structures in which they occur and without reference to the lithologic and paleontological features of the stratigraphic section. Individually a sedimentary structure may not be meaningful, or its significance may be equivocal, but collectively such structures may be important indications of depositional environments and the physical processes responsible for sedimentation. Broadly speaking, questions placed on the interpretation of sedimentary structures hinge on whether the sediments are marine or nonmarine. If they are marine, it must be asked whether the sediments were laid down as shoreline deposits or in deeper water; if they are nonmarine, it must be decided whether the sediments are of fluvial or lacustrine origin. This statement oversimplifies the issue as some sedimentary structures occur in a variety of depositional environments. Consequently, not only must the field descriptions of such structures be qualitatively complete and accurate with respect to their features and sequence, but they must also be quantitative insofar as possible.

A common deficiency in field notes describing outcrops of sandstone is the brief notation that strata exhibit cross bedding, without reference to the type, angle of foreset beds, approximate direction of slope of foreset beds, amplitude of individual sets, or two-dimensional and bedding plane views where these can be seen. This last point is emphasized by McDaniel (1968) and illustrated by views C, E, and F in Fig. 5-10. Similarly, notations recording the appearance but not the type or orientation of ripple marks on bedding planes fail to adequately record the information available. Quantitative studies of directional features require a statistical approach which may be precluded by the constraints of time and area to be mapped; but even a few scattered observations may be sufficiently consistent to be of help in interpreting the regional or local direction of sediment transport.

Individually or collectively, sedimentary structures may indicate a particular depositional environment, but it must be kept in mind that they do so indirectly. Physical structures such as cross bedding and groove casts are produced by currents. The key question is the geographic and depth

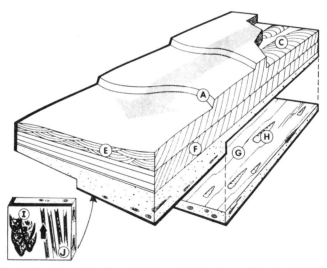

Fig. 5-10. Sedimentary directional features. A, foreslope of ripple; C, rib and furrow structure; E, festoon cross bedding; F, planar cross bedding; G, parting lineation; H, fossil lineation; I, flute casts; J, striations and groove casts. [After McDaniel (1968).]

limitations imposed on the hydrodynamic characteristics of such currents. Turbidity currents are known to commonly occur within the depth range encompassed by the continental shelf, rise, and abyssal plains. But this distribution does not necessarily preclude the possibility that turbidity currents may also occur at much shallower depths in large freshwater lakes. Conversely, fluvial-type cross bedding typical of river deposits can form in a wide range of water depths depending on the strength of current and availability of sand.

Sedimentary directional features are commonly measured without specific reference to the possible paleogeographic setting and depositional environment in which they were formed. Later evaluation and interpretation of the data may result in a reconstruction of the distribution of ancient landmasses, drainage systems, shoreline trends, continental shelves, and basins of deep-water sedimentation. Initially the study depends on the thoroughness and accuracy with which the sedimentary structures in outcrops are observed and recorded.

Fig. 5-10 illustrates some of the physical sedimentary structures commonly recorded. Biogenic sedimentary structures such as formed by crustaceans, molluscs, worms, and trace fossils have been variously interpreted with reference to ecological factors and depositional environments. They are not specifically discussed in this book as they are essentially biostratigraphic rather than lithostratigraphic and as such fall within a very comprehensive and specialized field. The origin of many trace fossils and the depth at which they were formed remain problematical. Some tubular structures, such as *Scolithus,* are similar to those formed by reef-forming marine worms living today in intertidal zones. Reference to these and other biogenic structures can be made to Seilacher (1964, 1967). With reference to common sedimentary directional features, Fig. 5-10 illustrates structures seen both on bedding planes and in sectional views of the strata. The former include ripple mark, rib, and furrow structures, flute casts, striations, and groove casts. The latter are various types of cross bedding depending on the size, relationship, and orientation of superimposed sets of ripples. Other directional features seen on bedding planes include the orientation of linear fossils parallel to the direction of flow of the current and parting lineation which reflects the depositional current direction and is probably related to the orientation of the long axes of sand grains parallel to the direction of flow.

Plots of such directional data are commonly superimposed on various types of geological maps. A common practice is to separate the data on cross bedding from those on striations, flutes, and grooves for the following reasons. The former are based principally on sectional views of the strata, whereas the later are seen on the bedding surfaces; also the former are predominantly formed in shallow water environments of the molasse facies, whereas the latter are characteristic of deep-water environments of the flysch facies. Studies of these directional features may be local or regional, depending on the availability of outcrops and sedimentary struc- tures and on the detail required. Fig. 5-11 illustrates a local study of the Pennsylvanian Smithwick Shale of central Texas. This stratigraphic unit comprises approximately 120 m of dark gray marine claystone grading upward into interbedded sandstone and claystone. The sequence is de- scribed as flyschlike by McBride and Kimberly (1963) who considered the lineaments, chiefly groove and flute casts, to have been formed by strong marine currents flowing southward down a paleoslope away from a region of uplift. Fig. 5-12 illustrates similar studies on a regional scale, one car- ried out in Arkansas, United States, and the other in the East Alps of Bavaria. Both extended over a distance of 100 km along the trend of outcrops. The study in Arkansas was made by Morris (1971) who plotted the mean directions of orientation of linear wood fragments, grooves, and

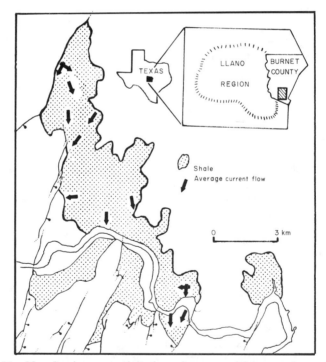

Fig. 5-11.   Map showing mean orientations of groove and flute casts (arrows) indicating directions of average current flow during deposition of flyschlike Pennsylvanian Smithwick Shale in central Texas. [After McBride and Kimberly (1963).]

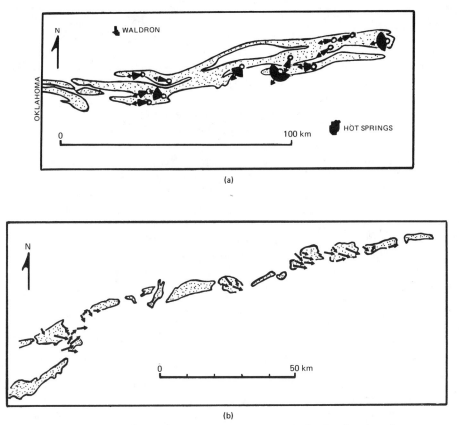

(a)

(b)

Fig. 5-12. (a) Map showing mean orientations of features indicating direction of transport of sediments in the Pennsylvanian Jackfork Group, Arkansas, United States. [After Morris (1971).] (b) Map showing mean directions of sediment transport in turbidites of the Early Cretaceous Gault Formation, East Alps, Bavaria. [After Hesse (1974).]

flute casts in shales and lithic arenites comprising flysch deposits of the Pennsylvanian Jackfork Group. Orientations indicated a regional westward to southwestward direction of sediment transport. The Alpine study, carried out by Hesse (1974), shows the mean orientations of sole marks in turbidites of the Early Cretaceous Gault Formation. These indicate a general easterly to southeasterly direction of sediment transport. Many other examples may be cited of directional features in flysch-type sequences, and of these a study by Scott (1966) is of particular interest in that it may have a bearing on the tectonism relating to crustal plate movements in the region of Tierra del Fuego in South America. The study showed that within flysch deposits of the Late Cretaceous Cerro Toro

Formation of Chile lineaments measured over a distance of 100 km indicate a predominant current direction from north to south.

The examples mentioned above are of linear features observed on bedding planes. Measurements of the directions of dip of sets of cross bedding are various because the foreset slopes are seldom exposed. The mean direction within the range of the arc measured is taken as the general direction of current movement and, consequently, of the direction of transport of the sediments. Fig. 5-13 illustrates an example of this type of study carried out by Schlee and Moench (1961). The Late Jurassic Jackpile Sandstone of New Mexico forms a river channel deposit that has a width of up to 20 km and a thickness of up to 60 m. It has been traced for more than 50 km in a northeasterly direction before dividing into three separate channels. The average direction of flow of the river which deposited the fluvial sands of the Jackpile is indicated by cross bedding, the mean directions of which also trend to the northeast.

Mapping the depositional trends of the Jackpile and other ancient river channels can have important economic applications. The Jackpile forms host rock for uranium mineralization in the area west of Albuquerque, New Mexico. Other river channels in the subsurface are known to form reservoirs for oil and gas where the configuration of the strata permits a structural closures. Data from dipmeter borehole surveys indicate the general current direction and, consequently, the depositional trend of the channel.

Another study of cross bedding as an indicator of regional current flow and directions of sediment transport was carried out by Goldberg and Holland (1973). They plotted the mean directions at various localities of cross bedding in sandstone beds of the Early Triassic Narrabeen Group in the Sydney Basin of New South Wales, Australia. The area covered by this study and the distribution of associated Permian and Triassic formations are shown in Fig. 5-14. The mean directions of dip of foreset beds of large ripples viewed as sets of cross bedding at various locations in the Sydney Basin are shown in Fig. 5-15. The sandstone beds of the Narrabeen Group are fluvial deposits, probably laid down by a river system. The current directions indicated in Fig. 5-15 suggest that the system drained southeast and northeast into the present area of the Sydney Basin. These converging branches are believed to have formed a major eastward-flowing river system in the coastal area north of Sydney.

In association with analyses of cross bedding in molasse-type sedimentary sequences, studies can be made of the distribution of pebble sizes within a particular formation or member. Measurements of the average maximum diameters of pebbles in separate outcrops within a limited locality can be plotted graphically on a map to show the direction of size

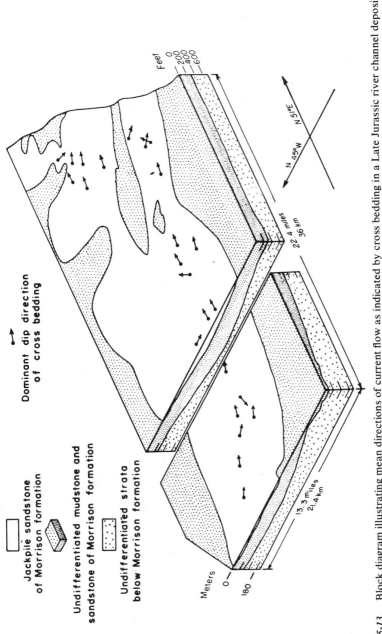

Fig. 5-13. Block diagram illustrating mean directions of current flow as indicated by cross bedding in a Late Jurassic river channel deposit, the Jackpile Sandstone of New Mexico. Note: Top of blocks is base of Dakota sandstone. [After Schlee and Moench (1961).]

313

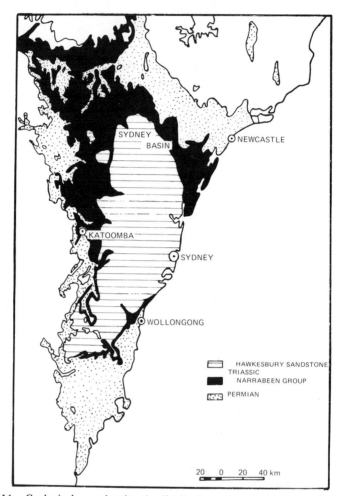

Fig. 5-14.   Geological map showing the distribution of Permian and Triassic formations, with particular reference to the Narrabeen Group, in the Sydney Basin, N.S.W., Australia. [After Goldberg and Holland (1973).]

gradation. Fig. 5-16, presented by Pettijohn (1962) and based on studies by Pelletier (1958), illustrates an example of pebble gradation within the Early Mississippian Pocono Sandstone of the Appalachian region in eastern United States. The Pocono, which ranges in thickness from 100 m in the west to 600 m in the east, forms a deltaic wedge of sediment that passes westward into marine beds. The distribution of average maximum pebble diameters shown in Fig. 5-16 indicates the eastward provenance of the sediments or, conversely, the westward direction of transport.

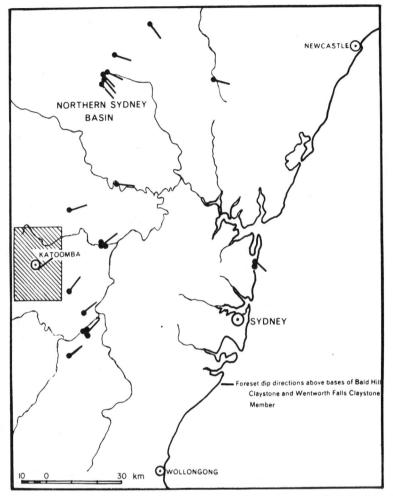

Fig. 5-15. Map of the central area of Fig. 5-14 showing mean current directions based on cross bedding in fluvial sandstone beds of the Early Triassic Narrabeen Group in the Sydney Basin, N.S.W., Australia. [After Goldberg and Holland (1973).]

Another example from the same region, studied by Yeakel (1962), is illustrated in Fig. 5-17. This shows the sediment dispersal pattern of the Early Silurian Tuscarora Quartzite as indicated by the distribution of average maximum pebble sizes. Sediment transport in a westerly to southwesterly direction was effected in large part by fluvial processes which terminated westward along a shoreline, as indicated by the presence of tubular structures referred to as *Scolithus*. With reference to the

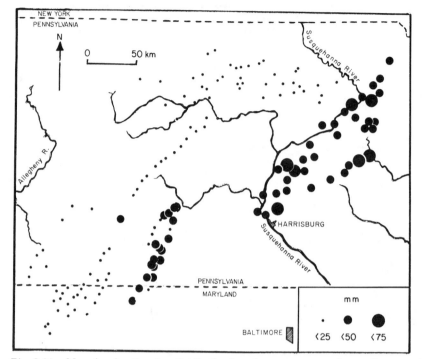

Fig. 5-16. Map showing the size distribution based on average maximum diameter of pebbles in the Early Mississippian Pocono Sandstone, Appalachian region, United States. [After Pettijohn (1962) and Pelletier (1958).]

application of pebble size studies to the analysis of the sedimentary basin, Yeakel states (p. 1515):

> Thickness, lithofacies, maximum pebble size, petrographic and fossil data, together with the cross-bedding, establish the Tuscarora depositional framework. These data enable one to locate the edge of the sedimentary basin and the shore line.

Sedimentary structures such as *Scolithus* and other similar features of bioturbation are indications of deposition in a body of water, commonly in a shallow paralic to intertidal beach environment. Other sedimentary structures, although formed by physical or chemical processes, may with reservation be ascribed to a particular depositional environment. For example, the sedimentary features formed by turbidity currents, including graded bedding, scours, pull-apart, and flame structures are commonly formed in deep-water marine environments. They also form in deep freshwater environments and may in some situations form in relatively

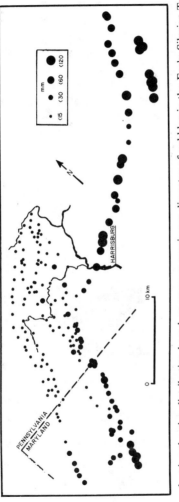

Fig. 5-17. Map showing the size distribution based on average maximum diameter of pebbles in the Early Silurian Tuscarosa Quartzite, Appalachian region, United States. [After Yeakel (1962).]

317

shallow-water environments. The structures themselves are indicative of a physical process which may operate in a variety of depositional situations but which are commonly a deep-water phenomenon. As such the structures must be considered in context with the lithologic and other characteristics of the stratigraphic sequence in which they occur. Structures similar in appearance to the breccia formed by turbidity currents pulling apart an underlying clay layer may be formed by wave and current action washing away fragments of sun-dried clay on a mud flat. Typical features formed by turbidity currents in flysch deposits are illustrated in Fig. 5-18 and include graded bedding, small-scale ripples, scours, and sedimentary breccia. The structures shown are those that would be seen in cores but which are drawn from polished slabs of rock taken from outcrops of clastic beds interfingering with pelagic radiolarites described by Schlager and Schlager (1973) as deep-sea sediments. Similar sedimentary structures in flysch sequences are commonly inferred to indicate a deep-water origin.

The stratigraphic distribution of turbidites tends to be in linear belts. Furthermore, the lithology and character of the contained sedimentary

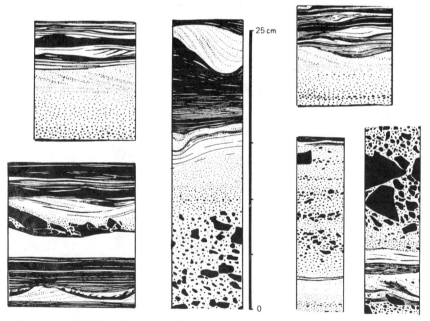

Fig. 5-18. Turbidity current structures in clastic sediments interbedded with pelagic radiolarite in the Late Jurassic Tauglboden–Schichten Formation of the Eastern Alps. [After Schlager and Schlager (1973).]

structures may vary considerably within a distance of a few kilometers. Correlation of individual tongues of turbidite flows with lateral tongues of the same flow is consequently not always possible. Other sedimentary structures may have a wide stratigraphic distribution as well as being indicative of sedimentary processes operating in a particular depositional environment. An interesting case in point is illustrated in Fig. 5-19, described by Kennedy and Garrison (1975) with reference to the genesis of nodular structures and features termed hardgrounds in Late Cretaceous chalk beds of southern England. Kennedy and Garrison say that the term hardground is used not only to describe a rocky sea floor but also for inter- or intraformational horizons in marine sediments which are inferred to have been exposed as rocky sea bottoms. This concept implies a hiatus in sedimentation, and stratigraphic horizons marked by features termed hardground can be placed in the category of paracontinuities (see Chapter 3).

The chalk beds, comprising bioturbated coccolith micrites, show rhythmic bedding and layers of hardground. Deposited in an outer shelf environment, in water depths of up to 300 m, it is inferred by Kennedy and

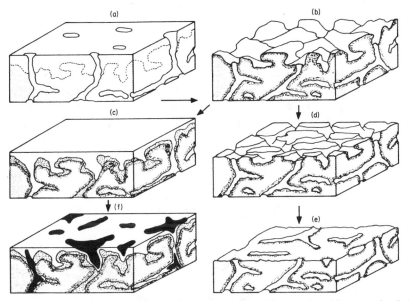

Fig. 5-19. Diagram illustrating possible genesis of nodular stratigraphic zones in chalk and other limestones. (a), lithification of nodules below sediment–water interface, organism burrows; (b), scour and exposure on sea floor, glauconitization and phosphatization. (c), temporary burial, cementation; (d), erosion of (b); (e), further erosion; (f), silicification (black) in burrows. [After Kennedy and Garrison (1975).]

Garrison that the hardground represents a pause in sedimentation and the operation of processes described as follows. With reference to Fig. 5-19 they say (p. 311):

> The products of sea floor cementation are widely represented in the sequence, and a series of stages of progessive lithification can be recognised. These began with a pause in sedimentation and the formation of an omission surface, followed by (a) growth of discrete nodules below the sediment-water interface to form a *nodular chalk,* erosion of which produced *intraformational conglomerates.* (b) Further growth and fusion of nodules into continuous or semicontinuous layers: *incipient hardgrounds.* (c) Scour, which exposed the layer as a true *hardground.* At this stage, the exposed lithified chalk bottom was subject to boring and encrustation by a variety of organisms, whilst calcium carbonate was frequently replaced by glauconite and phosphate to produce superficial mineralized zones. In many cases, the processes of sedimentation, cementation, exposure and mineralization were repeated several times, producing *composite hardgrounds* built up of a series of layers of cemented and mineralized chalk, indicating a long and complex diagenetic history.

Similar features are observed in outcrops of limestones other than chalk. Those in chalk may have a widespread stratigraphic distribution, covering up to 1500 km² of sea floor, and their recognition in outcrop is consequently of significance not only as a potential stratigraphic marker but as an indicator of seabed sedimentologic processes.

The examples of sedimentary and biogenic structures mentioned here are merely a few of those described in the extensive literature on the subject, and their inclusion is intended to serve only as a guide to the lithostratigraphic analysis of sedimentary basins.

MICROSTRATIGRAPHY

Microstratigraphy is an essential adjunct to macrostratigraphy where much of the strata are exposed and is the only approach to sedimentary basin analysis where the beds are entirely in the subsurface. Microstratigraphy employs the same concepts as macrostratigraphy but relies entirely on direct and indirect methods of determining the lithology, fossil content, and stratigraphic sequence in a borehole. Direct methods include examination of drill cuttings, conventional core, and sidewall core. Indirect methods require geophysical techniques and include electric logs, gamma ray–neutron logs, sonic logs, dipmeter surveys, and others. Supporting the data obtained from individual boreholes is the information obtained from seismic reflection and refraction surveys, particularly where the geophysical information can be tied-in to a particular borehole. Additional information is obtained from borehole surveys on the petrophysical characteristics of particular beds, such as porosity and permeability, and also on the nature of their fluid content.

Interpretations of seismic data and of data from geophysical borehole surveys depend on knowledge of specialized fields within the wide realm of geophysics and are mentioned here only insofar as they relate to lithostratigraphic methods. There are numerous general and specialized references to geophysical methods, including Dobrin (1976) with reference to geophysical prospecting, Moore (1964) and Pirson (1970) with reference to the application of borehole logs to the analysis of subsurface geology, and several detailed and specialized publications of Schlumberger Limited on the interpretation of various types of logs with reference to petrophysical data.

Lithology and Lithogenesis

Determination of the lithology and stratigraphic sequence in which beds occur depends primarily and principally on the examination of drill cuttings during the course of drilling a well. The information logged is subsequently tied-in to electric and other logs which show characteristics that can be used as markers to precisely define the depth of the upper surfaces of particular beds. This calibration is necessary in view of the fact that samples are commonly collected from the returning drilling mud at drilling intervals of approximately 3 m. Also the up-hole time required for the drilling mud to carry the rock chips to the surface and the rate of drilling are variable. Very precise correlations between geophysical and drill cutting logs can be made on the basis of particular zones or beds that have a marked petrophysical characteristic such as relatively high radioactivity or electrical resistivity. Widespread thin beds of bentonite rich in montmorillonite derived from the alteration of volcanic ash are good marker beds. Some thin beds of dark argillaceous limestone, rich in fish scales and other organic matter, show radioactivity exceeding that of the enclosing rocks and are consequently well defined by the curves of a gamma-ray log. Coal beds show a markedly high resistivity on electric logs as do tightly cemented zones in sandstone or silicified zones in limestone. Some knowledge of the stratigraphic sequence is therefore necessary to the interpretation of geophysical logs. It must be emphasized that final interpretation of subsurface data depends on a consideration and comparison of interpretations placed on all the available geophysical, geochemical, petrological, and petrographical data.

Initial examination of drill cuttings is made with the binocular microscope; subsequent examination of thin sections of rock chips from particular horizons can also be carried out using both binocular and petrographic microscopes. Collection of the samples of drill cuttings at 2- or 3-m intervals should always be carried out by the geologist responsible for sample examination and logging. Samples are bagged and labeled with the retained drilling mud which contains any microfossils present. Part of the sample will be thoroughly washed, dried, and placed in labeled plastic containers for examination.

Examination of these samples requires patience, thoroughness, a knowledge of various techniques, and experience. There are many pitfalls of omission and commission, some of which may later be rectified by comparing the drill cuttings sample log with geophysical logs. One source of possible error is the interval at which the samples are collected. If too large, drill cuttings from a thin intervening bed may be entirely missed. Another source is the ubiquitous presence of cavings picked up from higher intervals by drilling mud returning up the hole. The unwary geologist may either miss the lithology of beds below such a caving zone because the flood of cavings masks the presence of sparce cuttings from lower beds, or he may log a repetition of the upper caving bed. Such errors could be costly where the coring of an important zone depends on the recognition by drill cuttings of a thin marker bed.

Examination and description of drill cuttings viewed under a binocular microscope are done essentially in the same way as an examination of rock samples in the field, using a hand lens. The main difference is that examination in the laboratory affords more opportunity for additional testing. Two important considerations must be kept in mind when examination is made under a binocular microscope. First, samples under magnification commonly appear to have a different color or intensity of shade than the same rocks seen in hand specimen or outcrop. Second, it is essential that descriptions of color, grade, and texture be quantified by means of standard comparison charts and samples. The eye is not always a reliable judge. After examining a number of samples of a thick shale section, the geologist may find that variations of color from top to bottom are more apparent than real. In this respect it is helpful to have the samples laid out in clear plastic vials that show any such variations within a thick, monotonous shale section. Descriptions of color and shale should be made with clean, dry samples free of dust. Wet samples always appear much darker. On the other hand, wet samples more clearly show such textural features as the clasts in calcarenite. Qualitative tests of silt and clay content in limestone and of carbonate content in sandstone and shale are quickly made by crushing a small rock chip. In the first test stirring the residue in a watch glass with a rounded glass rod will indicate silt, and the nature of the insoluble residue will indicate the presence of clay. In the second test the application of dilute HCl will indicate the presence of carbonate.

The presence of microfossils or fragments of macrofossils is best detected when the samples are wet. Forams, ostracods, sponge spicules, shell fragments, fish scales, and tiny tests such as those of the pteropod *Tentaculites* may be seen. In some cases a particular characteristic in the internal structure of a macrofossil may, where only a fragment is seen, give a clue to the type of fossil. An example is the columnar structure of some pelecypod shells which in subsurface Cretaceous beds of Western Canada indicates the presence of *Inoceramus*. With all such observations of fossils in samples of drill cuttings, the recording of depth and rock layer in which the fossil first occurs is most important.

Interpretation of depositional environment, based on examination of drill cuttings and geophysical logs, is possible in some cases but must be attempted with reservations. Studies along these lines have been carried out by Davies and Ethridge (1975) who state (p. 239):

> The detrital mineralogy of sandstones is affected significantly by depositional environment. The imprint of the environment is recorded as the relative abundance and size of individual detrital species. Differences between environments within any one basin are accompanied by concomitant changes in detrital–sediment composition.

The approach used by Davies and Ethridge was based on the size and relative abundance in distributions of detrital fragments of quartz, feldspar, mica, rock, heavy materials, and other accessories. Interpretations based on such methods depend on the validity of certain premises and consequently tend to be equivocal.

Interpretations relating to lithogenesis may be difficult in view of the small size of the drill cuttings which are commonly less than 10 mm wide. Sidewall cores up to 2 cm in diameter and several centimeters in length offer more scope for petrologic and petrographic examination of carbonates and sandstones. Sidewall cores can be used to make accurate measurements of porosity and permeability which are related to the diagenetic history of the rock. In the absence of sidewall cores, estimates of the possible reservoir properties of a rock can be made from examination of the drill cuttings, as pointed out by Archie (1952) and Robinson (1966). In the case of sandstones it is possible to come to some conclusions as to the physical conditions and depositional environments in which they were deposited by examination of the composition, sorting, grade, and roundness of the sand grains. With carbonates, the texture (ranging from calcilutite through calcarenite to calcirudite), the fragmental or oolitic nature, the fossil fragments, the content of glauconite and other accessories, and the degree of crystallization are all features that may indicate the lithogenesis and depositional environment of the carbonate.

Large areas of a carbonate shelf may periodically be raised above sea level, as in the Bahamas, and become subject to the percolation of meteoric water which cements the carbonate by dissolution and precipitation of calcite. Such zones, including raised beaches, may subsequently be buried and become subject to further diagenetic alterations which fill the vugs or alter to micrite the fine carbonate debris that fills the vugs after resubmersion of the carbonate rock. An example of the type of alteration that may occur as the result of dissolution and precipitation of calcite is described by West (1973) and shown in Fig. 5-20. This figure illustrates the alteration of Pleistocene carbonate–quartz sands to calcareous quartz sandstones in north Devon and Cornwall, England. The type of alteration shown would be visible in large chips of drill cuttings, particularly when examined in a watch glass filled with water.

Examination and description of drill cuttings require considerable attention to details which may be of paramount significance in further investigations. The rate at which samples taken at intervals of 3 m can be logged will obviously vary with changes of rock type and bed thickness. Where a carbonate section is examined in detail, it may not be possible to log more than 100 m a day, whereas a thick homogeneous shale section may be logged at a rate of 300 m a day. It is essential that descriptive notes accompanying or embodied in the graphic sample log of a well be entirely objective and clearly defined from any interpretations regarding depositional environment or lithogenesis. The log constitutes a permanent record which is available for future reference.

Provenance of Lithic Sediments

Lithostratigraphic analysis of any unit within a sedimentary basin requires the formulation of a depositional and paleogeographic setting to account for the source of sediments and their mode of transportation. The problem of provenance of flysch sediments is more difficult than that of molasse sediments because in general they are much finer grained. Also the flysch sediments have been subjected to a greater degree of recycling and mixing from different sources, and their direction of transport within a

Beach sand

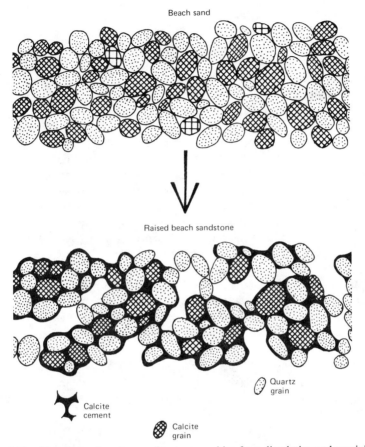

Raised beach sandstone

Quartz
grain

Calcite
cement

Calcite
grain

Fig. 5-20. Sketch showing change of texture resulting from dissolution and precipitation of calcite in conversion of a carbonate–quartz sand to a calcareous quartz sandstone. [After West (1973).]

trough may be more parallel than normal to the shoreline trend of the landmass from which they were derived. Molasse sequences, on the other hand, form thick wedges of sediment deposited on a paleoslope dipping away from the edge of an eroding landmass. These sequences comprise alluvial fans, fluvial and lacustrine deposits, and deltaic accumulations ranging from continental to paralic. Although directional features within a molasse sequence may show a wide range within an arc of 90° or more, they consistently trend away from the original source of the sediments. Studies of the distribution of the average maximum diameter of pebbles at various localities (Figs. 5-16, 5-17) have shown that the mean size of pebbles diminishes away from the source. Where the source is not remote and pebbles of granite, quartzite, volcanic rock, or chert are preserved, the source of these rock types may be determined. In such examples the period of sedimentation can be correlated with periods of tectonic uplift during which formations or igneous bodies of known age were eroded. In other examples the source rocks have been completely eroded away, or buried, leaving no trace on the surface. In these cases the determination of provenance is more difficult.

Where pebbles are not present in the sandstones, determination of provenance may be made on the basis of heavy mineral content or on grains of a particular rock type known to characterize a certain formation. An example of the application of heavy mineral determination to the problem of provenance and transport of sediment is seen in the Early Cretaceous basal quartz sandstone, termed the Dina Formation, of the eastern Alberta basin in Canada. The heavy mineral content of this sandstone includes tourmaline, zircon, rutile, sphene, garnet, sillimanite, monazite, and ilmenite, indicating a source in the igneous and metamorphic rocks of the Precambrian shield to the northeast. Quartzose sandstone beds of the tar-bearing McMurray Formation in northeastern Alberta are of comparable age, and are also derived from the Precambrian, but differ from the Dina in having an abundance of kyanite. Percentage differences in heavy mineral content, or the absence of certain minerals, result from differences in the source rocks underlying the areas which supplied the sediments. In the Precambrian of western Canada these source rocks include a wide range of leucocratic and basic igneous rocks, volcanics, metamorphic rocks such as granulite, gneiss, schist, quartzite, and amphibolite.

The percentages of varieties of heavy minerals can serve to indicate not only the provenance of a sediment but also the variety of a particular mineral such as garnet or zircon. Studies made on the basis of characteristics of particular minerals require more specialized techniques, and although they can be definitive, they are particularly subject to errors of premise. Similarly, studies on the distribution of grains of a particular rock type, with a view to relating that rock type to a possible source area, may depend on evidence that cannot readily be correlated. For example, deltaic basal sandstone of the Late Cretaceous Belly River Formation of Alberta is composed largely of grains of metamorphic quartzite, silicified argillite, chert, and volcanic rock, with minor grains of quartz and feldspar. The assemblage indicates that the source area was not the Precambrian shield to the east but rather an ancient Cordilleran area to the west. Examination under the petrographic microscope of thin sections of the Belly River Sandstone and comparison with silicified argillites, quartzites, and volcanic rocks of the Permian Cache Creek Series of

British Columbia, some 300 km to the west, indicate many similarities. It can be inferred, but not unequivocally demonstrated, that the basal sands of the Belly River Formation were derived, probably through several cycles of deposition, from uplifted and eroded highlands situated in the present area of south-central British Columbia.

An interesting study of the distribution of heavy minerals in the Pliocene Paso Robles Formation of the Salinas Valley area of west-central California was made by Galehouse (1967). The Paso Robles consists largely of continental conglomeratic sand, silt, and clay derived from the uplift of the Santa Lucia and La Panza ranges. Part of this study is illustrated in Fig. 5-21 which shows the relative distributions by percentage weight of combined sphene–apatite and hornblende–garnet. From these and similar mineral distribution maps and from studies of cross-bedding

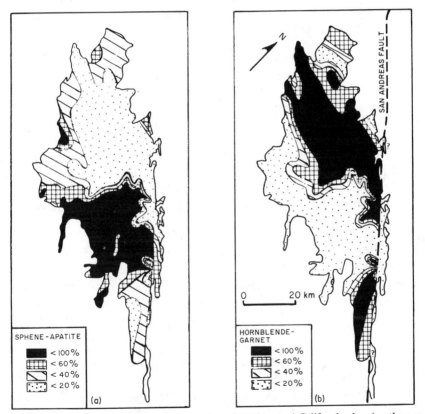

Fig. 5-21. Map of the Salinas Valley region of west central California showing the percentage distribution by weight of heavy minerals in the Pliocene Paso Robles Formation. [After Galehouse (1967).]

directions, Galehouse deduced that the provenance of the Paso Robles was probably principally to the south, as indicated by the pattern of distribution of sphene and apatite, but also to the northwest from which direction the hornblende and garnet were derived. It is of interest to note that vertical variations of heavy mineral content within the Paso Robles do not alter the patterns of distribution.

Interpretations of provenance must include the possibility that the sediments have been reworked through two or more cycles and that the original source rock may have been completely removed by erosion or widely separated from the sedimentary rocks derived from it by sea-floor spreading. With each cycle of erosion, sedimentation, and consolidation of the sediment, a sandstone becomes more quartzose as the grains of feldspar and rock fragments are crushed and decomposed. But where the distance of transportation is short, the composition and texture of a second-generation sandstone may closely resemble that of the first.

An example is mentioned by Conybeare (1949) with reference to Proterozoic arkosic sandstone interbedded with basalt flows north of Lake Athabasca in Canada. These reddish sandstone beds, which are lithologically similar to the Athabasca Sandstone south of the lake, include conglomerate lenses containing rounded pebbles, gneiss, basalt, quartzite, and sandstone. The sandstone pebbles are also reddish and have a composition and texture similar to those of the enclosing sandstone which is inferred to have been derived from a preexisting sandstone. Several cycles can be represented in the entire sequence of sandstones north and south of Lake Athabasca. Alcock (1936) believed these sandstone beds to have been deposited in broad, subsiding basins between mountains of considerable relief in a semiarid climate. Cross bedding, ripple mark, and mud cracks in the red shale interbedded with sandstone indicate a fluvial to lacustrine environment of deposition.

With reference to the role of sea-floor spreading in separating an area of provenance from the product of its erosion, mention should be made of the distribution of Quaternary diamond-bearing sediments in West Africa, Brazil, and the Guyana region of South America. These diamonds occur in fluvial gravel deposits derived from older sediments which in turn were probably derived from diamond-bearing Precambrian conglomerate. The Precambrian belt of metamorphosed sediments has been rifted apart by sea-floor spreading, but erosion of the diamondiferous conglomerate and recycling of its products have continued. Junner (1934) states that the diamondiferous conglomerate at the base of the Precambrian Tarkwaian sequence in Ghana was derived from older sediments of the Birrimian sequence. The end product of several such cycles of erosion and deposition probably consists mainly of quartz and heavy minerals, with varying proportions of sediment from other sources added during the cycles of sedimentation.

Subsurface Mapping

Subsurface mapping is based on geophysical data and rock samples obtained from boreholes situated at various points on the surface of a sedimentary basin. Mapping may be based on either geophysical data, lithologic data or on both. Mapping of the basement topography, and to some extent of its lithology, can be undertaken by seismic, gravity, and radioactivity methods. Mapping of the structure and thickness of sequences of strata lying above the basement depends on the use of seismic reflection or refraction data obtained from shotpoints which may form a grid, linear, or irregular pattern. Basement is here defined in a geological sense as an igneous–metamorphic complex beneath the surface of which no seismic reflections are obtained. The term economic basement refers to a zone of low-grade metamorphic or indurated rocks below which no oil or gas are likely to be found but from which seismic reflections may be obtained. Mapping of the lithology and thickness of stratigraphic sequences within a sedimentary basin, and to some extent (where control permits) of the basement lithology, can be done by means of drill cuttings and cores obtained from boreholes. The main constraints placed on mapping are those imposed by the spacing and depth of the boreholes. Direct observation of samples is supported by the use of various borehole surveys.

It is not intended to describe the techniques involved in surface or borehole geophysical surveys. These include several specialties in which technology effects continual changes in methods and interpretation. For general reference there are several textbooks available on the principles involved and their applications to geological interpretations, including Moore (1964), Pirson (1970), Garland (1971), and Dobrin (1976). Surface geophysical surveys relate in only a general way to the subject of this chapter, whereas borehole geophysical surveys are directly related to the lithologs obtained from examination of drill cuttings and core. To this extent some brief mention will be made of their application to microstratigraphy.

The configurations of curves seen on borehole geophysical survey logs are direct reflections of the petrophysical characteristics, determined by lithology and fluid content, of the strata penetrated. Petrophysical characteristics may be local or widespread and may have specific vertical limits that enable their curve characteristics to be used as stratigraphic markers.

An example given by Payne (1976) is illustrated in Fig. 5-22 which shows the radioactivity and sonic characteristics of some beds in the Late Permian Delaware Mountain Group of the Delaware Basin in Texas and New Mexico. The gamma curve indicates the degree of natural radiation coming from the various layers of rock, which is a reflection of their mineralogical content which in turn is a reflection of their chemical composition and variable content of radioactive isotopes such as that of potassium, a common element in silicate minerals. Curves indicating the radiation produced by bombardment with neutrons from a radium–beryllium source are commonly included with gamma-radiation curves on the same log. Measurement can be expressed in terms of micrograms of radium equivalent per metric ton of rock. The sonic curve is a measure of the sonic velocity characteristic of each rock layer. Sound waves moving through rock commonly have a velocity in the range 1000–7000 m/sec depending on the lithology. Dense rocks such as hard, dolomitic limestones or granite will have sonic velocities in the upper part of the range, whereas less dense rocks such as claystone and friable sandstone will have velocities in the lower part. Measurement is expressed as the reciprocal of the velocity, that is, in terms of microseconds per unit length of rock.

Correlation of gamma-ray and neutron logs with the self-potential and resistivity curves of electric logs and also with drill cuttings logs is another useful method of integrating the lithologic and petrophysical characteristics of the stratigraphic section penetrated. Fig. 5-23 (Moore, 1964) illustrates the gamma-ray, neutron-radiation, self-potential, and resistivity

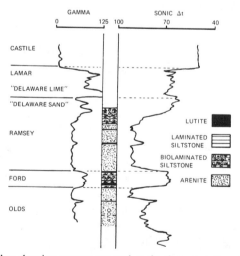

Fig. 5-22. Well log showing gamma-ray and sonic characteristics of some beds in the Late Permian Delaware Mountain Group, Texas and New Mexico. [After Payne (1976).]

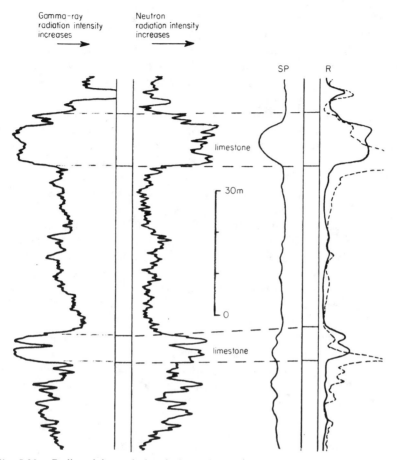

Fig. 5-23. Radioactivity and electric logs of a Pennsylvanian section in north-central Oklahoma showing correlation of limestone beds. [After Moore (1964).]

characteristics of limestone beds in a Pennsylvanian sequence in north central Oklahoma. The radiation log curves indicate a relatively low natural radioactivity for the limestone beds and a relatively high induced radiation caused by the fluid content. The self-potential curve of the electric log indicates a relatively high permeability, whereas the resistivity curve shows a relatively high resistivity caused by the presence of either fresh water or hydrocarbons in the limestone bed. In combination with a drill cuttings log, a preliminary assessment can be made of the hydrocarbon-producing potential of various beds. Other types and varieties of logs are used for specific purposes and borehole conditions, including laterologs, induction logs, micrologs, microlaterologs, nuclear magnetism logs, dipmeter logs, temperature logs, drilling mud analysis logs, caliper logs, and drilling-rate logs.

The configuration of self-potential curves on electric logs is not only a reflection of particular petrophysical characteristics but may also indicate the genesis of some sandstone bodies. Commonly, but not consistently, a bell-shaped self-potential curve is indicative of grain gradation from coarser below to finer above, a characteristic of river sand deposits such as point bars. Conversely, a funnel-shaped curve indicates grain gradation from coarser above to finer below, a characteristic of offshore sand bodies such as bars and barrier islands. Fig. 5-24 illustrates an example of the self-potential curves of Tertiary barrier bars in Texas and of inferred Early Cretaceous bars in Papua as interpreted by Conybeare and Jessop (1972). Where well log control is sufficiently close, as in the area of a producing field, it may be possible to trace the details of self-potential and resistivity curves with sufficient accuracy to show the relationships, such as offlap, between adjacent sand bodies. Examples have been cited by Conybeare (1976) and Evans (1970) with respect to the lower beds of the Late Cretaceous Belly River Formation of the Pembina oil field, Alberta, and the Early Cretaceous Viking Formation of the Eureka and Avon Hills oil and gas fields of Saskatchewan.

Over considerable distances the correlation of such details is not possible and so-called jump correlations are necessary. Where good marker beds, within a few hundred feet vertically from the target strata, are not available, miscorrelations are a potential hazard to subsurface mapping. These may arise simply because of facies changes in which the lithology

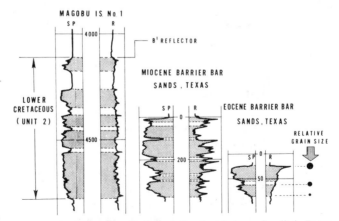

Fig. 5-24. E-log character of barrier bar sands: electric logs of wells in Papua and Texas showing funnel-shaped self-potential curves characteristic of barrier bars. [After Conybeare and Jessop (1972).]

and consequently the petrophysical characteristics of the particular bed being traced are altered. Or they may stem from thickening and thinning of the bed and from lensing out completely. A further consideration is the effect of compaction on stratigraphic sequences comprising variable volumes of sandstone and shale. Sandstone bodies within sections consisting predominantly of shale will be displaced vertically to a greater extent than sandstone bodies within a section consisting predominantly of sandstone. Consequently, a comparison of logs laid side-by-side may lead to miscorrelations. Similar problems may arise where the sequences were deposited at markedly different rates of sedimentation. Despite these problems which place constraints on more detailed correlations, it is usually possible to correlate sequences on a broader scale. For example, where a sequence predominantly of shale is overlain by a sequence containing much more sandstone or limestone, the demarcation of the two sequences can be arbitrarily chosen, keeping in mind that variations in facies will allow some degree of vertical migration. Where lithofacies or isopach maps are constructed, they will be limited in accuracy by the horizontal control supplied by boreholes and the vertical control supplied by stratigraphic markers. In this respect it is essential that geophysical well logs are compared closely with the lithologs and accompanying descriptions of drill cuttings and cores. Where necessary, reexamination of samples should be undertaken to fill in potentially significant data that may have been missed during the initial examination, particularly where the litholog was drawn during the drilling of a well and, consequently, before the running of geophysical logs.

Microstratigraphy, involving the correlation of observed features in drill cuttings with petrophysical features indicated by geophysical well logs, is the basic approach to construction of subsurface maps. In the following chapter various types of subsurface maps will be discussed.

PHYSICOCHEMICAL PROPERTIES

Porosity and Permeability

Porosity and permeability are measurable physical properties of a sedimentary rock. They may be expressed in qualitative terms such as good, fair, and poor but are commonly defined in quantitative terms based on measurements of side wall cores or conventional cores. Porosity is expressed as the percentage of volume of the whole rock occupied by its pore space. Sandstones and carbonate rocks commonly have a porosity ranging up to 20%, but some rocks may range up to 30%. A porosity of up to 10% is considered poor, 10–20% fair to good, 20–30% excellent. Permeability is expressed in darcies or millidarcies. A rock is said to have a permeability of one darcy if one cubic centimeter of a fluid, having a viscosity of one centipoise, flows through one cubic centimeter of the rock in one second under a pressure differential of one atmosphere. The permeability of most sandstones and carbonate rocks commonly ranges up to a few hundred millidarcies but in the case of some friable quartzose sandstones ranges up to a few darcies. Permeability of up to 10 millidarcies may be termed as poor, 10–100 as fair, and 100–1000 as good to excellent.

From an examination of drill cuttings one may obtain an approximate estimate, in qualitative terms, of the potential porosity and, consequently, of the possible permeability of a rock layer. Chips of sandstone may show intergranular pore space and differences in types and amount of intergranular material consisting of the original matrix, of introduced cementing material, and of minerals derived from the diagenesis of part of the origin matrix. Examination under a petrographic microscope of thin sections made from the largest available chips will assist in such estimations. Drill cuttings of carbonate rocks may also give some indications of the porosity and permeability of the rock from which they came. Pinpoint porosity may be visible in the rock chips. Where recrystallization has taken place, the sucrosic texture may show intercrystalline porosity. Vugular porosity may be indicated by the crystalline texture of a carbonate rock but is not visible as such in the rock chips of drill cuttings. From the apparent degree of recrystallization, dissolution, and cementation, as illustrated in Fig. 5-25, some approximation can be obtained of the probable porosity and permeability of the rock. Diagram (a) has been used by Wardlaw (1976) to graphically show in two dimensions the reduction of percent porosity with crystal growth in dolomite and to show also the reduction of mercury-ejection efficiency with reduction of pore space. Diagram (b) is taken from Choquette and Pray (1970). It shows the various combinations of progressive dissolution and cementation that may take place where a constituent of the rock, such as a fossil, is dissolved. In this example the original mould is that of a fragment of crinoid stem.

The relative size and distribution of pore space may be used to estimate in qualitative terms the approximate porosity of a rock. The diagram of Fig. 5-26a illustrates an example described by Robinson (1966). This shows idealized capillary–pressure curves for three rocks having varying degrees of visible pore space. In all of the three rocks the pore spaces are equidistant, but each rock has pores of a different size. The capillary–pressure curves show that attainment of the same mercury saturation in all

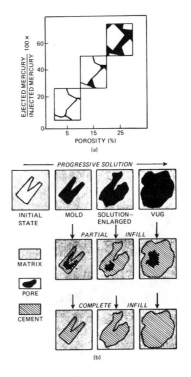

Fig. 5-25. (a) Diagram showing reduction of percent porosity and of mercury–ejection efficiency resulting from growth of crystals in dolomite [After Wardlaw (1976).] (b) Diagram showing stages of dissolution and cementation of a fossil mould. [After Choquette and Pray (1970).]

three rocks requires a progressively increasing capillary pressure with decreasing pore size. The diagram of Fig. 5-26b described by Schwarzacher (1961) illustrates a feature of prime importance to permeability, the interconnection of pore space. Pores or vugs may be discrete and, although filled with oil, may be unable to yield their contents effectively because of poor connections. Appearances may be deceiving in this respect. For example, a core of vugular dolomitic limestone may appear to have attractive reservoir characteristics but may in fact have a lower porosity and permeability than a limestone of more solid appearance with interconnected pinpoint porosity. For this reason a distinction is commonly made between the percentages of total porosity and effective porosity. The later is obviously essential to permeability.

The relationships between a rock's petrophysical characteristics of porosity and permeability and the physicochemical processes and changes to which the same rock has been subjected are complex. Various aspects have been studied by many workers, including Maxwell (1964), Shenhav (1971), Pittman (1974), and Sprunt and Nur (1976). Maxwell studied the effects of pressure and temperature on quartzose sandstones and found that porosity tends to decrease with increasing depth of burial.

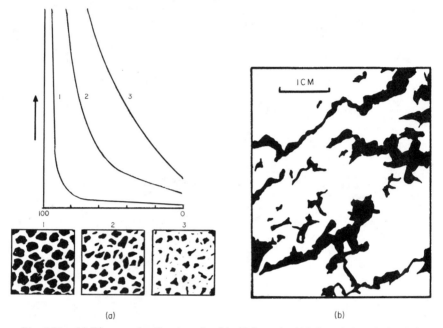

(a)   (b)

Fig. 5-26. (a) Diagram showing porosity (black) in rocks (1,2,3) and the relationship to idealized capillary–pressure curves (0–100% mercury saturation with increase of capillary pressure). [After Robinson (1966).] (b) Tracing of pores and interconnecting channels made from thin section of Early Carboniferous reef limestone in northwestern Republic of Ireland. [After Schwartzacher (1961).]

Shenhav investigated the porosity and permeability of sandstone beds of the Early Cretaceous Helez Formation in Israel. He found that the amount and mineralogy of cement are related to the depositional environment and are the main factors influencing porosity, whereas permeability is related to the amount of cement only. Pittman found that the porosity and permeability of Pleistocene coral limestone in Barbados depended on the transition of aragonite to calcite and the consequent growth of sparry calcite in voids. This diagenetic process depends on the infiltration of fresh meteoric water (Fig. 5-20). Sprunt and Nur experimented with samples of the Early Ordovician St. Peter Sandstone of the eastern United States and showed that pressure–solution was an important factor in the rapid reduction of porosity. These and many other studies relate to a very complex problem which is of prime significance to the production of oil and gas from any reservoir.

Preliminary qualitative estimates of the potential porosity and permeability characteristics of a rock layer penetrated by a borehole can be made by examination of drill cuttings; further quantitative determinations can be made by measurements of side wall or other cores, and calculations of the vertical distribution of porosity and permeability can be made on the basis of electric logs, as illustrated by Fig. 5-27 (Berg, 1975). This figure shows the relationship of electric log characteristics to the vertical distribution of permeability, porosity, grain size, and composition of the Late Cretaceous Sussex Sandstone of the Powder River Basin in Wyoming.

Determination of porosity and permeability based on measurements of core samples is spot determinations only. The values given may be purely local, or they may be applicable to the entire thickness of the sampled bed where it has been penetrated by the borehole, and to the lateral extension of the bed for a considerable distance form the borehole. Homogeneity of texture and composition within a sandstone unit facilitates uniformity of porosity and permeability, but variations in thickness and number of thin shale layers at various locations within a sandstone unit will affect the vertical permeability. Within a homogeneous unit of sandstone other factors such as preferred orientation of fabric, produced by depositional

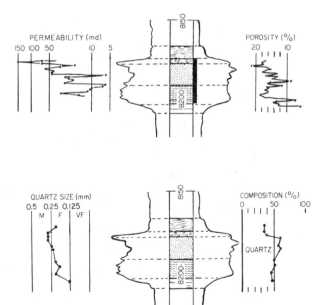

Fig. 5-27. Electric logs of the Late Cretaceous Sussex Sandstone, Powder River Basin, Wyoming, showing relationships to permeability, porosity, grain size, and composition (Woods Mandel 1). [After Berg (1975a).]

alignment of sand grains, will cause variations in permeability in different directions along a bed. These variations can be of significance to the production of oil from a well, and consequently the depositional pattern and current trends indicated by the mean direction of preferred orientation of sand grains may be of importance to the techniques of oil recovery.

Where a sandstone unit lacks homogeneity, such as in alluvial deposits which comprise lenses of sandstone and conglomerate with interfingering layers of shale, the porosity and permeability vary from lens to lens. Production problems include those caused by the infiltration of water into the oil-bearing sandstone, cutting off zones of oil in less permeable lenses. These problems, which may result in a lower ultimate recovery of oil, are particularly important with reference to the subsurface recovery of tarry oil from the alluvial sandstone beds of the Early Cretaceous McMurray Formation in Alberta. These oil sands are completely uncemented where they are oil-saturated, the grains being bound in a matrix of tar. Where they are not oil bearing, the sandstone, gritstone, and conglomerate lenses are poorly cemented and friable. These lenses may disappear within a few meters from a borehole and, consequently, the porosity and permeability characteristics may vary appreciably from one location to another.

Fracture porosity has so far not been mentioned. It cannot be seen in drill cuttings but is evident in conventional cores. This type of porosity is characteristic of dense carbonate rocks that have been shattered by folding and is the main factor in the formation of some reservoirs such as those producing gas from Devonian beds in the Waterton Lakes area of Alberta. Such fractured zones may also be highly permeable.

In evaluating the porosity and permeability characteristics of a sedimentary unit, with reference to its possible economic potential, it must be kept in mind that some beds may have a fair to good porosity but extremely poor permeability. Where such beds contain oil, a slight stain may be evident in the drill cuttings. This can be confirmed by placing a rock chip in a white porcelain dish and adding a few drops of solvent. On evaporation a brown stain remains on the porcelain, indicating the presence of oil. Where the bed being sampled by drill cuttings contains gas, there is no indication of hydrocarbon content in the rock chips. The presence of hydrocarbons is indicated by a gas detector which continuously analyzes the drilling mud. Tight zones are brought into production by artificially fracturing the formation by pumping fluid and sand into the hole at very high pressures until (at some pressure exceeding hydrostatic but less than lithostatic) the formation fractures (Nelson and Handin, 1977). This event is indicated by a sudden drop in pressure as the fluid is forced into the fractures. When the fluid is recovered, the sand remains to prevent the fractures from closing. The orientation of the fractures depends on the condition of gravitational or tectonic stress to which the formation is subjected.

Porosity and permeability are of prime importance not only to the production of oil and gas but also to the subsurface disposal of waste fluids and to the supply of water from aquifers. These aspects will be discussed in Chapter 7.

Hydrocarbon Source Rock Analysis

The origin of hydrocarbons in sedimentary rocks has occupied the attention of geologists and geochemists for many years. It is now generally agreed that oil and gas are generated by the transformation of organic matter buried with the sediment. The optimum conditions of temperature and pressure under which this transformation is believed to occur are obtained at depths exceeding 2000 m. From the point of view of origin and entrapment of hydrocarbons, the strata within a sedimentary basin may be regarded as potential source beds or reservoir beds. Some may serve both functions. Many beds contain too low an organic content to be considered as potential source beds, and others are sufficiently lacking in porosity and permeability to be considered as potential reservoir beds. These qualities are essential in any sedimentary basin considered to be potentially favorable for the discovery and production of petroleum. Most sedimentary basins contain beds that could be reservoirs for oil or gas, but not all contain beds that can be considered qualitatively and quantitatively as good source rocks.

These matters have generated numerous scientific papers, including those by White (1935), Brooks (1938), Trask (1939), Knebel (1946), Smith (1954), Philippi (1956, 1965), Sokolov (1957), Erdman (1961), Weeks (1961), Hunt (1961), Gehman (1962), Hedberg (1964, 1968), Larskaya and Zhabrev (1964), Bray and Evans (1965), Conybeare (1965), Louis and Tissot (1967), Baker and Claypool (1970), Tissot et al. (1974), McKirdy and Powell (1974), Bailey et al. (1974), Peters et al. (1977), and Harwood (1977).

The evaluation of potential source rocks for petroleum is undertaken by chemical analysis of unweathered outcrop and core samples to determine the content of soluble and insoluble (kerogen) organic matter and compounds. Analyses are made by crushing the rock sample and processing the powder in a sequence of operations that include the use of solvents, x-ray diffraction, thermogravimetric analysis, infrared analysis, elemental analysis of kerogen for carbon, hydrogen, and ash, carbon isotope analysis, and gas chromatograms to show the content of organic compounds including alkanes, n-alkanes, pristane, and phytane. Examples of x-ray diffractograms and gas chromatograms are shown in Fig. 5-28 (McKirdy, 1976; McKirdy et al., 1975). Analyses and experimentation have shown (Tissot et al., 1974) that increasing temperatures and pressure during burial of a sediment produce physicochemical transformations of the kerogen which define the evolutionary paths for each type of kerogen. The main products of evolution are carbon dioxide, water, and the hydrocarbons collectively called oil and gas.

The results of such analyses can be used to estimate on a relative basis the value of various rock units, such as formations and members, as source rocks. This does not imply that rock units having higher values have necessarily yielded more hydrocarbons to the surrounding rocks because the yield depends also on permeability within the bed of source rock. A distinction is therefore made (Conybeare, 1965) between hydrocarbon-generation potential, which refers to the organic and hydrocarbon content that theoretically could produce petroleum, and the hydrocarbon-yield capacity which refers to the ability of the bed of source rock to expel the petroleum produced within it. Estimations of both factors suggest that in all petroleum-producing sedimentary basins only a very small amount, possibly a fraction of one percent, of the hydrocarbons generated has been trapped in reservoirs, the recovery from which is commonly only 20–40%.

Source rock analysis is of value not only in giving some approximation of the petroleum potential of a sedimentary basin but also as an indication of the possible source of oil in a reservoir. Where petroleum-bearing beds overlie an unconformity, it may be important to exploration to know whether the oil was derived from beds below or above the unconformity. This problem arises in western Canada where oil accumulations in Cretaceous sandstone beds overlie Devonian carbonates and shales, and in Queensland, Australia, where oil accumulations in Jurassic sandstone beds overlie Permian carbonates and shales. Hedberg (1968) showed that a high wax content in oils indicated an origin from plant material deposited with the sediment in fresh to brackish water. He further noted that although not all oils of freshwater origin are waxy, most of them are. The study suggests that a high wax content precludes an origin from marine carbonates. Methods including sulfur isotope (Thode et al., 1958) and trace element (Scott et al., 1954) analyses of possible source rocks and oils may be of use in tracing the provenance and migration path of oil. Other aspects related to analysis of the petroleum potential of a sedimentary basin will be discussed in Chapter 7.

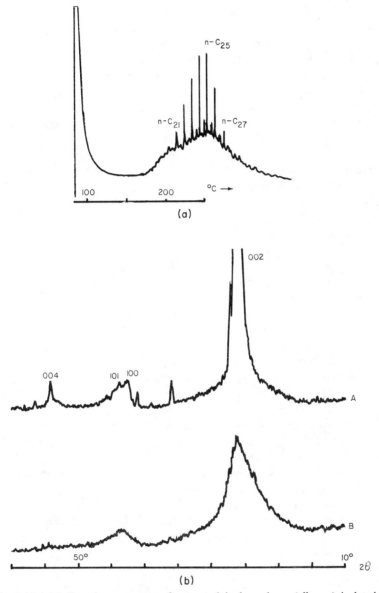

Fig. 5-28. (a) Gas chromatogram of saturated hydrocarbons (alkanes) isolated from stromatolites of the Early Devonian Cavan Limestone, New South Wales, Australia. [After McKirdy (1976).] (b) X-ray diffractograms of graphitic (A) and subgraphitic (B) kerogen from Proterozoic shales in South Australia. [After McKirdy *et al.* (1975).]

Clay Mineral Content Analysis

X-ray diffraction and electron microscope analyses of the clay mineral content of sedimentary rocks obtained from outcrops and subsurface drill cores are commonly undertaken with a view to determine one or more of the following problems: first, the provenance of the sediment, second, the degree of burial metamorphism to which it has been subjected, and third, the effect of clay mineral content on porosity and permeability. Provenance may be indicated by the known character of clay minerals washed from particular rock types into drainage systems. The depth to which the sediment has been buried, regardless of present depth, may also be indicated. This point is of particular significance to interpretation of the geologic history of a basin. Porosity and permeability are influenced to a large extent by the nature and amount of clay minerals within the matrix of a sandstone (Kharaka and Smalley, 1976).

Studies of the distribution of clay mineral types in modern sediments can certainly be related to areas of origin, as described by Potter *et al.* (1975), but whether such distributions in the subsurface can be related to sources (in view of the effects of burial metamorphism) is open to question. Potter *et al.* have shown that modern alluvial muds within an area of more than 3,000,000 km^2 in the Mississippi River Basin have two major clay mineral associations. One mud type is characterized by smectite, and term synonymous with the montmorillonite group of clay minerals that expand in water, some of which are derived from the alteration of volcanic ash. The smectite-bearing mud is derived from Cretaceous and Tertiary sedimentary rocks lying westward toward the Rocky Mountains. The other mud type, containing an abundance of illite, chlorite, and kaolinite, is derived from Paleozoic rocks to the east. Mineralogic changes with increasing depth of burial are reported by Hower *et al.* (1976) who note that smectite is largely converted to illite within the depth range 2000–3700 m. Other changes include loss of calcite, mica, and potassium feldspar, and an increase in the content of chlorite and quartz. They note also that compositional changes in shale, related to metamorphic grade resulting from depth of burial, are matched by compositional changes related to geologic age, a relationship illustrated by Fig. 5-29. This figure shows the distribution of clay mineral content is sedimentary rocks of North America. A marked feature is the decrease of expandable clay minerals, such as the smectite group, and of kaolinite with age and the increase of illite and chlorite.

The ability of water to move through a sedimentary rock can be not only a function of effective porosity but also of the clay mineral content within the matrix. This factor is of particular significance to the initial migration

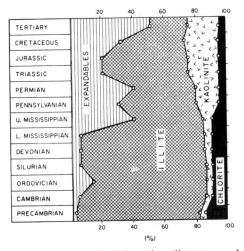

Fig. 5-29.   Distribution of clay minerals with age in sedimentary rocks of North America. [After Potter *et al.* (1975) and Weaver (1967).]

of oil or its precursors from source rocks into potential reservoir rocks. Kharaka and Smalley (1976) point out that the clay content of sedimentary rocks in the subsurface can act as semipermeable membranes and that movement of water is particularly effective within shales containing a high content of smectite. Of probable significance with reference to the transformation and migration of petroleum precursors is the finding of Johns and Shimoyama (1972) that certain clay minerals in the smectite group act as catalysts in promoting the decarboxylation of fatty acids to form long-chain alkanes and in promoting the subsequent cracking of these alkanes to shorter chain alkanes with a molecular distribution similar to those of petroleum.

With reference to exploration for petroleum in Cretaceous sandstones in Wyoming, an interesting application of clay mineral content in petroleum reservoir rocks is pointed out by Webb (1974). He says (p. 2245):

Thin-section, sieve, and clay-mineral analyses indicate that many of the clay-rich sandstones in Wyoming originally were deposited as clean well-sorted sands, the pore spaces of which were filled with either hydrocarbons or secondary clay cement after deposition and burial. Clay precipitation and hydrocarbon migration began after burial. Those sands not filled with hydrocarbons were subjected to continued precipitation of clay from the formation waters until effective porosity and permeability were eliminated. Later structural movements may position these "clay-sealed oil traps" in seemingly unfavourable structural positions. A thorough understanding of the depositional and tectonic development of the basin is necessary to prospect for this type of trap.

Carbonate Rock Analyses

Chemical, physical, and mineralogical analyses of limestones, dolomitic limestones, and dolostones can be undertaken for a variety of reasons. The results may be indicative of the depositional environment in which the carbonate beds were laid down or of the diagenetic processes to which they were subsequently subjected. Analyses may indicate whether the original carbonate was largely of aragonite, and to what extent and at what stage of burial of the sediment alteration to calcite and dolomite occurred. They may also indicate the original constituents of a carbonate rock such as abundant coccolith plates in chalk, the content of sand, silt, and clay minerals, and the presence of trace elements and diagenetic minerals. The conclusions reached from these analyses are commonly based on premises which may either be generally accepted or equivocal such as those deriving from the presence and concentration of certain trace elements. Where a certain relationship has been demonstrated to obtain, the validity of analysis is not often challenged; but the inferences placed on such analyses may be open to question. This has been stated succinctly by Davies (1977) with reference to his own work on the diagenetic sequence observed in Paleozoic reefs of the Canadian Arctic. He says (p. 11):

> By combining field observations and petrography with isotopic and elemental analyses and by comparing the results with modern submarine cement analogues, a reasonably convincing sequence of submarine cementation by magnesian calcite and aragonite, interupted by fracturing, has been constructed for the Artic Paleozoic reefs.

Conflicting interpretations from valid analyses are pointed out by Allan and Mathews (1977) who discuss the application of carbon and oxygen isotope analyses of limestones to problems of stratigraphy and diagenesis. With reference to Pleistocene surface limestones in the West Indies, they point out that the increase in ratio of C^{12} to C^{13} isotopes accompanying conversion of aragonite and high-magnesium calcite to low-magnesium calcite takes place on exposure to fresh water in the presence of carbon dioxide within the soil. These changes have previously been considered to be characteristic of diagenesis under subaerial conditions. The distinction is a fine one where diagenesis takes place above sea level but may be significant where meteoric water percolates downdip through carbonate beds and emerges as submarine freshwater springs.

The temperature at which carbonate sediments were deposited may have some bearing on interpretations relating to depth of the sea floor, ocean currents, latitude, and climate. Analyses of the oxygen isotope ratios in carbonates can be used as guides to the temperature range in which they formed. A modern example that illustrates agreement between such analyses and present-day water temperatures is cited by Neumann *et al.* (1977, p. 8) who analyzed samples from Pleistocene limestone mounds in the Straits of Florida. They say:

> Some preliminary geochemical data are available and are helpful. Oxygen isotope analyses of bulk biomicritic rock yield ratios that correspond to paleotemperatures of 5.6° and 8.4°C, and these are within the 5.5° to 10°C range of bottom temperatures recorded in the Straits of Florida today (Moore, 1965).

The limestone mounds sampled are indicated by radiocarbon dating to be 26,000–28,000 years old.

The analyses of sedimentary basins may also involve the problem of determining whether a particular sequence of beds was deposited in a marine or nonmarine environment. Where illite-bearing limestone or shale beds are present in the sequence, and in the absence of definitive fossils or stratigraphic features, the equivalent boron content of the illite has been used (Walker, 1964, 1968) as an indictor of salinity of the water in which the sediments were deposited. Lower equivalent boron values reflect lower salinity. He points out, with reference to the Early Carboniferous Yoredale Formation of northern England, that many zones within the limestone and calcareous shale sequence have variations of equivalent boron content which appear to reflect changes in salinity. He suggests that shale intercalations indicate temporary incursions of mud from a freshwater deltaic source. Such a conclusion might be drawn from other considerations of the stratigraphic sequence but may be substantiated by equivalent boron analyses.

Other trace element analyses may serve different purposes. An interesting application of strontium and barium analyses to exploration for petroleum is described by Weaver (1968) with reference to the oil- and gas-bearing Late Devonian stromatoporoid and algal reefs of western Canada. These reefs are flanked by widespread greenish gray shale beds of the Ireton Formation and underlain by the brownish calcareous shale and argillaceous limestone beds of the Duvernay Formation. Weaver analyzed selected drill cuttings from wells at various distances from reefs and found that the strontium and barium contents increased markedly within 8 km of the reefs. The strontium content in particular is a good indicator of proximity to reefs. Weaver noted that the strontium is related to the calcium content which he thought reflected the presence within the Ireton shale beds of fine, reef-derived carbonate material. This seems to be a reasonable explanation, as seismic surveys indicate that the shale beds adjacent to the flanks of some reefs are sufficiently calcareous to show very little velocity contrast to the reefs.

Physical analyses of carbonate rock cores may reveal a hitherto unsuspected degree of compaction. Carbonate reefs, calcirudites, and calcarenites are generally thought to be only slightly compactible, but calcilutites appear to be readily compactible. An interesting feature of the compaction of carbonate muds has been noted by Shinn *et al.* (1977, p. 21) who says:

> Compression of an undisturbed carbonate sediment core under a pressure of 556 kg/cm^2 produced a "rock" with sedimentary structures similar to typical ancient fine-grained limestones. Surprisingly, shells, foraminifera, and other fossils were not noticeably crushed, which indicates that absence of crushed fossils in ancient limestones can no longer be considered evidence that limestones do not compact.

Compaction and Diagenesis

Lithostratigraphic analysis of a sedimentary basin involves not only the stratigraphic sequence and framework present today but also the evolution of these features during the development and subsequent history of erosion of the basin. These studies are concerned with compaction and diagenesis which are prime factors in the history of paleopressure and paleothermal gradients within the basin. An understanding of the changes which these gradients have undergone during the periods of growth and erosion of a basin is fundamental to the interpretation of hydrocarbon generation and migration within a basin. With these considerations in mind, interpretations of paleothermal gradients can be based on analyses and optical determinations of diagenetic minerals present in rock samples and on the homogenization temperatures of liquid-gas inclusions in quartz veins cuttings of the rock samples (Currie and Nwachukwu, 1974).

The technique used by Currie and Nwachukwu involves microscopic examination of an inclusion, while the thin section in which it is contained is gradually heated. The homogenization temperature is that at which the gas bubble disappears. When this temperature is equated with the same temperature obtaining today in similar sedimentary basins having average geothermal gradients, an approximate estimate can be made of the maximum depth at which the quartz vein formed and presumably to which the rock sample was subjected. Interpretations of paleopressure gradients can be based on data from sonic surveys of boreholes, used with reference to shale–compaction data and the geothermal data obtained from studies of homogenization temperatures.

With reference to the Mesozoic sedimentary basin of western Canada, Magara (1976) has shown that integrated studies such as these have indicated removal by erosion of up to 1400 m of sedimentary rocks and that during the period of maximum burial of the Late Cretaceous Cardium Sandstone, from which samples were examined, the paleogeothermal gradient of the basin probably varied only slightly from the present geothermal gradient. The effects of removal by erosion of the upper layers in a sedimentary basin, resulting in lower pressures at depth and the development of fracture porosity, have been pointed out by Currie (1977). On the other hand, as discussed by Stephenson (1977), the systematic reduction of pore porosity with increasing depths of burial, and possibly also with time, cannot be quantified with respect to the relative effectiveness of any of these parameters. The significance of paleogeothermal gradients with respect to the generation of hydrocarbons in a sedimentary basin has been discussed for many years and is reiterated by Pusey (1973, p. 1836) who says:

> Concentrations of liquid hydrocarbons in modern sediments are only a few parts per million, but hydrocarbons in subsurface shales reach several thousand parts per million. The increase in hydrocarbons is the result of the alteration of insoluble organics called kerogens. The primary energy for the reaction is heat associated with burial. Significant hydrocarbon generation begins at 150°F, but liquid hydrocarbon destruction dominates at temperatures greater than 300°F.

The effects of diagenesis on porosity in hydrocarbon-bearing sandstones of the Pennsylvanian Morrow Formation of the Anadarko Basin in Oklahoma have been studied by Adams (1964). He describes several types of fabric recognized in thin sections of quartz sandstones illustrated in Fig. 5-30 and concludes that a knowledge of diagenesis and its relationship to the fabric of potential reservoir rocks has significant application to the search for and production of hydrocarbons. Adams discusses the effects of pressure–solution on the reduction of porosity in a quartz sandstone and relates different types of fabric to diagenetic alterations of different matrix compositions that reflect variations of offshore depositional environments. The clean, well-sorted quartz sandstones were

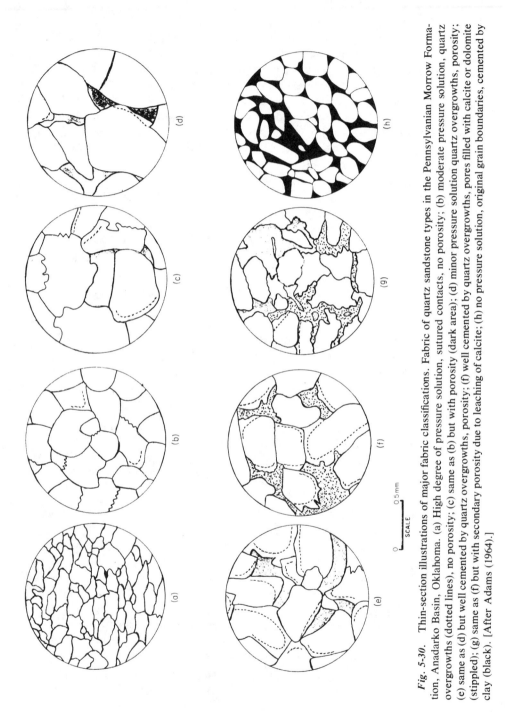

Fig. 5-30. Thin-section illustrations of major fabric classifications. Fabric of quartz sandstone types in the Pennsylvanian Morrow Formation, Anadarko Basin, Oklahoma. (a) High degree of pressure solution, sutured contacts, no porosity; (b) moderate pressure solution, quartz overgrowths (dotted lines), no porosity; (c) same as (b) but with porosity (dark area); (d) minor pressure solution quartz overgrowths, porosity; (e) same as (d) but well cemented by quartz overgrowths, porosity; (f) well cemented by quartz overgrowths, pores filled with calcite or dolomite (stippled); (g) same as (f) but with secondary porosity due to leaching of calcite; (h) no pressure solution, original grain boundaries, cemented by clay (black). [After Adams (1964).]

348

deposited under higher-energy conditions in an inner neritic environment, whereas the calcareous, poorly sorted, glauconitic, and clayey sandstones were deposited under lower-energy conditions in a middle neritic environment. Diagenesis of different sandstone types, under the same conditions of pressure and temperature, produces various textures and degrees of porosity. Interpretations based on microscopic examination of rock samples and of thin sections made from the rock samples must be accepted with some reservation. In general, interpretations of carbonate rocks are more reliable, for although they are more easily altered by diagenesis than quartzose or lithic sandstones, their original fabric is more diagnostic of depositional environment. Further discussion on the efficacy of this approach to interpretations of depositional environment is offered by Selley (1970).

Grain Orientation

Experimentation in the laboratory and sampling in the field have shown that the long axes of sand grains tend to be aligned parallel to the direction of flow of the current. On a beach, for example, the main direction of water movement results from the action of waves breaking in the shallows and producing a strong current down the slope of the beach. Oriented samples of beach sand can be obtained by means of a short cylinder which is thrust into the sand. A liquid plastic that hardens quickly is then poured into the cylindrical mould, the orientation of which is marked. On hardening, the core of sand can be examined by thin section or other means (Conybeare and Crook, 1968). Statistical analyses of the preferred orientation of sand grains in samples of beach sand have shown a mean alignment normal to the beach. This orientation is confined to the narrow zone of wave action which may not be commonly preserved in the geological record. Farther from shore, where longshore currents control dominant direction of flow, the preferred orientation of sand grains tends to be parallel to the coastline. This internal feature is known to be preserved in sandstones (Nanz, 1960; Shelton and Mack, 1970; Winkelmolem, 1972) and is obviously a valuable indicator of paleocurrent directions which determine the geometry and depositional trends of sandstone bodies.

Early investigations of the preferred orientation of grains in sandstones were carried out by means of a petrographic microscope; but these tedious optical methods were supplanted by rapid analyses that measured the dielectric anisotropy of the sandstone. It was found that this was related to the preferred orientation of grains within the sandstone sample. Shelton and Mack (1970, p. 1119) say:

Dielectric anisotropy measurements of grain orientation have been favorably calibrated to (1) palecurrent indicators, such as parting lineation, rib-and-furrow structures, sole marks, and cross-bedding, (2) upcurrent grain inbrication, (3) local current trends of Cimarron River in partly impregnated samples, and (4) sandstone trend of the lowermost unit of the Eagle Sandstone at Billings, Montana, and of three lenticular Permian sandstone beds.

The Eagle lies within the Upper Cretaceous sequence. Later, reservations were placed on this method, as stated by Winkelmolen (1972, p. 2159) who says:

Although the grain long-axis fabric is important in establishing the dielectric anistropy of cylindrical sediment plugs, the observed anistropy is not related exclusively to the grain orientation. Crystal lattice effects also are present and the influence of bedding phenomena is marked wherever this bedding makes an angle with the plane of rotation of the plug.

These reservations do not negate the usefulness of the method in its application to interpretations of depositional trends but emphasize the necessity for caution and for the integration of laboratory data with field data.

6

Construction of Maps and Sections

SEDIMENTARY BASIN CONFIGURATION

The principal aim of sedimentary basin analysis is to depict the solid geometry, structural, and stratigraphic framework of a basin as it is today. In order to do this, it is necessary to rely on the available geological and geophysical data. The quality and quantity of control provided by areal coverage of a basin, survey methods employed, closeness of boreholes, and applicability of surface geological mapping place constraints on the validity of interpretations. As new data come to hand, modifications can be made to the model constructed. This model embodies a consolidation of all the features and parameters that show agreement, or appear to be compatible, in a variety of maps and sections of sequences within the basin. These commonly illustrate structure, thickness, growth, unconformity features, basement topography, and various facets of lithofacies and biofacies but may also include specialized maps such as the distribution of trace elements, maturation of kerogen, radioactivity of particular beds, and many other characteristics or diagnostic features. Many such maps may prove to be interesting but not particularly helpful from the point of view of explaining stratigraphic relationships, depositional environments, or the history of the basin. This lack of a definitive nature may stem from poor control, too broad a treatment of the data, or insufficient knowledge, experience, or imagination on the part of the person whose responsibility it is to analyze the information. Such maps or sections should never be discarded, as their relevance may become apparent in the light of future investigations. The construction of a model illustrating the present-day structural and stratigraphic parameters and relationships within a sedimentary basin has direct applications to exploration for hydrocarbons, coal, and minerals, to water supply from subsurface aquifers, and to the disposal of radioactive and other chemical wastes. These matters will be discussed in Chapter 7.

A secondary aim, which from an academic and in some cases from a pragmatic, point of view also may be no less important, is the reconstruction of the tectonic, structural, and stratigraphic development of a basin from the time of its origin to the present. The history of a basin may involve periods of uplift and erosion followed by subsidence and accumulation of sediments. In tectonically active regions lateral thrusting and uplift of the earth's crust flanking a basement may also be a significant factor in the solid geometry, structural framework, and sediment provenance of the basin. What has gone on in the past will in one way or another be reflected in the present. This is particularly true with reference to the migration of hydrocarbons. An example can be made of the folded structures west of the oil- and gas-producing structures of the Turner Valley Field in southern Alberta. When westerly folds were drilled, they proved to dry, presumably because they had not obtained closure at the time of hydrocarbon migration into the nearby easterly structures (Link, 1949). The growth of folds and faults in consolidated and unconsolidated sedimentary strata, the tilting of strata resulting from tectonic movements or compaction of a sedimentary pile, and the uplift and erosion of strata are of particular significance to the migration and entrapment of oil and gas during the history of development and denudation of a basin.

In order to interpret the present internal structural and stratigraphic features of a sedimentary basin, it is obviously necessary to make extensive use of geophysical data, particularly in areas that have been sparcely drilled. Initial investigations of comparatively unexplored areas include surface mapping, where applicable, and airborne geophysical surveys. The latter include gravity, magnetic, radioactivity, and infrared line scanning surveys. Later exploration for specific structural features will include reflection and refraction seismic surveys, the latter being particularly applicable in folded and faulted belts such as those of the surbsurface in the region of foothills flanking a mountain range. Use may also be made of remote sensing methods employed in satellite surveys. Geological, petrophysical, and geophysical information obtained from boreholes affords direct contact with the subsurface and offers a means of checking and coordinating the results obtained from surface and airborne geophysical surveys. It is consequently of obvious advantage to the geologist working on subsurface problems relating to exploration, or to more esoteric and academic projects, to have a firm understanding of the application of geophysical methods to geological problems. The reverse is equally true, that geophysicists working on subsurface investigations should have a good understanding of geology with respect to the area being studied. As stated earlier in this book, the ideal approach incorporates a basins study group comprising both geologists and geophysicists who work cooperatively.

The construction of interpretative maps and sections commonly requires some degree of vertical exaggeration in order to accentuate the structural and stratigraphic features. It is always advisable to keep this exaggeration to a minimum as excessive exaggeration disproportionately distorts the structures and stratigraphic relationships. Such distortions can cause gently dipping faults to appear nearly vertical, and reefs, which in their true perspective may be tabular and thin like an irregular pancake, are shown as pinnacles. As illustrated in Fig. 6-5, which has a vertical exaggeration of approximately four, most sedimentary basins viewed in sections drawn to their true proportions are remarkably shallow with respect to the earth's crust; also, their internal stratigraphic features may appear to be lamellar and flat. Interpretations that depend to some extent on the attitude of strata within a basin, such as those referring to hydrocarbon migration or gravitational sliding, must be seen in context with the true proportions of the sedimentary basins.

Structural and Stratigraphic Framework

Most sedimentary basins have not developed from simple downwarping of the basement but rather from downwarping accompanied by downfaulting which is commonly of the step-faulted variety (Fig. 6-1). The extent to which each type of movement has contributed to tectonic subsidence beneath the basement, and to the consequent structural framework within the lower part of the basin, is not always easy to determine. Continual movement, particularly of basement fault blocks, may have a significant influence on the sea-floor topography and consequently on the pattern and rate of growth of biohermal development. These will in turn be reflected in the stratigraphic framework of the basin.

An example of how the structural framework of the basement is reflected in the stratigraphic framework and geophysical characteristics of

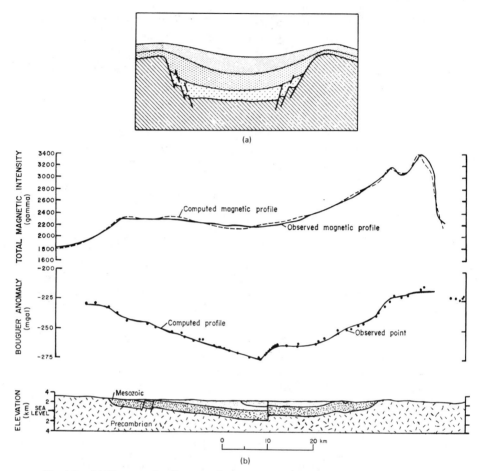

Fig. 6-1. (a) Diagram of sedimentary basin overlying hypothetical graben showing possible effects of faulting on the thickness of strata in basin. [After Whittaker (1975).] (b) Section across North Park Basin, Colorado, based on magnetic intensity and gravity survey traverses showing relationships of geophysical data to basement topography. [After Behrendt *et al.* (1969).]

the Illinois Basin of the United States is given by McGinnis and Ervin (1974, p. 517). They say:

> Historic earthquakes in Illinois occur on faults that have been intermittently active since Precambrian time. Vertical deflection of the lithosphere continues, but at an abbreviated pace. Evidence for this statement follows: (1) blocks of underformed and coherent Precambrian lithosphere are outlined by gravity closures; (2) Paleozoic strata

are folded and, in most cases, unfaulted over Precambrian blocks that have undergone vertical, differential displacements; (3) gravity closures, which contain negative residuals, overlie crustal blocks that have moved upward relative to blocks that are overlain by gravity closures, which contain positive residuals) and (4) historic earthquakes are not located within closures but are in high gravity gradient areas on the edges of closures or in the lower gradient areas lying between closures.

The principal methods employed in determining the configuration of basement are gravity and magnetic surveys, both of which can be carried out on land or by means of airborne instruments. The data require expert interpretation by geophysicists familiar with the lithologic nature of the basin, as particular geological relationships may be variously interpreted. For example, a thick layer of basalt within a sequence of sediments may be incorrectly interpreted as basement on maps showing the contoured variations of magnetic intensity.

Gravity surveys employ gravity meters that measure the variations of the earth's gravity at different locations. These variations result from differences in total mass between the instrument and the earth's center, the differences depending largely on the thickness of strata overlying basement at various localities. Magnetic intensity surveys employ magnetometers which measure variations in the earth's magnetic field which is influenced by the lithology of the basement and its proximity to the surface. The control points derived from both of these surveys can be plotted and presented as maps showing contoured values. From these maps sections may be constructed to illustrate the structural framework, as shown in Fig. 6-1. Diagram (a) in this figure is of a hypothetical graben drawn by Whittaker (1975) to show the possible effects of basement faulting on later sedimentation. Sections (b), based on magnetic intensity and gravity surveys, were presented by Behrendt et al. (1969) to show the basement profile of the North Park Basin in Colorado. A major fault within the basin is particularly well defined by the gravity survey.

Gravity surveys may also clearly show the regional relationships of rock-type distribution, as illustrated in Figs. 6-2 and 6-3, both of which are taken from St. John (1970). Fig. 6-2 is a gravity map of Papua New Guinea showing free air anomalies contoured in intervals of 10 mgal. The contours relate closely to the geological structures map shown in Fig. 6-3. This map shows the distribution of rock types, within the upper crust, that are exposed or near the surface. Throughout the center of the peninsula, extending northwest into the highlands of New Guinea, there is a basement ridge flanked by basic and ultrabasic rocks. This separates the

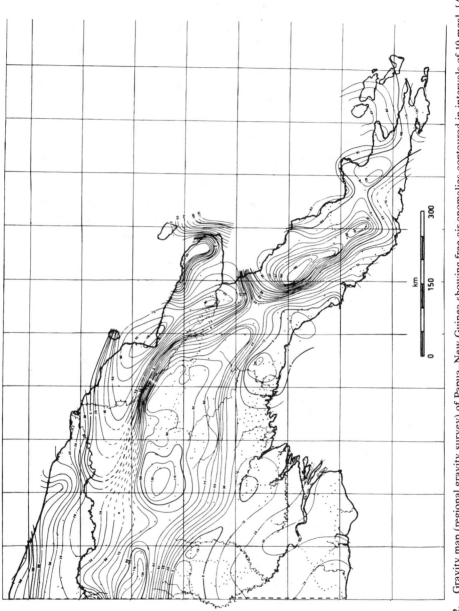

Fig. 6-2. Gravity map (regional gravity survey) of Papua–New Guinea showing free air anomalies contoured in intervals of 10 mgal. [After St. John (1970).]

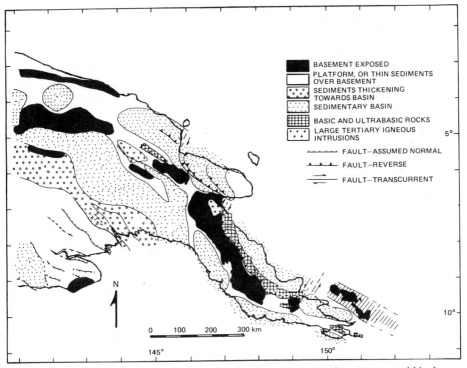

Fig. 6-3. Geological structures map of Papua–New Guinea showing structures within the upper crust of eastern New Guinea and ridge of exposed basement separating the eugeosynclinal New Guinea Basin on the northwest from the miogeosynclinal Papua Basin on the southwest. [After St. John (1970).]

eugeosynclinal New Guinea Basin on the north from the miogeosynclinal Papua Basin on the south. The Oriomo promontory to the south of the Papua Basin is underlain by granitic basement on which gently dipping Mesozoic strata are truncated. These major geological structures can be outlined by geophysical surveys, as stated by St. John (1970, p. 41) who says:

> Interpreted in the light of existing geological knowledge a regional gravity survey of Eastern New Guinea defines the major tectonic units of the area and clarifies its tectonic history.

The major stratigraphic and structural famework of the Papua Basin, thickening off the exposed Oriomo basement and terminating abruptly in a folded and faulted belt flanking the basement ridge to the northeast, is similar to that of many sedimentary basins throughout the world, including the much smaller Sacramento Valley Basin of California, shown in

Fig. 6-4. The latter has a maximum thickness of more than 10,000 m of Jurassic and Cretaceous strata comprising lithic sandstone, siltstone, and mudstone with minor conglomerate and bentonitic layers derived in large part from the east. According to Ojakangas (1968), graded bedding is common and, with other paleocurrent structures, suggests that many of the beds were deposited as turbidites in outer neritic to upper bathyal depths. Subsequent crustal compression resulting from plate tectonics has folded and faulted the western fringe of the basin during a period of uplift and erosion.

Present-day sedimentary basins underlain by Tertiary beds are of particular interest as models for the study of relationships between sedimentation, stratigraphic configuration, and structural framework. Many studies have been made of the Mississippi Delta, including those of Fisk *et al.* (1954) and Bernard *et al.* (1970). These have shown some of the relationships between present-day sedimentary features and subsurface stratigraphic features revealed by drilling and geophysical surveys. More recent studies of the Niger Delta, including those of Weber (1971), Merki (1972), Mascle *et al.* (1973), and Evamy *et al.* (1978) have shown the contemporaneous development of growth faults, bed thickening, and other features characterizing the structural and stratigraphic framework of the Niger Basin. The major tectonic feature, shown in Fig. 6-5, is the development of the basin within a major and complex graben-type rift underlying the adjoining edges of the continental and oceanic crusts. Evamy *et al.* (p. 1), say:

> The development of the delta has been dependent on the balance between the rate of sedimentation and the rate of subsidence. This balance and the resulting sedimentary patterns appear to have been influenced by the structural configuration and tectonics of the basement.

The structural framework of a sedimentary basin is related to the regional tectonic framework in which it develops. Separate basins may be formed along a trend of tectonic subsidence, or subbasins may be developed within a broad regional basin. In the latter case the lower strata of the subbasins may not be of the same age whereas the upper beds may be correlated. An example is illustrated in Fig. 6-6 which shows the Devonian Black Creek Basin of northern Alberta underlying the much more extensive Devonian Elk Point Basin. According to McCamis and Griffith (1968), sedimentation in the Black Creek Basin began with the development of reefs and the deposition of associated open marine sediments. With uplift and withdrawal of the sea, the basin became restricted and a salt bed was formed. Renewed subsidence between the landmasses of the

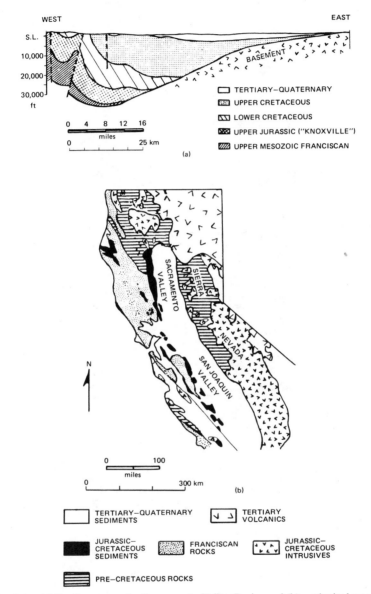

Fig. 6-4. (a) Section across the Sacramento Valley Basin, and (b) geological map of the region surrounding Sacramento Valley, California. [After Ojakangas (1968).]

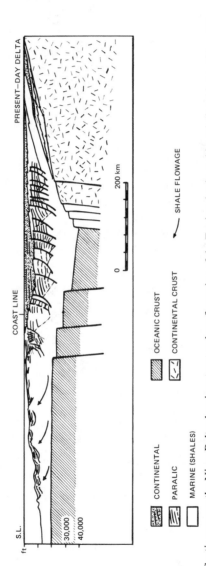

Fig. 6-5. Section across the Niger Delta showing structural configuration of the Tertiary basin and its relationship to rifting of the continental and oceanic crusts. Vertical scale in feet. Vertical exaggeration approximately 4. [After Evamy *et al.* (1978).]

360

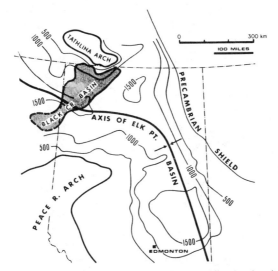

Fig. 6-6. Isopach map of the Devonian Elk Point Group, Alberta, showing the Devonian Black Creek Basin which underlies and forms a subbasin of the Elk Point Basin. [After McCamis and Griffith (1968).]

Precambrian Shield and the Peace River Arch resulted in widespread deposition of evaporites and associated sediments within the Elk Point Basin. This example illustrates the development of separate basins or subbasins on a continental shelf environment. Recent development of separate but adjacent or contiguous sedimentary basins is illustrated in Fig. 6-7 (Kesler *et al.*, 1974) which shows the Yucatan Basin, Columbia Basin, and Venezuela Basin underlying the Caribbean Sea. These basins, partly separated from each other by marine rises, are enclosed between Central and South America to the south, and the Greater Antilles to the north. Sedimentation within these basins is commonly taking place at depths greater than 2000 m in bathyal to abyssal environments. By way of comparison in size the present distribution of the Late Mississippian Chester Series (Fig. 6-8) in the Illinois Basin would easily fit into any of these Caribbean basins. By way of contract in depositional environments, the present-day sediments of the Yucatan, Columbia, and Venezuala basins comprise deep-sea deposits including turbidites, whereas the Chester Series comprises rhythmic alternations of paralic and inner neritic sedimentary units consisting predominantly of sandstone and limestone respectively. These units were probably deposited in environments which fluctuated from deltaic to inner continental shelf (Swann, 1964).

A further comparison in size can be made between the area of the Nicaragua Rise (Fig. 6-7) and the several subbasins of depositions on the

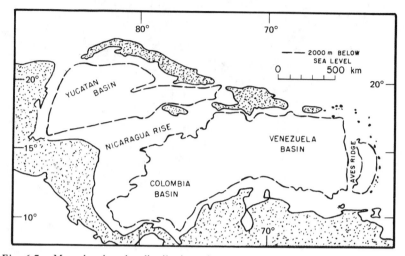

Fig. 6-7. Map showing the distribution of recent sedimentary basins underlying the Caribbean Sea. [After Fox *et al.* (1971) and Kesler *et al.* (1974).]

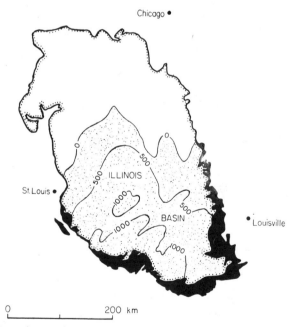

Fig. 6-8. Isopach map showing distribution of the Late Mississippian Chester Series in the Illinois Basin, United States. Area of outcrop shown in black. [After Swann (1964).]

continental margin off the coast between Santa Barbara and San Diego, California, as shown in Fig. 6-9. These subbasins are irregular, up to 100 km in length, and have somewhat linear trends parallel to the coast. They form centers of deposition, predominantly of clay and silt interbedded with turbidites, situated between rises in a topography controlled by block faulting. Deposition on the rises is very slow, consisting largely of pelagic debris, but in the deep areas rates exceed 20 cm per thousand years. According to Emery and Bray (1962), these rates are comparable to those which occurred during the Pliocene Epoch in the Ventura and Los Angeles Basins.

Fig. 6-9. Centers of deposition off the coast of California between Santa Barbara and San Diego. Rate of deposition is given in centimeters per thousand years. Centers are low areas in a submarine topography controlled by block faulting. [After Emery and Bray (1962).]

Structural Maps and Sections

Structural maps are analogous to contoured topographic maps in that they show the relief of a particular surface, or top of a stratigraphic zone, with respect to a horizontal datum. This datum is commonly sea level but may be any elevation parallel to sea level. The surface contoured may be basement topography, an unconformity, or the top of a stratigraphic unit such as a bed or bioherm. A stratigraphic zone may be defined by the first occurrence in drill cuttings of characteristic minerals such as glauconite, chamosite, and siderite or of fossil assemblages and particular fossils such as ostracods, forams, fish scales, and conodonts. With the exception of recognizable internal columnar structure seen in fragments of bivalve shells, the macrofossils are commonly not readily identified in drill cuttings. Structure maps are also based on seismic data tied-in where possible to sonic surveys of wells and consequently to known stratigraphic horizons. These maps are contoured in milliseconds, one-way or two-way time, and show the variations in depth to a particular reflecting bed. Some beds may be widespread and have a marked velocity contrast with the overlying bed, in which cases there is little difficulty in tracing them on seismic records and constructing a map to show their structure. Other reflecting beds may cause problems because they are not sheetlike within the area of study but lens out or change facies. In these cases miscorrelation of records may occur. Also as with structural maps based solely on well control, lack of data may seriously impair the validity of interpretation, particularly where questionable contours involve the possibility of closure.

Structural maps may be regional or local and may show the major tectonic features and configuration of a sedimentary basin or the local structures associated with a fault system or reef complex. An example of a structure map that delineates the general shape and tectonic relationships of a basin is illustrated in Fig. 6-10 (Harms, 1966) which shows the structure of the Denver Basin of Colorado, Wyoming, and Nebraska. The western flank of the basin is terminated by a folded and faulted belt of the mountain system, the deepest part of the basin forming a linear trough parallel and adjacent to the fold belt. The eastern flank gradually rises so that from east to west the basin is markedly asymmetrical.

The main structural features of a sedimentary basin can be mechanically defined by computers which are necessary tools for modern geomathematical approaches to geological problems (Davis, 1975; Broin and Schenck, 1977; Szumilas, 1977). But there is no substitute for contouring by geologists who understand the various geological situations that may obtain and who emphasize the possibilities favorable to exploration.

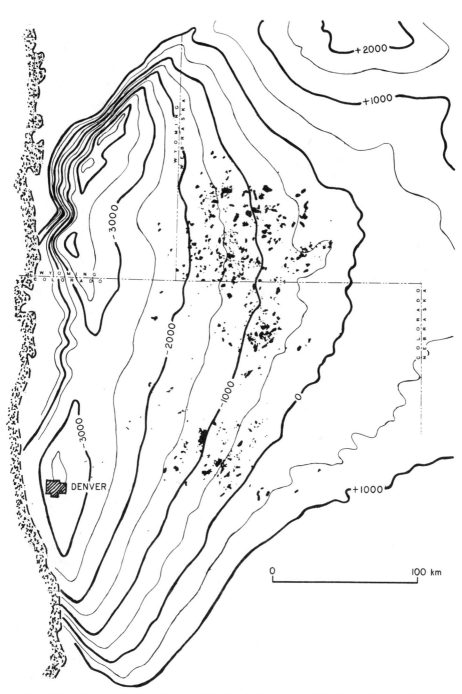

Fig. 6-10. Structural map of the top of the Early Cretaceous J Sandstone, Denver Basin, United States. Stippled area is folded and faulted belt. Oil fields shown in black. Contours in feet above and below sea level. [After Harms (1966).]

365

Rapid analysis by means of computer is normally a prelude to more detailed and imaginative interpretation by individuals. These later maps should not invalidate the data, or press interpretations beyond reasonable bounds, but must highlight both probable and possible prospects.

A structural section shows the configuration of strata, with reference to a datum, in a particular slice of a sedimentary basin. As with structural maps the datum is commonly sea level but may be any horizon parallel to sea level. The section may be based entirely on subsurface information obtained from boreholes such as oil wells or it may be constructed from a series of structure maps of various beds within a stratigraphic sequence. Commonly it is based on both, the contour maps serving to fill gaps between the proven control of boreholes. Conversely, the borehole data may require the contours to be modified, so the process of adjusting structural sections to structural maps on the basis of borehole data and interpretive contours is one requiring the best-fit method. It is essential, unless the section is to be illustrated as a fence diagram, that the section be as straight as possible, otherwise changes of azimuth will introduce changes in depth that will distort the apparent structure as seen in two dimensions. For this reason it is necessary to show the point-to-point line of section in cases where it deviates appreciably from a straight line. Another point is that excessive vertical exaggeration can seriously distort the structure shown. Some exaggeration is necessary, both for emphasis and convenience of drafting, but it should be minimal. This is a consideration of particular importance in viewing structures such as folds, faults, and reefs. Viewed in their natural scale, most sedimentary basins appear to be extremely shallow, which may prompt one to speculate further on the mechanisms of migration of petroleum. An example of the effect of vertical exaggeration is shown in Fig. 6-11 which is a nearly straight line section across oil-bearing Miocene reefs of the Salawati Basin in the western extremity of Irian Jaya, Indonesia, as described by Vincelette and Soeparjadi (1976). These reefs are bounded by a normal fault shown to be steeply dipping at about 60°. The vertical exaggeration is five which is comparatively small compared with many sections in which the factor is ten or more. Shown to natural scale, the reefs would appear to have shapes like mesas or buttes, and the fault would dip about 20°.

In the early stages of exploration of a sedimentary basin, where geophysical coverage is inadequate for interpretation of the basin as a whole and where borehole data is scanty or nonexistent, it is necessary to construct structural maps and sections in a more generalized and imaginative way. Structural maps can be drawn by means of outlining the general distribution of areas above or below an arbitrary datum. This practise may depend heavily on extrapolation but broadly shows the trends of major

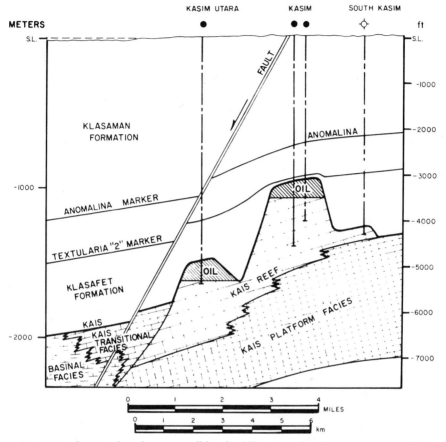

Fig. 6-11. Structural section across oil-bearing Miocene reefs in the Kasim Field of Irian Jaya, Indonesia. [After Vincelette and Soeparjadi (1976).]

structural features. The Exmouth Plateau, situated off the northwest continental shelf of Western Australia in depths to 4000 m (Fig. 6-12), is an example of a region where this approach has been useful, as shown by Exon and Willcox (1978). This figure illustrates the general features of the structure of seismic horizon C which is a slight angular unconformity at the base of the Late Cretaceous carbonate sequence. This horizon is a reliable tie to several wells of the Rankin Platform on the outer fringe of the continental shelf.

Structural sections may also be mainly interpretive in the absence of subsurface control. An example is shown in Fig. 6-13 which illustrates in a block diagram the surficial and subsurface structural features of the

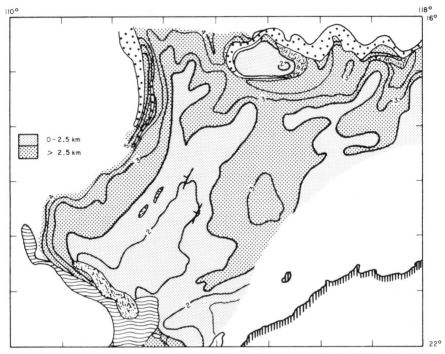

Fig. 6-12. Generalized structural map of Late Cretaceous seismic horizon C within offshore Exmouth Plateau, Western Australia. Zero datum is sea level. Cross-hatching shows Barrow Island oil field and mainland. [After Exon and Willcox (1978).]

Blatherskite Nappe in the Northern Territory of Australia, as interpreted by Stewart (1967). This structure occurs in Late Proterozoic beds of the basal sequence in the Amadeus Basin. The section has no vertical scale because there is no subsurface control, and consequently the interpretation of detail is somewhat equivocal. Other types of structural sections depict not only the structure as it may be today but also the possible evolution of the structure, as seen in Figs. 5-1 and 6-14. These types of structural sections serve a useful purpose in explaining not only how a particular structure may have formed but also its relationship to other events such as the migration of petroleum. An older fold belt may contain hydrocarbons, whereas a younger adjacent belt may be barren. The Turner Valley field of southern Alberta is a good example of this relationship (Link, 1949). Fig. 6-14 offers an explanation for the structure shown in Fig. 6-13. The Heavitree Quartzite, which rests on basement and is the basal member of the Proterozoic sequence, forms a prominent ridge along the MacDonnell Ranges of central Australia. This competent quartzite

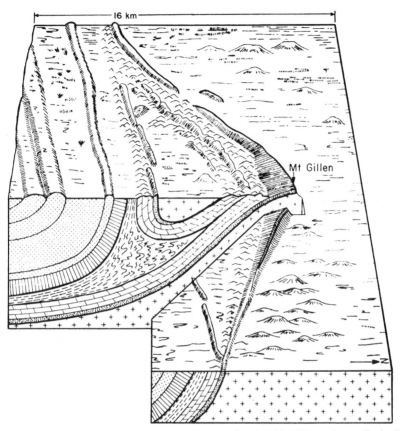

Fig. 6-13. Diagrammatic block diagram of the Blatherskite Nappe in an Early Proterozoic sequence, Amadeus Basin, Australia. [After Stewart (1967).]

formation is overlain by less competent dolomite beds of the Gillen Member and by incompetent shaly siltstone beds of the Loves Creek Member, both of which form the Bitter Springs Formation.

Isopach Maps

Isopach maps show the contoured variations in thickness of a stratigraphic unit which may be a member, formation, or any sequence of rock layers defined with reference to its upper and lower limits. Control may depend entirely or in part on measured sections of outcrops, on borehole data, or on seismic reflection surveys. Where possible all such information may be incorporated in the one map with palinspastic adjustments for

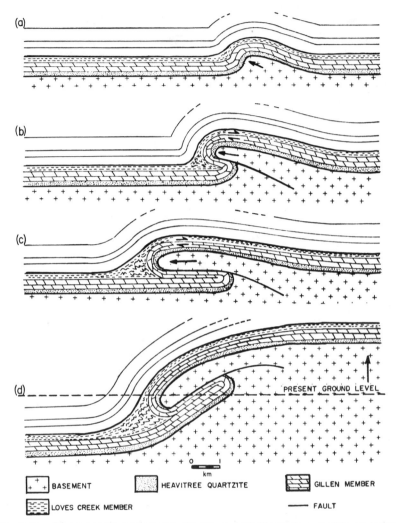

Fig. 6-14. Diagrammatic sections (a), (b), (c), and (d) illustrating possible evolution of Blatherskite Nappe (Fig. 6-13) in the Early Proterozoic sequence, Amadeus Basin, Australia. [After Stewart (1967).]

folding, faulting, and regional tilt. The latter is commonly not a significant factor, although it may appear so in sections showing considerable vertical exaggeration. As with structural maps, the contouring can be done mechanically or with imagination to illustrate the possible relationships between the stratigraphic interval mapped and contemporaneous tectonism, depositional environment, or sedimentation. For example, thicker sections of sediment may be deposited along a linear trend overlying the downthrown side of a normal fault or within basinlike depressions overlying sinking blocks of basement rock. Conversely, where the measured sequence overlies an unconformity, the thinner sections may overlie upthrown fault blocks, buried hills, and uplands of the topography preserved as an unconformity.

An isopach map may also indicate the shape of a stratigraphic unit and consequently its probable origin and depositional environment. Where the contoured unit is discrete and of limited extent, its shape may indicate or suggest its original geomorphology as a carbonate bioherm or sandstone body. Some caution is needed in such interpretations since, in the absence of definitive petrologic or petrophysical data, narrow and linear sandstone trends may be interpreted either as ancient river tributaries or barrier bars. Also reef-shaped sections of carbonate may be erosional remnants. In these respects the validity of interpretations depends very considerably on the background of geological knowledge, with particular reference to the local stratigraphy, available to the person constructing the isopach map. Essentially, an isopach map should accentuate the positive and the negative by closing contours where it is reasonable to do so thereby highlighting the distribution and shape of thick and thin parts of the stratigraphic unit.

On a regional basis series such as those of the Late Cretaceous Epoch, systems such as those of the Cretaceous Period, and even giant megaunits such as Erathems of the Mesozoic Era may be contoured to show the distribution and thickness of sedimentary basins and subbasins. In this respect it is essential to clearly differentiate depositional from erosional edges. This may be difficult or impossible to demonstrate in the internal stratigraphic framework of a basin for although the exposed or uppermost subsurface layers have suffered only minor erosion, there may be major unconformities within the sequence. In such cases further light may be shed on the possible thickness and consequent regional distribution of major stratigraphic units by investigations of compaction and diagenesis of the subsurface layers (Magara, 1976). Another point is that the regional direction of thickening of a major stratigraphic sequence is not necessarily in a seaward direction, as shown in Fig. 6-5 of the sedimentary pile underlying the Niger Delta. The pile does indeed thicken seaward to the region of maximum crustal subsidence of the basement but then thins seaward over oceanic crust.

An example of an isopach map illustrating regional distribution of the Mesozoic Erathem in northwestern United States is shown in Fig. 6-15 (Claypool *et al.*, 1978). This map clearly shows the distribution of three separate Mesozoic basins. It will be noted that in the region of Wyoming

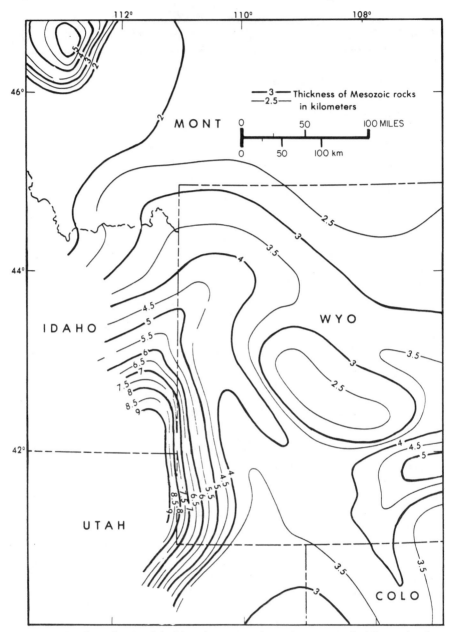

Fig. 6-15. Isopach map of the Mesozoic sequence in northwestern United States showing three main basins. [After Claypool *et al.* (1978).]

the contours could have been drawn to accentuate the distribution of areas where the Mesozoic sequence is either thick or thin. As illustrated, the contours do neither but allow room for interpretations either way. This cautious approach is warranted where control is insufficient to justify a decision one way or the other and where other geological factors have no apparent bearing.

At the other end of a scale of magnitude are isopach maps of individual bodies such as sand bars, river distributary channel sands, and carbonate reefs. This type of isopach map requires very close control, whereas a regional map may be constructed by extrapolation between scattered points of control. An example of an isopach map based on closely spaced subsurface control of a sandstone body is shown in Fig. 6-16. This illus-

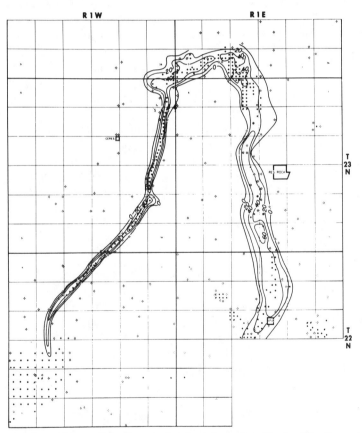

Fig. 6-16.   Isopach map of an oil-producing river channel sand body of the Pennsylvanian Red Fork Sandstone (South Ceres pool area) in Oklahoma. Contour interval equals 20′. [After Lyons and Dobrin (1972).]

trates the distribution of one of the oil-producing sandstone bodies of the Pennsylvanian Red Fork Sandstone in Oklahoma, as shown by Lyons and Dobrin (1972). The sandstone body is known from its lithology to have been deposited as a river channel sand, and Conybeare (1976) referred to it as horseshoe-shaped. This comment has been corrected by Busch (1977, p. 451) who says:

> The two limbs of this 'horseshoe-shaped channel' are actually the two major distributaries of a delta which had its apex on the north.

Sandstone bodies of this type are commonly referred to as shoestring sands and have frequently been interpreted as sand bars. Distinguishing characteristics of such river channel sands include lithic composition, carbonized wood fragments, and grain gradation from coarser below to finer above.

The isopach maps shown in Figs. 6-15 and 6-16 are of gross intervals. Special isopach maps of selected layers within the gross interval can also be plotted as isopach maps. For example, in a gross section comprising alternating layers of sandstone and shale, the cumulative thickness of the sandstone layers at every control point can be plotted and contoured to construct a net sandstone isopach map. Such maps may be more informative with regard to the geometry, depositional environment, and source of sediments than gross isopach maps. Net isopach maps of carbonate can also be constructed of gross limestone and shale sequences. These are commonly used with isolith and other types of lithofacies maps to interpret the genesis and depositional environment of a stratigraphic sequence.

Stratigraphic Sections

Stratigraphic sections are constructed to show the relationship of stratigraphic members and sequences both parallel and normal to the bedding planes. The emphasis may be either on lateral correlation such as the tracing and delineation of geometry of a sandstone or limestone bed or on vertical correlation such as the construction of a cumulative sequence from separate outcrops. Where it is essential to correlation and reconstruction of the original shape of a stratigraphic unit such as a bioherm, sand bar, or channel-fill deposit, a stratigraphic market may be selected as a datum. This is commonly an electric log marker above but close to the unit or in some cases the top of the unit itself (Fig. 6-17). The use of either the top or bottom of a unit as a datum can result in conflicting interpretations, as is evident from shoestring sands which may appear as river distributary channel sands where the upper surface is used as a datum or

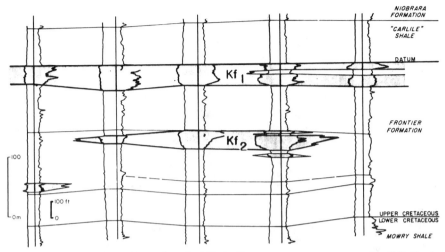

Fig. 6-17. Stratigraphic section of an Early Cretaceous sequence across the Salt Creek oil field, Powder River Basin, Wyoming. Length of section approximately 30 km. [After Barlow and Haun (1966) and Curry (1977).]

as sand bars where the lower surface is used. The best marker to use as a datum is some feature within a marine shale sequence which can be inferred to have had a fairly uniform and gentle depositional slope, although it is recognized that such a slope will rarely be a plane but will commonly be an undulating surface.

Correlations of this type, designed to show both the external and internal relationships of a stratigraphic unit, depend primarily on close subsurface control from boreholes but may also be assisted by seismic data (Fig. 6-18), as pointed out by Sheriff (1976). Experimentation with the filtration, stacking, digital recording, and processing of seismic data may yield records of sufficient resolution to delineate lensing and interfingering of sand bodies. Such details of shape are important to the understanding of the original geometry and depositional environment of sandstone bodies and consequently to the search for petroleum within them. They may reveal that a stratigraphic unit referred to as a blanket-type sandstone is a composite body of several units with distinct boundaries and depositional trends.

Interpretations of the depositional history and development of the stratigraphic framework of a sedimentary basin can be portrayed by a series of stratigraphic sections showing a reconstructed view of the basin at various stages of its growth. This involves the selection of several stratigraphic markers, each of which is successively used as a datum. Such a sequence of stratigraphic sections illustrates the progressive de-

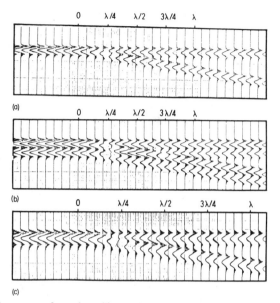

Fig. 6-18. Sequence of stratigraphic sections (a), (b), and (c) illustrating response to wedge of intermediate velocity. Seismic profile shows response to upper and lower surfaces of a stratigraphic pinchout at 0. Wedge angle is equivalent to 50 m/km. [After Sheriff (1976).]

formation of buried layers and marker bed surfaces resulting from compaction.

Stratigraphic sections may also be constructed with a view to representing the succession of rock layers exposed in outcrops and to show the correlation of sequences from one area to another. This method is particularly valuable for use in field guidebooks and field notes which are used for direct comparison with rock exposures. An example of sequences in the Bighorn Basin of Wyoming is illustrated by Todd (1964) in Fig. 6-19. This figure shows the Mississippian Sacajawea Formation overlain by the Pennsylvanian Amsden Formation in one outcropping sequence and the Amsden overlain by the Pennsylvanian Tensleep Formation in another. The strata are drawn to scale and represented by symbols that portray the layers in a realistic manner. Similar sections of outcrops at different localities can be drawn to show particular features and relationships such as changes of facies, thinning of stratigraphic members, or variations in the nature of an unconformity. An example is illustrated in Fig. 6-20 which shows an erosional surface overlain by Early Mesozoic carbonates and a mélange in the Dinaric region of Yugoslavia. The age of the mélange is considered by Dimitrijevič and Dimitrijevič (1973) to be Late Triassic. This type of stratigraphic section can be drawn in the field from direct

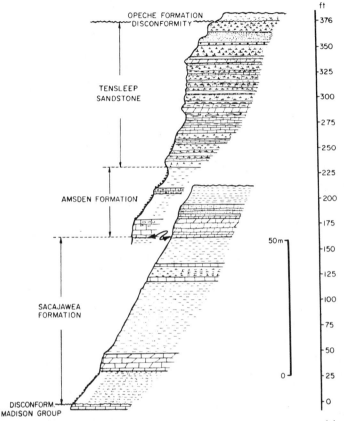

Fig. 6-19. Stratigraphic section showing correlation of sequences exposed in separate outcrops in the Bighorn Basin, Wyoming. Formations are the Mississippian Sacajawea and the Pennsylvanian Amsden and Tensleep. [After Todd (1964).]

observation of outcropping sequences of strata and may subsequently be an invaluable record where written descriptions prove to be an inadequate basis for interpretation.

Sections may be constructed to emphasize the sedimentary features and structures, as well as the lithology, and also to show the relationship of these to changes in depositional environment of the sequence. Fig. 6-21 illustrates this type of representation and shows the cyclic deposition of deltaic sediments that formed a sandstone within the Late Cretaceous Mancos Shale of Wyoming. Sections of this type are useful for comparison with rock sequences exposed in outcrops or road cuttings, particularly where it is necessary to interpret and illustrate details in the depositional

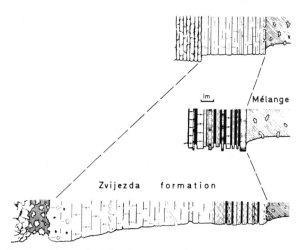

Fig. 6-20. Stratigraphic section showing correlation of exposed sequences of Triassic strata at three localities in the Dinaric region of Yugoslavia. [Reprinted from Dimitrijević and Dimitrijević, *J. Geol.* **81**(328–340). Copyright 1973 by the University of Chicago Press.]

history of sequences. In this respect these sections are particularly appropriate as illustrations in field guidebooks.

Logging of sequences seen in an outcrop or series of outcrops, or of sequences penetrated in a borehole, may be accomplished by detailed written descriptions accompanied by a strip log representing diagrammatically the sequence of rock layers, as illustrated in Fig. 6-22. This type of stratigraphic section shows by means of symbols not only the thickness and lithology of each layer but also the fossils present and the mineralogic content such as glauconite, siderite, and pyrite. It is a useful form of representation, particularly where the rock types are colored on strip logs which can then be compared for purposes of correlation. Standardization of colors is essential, the following having proved eminently suitable: igneous rocks (red), dolostone (mauve), limestone (blue), shale and claystone (brown, green), quartzose sandstone (yellow), lithic sandstone (orange), coal (black). Other color standards may be preferred, and any arbitrary choice is satisfactory provided it is adhered to.

Another type of stratigraphic section is drawn as a strip log to quantify the lithology. This type is commonly used for oil well logs as it combines several parameters that are of significance in correlation and in the evaluation of reservoir characteristics of particular layers. For example, sections of thin-bedded sandstone and shale, or of shaly sandstone are drawn to show the quantitative relationships of sand and argillaceous content in each unit within the section. This is done by a volumetric representation,

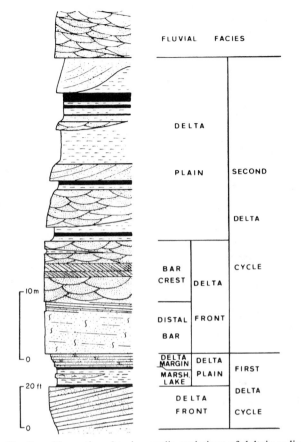

Fig. 6-21. Stratigraphic section showing cyclic variations of deltaic sedimentation in a sandstone tongue of the Late Cretaceous Mancos Shale, Wyoming. [After Geol. Soc. Amer. Coal Geol. Div. (1975).]

as illustrated in Fig. 6-23, which not only quantifies the lithology but qualifies the characteristic of individual layers by defining the natural gamma radiation, electrical resistivity, rock type, texture, mineral content, and diagnostic fossils.

Paleogeologic Maps

Paleogeologic maps are geologic maps of unconformities. As such they portray intervals of time in the history of a sedimentary basin during which there is a hiatus of deposition. Consequently, they complement

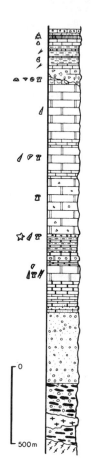

Fig. 6-22. Stratigraphic section of the qualitative strip log type showing lithology and fossil content of a sequence in the Triassic Engadin Dolomite, Switzerland. [After Dösseger and Müller (1976).]

maps showing the distribution of lithofacies, depositional environments, and paleogeography and form an essential link in any series of maps illustrating the geologic history of a sedimentary basin.

The construction of a paleogeologic map depends essentially on subsurface control but may be facilitated by surface exposures where the unconformity is visible in outcropping sequences or rarely where is has been exhumed. Subsurface control consists primarily of lithologic data supplied by drill cuttings and cores but may depend on seismic records where the reflecting surfaces truncated on profiles can be identified by reference to a borehole. Depending on the degree of control, a paleogeologic map may show the distribution of rock types and formations.

Regional maps covering very large areas obviously cannot show the distribution of rock types truncated by the unconformity mapped. Control

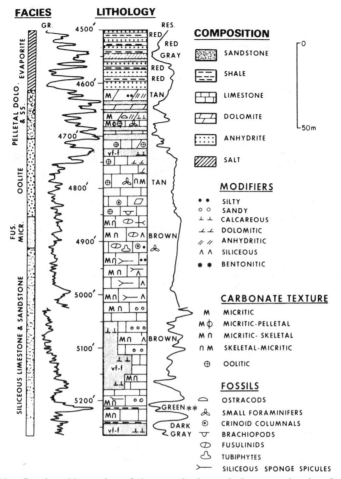

Fig. 6-23. Stratigraphic section of the quantitative strip log type showing the natural gamma radiation, electrical resistivity, lithology, texture, mineral and fossil content of a section of the Late Permian San Andreas Formation, Midland Basin, Texas [After Todd (1976).]

may be sufficient to show the general distribution of formations, but commonly such a map indicates only the distribution by age. An example is illustrated by Fig. 6-24 which is a paleogeologic map of the unconformity at the base of the Triassic in western Canada and the northwestern United States. This map shows the general distribution of Cambrian, Ordovician, Silurian, Devonian, Mississippian, Pennsylvanian, and Permian beds as they are believed to have been (Carlson, 1968) prior to Triassic deposition.

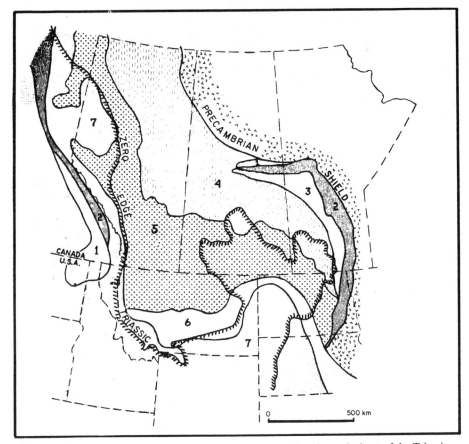

Fig. 6-24.   Pre-Triassic paleogeologic map of the unconformity at the base of the Triassic in western Canada and northwestern United States showing distribution of Cambrian sediments (1), Ordovician (2), Silurian (3), Devonian (4), Mississippian (5), Pennsylvanian (6), and Permian (7). [After Carlson (1968).]

The interpretation is subject to reservation in that the present erosional edge of the Triassic lies well within the area mapped, and consequently it is not known to what extent the present distribution of Paleozoic rock is the result of post-Triassic erosion. This uncertainty is always a problem in situations where unconformities converge.

On a broad but less regional scale, it may be possible to map the distribution of formations truncated by an unconformity. Where this is possible, the distribution of rock types is also indicated in that a formation is by definition a mappable lithologic unit. An example of this type of paleogeologic map is illustrated by Fig. 6-25 which shows the distribution

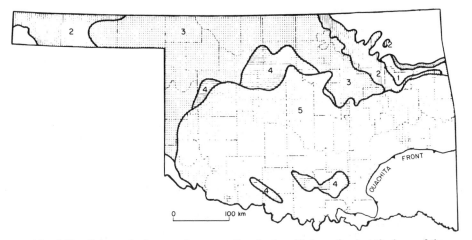

Fig. 6-25. Paleogeologic map of an unconformity (pre-Viola surface) at the base of the Late Ordovician Viola Formation in Oklahoma, showing distribution of Ordovician formations including the Oil Creek (1), McLish (2), Bromide (3 and 4), and Corbin Ranch (5). [After Schramm (1964).]

of formations of the Ordovician Simpson Group truncated by an unconformity underlying the Late Ordovician Viola Formation in Oklahoma, as described by Schramm (1964). This study stemmed from an appreciation of the possibility that formations within the Simpson Group might contain oil not only in reservoirs that are predominantly structural but also in stratigraphic traps. Delineation of belts of strata that include porous and permeable layers truncated by an unconformity and overlain by a suitable nonpermeable cap rock may point out trends along which hydrocarbon reservoirs may be found. Such reservoirs include the giant East Texas oil field which produces from the Late Cretaceous Woodbine Sandstone. In this case the Woodbine is underlain by an unconformity and overlain by the Engleford Shale which is truncated. The intersection of the two unconformities forms the eastern up-dip edge of the reservoir.

Paleogeologic maps may be significant in the interpretation of the tectonic history of a sedimentary basin, indicating trends of fold belts, directions of subsidence, tilt of strata, and areas of regional uplift. An example is illustrated by Fig. 6-26 which is a paleogeologic map of the unconformity at the base of the Late Devonian to Early Mississippian Chattanooga Shale showing the Nashville Dome in central Tennessee. The dome, which comprises Late Ordovician beds of the Nashville, Maysville, and Richmond Groups, forms a swelling on a major tectonic feature termed the Cincinnati Arch, according to Wilson and Stearns (1963). This

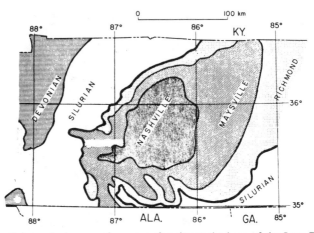

Fig. 6-26. Paleogeologic map of an unconformity at the base of the Late Devonian to Early Mississippian Chattanooga Shale in central Tennessee showing the Nashville Dome comprising the Late Ordovician Nashville, Maysville, and Richmond groups. [After Wilson and Stearns (1963).]

paleogeologic map reveals a major structure that not only helps to elucidate the tectonic history of the region but also offers the prospect of finding stratigraphic traps for oil in the updip truncated edges of permeable zones beneath the Chattanooga Shale or in other structurally high positions within the dome. Conversely, the disclosure of truncated structures may negate the prospects for hydrocarbon accumulations which may have been dissipated during the period when the structure was breached and eroded. Prospects will depend not only on the availability of permeable zones that structurally could be potential reservoirs but also on the time of migration of the hydrocarbons.

Paleotopographic Maps

Paleotopographic maps are contoured maps depicting the topography of an unconformity as it was prior to burial. As such they reveal the trends of buried hills and valleys and consequently of possible drainage patterns and channel sand deposits, of cuesta sands along buried ridges, and of mesa-shaped outlier of carbonate beds where overlying shale may form potential traps for hydrocarbon accumulations. The construction of a paleotopographic map depends on subsurface control, and the accuracy with which contours can be drawn and the topography defined is a function of both the closeness of control and the contour interval. Where the distance between control points is too large, or the contour interval too great, it is not possible to show the topography in sufficient detail to delineate features of significance to the discovery of individual oil or gas fields.

The initial step in constructing a paleotopographic map is to prepare a structure contour map of the present surface of the unconformity, using sea level or any other level as a datum. The next step is to prepare a set of structure contours which show the variations of direction and dip of the regional tilt to which the original unconformity has been subjected. This set of contours is superimposed on the first. At each point where the contours intersect, a new value is calculated to compensate for the estimated vertical distance the point has been depressed by the regional tilt. This will result in some slight displacement of the original position of points; but as viewed in a section drawn to natural scale, this displacement is commonly insignificant. These residual values are then contoured to reconstruct the original topography of the unconformity. This graphic method, described by Rich (1935), is somewhat tedious and slow but can be done in the field with a minimum of equipment. Where electronic computers are available, it is possible to devise a progrem that will rapidly carry out such a trend analysis, as described by Schramm (1968) and Robinson *et al.* (1969).

The examples shown in Figs. 6-27 and 6-28 are given by Schramm who says (p. 1655):

> Trend-surface analysis is the statistical approach to structural or topographic surfaces. The technique is designed to separate a contour map based on observed data into two components: (1) regional trend and (2) departure of the actual surface (observed) from the regional trend. The departures, or residual values, are the difference between the actual value and the calculated regional surface value at each data point.

The regional trend referred to by Schramm may be a regional tilt, and the observed structure of an unconformity, as shown by a structure contour map (Fig. 6-27), will be the cumulative result of original elevations (topography) and regional tilt. Fig. 6-27 is a structure contour map of the unconformity at the base of the Early Pennsylvanian Morrow Formation in northwestern Oklahoma showing the present configuration of that surface. Fig. 6-28 shows the first-degree residual map resulting from subtracting values for the regional tilt where the tilt is assumed to be a flat surface or plane. In some regions this would be a close enough approximation, particularly if the area studied is comparatively local. But commonly the surface of tilt is curved and forms a paraboloid so that a realistic recon-

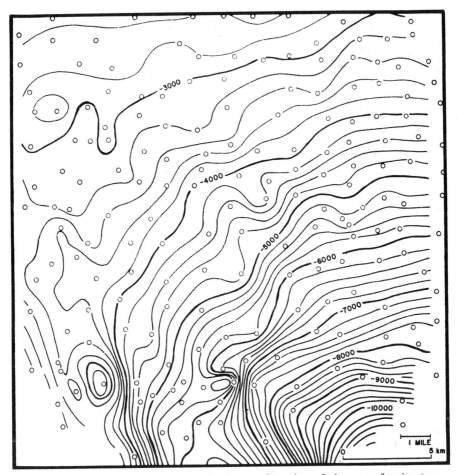

Fig. 6-27. Structure-contour map showing configuration of the unconformity (pre-Morrow surface) at the base of the Early Pennsylvanian Morrow Formation, Anadarko Basin, northwestern Oklahoma. [After Schramm (1968).]

struction of the original topography of an unconformity requires a less simplistic application of correction factors. Assuming that the surface of tilt is approximately a plane, the unconformity at the base of the Morrow can be shown by the first-degree residual map as in Fig. 6-28. This reconstruction shows the general configuration of the original topography but lacks the detail that could be shown in a second-degree residual map by the use of a best-fit paraboloid to represent the regional tilt.

The Morrow was deposited on the eroded surface of the Mississippian

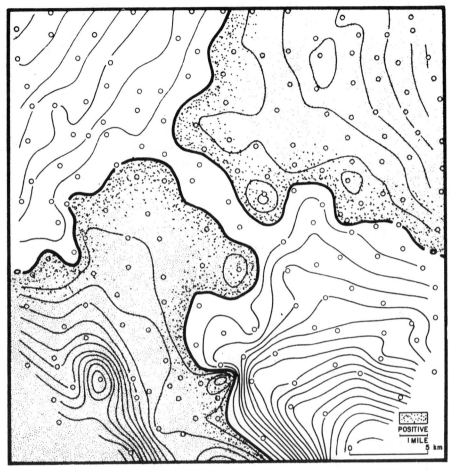

Fig. 6-28.   First-degree residual map (assuming a plane to represent regional tilt) showing general configuration of the original topography of the unconformity (pre-Morrow surface) at the base of the Morrow Formation. See Fig. 6-27. Here o is control well. Contour interval equals 200'. [After Schramm (1968).]

the topography of which is also reflected in an isopach map of the Lower Morrow. Forgotson *et al.* (1966, p. 532) say:

> The trend technique of map analysis emphasizes the inverse relation between the topography of the pre-Pennsylvanian surface and the thickness of lower Morrow sandstones. It also provides supporting evidence that most relief on this surface was developed prior to, and during, deposition of the lower Morrow.

The lower Morrow sandstones formed a transgressive unit, and the geomorphology of individual sandstone bodies was controlled or influenced by the local topography of the unconformity on which the sandstone was deposited (Busch, 1959, 1977; Conybeare, 1976).

Hydrodynamic Maps

Hydrodynamic maps of a sedimentary basin are constructed to show the movement of water within individual stratigraphic units of the basin. These movements may vary in different units with reference to their vectors and rates of flow. Consequently, sedimentary basins do not function as homogenous volumes to the flow of water through them and must be studied and measured as composite sequences containing separate aquifers having distinct hydrodynamic characteristics.

The flow of water through a permeable sandstone or limestone bed is slow, commonly amounting to a few meters but ranging up to 30 m per year, according to Meinzer and Wenzel (1942), Roof and Rutherford (1958), and Hodgson *et al.* (1964). The water will flow from its head or intake to its outlets and will consequently move through structures, such as folds, from points of higher pressure to points of lower pressure, or the reverse. But regardless of whether the direction of flow is from higher to lower, or from lower to higher pressures, it is always from a higher potential to a lower potential as measured from the intake to the outlet. This principle has been enunciated by Hubbert (1953, p. 1962) who said:

> In the flow of underground water only rarely does **E** have a magnitude greater than a small fraction of that of the vector **g**. Under these conditions the negative pressure-gradient vector can depart but slightly from the vertical and the equipressure surfaces will be nearly horizontal with the pressure increasing downward. The resultant vector **E**, however, can have any direction in space—vertically downward or upward, horizontal, or any arbitrary inclination. Since the fluid will tend to flow in the direction of **E**, it is clear that the flow can be in any direction whatever with respect to the negative gradient of the pressure: from lower to higher or from higher to lower pressure, parallel with the equipressure surfaces or in any other direction—a fact of great importance in view of the widely held opinion that fluids flow only from higher to lower pressures.

In pointing out that the direction of movement of a fluid in a subsurface aquifer tends to be that of a vector **E**, Hubbert defined **E** as a force–intensity vector which is the sum of two independent forces, gravity **g** and the negative gradient of the pressure divided by the density of the fluid.

The weight of a column of water that would rise within a well penetrating aparticular aquifer is a measure of the pressure of water within the aquifer at that point. In other words, the well acts as a manometer. The elevation of the top of the column of water is a measure of the potential of the aquifer at that point. The surface connecting such elevations in different wells is referred to as the potentiometric surface and is shown by means of a contoured potentiometric map.

Fig. 6-29 illustrates the general features of the potentiometric surface with reference to the hydrocarbon-bearing Early Jurassic Precipice Sandstone of the Surat Basin in Queensland, Australia. Scattered control precludes, except locally, the delineation of contours on the basis of

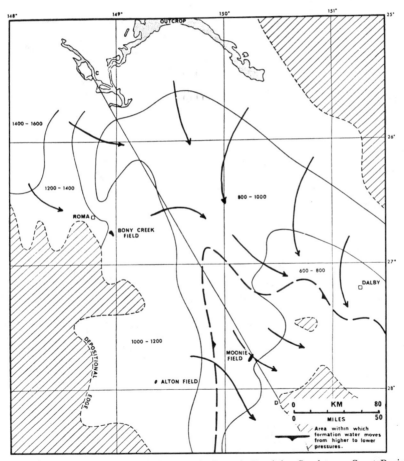

Fig. 6-29. Potentiometric map of the Early Jurassic Precipice Sandstone, Surat Basin, Queensland, Australia, showing directions of flow (arrows) of water through the formation. Ranges of values indicate potential in feet above sea level. [After Conybeare (1970).]

point-to-point control, but allows the contouring of areas containing ranges of values. The values expressed on this potentiometric map are feet above sea level to which columns of water would rise in each well. Naturally, water is not allowed to so rise, but the height can be determined by conversion of the initial shut-in pressure measured when the well penetrated the aquifer. The arrows shown in Fig. 6-29 indicate the direction of movement of the water which is always down the potentiometric slope and consequently normal to the contours at any point. Also defined is an area within which the water is moving updip from higher to lower pressure, as illustrated by the structure of the top of the Precipice Sandstone shown in Fig. 6-30.

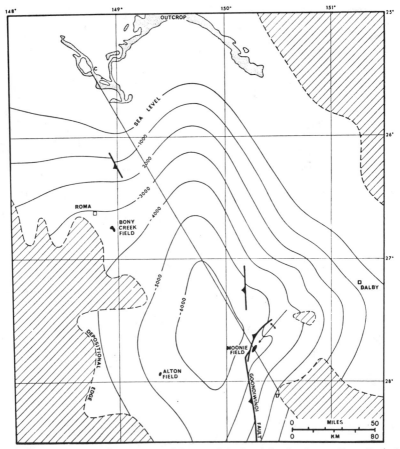

Fig. 6-30. Structure contour map of the top of the Precipice Sandstone, Surat Basin (see Fig. 6-29). [After Conybeare (1970).]

On a local basis potentiometric maps of oil fields and contiguous areas may indicate targets for drilling where hydrodynamic conditions may have tilted the oil–water contact sufficiently to form a secondary trap within a minor flexure on the flank of the major structure. On a regional basis an understanding of the movement of water through a hydrocarbon-bearing sequence may assist in estimating the direction and extent of oil and gas migration and consequently facilitate the planning of an exploration program. Hydrodynamic studies of sedimentary basins are also essential to the planning of subsurface water supply and disposal of liquid wastes.

Potentiometric maps may indicate anomalous hydrodynamic situations within a sedimentary basin arising from features of the structural or stratigraphic framework. Levorsen (1967) cites an example with reference to the Cretaceous Point Lookout Sandstone in the San Juan Basin of New Mexico and Colorado. Water flows through the Point Lookout from all sides of this semicircular basin to form a sink in the middle and structurally low region of the basin. As the water must flow somewhere, it is obvious that it leaks across the bedding, flowing upward or downward into other permeable beds.

LITHOFACIES

Lithofacies maps are designed to show the variations and distribution of stratigraphic units within a sedimentary basin. Variations consist of differences in rock types and of their percentages, ratios, or volumes in stratigraphic units and also of differences in color and grade, or in the contents of particular constituents such as clays, other minerals, and coal. The manner in which such differences are portrayed on a map depends on the aspects to be accentuated (Forgotson, 1960), but commonly the differences are shown by means of colors and contours.

The most general type of lithofacies map shows areas of distribution in which the stratigraphic unit mapped is characterized by, and commonly composed mainly of, a particular type of rock. These may be termed rock-type distribution maps as they show the distribution of predominant lithologic types such as sandy, shaly, or carbonate facies. This type of lithofacies map may be constructed with scattered control by extrapolation over considerable distances from the subsurface to outcropping sections. Such maps give a general impression of the regional distribution of rock types and consequently of the sedimentary facies within a stratigraphic sequence. Maps of a more definitive nature can be constructed as control permits. Complementary to a regional map showing the general distribution of lithofacies are isolith maps of individual facies. These are isopach maps which show not only the distribution but also the variations in thickness of a particular facies. Other maps include net rock-type maps which are contoured to show the cumulative thickness, within the stratigraphic unit mapped, of a particular rock such as shale, sandstone, or limestone. Rock-type ratio maps show the ratio of cumulative thickness of sandstone to shale, or of shale to limestone, within a stratigraphic unit. Another type in common usage is a percentage rock-type map showing the percentage of total thickness of a stratigraphic unit occupied by the cumulative thickness of shale, sandstone, or limestone.

Additional to these basic types of lithofacies maps are special maps designed to emphasize features that may be significant or diagnostic. Included among these are maps to show the distribution of clay mineral types, such as illite or montmorillonite, and of the other minerals including glauconite, pyrite, and siderite. Maps showing the distribution and content of kerogen, fixed carbon, or of coal within a stratigraphic unit are significant in studies of the maturation of organic matter, with reference to the origin and migration of oil, and to exploration for deposits of coal. Isograde maps showing variations in mean grain size may be of use to indicate the direction of transport of sediments within a stratigraphic unit. Variations in the color of shale units may also indicate changes of depositional environments or of proximity to features such as reefs or deltas. Many other types of maps illustrating variations of lithologic or petrophysical characteristics which may be related to lithofacies, such as porosity and permeability, natural radioactivity, electrical resistivity, and trace element content, can also be constructed for special purposes. The construction of all such maps, directly or indirectly related to the distribution of lithofacies, is commonly undertaken on an experimental basis, as it is not always known which ones may be meaningful or diagnostic. Taken individually, a map may not be useful because its significance is not understood; but examined in the context of other maps, the relationships shown may become apparent.

Rock-Type Distribution Maps

Rock-type distribution maps show the areas of a sedimentary basin in which the beds of a particular stratigraphic sequence are characterized by a predominance of limestone, dolostone, shale, sandstone, or some other rock type. The stratigraphic sequence may range from a member to a series but commonly has the status of a formation. By definition, a member is a division of a formation, separated by distinct lithologic features or composition; a formation is commonly the basic stratigraphic unit defined and mapped by its lithologic character; and a series is a stratigraphic sequence comprising several formations which may be placed in two or more groups deposited during a particular interval of time. In practice, there has commonly been some disparity in nomenclature between adjacent or separated areas mapped by different persons over various periods of years. With increasing knowledge of the details of a stratigraphic sequence, such differences are resolved by the consensus of a stratigraphic nomenclature committee. Rock-type distribution maps may consequently have a wide stratigraphic and areal range. On a very broad regional scale the distribution of late Early Mississippian to early Late Mississippian marine rock types in western Canada and the United States is illustrated by Johnson (1971) in Fig. 6-31. On this scale the assemblage

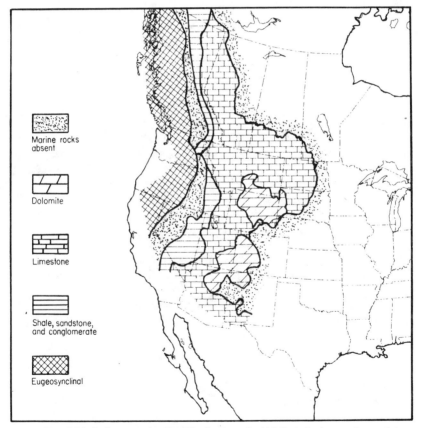

Fig. 6-31. Lithofacies map showing distribution of mid-Mississippian marine rock types in western North America. [After Johnson (1971).]

of rock types shows the regions characterized by eugeosynclinal and miogeosynclinal development and delineates the major tectonic and depositional features that obtained during this interval of time.

On very much smaller areal and time-stratigraphic scales, Fig. 6-32 illustrates the distribution of lithofacies in the upper part of the Devonian

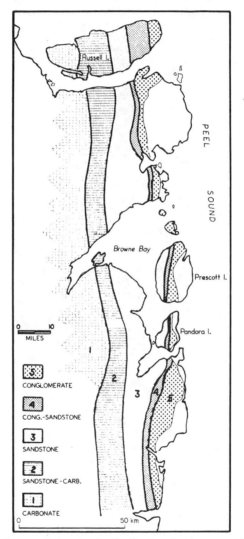

Fig. 6-32.   Lithofacies map showing the distribution of rock types within the upper part of the Devonian Peel Sound Formation of Prince of Wales Island in the Canadian Arctic. [After Miall (1970).]

Peel Sound Formation of Prince of Wales Island in the Canadian Arctic. According to Miall (1970), paleocurrent evidence indicates an easterly source for the clastic sediments. These range in facies from coarse conglomerate to reddish sandstone near the source area, and farther west to gray, nonmarine clastics which grade into marine carbonates. The interpretation of geomorphic relationships between lithofacies and sedimentary processes that deposited the upper section of the Peel Sound Formation is shown in Fig. 6-33. Compared with the study of Johnson (1971), that of Miall (1970) is on a microscale; but lithofacies maps may be constructed on still smaller scales for particular purposes such as the delineation of carbonate reef facies, or of barrier bar and lagoonal facies. Rock-type distribution maps on whatever scale are useful in interpreting the geologic history, paleogeography, and sedimentary processes of a particular time-stratigraphic unit.

The interpretation of lithofacies maps showing rock-type distribution must be carried out with some caution where the reconstruction of paleogeography is involved, and an appreciation of the possible sedimentary processes operative at the time of deposition of the interval studied is essential. For example, the distribution of sediment types with respect to a shoreline may show a wide variation depending on depth ranges, current patterns, and volume and grade of sediment reaching the coastline. As mentioned previously, mud carried shoreward by upwelling currents off the mouth of the Amazon River is deposited nearer the coastline than the sand. Carbonate shelf deposits may also fringe a coastline and be flanked on their seaward sides by deposits of mud. This is the reverse of the sedimentary pattern shown in Figs. 6-32 and 6-33. The usefulness of a rock-type distribution map depends largely on the interpretation placed on it.

Isolith Maps

Isolith maps are isopach maps of the cumulative thickness of a particular rock type within a defined stratigraphic interval. Where, for example, a stratigraphic interval consists of interbedded sandstone and shale, the cumulative thickness of sandstone at each control point is plotted and isopach contours are drawn. The resulting map shows areas of thicker and thinner cumulative sandstone layers and consequently of concentration or paucity of sand deposition during the interval of time represented by the stratigraphic interval. It does not necessarily define the geometry of sandstone bodies within the interval unless most of the sandstone is present within a single bed. In cases where distributary channel sand bodies

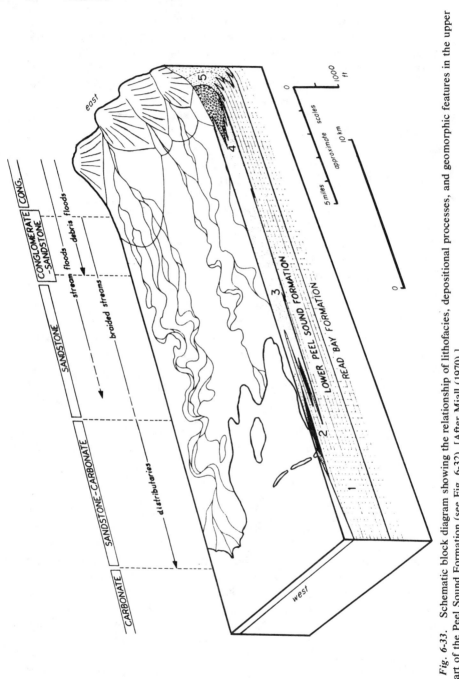

Fig. 6-33. Schematic block diagram showing the relationship of lithofacies, depositional processes, and geomorphic features in the upper part of the Peel Sound Formation (see Fig. 6-32). [After Miall (1970).]

cut into shoreline sand bodies in a delta environment, both channel and shoreline sands may lie within the same narrow interval or bed and appear to form a single body of sand. If this bed of sand comprises most of the sand within a stratigraphic interval, then an isolith map will reveal the areas of greatest sand deposition but will not show geomorphic features or the details of depositional trends. But where most of the sand within an interval is concentrated in a bed that has been deposited as a bar sequence or distributary system, an isolith map will clearly show, provided there is sufficient control, the geomorphic features and depositional trends.

Similarly, isolith maps of an interval consisting of limestone and shale can be constructed to show the distribution of limestone. Where the interval is not too thick, such a map will also reflect the distribution of limestone as shown on lithofacies maps. Also where the intervals are sufficiently narrow, isolith maps of contiguous intervals may show lateral shifts of the main areas of limestone formation. Such shifts would not be detected by the use of larger stratigraphic intervals. In other words, the resolution obtained in an isolith map depends not only on the thickness of the interval chosen but on the nature of the sedimentation patterns that were formed during the time of deposition of the interval. As with many types of maps, the end results may be immediately meaningful or obscure. In the latter case, obscurity may stem from the use of an interval that is too thick or too poorly defined, or may result from the lack of other data.

An example of a depositional pattern outlined by an isopach map based on the thickness of a unit of sandstone within a stratigraphic interval is given by Swann (1964) and illustrated in Fig. 6-34. This is an isopach map of distributary channel sandstone bodies within the Late Mississippian Degonia Formation in Illinois. It is also an isolith map of the Degonia in this area. An isolith map of sandstone within the interval between the limestone beds indicated as 1 and 2 would also show a similar depositional trend but with less resolution because it includes sandstone beds of the Clore Formation underlying the Degonia. The sandstone channels within the area defined by Fig. 6-34 are traced for a distance of 15–20 km. Fig. 6-35 illustrates another example of distributary channel sandstone bodies which, in the area shown, are traced for a distance of 55 km. These so-called shoestring sands described by Shelton (1967), lie within the Early Cretaceous Cutbank Formation of Montana. The sandstone fills channels cut into the eroded surface of the Late Jurassic Swift Formation and Rierdon Formation and forms the basal member of the Kootenai Formation. These channel-fill sandstone beds are locally oil-producing. Another example of an isolith map that delineates the depositional trend of an oil-bearing sandstone is illustrated by Fig. 6-36. This figure shows the distribution of the Early Cretaceous Gas Draw Sandstone member of the

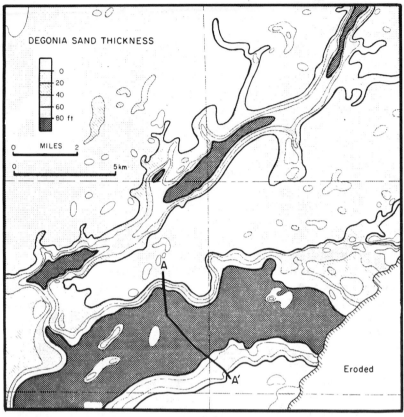

(a)

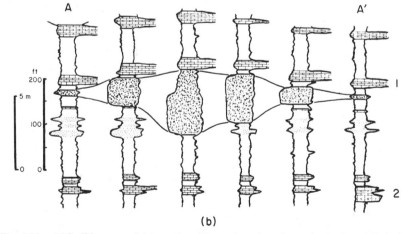

(b)

Fig. 6-34. (a) Isolith map and (b) stratigraphic section of sandstone in the Late Mississippian Degonia Formation, Illinois, showing the depositional trends of distributary channel sandstone bodies. [After Swann (1964).]

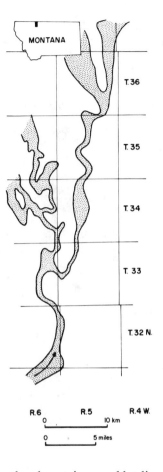

Fig. 6-35. Isolith map of oil-bearing distributary sandstone where the thickness exceeds 15 m within the Early Cretaceous Cutbank Formation, Montana. [After Shelton (1967).]

Muddy Formation in Wyoming. In this case the shoestring sand bodies are thought by Stone (1972) to have been deposited as beaches and offshore bars.

With reference to predominantly carbonate sequences, isolith maps can be constructed to show the variations in content of sandstone and shale or dolostone within a totally carbonate section. The distribution of limestone and dolostone, and of evaporites within a dolostone section, can be quantified by means of isolith maps which are used in conjunction with lithofacies maps. Such maps can serve to indicate areas that once were landmasses, or carbonate shelves, and may be useful to define the paralic range of a sabkha environment or the possible trend of a reef complex.

Some isolith maps are also net rock-type maps in that they are a measure of the distribution of the cumulative thickness of a particular rock type such as sandstone, shale, or limestone within a determined strati-

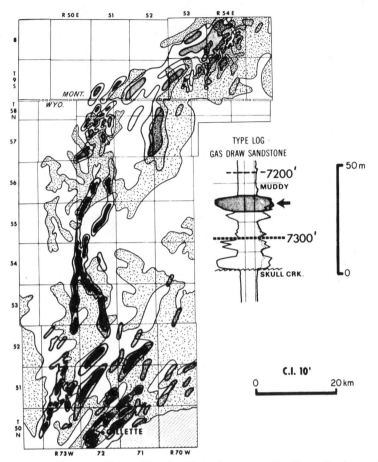

Fig. 6-36. Isolith map of the oil-bearing Early Cretaceous Gas Draw Sandstone zone, Wyoming, showing trends of sandstone bodies deposited as beaches and offshore bars. [After Stone (1972).]

graphic interval at any point. Other isolith maps may be a measure of the cumulative thickness of sections that are predominantly of one rock type but include others. For example, where a stratigraphic sequence includes thick alternating sections of shale, and of limestone with minor shale, the gross thickness of the limestone sections at each control point may be used to construct an isolith map of the limestone facies. For practical purposes the minor content of shale in the limestone sections may not be of significance. But if further differentiation is required, then the cumulative thickness of the shale beds within the limestone sections must be subtracted in order to construct a net rock-type map. Each problem will pose its own methods of attack, and the number and types of maps constructed will depend on the scale of the maps, whether regional or local, the control available, the definition required, and the exigencies attending the study.

Rock-Type Ratio Maps

Rock-type ratio maps commonly show the distribution by ratio of two components such as sandstone and shale, shale and limestone, limestone and dolostone, or of the ratio of total clastics to nonclastics. This distribution can be shown by contours or by areas in which certain ranges of ratios are predominant. The latter method is preferable in that it smooths out the minor anomalies and is more easily viewed. Where it is desirable to show the ratio of three components such as sandstone, shale, and carbonate, it is necessary to construct a triangle in which each apex represents one hundred percent of one of the three components, and the area of the triangle is divided into blocks representing ranges of composition and consequently of ratio of the three components. Each block is shown as a separate symbol which can then be used on a map to show the distribution by areas of ranges of composition expressed as ranges of ratio. This three-component method may show a closer correlation with other lithofacies maps if correctly plotted but is subject to arbitrary decisions and inconsistency which may tend to detract from the meaningfulness of the map and so add further difficulty to interpretation.

Where a rock-type ratio map is constructed of an interval 100 m or more thick, the common practice is to use the gross thickness of each sandstone interval, as illustrated in Fig. 6-37. This figure shows a section of an electric log of a sandstone and shale sequence. The shale base line and the 10-mV sand line are indicated. Curves to the left of the sand line include sandstone intervals which are also shown as columnar blocks. The two thickest sandstone intervals A and B also include minor shale partings which are inconsequential in calculating the gross cumulative thickness of sandstone intervals within a section 100 m or more thick. If higher resolution is required, such as the construction of a sand–shale ratio map of sandstone unit A or B, then each shale parting must be measured by the use of micrologs or microlaterologs on a larger scale. These calculations provide the net sandstone thickness rather than the gross thickness of sandstone intervals A or B. This type of application is commonly required in studies of the production characteristics of oil and gas reservoirs; whereas the gross sandstone interval only is significant in studies of the regional distribution of strata within a sedimentary basin.

It is essential in constructing a rock-type ratio map to also incorporate an isopach map with sufficient contours to establish the trend. Otherwise the rock-type ratio map is of limited value for the reason that in sandstone

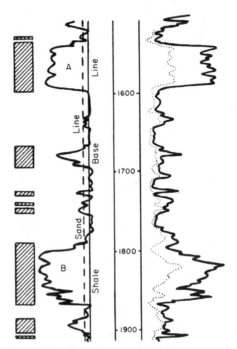

Fig. 6-37.   Electric log section of a sandstone and shale sequence showing shale base line and 10 mV sand line. Columnar blocks indicate gross thickness of sandstone intervals. A and B include shale partings. Depths in feet. [After Potter (1962a).]

and shale sequences a one-to-one ratio in a 100 m-section represents 50 m of sandstone, whereas the same ratio in a 300-m section represents 150 m of sandstone. The same treatment is required in the construction of rock-type percentage maps.

Fig. 6-38 is an example of a rock-type ratio map showing the distribution of areas in which the sandstone-to-shale ratios within the upper part of the Late Cretaceous Eagle Ford Formation of the East Texas Embayment are within the ranges 1/1–1/2 (black), 1/2–1/4 (dark), and less than 1/4 (light). The superimposed isopach contours indicate that with increasing thickness the sandstone-to-shale ratio decreases regionally but not necessarily locally. Referring to this figure, Stehli *et al.* (1972, p. 44) says:

> The map indicates that the largest sand accumulations are in the northeast, whereas the northwestern region is characterized by shaly rocks.

In the quantification of a section comprised entirely of sandstone and shale, an expression of ratio is also an expression of percentage. But in a section comprising sandstone, shale, and limestone the ratio obviously

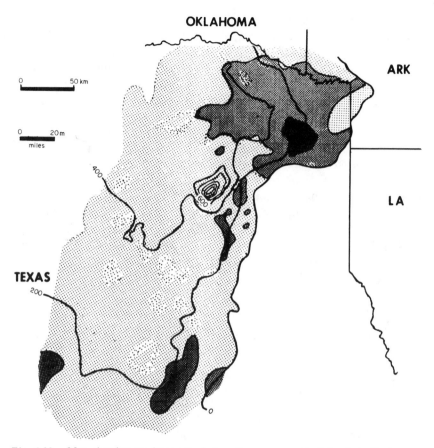

Fig. 6-38. Map showing sandstone to shale ratios, upper part of the Late Cretaceous Eagle Ford Formation, East Texas Embayment. Ratio ranges 1/1–1/2 (black), 1/2–1/4 (dark), less than 1/4 (light). Isopach contours in feet. [After Stehli *et al.* (1972).]

does not relate to the percentage as it is quite independent of the cumulative thickness of limestone in the section. What it does relate to is variation in facies or microfacies, the influx of clastic sediments into an environment of carbonate desposition, and consequently paleogeographic changes and fluctuations of coastal features such as migrating sheets of sand or mud moving from the estuary of a mobile river system.

Rock-Type Percentage Maps

Rock-type percentage maps show the distribution by percentages of volume of the rock-type constituents within a defined stratigraphic sequence. The map may show the percentages of sandstone, limestone, total clastics, or any other component. As with rock-type ratio maps, isopach contours must be superimposed to show the variations in total thickness of the stratigraphic interval in order to view the percentage distributions in perspective. Percentage maps are a type of quantified lithofacies map and correlate well with rock-type distribution maps. Anomalous results may arise in the case of percentage, ratio, and isolith maps of evaporite sequences where mobilization or removal of salts or anhydrite have occurred. These anomalies may adversely affect interpretations of the original distribution of sediments and evaporites and consequently of the depositional environments associated with the paleogeography. On the other hand, recognition of the total or partial removal of the evaporites within a stratigraphic section may be the clue to depositional and structural features caused by collapse and reflected in the distribution and internal deformation of the overlying strata. Structures resulting from differential dissolution of salts can form potential traps for oil and gas and are known to occur in association with the early Late Devonian Prairie Formation, comprising halite, anhydrite, and limestone, in southern Saskatchewan (Carlson and Anderson, 1965).

Rock-type percentage, ratio, and isolith maps are commonly constructed on the basis of subsurface control but may also incorporate the data from measured sections of outcrops. Subsurface data are obtained chiefly from electric and other logs used in combination with sample logs based on drill cuttings and cores. Depending on the definition, required sample logs may be sufficient, but as these are commonly constructed from samples representing intervals of 3 m, they do not always show the detailed relationships of interbedding. Also in the case of sections containing anhydrite, it may not be possible to determine, on the basis of drill cuttings only, whether the anhydrite is interbedded with the carbonate and shale layers or present as veins filling preexisting fractures. The latter situation is common in areas where dissolution of evaporite deposits has caused collapse of overlying beds and development of a complex pattern of fractures.

Mineral Distribution Maps

Mineral distribution maps show, on a qualitative or quantitative basis, the distribution by area of specific minerals contained in a particular stratigraphic sequence. Such minerals commonly include glauconite, siderite, pyrite, collophane, and clay minerals. Some of these, in particular glauconite, may characterize widespread zones and constitute important marker beds. Others, such as the clay mineral montmorillonite in bentonite beds, may characterize marker zones that are distinctive in electric logs but have a more restricted distribution.

The present distribution of minerals on the ocean floors appears to be more local than regional, although numerous patches of manganese nodules are spread over vast regions of the Pacific abyssal plains. In general, the distribution of minerals in sediments is related to local features such as volcanic activity, hot brines emanating from faults beneath the ocean floor, upwelling of cold currents from ocean depths, and variations in the water depths and salinity of neritic and paralic environments. The distributions of clay minerals, such as those observed today in sediments of the Gulf of Mexico, may not be preserved in future sedimentary basins in view of the fact that smectite is converted to illite with burial to the depth range 2000–3700 m (Hower et al., 1976). Consequently, the distribution of illite in many sedimentary basins probably represents the cumulative results of original distributions of illite deposited with the sediments and of smectite converted to illite by burial metamorphism. Such problems are not readily resolved, and the application of present-day mineral distributions in marine sediments to the genesis and significance of minerals in stratigraphic zones of sedimentary basins is not always certain.

An interesting example of a mineral distribution map showing the percentage weight of illite in bulk sediment samples from bathyal and abyssal zones of the Gulf of Mexico is shown in Fig. 6-39. As pointed out by Devine et al. (1973), the marked concentration of illite in the Mississippi cone clearly indicates the input of sediment from the Mississippi River which is known to have a high content of illite (Grim and Johns, 1954). The other main area of illite concentration extends from the delta of the Rio Grande and appears to be related to the present and previous load of sediment transported by that river.

The content of other minerals may signify the extent of certain depositional environments. In particular, the association of pyrite with black shales may indicate deposition in a euxinic environment, whereas an abundance of glauconite in carbonate or sandstone may signify a shallow water, marine shelf environment, particularly where the glauconite is pelletoid and appears to have been derived by the alteration of clay minerals in faecal pellets.

Isocarb and Coal Distribution Maps

Fixed carbon is the carbon content of coal or carbonized matter. It is obtained as the solid residue, after extraction of ash and moisture, of such

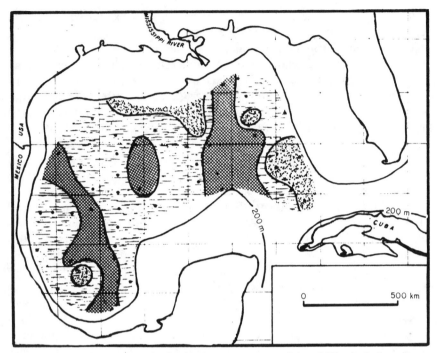

Fig. 6-39. Map showing distribution by percentage weight of illite in bulk sediment (carbonate removed), Gulf of Mexico: 35%–42% (dark), 30%–35% (hatching), less than 30% (stippled). [After Devine *et al.* (1973).]

matter subjected to destructive distillation. The content of fixed carbon, expressed as a ratio, has long been used as a measure of the degree of low-grade regional metamorphism to which the beds in various parts of a sedimentary basin have been subjected. The distribution of these ratios, shown on maps in the form of isocarb contours, has been used to upgrade or downgrade the prospects for discovery of hydrocarbons. Based on the assumption that the higher the content of fixed carbon, the lower the chances for finding oil and that the ratio of gas to oil increases directly as the fixed carbon content, an arbitrary scale of exploration prospects was proposed by Fuller (1920). Objections to such guidelines can be raised on the basis of analytical methods, sampling methods, and geological considerations. It can be argued that with respect to moisture content, the premises on which analytical methods are based are not always consistent. It has also been pointed out that isocarb maps may be inconsistent in that the sampling has been from different stratigraphic zones or from weathered exposures of rocks. Furthermore, an arbitrary scale of exploration prospects, based on fixed carbon ratios, does not take into consideration the possibility of petroleum migration subsequent to the period of metamorphism. The application of isocarb maps to petroleum exploration must be considered as a tentative guide which, in general, tends to downgrade prospects for oil in zones having comparatively high fixed carbon contents but which cannot be considered as definitive in specific zones where potential structural or stratigraphic traps may exist.

Coal distribution maps are commonly of three types: (a) isopach map of a single coal seam; (b) isopach map of the net coal thickness in a coal-bearing zone of interfingering coal seams, shale layers, and sandstone lenses; and (c) isopach map of the net coal thickness in a defined stratigraphic sequence. The first type of map shows a limited areal distribution and depends on control sufficiently close to allow positive correlation of the coal seam. The second type may cover a much larger area, does not depend on such close control, but does depend on the correlation of marker beds or recognizable stratigraphic relationships in proximity to the coal-bearing zone. The third type is commonly regional and shows the distribution of areas underlain by coal and the ranges of cumulative thickness of coal within these areas. Maps of this type can be constructed with limited subsurface control and are the most useful from the point of view of sedimentary basin analysis and evaluation of the regional coal prospects.

Coal distribution maps, particularly of types (a) and (b) where the beds are nearly flat-lying and close to the surface, may be based on cores from shallow boreholes. Maps of cumulative thicknesses of coal in zones several hundred to thousand meters below the surface are best constructed on the basis of electric and other logs rather than on logs of drill cutting only. The reason is that coal seams tend to cave badly, and consequently the drilling mud returning with cuttings up the hole is continually picking up fragments from the coal seams penetrated. This flood of coal in the samples of drill cuttings is likely to result in overestimating the number and thickness of seams. Experience may show that certain fragments are too large to be drill cuttings, but small fragments of cavings may be sufficiently numerous to cause problems.

Coal distribution maps of type (b) commonly show a close relationship to the distribution of deltaic sediments and consequently are of use in the delineation of paleogeographic features. Maps of type (c), which are a measure of the cumulative thickness of coal seams within a defined stratigraphic interval, commonly show a wider spread of coal distribution which reflects both the lateral and vertical migration of deltaic sedimentation.

Isograde Maps

Isograde maps show, by means of contours, the distribution of mean or maximum size clastics within a defined stratigraphic interval. Where control is too scanty to allow the meaningful construction of isograde contours, the generalized distribution of grades can be shown by arbitrarily outlining areas designated by colors or symbols. These maps are constructed on statistical data based on measurement of a number of grains, pebbles, or boulders. Consequently, they depend very largely on examination of outcrops in the field. Where the sizes range from small pebbles to sand, and where control is sufficiently close to obtain representative samples or to be certain of correlation, it may be possible to construct an isograde map on the basis of subsurface data only provided core is available. Drill cuttings cannot be used for this purpose because the sizes of the pebbles crushed by the drill cannot be accurately determined.

Determination of the maximum size of pebbles or boulders in outcrops is somewhat uncertain as it depends on the size of the outcrop and the extent to which it is weathered or obscured by vegetation. Also the maximum size estimated is that of the largest pebble, cobble, or boulder observed, which may not be representative. But provided that the outcrops are sufficiently large and well exposed, the errors so resulting are probably not significant. Similarly, in estimating the mean grade size, the question will arise as to the adequacy of the sample measured. Theoretically, 1000 measurements will give a more accurate figure than 100, but in view of the constraints of time and other considerations imposed on most field work, measurements of 10 pebbles or boulders will give sufficiently good results. Better definition resulting from larger samples is commonly insignificant and not worth expenditure of the required resources and time. Where doubt may exist, it is best to carry out a trial run of 10, 50, or 100 pebbles to determine the variations of mean grade size in the three samples and then to plan the field sampling operation in view of the results taken in context with other priorities.

Isograde maps that are contoured or constructed to show distribution of grade size by area or location sampled (Figs. 5-16, 5-17) are particularly useful on a regional basis as indicators of sediment provenance. They may also be useful in the interpretation of paleogeographic and paleogeomorphic features such as coastal cliffs fronted by boulder beaches, large alluvial cones fronting belts of uplift, and river system deposits covering tens of thousands of square kilometers. Of particular interest is the application of boulder size analysis to paleogeography, as illustrated in Fig. 6-40. This figure shows the size distribution pattern of boulders in the transitional Late Mississippian to Early Pennsylvanian Johns Valley Formation of

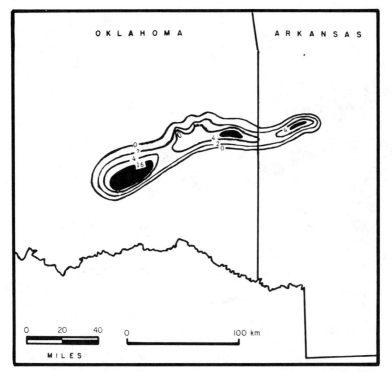

Fig. 6-40. Isograde map showing distribution of boulders in the Late Mississippian to the Early Pennsylvanian Johns Valley Formation, Oklahoma and Arkansas. Contours show mean diameters of 2, 4, and greater than 16 feet (approximately 5 m). Black areas are interpreted as island loci for boulders. [After Shideler (1970).]

Oklahoma and Arkansas. With reference to this distribution, Shideler (1970, p. 801) says:

> The pattern delineates four areas in which boulder diameters exceed 16 ft. These areas are believed to represent the paleogeographic loci of, areas adjacent to prominent islands along the geanticlinal ridge. The islands functioned as the most prolific sources of boulders, and therefore constituted major boulder dispersal centres during Johns Valley deposition.

Another type of map that can be used in conjunction with isograde maps and one that can be constructed on the basis of both outcrop and subsurface data is a clastic percentage map such as illustrated in Fig. 6-41. This map shows the distribution, by percentage of volume, of coarse clastics in Pennsylvanian strata of southeastern Colorado. The westward increase in

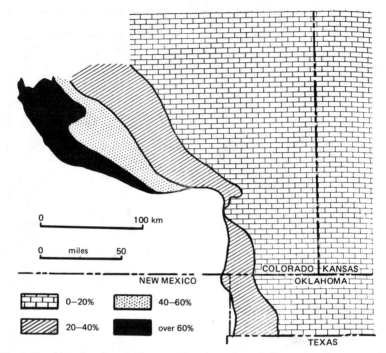

Fig. 6-41. Map showing clastic percentage by volume in Pennsylvanian strata of Colorado. [After Maher (1953).]

the content of coarse clastics is interpreted by Maher (1953) as indicating the general direction of the landmass from which the clastics were derived.

Shale Color Maps

Shales have a variety of shades of gray, brown, green, and red. Descriptions vary depending on whether the rock is examined wet or dry; in outcrop, hand specimen, or under the microscope; in natural or artificial light. Descriptions depend not only on the color sense of the person examining the rock but also, in the case of sample logging of drill cuttings, on the state of eye fatigue. Cuttings of shale logged as greenish gray in the upper part of a thick section may be logged as light gray in the lower part although comparison of cuttings from the lower and upper part of the section may show no change in the shade of color. The eye is not always reliable as a color guide, and where shale color maps are constructed, it is necessary to calibrate the shades mapped by means of a standard color chart.

The color of some thick shale sections, particularly marine sections, remains remarkably constant throughout their entire thickness. Others, particularly those deposited in paralic and nonmarine environments, may comprise interbedded shales of various shades. In the latter case it may be necessary to designate the shale unit mapped by a color range or by its predominant shade of color. The method chosen and the source of calibration should be stated.

With some exceptions, shale colors are commonly associated with the content of organic matter or iron compounds and consequently with depositional environments. Black pyritic shales are associated with marine or nonmarine euxinic conditions, brown carbonaceous shales with deltaic environments, light brown to buff and variegated shales with river floodplain deposits, greenish gray shales with open marine environments, and red shales with derivation from a red-weathered regolith and floodplain deposition as well as with deep-sea deposition of volcanic ash. The color of shales may vary, in places almost imperceptibly, with a slight change of lithofacies or of biofacies which may reflect proximity to a delta system, a reef trend, or a carbonate shelf. In some places variations of color or shade may have no relationship to such environmental features but may depend on depositional conditions such as rapid burial of the sediments that precludes oxidation beneath the sediment–water interface and on diagenesis caused by burial metamorphism.

An example of a shale color map is illustrated in Fig. 6-42 which shows the isopach contours and color distribution of a shale unit within the late Early Cretaceous of northeastern Texas. The map shows a general relationship between thickness and color, the unit becoming darker away from the landmass and towards deeper water in the basin. With reference to this relationship, Stehli *et al.* (1972, p. 56) say:

> The common occurrence of black shales in the southern part of the area in a region probably characterised by law energy, deep water, and great distance from a major source of terrigenous clastics is consistent with the interpretation that low permeability in fine-grained, organic-rich sediments was responsible for the production of reducing conditions and, thus, of the black shales.

Commenting further on the significance of color distribution in Cretaceous shales of the area, Stehli *et al.* say that the distribution of red shales appears to be limited to areas near to terrigenous sources and that gray shales are common to areas of relatively shallow-water deposition such as the middle to inner neritic environments of a continental shelf.

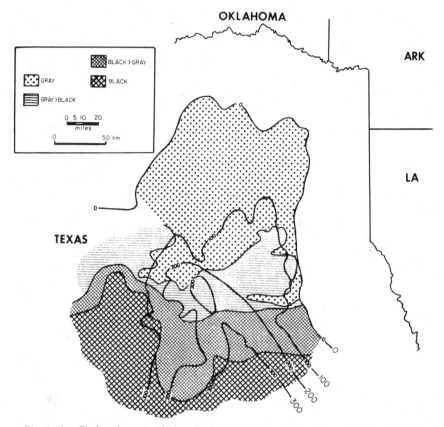

Fig. 6-42. Shale color map of a late Early Cretaceous stratigraphic unit in the East Texas Embayment showing gradation from black to gray. Isopach contours in feet. [After Stehli *et al.* (1972).]

BIOFACIES

The distribution of fossil biota and the construction of various biofacies maps showing the relationships of various genera or species to stratigraphic horizons or intervals is not within the province of this book; but brief mention will be made of certain biofacies maps insofar as they relate to lithofacies. The distributions of genera are broadly related to the environment in which they live, that is to say they thrive best under certain living conditions but may be able to tolerate a broad range of environmental conditions. Insofar as some benthonic genera are selective, preferring sand to mud, or shallow water to deep, their distributions can be related generally to geographical features and depositional environments commonly characterized by particular sediment types. Associations of living forms preferring particular environments constitute with their physical surroundings an ecosystem which may be subsequently reflected in the lithified sediments as a faunal assemblage. The distribution of such an assemblage, shown on a biofacies map, may have a direct correlation with the lithofacies shown on a rock-type distribution map. Broadly, these relationships may indicate that where the facies of a rock-time interval changes from limestone to shale, the assemblage of fauna also changes, so that limestone is characterized by one assemblage and shale by another.

Certain genera included in a particular assemblage may be more sensitive than others to changes of environment such as variations in the volume or type of sedimentation. For example, an increase in the influx of mud into an environment of carbonate sedimentation may depress or kill off the growth of certain genera, whereas others can tolerate the mud and survive. In such cases a biofacies map showing the distribution of a particular genera or species, as a percentage by number of specimens counted in a faunal assemblage, may show a correlation with rock-type ratio or percentage maps. Other biofacies maps showing the distribution of a particular genera or species may indicate a restricted habitat. Some fossils are found only within a specific stratigraphic zone deposited in a particular depositional environment subject to definite physical constraints imposed by depth and temperature of water, ocean currents, wave action, sedimentation, and food supply. These constraints are operative in the limestone facies of reef complexes and their associated lagoons. Modern and fossil reefs, viewed in section and in plan views, show local distributions of assemblages and genera. Limited distributions of this type are known as microbiofacies, and where control is close enough to plot such distributions they can be an invaluable aid to interpretation of the geomorphology of ancient carbonate environments.

The lateral distribution, within a stratigraphic unit, of a specific pelagic form also constitutes a biofacies and may be of particular value as a time marker within a thick stratigraphic sequence otherwise devoid of widespread definitive markers. An example may be cited in the first occurrence of *Tentaculites* in shales of the Late Devonian Ireton Formation of Alberta. These fossils are readily recognized in drill cuttings. In some areas the lateral and vertical distribution of microflora may constitute biofacies which also serve as important time-rock units, particularly within thick sequences of prodelta sediments in which it is otherwise difficult to correlate sandstone beds. Considerable use of microflora, in the stratigraphic zonation and identification of faulted strata, has been applied to the Tertiary of the Maracaibo Basin of Venezuela. These and other biofacies maps should be considered as valuable adjuncts to lithofacies maps and sections with reference to interpretation of the geologic history of a basin.

COLLATION AND REPRESENTATION

The collation of various types of maps and sections of a particular stratigraphic interval commonly requires a sifting process to sort out the key elements to be illustrated in some form of representation. Taken individually, and out of context with others, such maps and sections may not appear to be meaningful. The significance placed on particular relationships depends on interpretation; the emphasis shown depends on the form of representation in one or more maps, sections, or block diagrams. Some examples are given in the following pages.

Depositional Environments of Rock-Time Units

The depositional environment of a rock-time unit may possibly be determined by examination of outcrops. Where the unit is not exposed, its depositional environment may be indicated by the character of its electric log, by cores, or by a combination of both. Commonly, it is necessary to construct a three-dimensional model of the unit, using isopach maps, facies maps, and stratigraphic sections. This model must then be considered in the light of core and electric log data and in context with the lithology and stratigraphic sequence of overlying and underlying beds. Figs. 6-43–6-45 show examples of the geometry and depositional environment of a stratigraphic unit placed in the context of its regional tectonic and paleogeographic setting. The first two figures are based on studies by Katich (1958) and Quigley (1965), the last figure on studies by Mallory (1960) and Tillman (1971).

Fig. 6-43 is a schematic section extending northeast from Glenwood Springs across the Maroon Basin of northwestern Colorado. It shows the stratigraphic relationships of the Early Pennsylvanian Eagle Valley Evaporite consisting of gypsum, anhydrite, and salt. The Eagle Valley is underlain by dark gray shale and limestone of the Belden Formation which overlies an unconformity and is overlain by buff to reddish sandstone and thin beds of gray shale and limestone of the Minturn Formation. Quigley (1965) describes the Eagle Valley as evaporites of the basin-center type

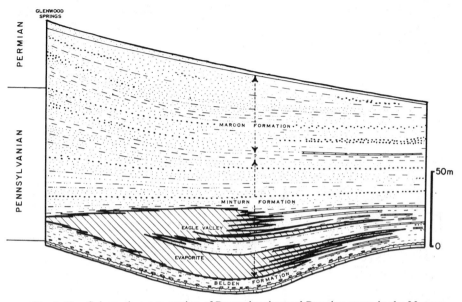

Fig. 6-43. Schematic cross section of Pennsylvanian and Permian strata in the Maroon Basin, northwestern Colorado, showing stratigraphic relationships of the Eagle Valley Evaporite. [After Katich (1958) and Quigley (1965).]

deposited with thin layers of dolomite, shale, and sandstone in a rapidly sinking trough flanked by rising highlands. The trough was subsequently filled during the Late Pennsylvanian to Early Permian with up to 700 m of fluvial and paralic red beds of the Maroon Formation.

Fig. 6-44 is an isopach map of the Eagle Valley Evaporite and shows the unit to have a thickness of up to 400 m. It is distributed in a somewhat elongated basin-shaped area surrounded by uplifted blocks of basement rocks. The regional distribution of these areas of exposed basement rocks is illustrated in Fig. 6-45, which also shows the distribution of fluvial and paralic arenites and lutites. These sediments were derived from erosion of highlands during the Pennsylvanian and swept out along flanks of the uplifted belts as alluvial fans and outwash plains of gravel, grit, askosic sand, and silt. These sediments comprise the Fountain Formation and correlative units.

The composite picture suggests that during the Pennsylvanian the region of western Colorado was one of basin and range topography, with rapid erosion of the uplifted basement blocks, and of drainage into downfaulted basins that periodically were either totally or partially closed. The paleogeographic situation may have been similar to that of the

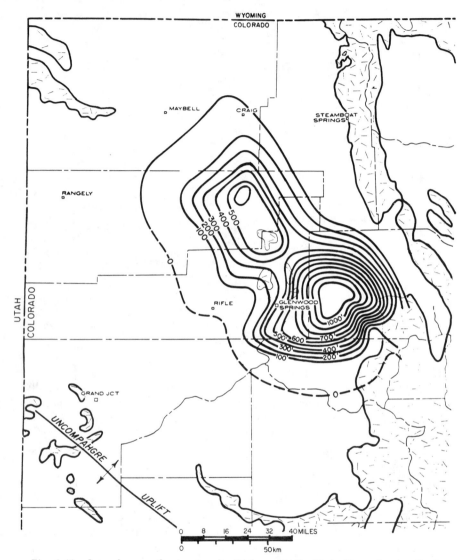

Fig. 6-44. Isopach map of net evaporite thickness in the Early Pennsylvanian Eagle Valley Evaporite, Maroon Basin, northwestern Colorado, showing areas (hatched) where Precambrian rocks are now exposed. [After Katich (1958) and Quigley (1965).]

region surrounding the northern end of the present-day Gulf of California, as shown in Fig. 4-20. Such inferences, on the basis of evidence supplied by the three figures mentioned, must be considered as tentative; but they serve to illustrate the way in which information can be collated and

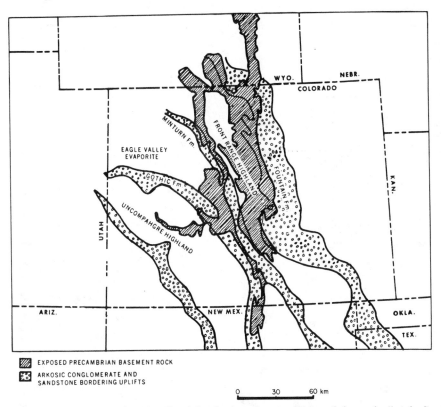

EXPOSED PRECAMBRIAN BASEMENT ROCK

ARKOSIC CONGLOMERATE AND
SANDSTONE BORDERING UPLIFTS

0   30   60 km

Fig. 6-45. Map showing (a) regional distribution of exposed Precambrian rocks (hatched) and Pennsylvanian to Permian arkosic arenites derived from their erosion (stippled), (b) belts of tectonic uplift, and (c) Maroon Basin partly filled with the Early Pennsylvanian Eagle Valley Evaporite in northwestern Colorado. [After Mallory (1960) and Tillman (1971).]

stratigraphic and tectonic relationships highlighted. Positive representation emphasizes interpretation of the paleogeographic setting in which the indicated depositional environments could have existed.

Another example of composite representations that show the paleogeographic framework and depositional environments in which different sandstone trends were deposited is illustrated by Haun (1962) in Figs. 6-46 and 6-47. The former is a schematic section across the northern part of Wyoming. It shows the Jurassic and Early Cretaceous sequence, with particular reference to stratigraphic relationships of the Early Cretaceous Muddy Sandstone, Newcastle sandstone, and J sandstone. The latter illustrates the distribution in Wyoming of the above-mentioned sandstone units and shows the Muddy to have a northwesterly and the

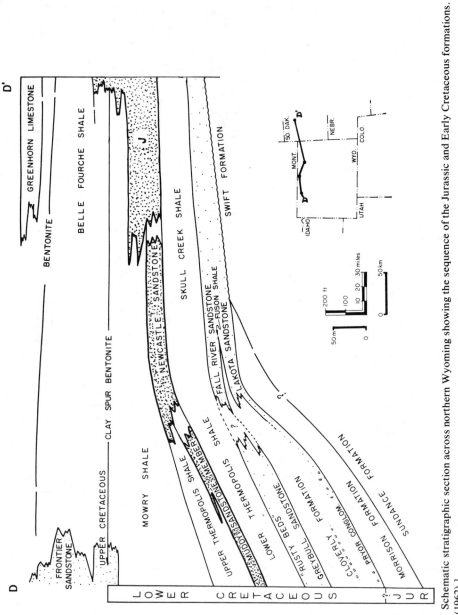

Fig. 6-46. Schematic stratigraphic section across northern Wyoming showing the sequence of the Jurassic and Early Cretaceous formations. [After Haun (1962).]

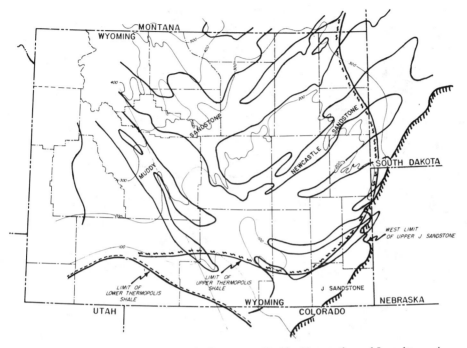

Fig. 6-47. Distribution of the Early Cretaceous Muddy, Newcastle, and J sandstones in Wyoming showing main depositional trends of these composite units comprising river distributary channel and shoreline sand bodies (see Fig. 6-46). [After Haun (1962).]

Newcastle a northeasterly depositional trend. Both of these sandstone units comprise complex systems of river channel, delta distributary, and shoreline sand bodies. The Thermopolis shale units underlying and overlying the Muddy thicken to the northwest, as does the younger Mowry shale. The upper unit of the J Sandstone, which is a blanket-type unit consisting of regressive marine sandstone bodies deposited in the front of a northwesterly prograding delta, is shown to have its western limit in the southeast corner of Wyoming. The map outlines the main depositional trends and paleogeographic setting of these sandstone units but does not differentiate the paleogeomorphic features. It is evident that the fluvial, deltaic, and shoreline depositional complexes represented by the Muddy and Newcastle were drowned by a marine incursion that deposited the Mowry Shale during the period of deposition of the penecontemporaneous J Sandstone.

One further example of representation is given in Fig. 6-48 which is a block diagram illustrating the thickness, provenance, and paleogeographic setting of the Late Devonian Cleveland Shale in Ohio. This unit is a silty

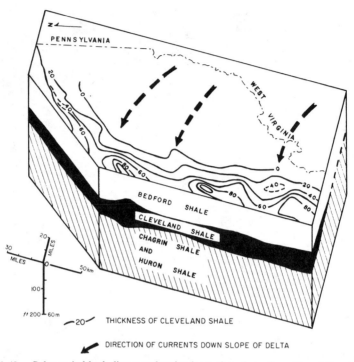

Fig. 6-48. Schematic block diagram showing isopach and stratigraphic relationships of a segment of the Late Devonian Cleveland Shale in Ohio. Arrows indicate direction of transport of silt content. [After Lewis and Schwietering (1971).]

and organic-rich black shale in which the silt content increases eastward and the calcareous content increases westward. Lewis and Schwietering (1971) concluded that the silty component of the black shale unit was introduced by currents flowing westward from the Catskill Delta into the restricted depositional basin of an inland sea. The features shown in this block diagram do not in themselves constitute or indicate evidence, but they effectively illustrate the conclusion stated.

Paleogeographic Maps of Rock-Time Units

A rock-time unit is a stratigraphic unit that can be defined as having been deposited within a particular range of time. It may be a lithostratigraphic unit in which the rock type is (a) fairly uniform in composition and character, such as sandstone or limestone; or (b) variable in composition and character such as interbedded sandstone and shale, or limestone and shale, but readily defined as a stratigraphic unit distinct from overlying and underlying sequences. It may also be a unit that is either uniform or variable in composition and character at any particular location but which changes facies laterally. Provided the rock-time unit is not too large in terms of time range, thickness, or both, the paleogeography that existed during the period of deposition of the unit may be reconstructed on the basis of several criteria. These include the following: (a) depositional limits of lithostratigraphic units; (b) distribution of lithofacies and biofacies in a lithostratigraphic unit; (c) distribution of outcrops or subsurface occurrences of a lithostratigraphic unit known from its faunal content to be either marine or nonmarine; (d) distribution of stratigraphic units and sequences containing faunal assemblages that define age ranges; (e) distribution and depositional trends of stratigraphic units; and (f) evidence based on sedimentary structures and petrological analyses that indicate directions of transport and provenance of sediments. Interpretations of paleogeography may be based on any one or a combination of these stratigraphic situations or parameters, some of which may themselves be to some extent inferential and based on interpretation. Some examples are given in what follows.

Fig. 6-49 illustrates an example of a landmass subsiding under a transgressive sea. It shows the edges of the early Late Jurassic Rierdon Formation and of the successively overlying Late Jurassic Swift Formation and Morrison Formation in southwestern Montana. The edge of the Rierdon is probably in part an erosional edge, but no marked hiatus in deposition or unconformity appears to exist between the Swift and Morrison. The decrease in size of the area in which these formations are not present is interpreted by Suttner (1969) as evidence of an encroaching Late Jurassic sea. One of the main difficulties with interpretation based on such evidence is the problem of defining the extent to which the edge of a stratigraphic unit is depositional or erosional. In this respect it may be possible, provided there is sufficient control, to determine the answer on the basis of lithofacies or biofacies. The combination of an edge and a nonmarine facies that grades away from the edge to a marine facies strongly supports the interpretation that the edge is depositional but does not preclude the possibility that a considerable part of the nonmarine facies may have been eroded. On the other hand, the combination of an edge and a marine facies suggests that the edge is probably erosional but does not necessarily indicate erosion of a preexisting nonmarine facies.

Interpretation of paleogeographic relationships, based on facies variations of correlative stratigraphic units, may be possible in a generalized way where the control is scarce and widespread. An example of a generalized reconstruction of the main paleogeographic features that existed

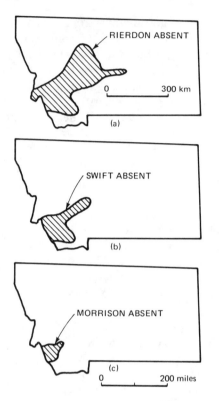

Fig. 6-49. Paleogeographic maps showing areas of decreasing size in which the successive (a) Rierdon, (b) Swift, and (c) Morrison formations of southwestern Montana are absent and presumably not deposited, indicating transgression of the Late Jurassic sea. [After Schmitt (1953) and Suttner (1969).]

during the early Late Devonian in the southwestern Arctic Islands of Canada is illustrated in Fig. 6-50. Based on a few scattered control points, the map is tentative but serves as a base to which additional data can be incorporated to change or modify the interpretation. The Weatherall Formation facies, comprising fine-grained quartzose sandstone, siltstone, shale, carbonaceous shale, and minor coal, grades laterally into the Orksut Formation which consists mainly of calcareous shale, with minor silty shale and argillaceous limestone. The lateral relationships between strata of the Orksut Formation and the limestone beds of the Hume Formation are not well understood, but it is considered by Miall (1976) that the Hume in this region represents a shelf facies flanking an area of deltaic sedimentation prograding to the south and southeast from an inferred landmass.

A generalized paleogeographic map showing the distribution of land and sea masses during the interval of time represented by a rock time unit can be constructed on the basis of the known occurrence and depositional environment of the unit. An example is given by Day (1969) in Fig. 6-51

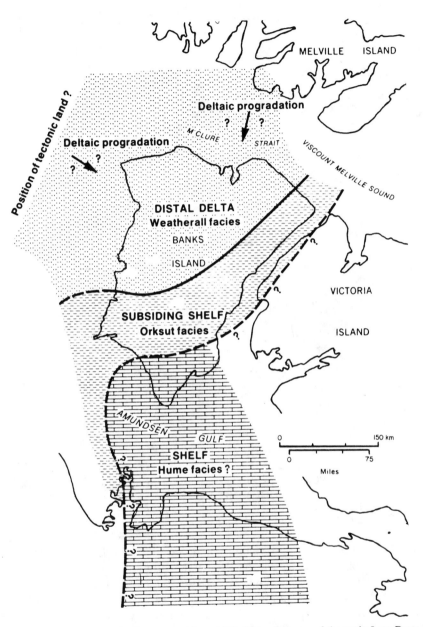

Fig. 6-50. Generalized paleogeographic and lithofacies features of the early Late Devonian Weatherall, Orksut, and Hume formations in the southwestern Arctic Islands of Canada. [After Miall (1976).]

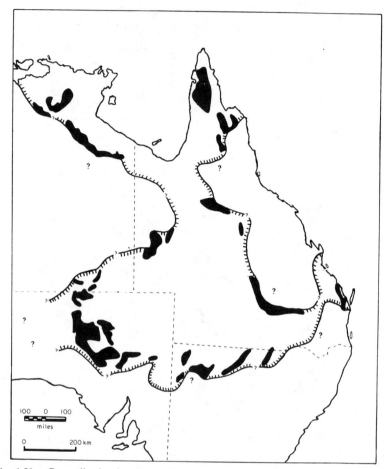

Fig. 6-51.   Generalized paleogeographic map of eastern Australia showing the distribution of the Late Aptian sea, based on known distribution of strata containing a marine fauna. [After Day (1969).]

which illustrates the approximate configuration of the Late Aptian sea in eastern Australia. Based on the known distribution of outcrops and subsurface strata containing a marine fossil assemblage, the map shows the area of the Great Artesian Basin open to the sea in the north and probably also connecting to the west with another embayment of the sea which transgressed eastward from the present Northwest Shelf of Western Australia. The shoreline indicated on the map shows the minimal area that the sea could have occupied, as it discounts the possibility of erosion of preexisting outcrops. In such cases interpretation hinges to a significant extent on the degree to which a particular erosional edge approximates a depositional edge.

The limits of stratigraphic distributions of fossil assemblages indicating particular time ranges can also be plotted to show the transgressive or regressive migration of a coastline. Fig. 6-52 illustrates the delta of the Niger River, Nigeria, and shows the depositional limits of Tertiary rock-time units spanning the interval from Middle Eocene to Pleistocene. The seaward progression of the units represents the growth of the Niger Delta during the Tertiary and serves to illustrate the progressive changes in paleogeography. By the end of the Pliocene, according to Short and Stäuble (1967), the delta shoreline had extended as far as 80 km seaward of the present coast. The postglacial rise of sea level resulted in a marine transgression that drowned the old river systems and formed the existing estuaries.

On the basis of the distribution, depositional trends, and depositional environments of various parts of a rock-time unit, the gross elements of major paleogeographic features such as deltas may be outlined. This has been illustrated by Asquith (1974) in Fig. 6-53 which shows the distribution of various stratigraphic units that comprised a delta system during the Late Cretaceous in Wyoming. Within the rock-time unit determined, the Upper Erickson Sandstone is fluvial and the Williams Fork Formation and Mesaverde Formation are paralic. The Parkman sandstone is of inner neritic origin and includes shoreline sand bodies and barrier islands. The Pierre Shale was deposited in a deeper marine environment. Reconstructions of the paleogeography based on such evidence are approximations only as in any deltaic environment transgressive and regressive phases are common. They serve to indicate the general configuration of the coastline during an interval of time and to relate the progradation of deltaic sequences to tectonic events such as the uplift of crustal belts. Another example given by Haun (1963) is illustrated in Fig. 6-54 which is an isolith map of the Early Cretaceous J Sandstone in northeastern Colorado showing the general trend of the distributary system and seaward fringe of the delta in which the sand bodies were deposited.

Paleogeographic maps can also be constructed on the basis of sedimentary structures and provenance of sediments although these features are seldom the only criteria. Within a rock-time unit, the assemblage of sedimentary structures may grade laterally from those associated with fluvial environments to those typically formed in paralic or neritic environments, as described by Conybeare and Crook (1968), Reineck and Singh (1973), and many others. The sediments may also show a gradation in mean grade size (Figs. 5-16 and 5-17). Furthermore, it may be possible to relate the rock types seen in pebbles, or in grains viewed under the

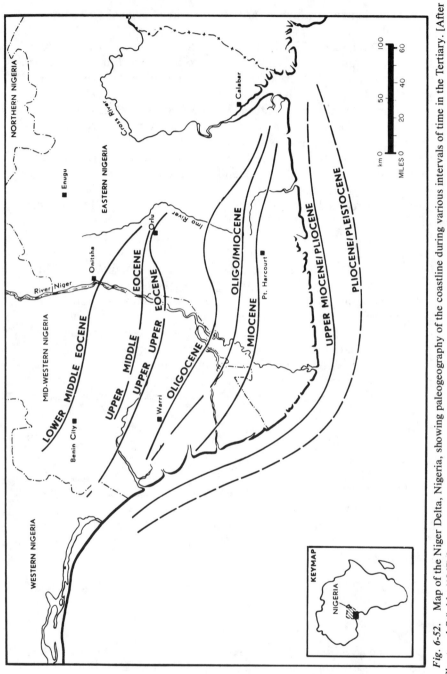

Fig. 6-52. Map of the Niger Delta, Nigeria, showing paleogeography of the coastline during various intervals of time in the Tertiary. [After Short and Stäuble (1967).]

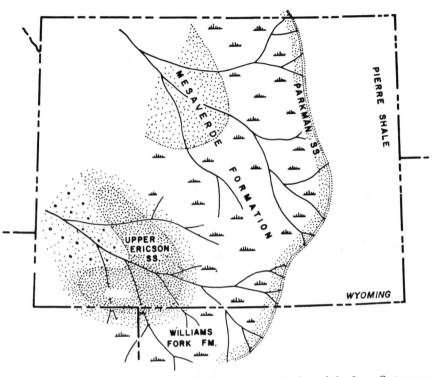

Fig. 6-53. Paleogeographic map showing general distribution of the Late Cretaceous delta system in Wyoming. [After Asquith (1974).]

microscope, to rock sequences from which they were derived. A case in point is the Late Cretaceous Belly River Formation of Alberta, the sandstones of which are indicated by thin section analysis to have been derived mainly from sources to the west, probably in part from the Permian Cache Creek Series presently exposed some 300 km to the west in British Columbia.

Paleogeomorphic Maps of Rock-Time Units

Paleogeomorphic maps are commonly isopach maps of rock-time units that delineate the solid geometry, trends, distribution, and where relevant the spatial relationships to each other of particular sandstone or limestone bodies such as barrier bars of lithic sand, or algal and coral reefs. These maps can be constructed from outcrop data only, but commonly this is insufficient because of lack of continuity or exposure. Construction of

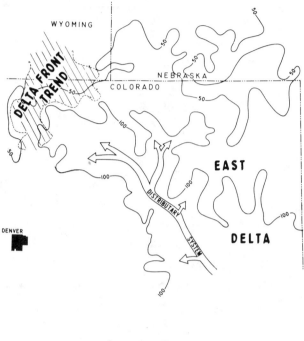

Fig. 6-54. Paleogeographic map showing the main depositional features of the Early Cretaceous J Sandstone (J delta sand isolith) in Colorado (Denver Basin). [After Matuszczak (1973) and Haun (1963).]

paleogeomorphic maps based on subsurface data depends on fairly close control such as is found in oil fields and adjacent areas, although it may be possible to rely on extrapolation between widely spaced control points in the delineation of linear trends such as barrier bars and reefs. Many examples of ancient river valley channel sands, cuesta strike-valley sands, delta distributary sands, delta-front sheet sands, bars, barrier islands, sand dunes, and other features have been described in the literatures. Some of these features have formed stratigraphic and structural-stratigraphic traps for oil and gas, as described by Conybeare (1976).

It must be emphasized that the basis on which interpretation is made should not be taken out of context with other available information, although differences of interpretation may arise. For example, in the case of a sandstone body, if an interpretation is based on isopach data only, the resulting map could indicate its origin as either a bar or distributary. In the absence of fossils, the problem may be resolved by the sequence of grain gradation as seen in cores or as indicated by the self-potential curves of electric logs. Where a range of grain sizes is present, the coarser sand was deposited in the upper part of offshore bars and in the lower part of river distributaries. Other criteria such as electric log characteristics indicating offlap relationships within a composite sandstone body, sedimentary structures, and stratigraphic relationships may be definitive or indicative.

Paleogeomorphic maps showing the topography of an unconformity can be constructed as either structural or isopach maps. The former is a topographic map based on a datum such as sea level; the latter depends on the selection of a stratigraphic marker and the construction of an isopach map of the interval lying between this marker and the unconformity. The marker selected should be widespread and as nearly as possible represent deposition on a flat surface. Obviously, this latter method is not precise, and the smaller the range of thickness selected for the isopach map the greater the accuracy obtained. In such cases the map is not only an isopach map of the rock-time unit but also a paleogeomorphic map showing the configuration of hills and valleys of the unconformity. In some structural-stratigraphic situations where the erosional margin of permeable limestone beds is overlain by impermeable shale, outliers of limestone form buttelike hills and may be traps for oil and gas. Examples in Alberta and Saskatchewan have been described by Martin (1966).

An isopach map of a portion of a sandstone body interpreted by Forgotson and Stark (1972) as a barrier-bar complex is illustrated in Fig. 6-55.

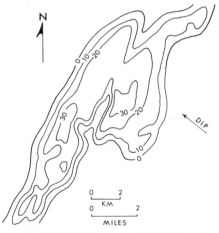

Fig. 6-55.   Isopach map of the Early Cretaceous Muddy Sandstone, Bell Creek oil field, Montana, showing the paleogeomorphology of this barrier-bar complex. Thickness in feet. [Modified after McGregor and Biggs (1968) and Berg and Davies (1968).]

This shows the configuration of the oil-producing Early Cretaceous Muddy Sandstone of the Bell Creek Field in the Powder River Basin, Montana. The basal unit of this complex body exhibits extensive bioturbation. Another example of a paleogeomorphic map of a barrier-bar complex is illustrated in Fig. 6-56 of an oil-producing sandstone unit within the Late Cretaceous Gallup Formation of the San Juan Basin in New Mexico. This illustration is based on isopach maps of the three sandstone units comprising the bar complex and shows not only the distribution of these units but also their spatial relationships to microfacies within an inner neritic environment, as interpreted by Sabins (1963).

Paleogeomorphic maps of organic carbonate reefs may be constructed from either subsurface isopach or structural data. In the majority of cases the latter method is employed, the depths from sea level to the top of the reef being contoured as a structure map which is then adjusted for regional tilt or folding to conform to the original configuration of the reef surface. In oil or gas fields producing from reefs, few of the wells penetrate sufficiently deep to provide isopach data. Isopach and structural maps of seismic intervals may also be used to outline the structure, configuration, and thickness of carbonate reefs provided there is sufficient velocity contrast between the reef and its overlying and underlying rock layers.

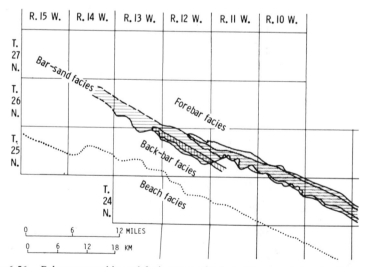

Fig. 6-56. Paleogeomorphic and facies map of a Late Cretaceous barrier–bar complex (Bisti Oilfield) in the Gallup Formation, Sand Juan Basin, New Mexico. [Modified after Sabins (1963).]

Paleocurrent and Paleowind Pattern Maps

There is a vast amount of literature on paleocurrent indicators, much of which has been summarized by Potter and Pettijohn (1963), Conybeare and Crook (1968), Selley (1970), Pettijohn *et al.* (1972), Reineck and Singh (1973), and others. Some of the information that can be obtained from these indicators has been discussed in this book. Paleocurrent indicators commonly include both linear features, such as groove or flute moulds, and cross bedding. These are indicative of the direction of flow of the current that deposited the sediment at a particular location. Vertically or laterally within the same or adjacent outcrop, the linear features or cross bedding may indicate variations of current direction which, in the case of tidal channels, can reverse direction. In the case of fluvial sediments deposited in a meandering channel, the directions may range within an arc of 180°. A rose diagram based on a number of measurements of cross bedding will show the mean direction of flow of the river system. Consequently, the construction of a paleocurrent map depends on statistical methods of analysis.

The difference in results obtained from a few or a large number of readings may not be significant but can only be determined empirically. In many locations the constraints placed by lack of outcrop, or difficulty of access, preclude the collation of a large collection of data, and results must be obtained from a few dozen control points. In all such cases it is essential, in reporting the results of an investigation, to state the number of readings and the nature of the indicators on which they are based. Current directions can be represented on a map by means of arrows. Where each arrow represents several measurements, numbers can be attached referring to the quantity, quality, and type of data on which the directions are based. It is also essential to stipulate the rock-time unit within which the measurements have been made. Investigations of current directions based on measurements of various sedimentary structures are fundamental to the determination of sediment provenance and major paleogeographic features, including landmasses and abyssal trenches or plains. An interesting example of a study of paleocurrent directions is given by Bennacef *et al.* (1971) with reference to sedimentation in the Paleozoic cratonic basin of central Algeria. In this basin Late Cambrian to Early Devonian sandstone beds exhibit a remarkable uniformity of cross bedding indicating northward transportation from the present area of the Precambrian Hoggar basement. These cross-bedded sandstone units are separated by beds of Early Silurian marine shale and by Late Ordovician glacial strata, a relationship that suggests periodic uplift of the Precambrian basement fringing the southern margin of the cratonic basin during a long period of time.

Paleowind pattern maps are constructed from measurements of eolian cross bedding. Among the early students of windblown sands, Bagnold (1941) made notable contributions. His work was followed by many others. In particular, the work of McKee (1966, 1969) has been instructive in understanding the development and stratigraphic relationships of sand dunes.

Sand dunes can be preserved in the geological record only where they are buried by sedimentary layers. These can be of marine sediments deposited in a neritic environment by a transgressive sea or of fluvial sediments commonly deposited by braided streams flowing over alluvial fans into cratonic basins. Concordance of directions of transport of the eolian sand, determined by measurements of cross bedding at various localities, is indicative of the direction of the prevailing winds that formed the sand dunes. Interpretations of prevailing wind directions are useful, in conjunction with those pertaining to paleogeography, paleoclimates, and paleolatitudes, in unraveling the complexities of earth's geologic history.

Reconstruction of Basin Growth

Reconstruction of the stages of growth of a sedimentary basin shows the evolution and relationships of its main structural and stratigraphic elements during successive intervals of time represented by rock-time units. This form of representation, based on the collation of a number of structural and stratigraphic maps of individual rock-time units, is useful in depicting the contemporaneity of tectonic and sedimentologic events and in illustrating the sequence of development of a basin.

Illustrations commonly are drawn as cross sections of the basin showing its structural and stratigraphic configuration at the times of deposition of various successively younger marker beds. This method can be more easily employed where the basin is smaller and stratigraphically less complex. Larger basins comprising a complex of two or more sedimentary piles deposited under different tectonic and environmental conditions may initially require analysis of their separate parts rather than of the whole basin. Subsequently, the development of the whole basin may be represented by collating the various stages of its parts. This process presumes the recognition of time marker beds over much of the basin, or of the determination of rock-time units common to two or more sedimentary piles. Falls of airborne volcanic ash form excellent time marker beds of bentonite and commonly are easily recognized in electric logs. The recognition of rock-time units that may be lenticular, or otherwise not laterally traceable one to the other, depends on the presence of fossils known to define a particular range of time. In some situations it may be possible to make broad correlations on the basis of age dating of intercalated volcanic flows. These problems commonly arise in reconstructing the geologic history of a complex basin comprising one or more subbasins.

Representation of cross sections illustrating successive stages of growth of the whole or part of a sedimentary basin must be drawn with due consideration to the distortions resulting from vertical exaggeration of scale. Where the cross sections are hung from a marker bed which is used as a surface of reference, the bed may be shown as a horizontal datum, an inclined plane, or a warped surface depending on the horizontal scale of the cross section and the available information and concepts as to the depositional configuration of the marker bed. It is useful to keep in mind that where little or no vertical exaggeration is employed, the original configuration of the marker bed may be of little consequence to the shape of the marker bed trace as drawn on the cross section and that, for practical purposes, the trace may be shown as a datum. Where considerable vertical exaggeration is employed, the delineation of the trace is a matter of judgment and interpretation. Sequences of cross sections are particularly useful as a graphic method to illustrate the stages of development and stratigraphic and structural relationships within a sedimentary basin of thrust sheets, growth faults, bioherms, salt domes and other features as shown in Figs. 2-34, 2-46, 2-48, 3-5, 3-22, 3-23, 4-19, and 4-43.

Palinspastic and Regional Tilt Adjustments

The construction of lithofacies, biofacies, and paleogeographic maps of folded and faulted strata poses a problem in that the geographic locations of the control points do not have the spatial relationships that existed at the time of deposition. Sediments deposited at sites that were originally many kilometers apart may have been brought into close proximity by folding and thrusting, and, conversely, sediments deposited at adjacent sites may have been separated by many tens of kilometers by thrust faulting. In order to plot the control points in positions approximating their original distribution with respect to one another, it is necessary to calculate the amount and direction of displacement of each control point. It has been found empirically that such calculations need not be precise, as there is a range of latitude within which variations in the plotting of lithofacies and paleogeographic data does not materially affect the interpretation. Where this procedure of palinspastic adjustment has been carried out, the control points replotted, and the original distribution of lithofacies or paleogeographic features reconstructed, many previously irreconcilable features and apparent stratigraphic ambiguities may be explained. Also the original depositional trends of features within fold belts, such as carbonate reefs and sandstone barrier bars, may be delineated and thereby indicate areas within a basin in which the continuation of these potential hydrocarbon-bearing features may be found.

Palinspastic adjustment is applicable to the folded and faulted belts on the flanks of many sedimentary basins. Within the basins, and particularly on the flanks overlying stable basements subject to periodic regional uplifts, adjustments may be made for regional tilt. This procedure is particularly useful to restore the topography of unconformities where erosional edges and outliers, or sandstone bodies deposited as fluvial sand in valleys, may be of interest as reservoirs for oil and gas. Tilting of strata has the effect of bringing closer together on a vertical plane the projections of original depositional sites located on a line normal to the axis of tilt. Where tilted to a vertical position, the projections converge to a single point. In many sedimentary basins the necessary corrections involve regional tilts of less than 10 m/km. Although slight, these can be significant in the construction of structure contour maps of buried hills or reefs.

It has been pointed out by Dahlstrom (1969) and others that palinspastic adjustment of control points along a folded and faulted belt must take into consideration the variations in amount of displacement of a particular belt by different thrusts and also that calculations of shortening must be made on one stratigraphic horizon only. These concepts are illustrated in Fig. 6-57. Diagram (a) shows a stratum folded and displaced by thrusting. Original depositional sites from left to right are numbered 1–7. These have been moved from their original positions relative to one another so that sites 4, 5, and 6 are presently located between sites 1 and 2. Obviously, a facies map based on data obtained from the present spatial relationship of the outcrops would be incorrect as the facies are in the wrong sequence and would appear to be anomalous. Diagrams (a) and (b) illustrate distortion produced in the apparent depositional trend of a reef as a result of displacement by three overriding thrust sheets. Differential displacement by different thrust sheets, and shifts of displacement between thrust sheets, as shown in diagram (d), further complicate the problem of palinspastic reconstruction.

Computer Synthesis of Quantitative and Qualitative Data

The use of computers to synthesize and evaluate masses of geological data has become an important adjunct to the imaginative approach of

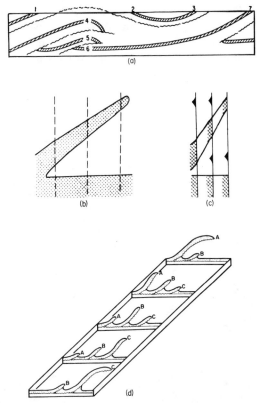

Fig. 6-57. Diagrams illustrating problems in palinspastic reconstruction involving (a) displacement of depositional sites 1–7 (ordinary map presentation would show facies data in wrong sequence), (b) and (c) displacement of depositional trend (trends before thrusting and apparent trends after thrusting, respectively), and (d) shift of displacement between thrust sheets in a fold belt. [Modified after Dahlstrom (1969).]

individual geologists and as such is a significant tool in the solution of geological problems. The great advantage of computerized synthesis is the speed with which it can process data and present the evaluated results in various tabulated or graphic forms.

The importance of evaluating all the quantitative and qualitative parameters of a sedimentary basin has been pointed out by Forgotson (1960, p. 99) who said:

A variety of stratigraphic attributes can be expressed quantitatively and illustrated in map form. The geometrical configuration of a stratigraphic unit, the areal variation in composition, degree of mixing of components, differentiation into discrete units, and the number,

vertical position, and arrangement of discrete units within a strati-
grafic interval can be expressed quantitatively and presented on
maps that serve as useful exploration tools. Each design conveys a
certain type of information and two or more designs can be combined
on one map to show the relationships between the aerial variation of

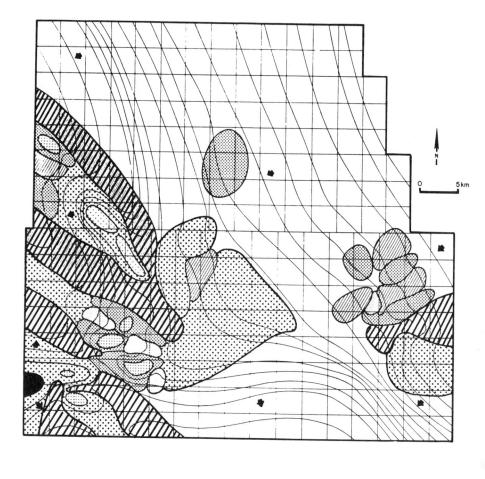

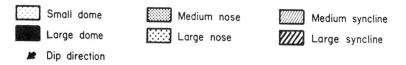

▨ Small dome   ▨ Medium nose   ▨ Medium syncline

■ Large dome   ▨ Large nose   ▨ Large syncline

🐝 Dip direction

Fig. 6-58. Form-line map showing main structural features at the base of the Pliocene,
San Joaquin Basin, California. [Modified after Grender *et al.* (1974) and Hoots *et al.* (1954).]

several attributes. Maps showing different kinds of information can be interpreted jointly to present a comprehensive three-dimensional picture of a stratigraphic unit. The variation in thickness, composition, and vertical variability of a stratigraphic unit can form the basis for paleogeographic and paleotectonic interpretations when the components selected for mapping are chosen properly.

The effective design of computer programs can greatly facilitate implementation of the various types of analysis discussed by Forgotson, particularly in outlining the major stratigraphic and structural components of the area reviewed (Fig. 6-58). In this regard the combination of computers and automatic plotters, as pointed out by Forgotson and Iglehart (1967), is invaluable for rapidly selecting and displaying key features that can subsequently be reviewed and evaluated. Additional studies of areas of interest can be carried out with greater definition by the further use of computer data, as described by Grender *et al.* (1974), to construct various maps and sections. In particular, computer data can be used to construct models which can serve as a basis for estimating the oil and gas potential of a sedimentary basin.

Application of Basin Analysis to Economic Geology

INTRODUCTION

Lithostratigraphic, biostratigraphic, and structural analyses of sedimentary basins are commonly undertaken by individuals or groups within government geological surveys, oil and mining companies, and universities or research institutions in order to assist in evaluating the petroleum, mineral, and water supply potential. Studies may place various degrees of emphasis on aspects regarded as pure or applied research and may direct efforts with particular reference to certain fields of interest, basin areas, or stratigraphic intervals. The collation, review, and interpretation of all available geological and geophysical information may be a massive task beyond the ability or scope of any individual and, consequently, require the combined effort of several skilled people in various disciplines. A further constraint, ever present, is time. The result is a compromise based on available personnel, time allocated for the study, and the emphasis to be placed on the direction of the study. These constraints apply in particular to projects of applied research where it is hoped that useful results may be obtained within a specified period of time. During the course of such a study, several interesting lines of investigation peripheral to the main thrust of the project may emerge. These cannot be pursued at the time but may be set aside for further investigation in the future. The end result of the study may not produce as many answers as might be hoped for but does require the resolution of data and ideas into interpretations subject to future modifications and thereby offers a working basis on which to proceed with an exploration program.

EXPLORATION FOR OIL AND GAS

Source Rock Potential of Basins

Much has been written on the subject of source rock analysis, some of which has been referred to in Chapter 5. Studies concern the hydrocarbon and organic content of shales and limestones, with reference to the probable nature and quantity of original organic matter contained in the sediment and to the physicochemical conditions and processes under which maturation and conversion to hydrocarbons occur. It is generally agreed that oil and gas are formed from organic matter (Watson and Swanson, 1975) buried in sediments and subjected over long periods of time to heat and pressure, but the parameters are not precisely known. The work of Tissot *et al.* (1974) indicates that within the range of geothermal gradients that commonly obtain the optimum depth range for the formation of oil is 2000–3000 m (Fig. 7-1). This does not preclude the formation of oil in shallower or deeper zones. Also the optimum depth range for the formation of natural petroleum gas is 3000–4000 m, although considered quantities of gas may be generated at shallower depths. These optimum depth ranges may vary, depending on the nature of the original organic matter, the geothermal gradient, and the length of time the hydrocarbon-generating stratigraphic unit is buried within various ranges of depth. These factors involve a knowledge of the geologic history of the sedimentary basin based on analyses previously discussed in this book. For example, the curves shown in diagram (a) of Fig. 7-1 apply only to sedimentary piles that have not been uplifted, eroded, and reburied but have been continually if not continuously deposited. Also the original nature of organic matter derived from terrestrial or marine sources appears to be a factor in the type of hydrocarbons derived by maturation (Ishiwatari *et al.*, 1978).

Referring to this diagram, Tissot *et al.* (1974, p. 500) note that during a certain period of time and within a depth range to about 1000 m the hydrocarbon content remains low. They say:

> Then, beyond a certain threshold, new hydrocarbons are generated, first gradually, then more rapidly, by thermal breakdown of the insol-

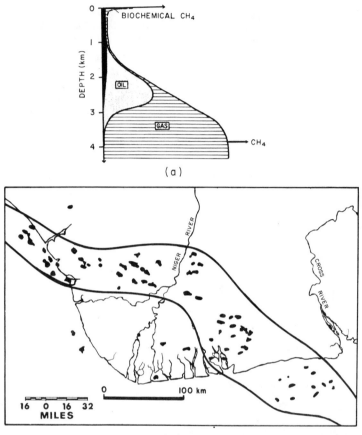

Fig. 7-1. (a) Diagram illustrating effect of increasing burial of source bed on the generation of hydrocarbons. [Modified after Tissot *et al.* (1974); and Hood and Castaño (1974).] (b) Map of Niger River Delta, Nigeria, showing distribution of oil fields forming a belt of prolific production. The belt is presumed to be a zone of greater subsidence. [After Dailly (1976).]

uble organic matter ('kerogen'). These hydrocarbons have no characteristic structure, and the previous hydrocarbons (the geochemical fossils) are diluted progressively by the new ones. This is the principal phase of oil formation, as designated by Vassoyevich *et al.* (1969).

At greater depth, increased cracking of carbon-carbon bonds occurs and generates light hydrocarbons from kerogen and from previous oil as well. This is the phase of condensate and gas formation. Later, under epimetamorphic conditions, methane, carbon dioxide, and a carbon-rich residue would be the end point of organic matter.

The geochemical fossils referred to are inherited from the organisms, algae, and plant matter deposited with the sediments and include naturally occurring fatty acids and waxes.

Assuming that the generation of hydrocarbons in a sedimentary basin can be illustrated by a set of curves similar to those in Fig. 7-1, it follows that an evaluation of the petroleum generating potential of a basin can be made provided the lithology, thickness, and maximum depth ranges of stratigraphic units are known. In other words, where unconformities occur within the sedimentary pile, the thickness of strata removed by erosion must be estimated. Calculations based on these parameters give estimates of the possible amount of hydrocarbons generated from potential source rocks but do not take into consideration the relative quantities that may have escaped, be dispersed within the rock layers, or concentrated in small unrecoverable or uneconomic pockets.

Within the depth range 2000–3000 m, the temperature range may commonly be 40°–55°C but can be lower or considerably higher as indicated by graph (a) of Fig. 7-2. The temperatures and pressures that obtain or have obtained within a sedimentary basin are important factors in the generation of petroleum and can be measured with a fair degree of accuracy by studies of mineral diagenesis within the sediments and by measurements of the reflectance of vitrinite in coal or other carbonaceous material, as illustrated in graph (b) of Fig. 7-2.

Where an organic-rich stratigraphic unit has been buried in a sedimentary basin to a depth of more than 2000 m, it will probably have generated oil and gas which have migrated updip to structurally and possibly stratigraphically higher zones in which a small percentage may have been trapped in oil and gas accumulations. Where subsequent uplift occurs, these accumulations of oil and gas may be breached by erosion or may be lifted so close to the surface that the hydrocarbons escape through joints, faults, or by tilting sufficient to destroy the structural closure. Reburial of the hydrocarbon-generating beds to depths in excess of 3000 m may result in further generation of gas only. These considerations are of prime importance in evaluating the hydrocarbon potential of a sedimentary basin, with particular reference to whether it is likely to be an oil or gas basin.

The process of evaluating a sedimentary basin with respect to its past capacity to generate oil and gas can be summarized as follows: (1) analysis of kerogen and other organic and hydrocarbon content of selected shale and limestone beds; (2) calculations of the volumes of these beds; (3) determination of the depth ranges, from drilling or seismic data, of various parts of these potential hydrocarbon source beds; and (4) estimation of the maximum depth ranges, from vitrimite reflectance or data on diagenesis, within which these parts have been buried. Analyses of the hydrocarbons,

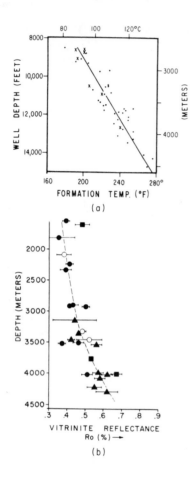

Fig. 7-2. (a) Temperature–depth plot, Tejon area, California. Crosses represent long-duration tests, dots represent determinations from electrical logs. (b) Vitrinite reflectance–depth plot, Tejon area. Symbols represent four wells. [Modified after Castaño and Sparks (1974) and Hood and Castaño (1974).]

kerogen, and organic matter contained in a shale or limestone can be used on a relative basis for comparing the probable potential capacity or past performance of a rock to yield oil and gas, provided the comparison is made with other rocks that have been subjected to the same conditions of temperature and pressure during a period of geologic time. Such analyses are a qualitative rather than quantitative measure. The total content of kerogen is rarely converted to oil and gas, and the problem arises of estimating the quantity of organic matter originally preserved in the buried sediment and the percentage of this matter that may have been converted to hydrocarbons.

Investigations along these lines have been carried out for a long time. Trask (1939) estimated that the average content by weight of organic matter in samples of recent marine sediments from various regions in the world was 2.5 percent. Stetson and Trask (1953) found the organic content of clays in the Gulf of Mexico to range up to 2 percent; and Rao (1960) notes that silty clay muds off the east coast of India have an organic content of 1–2 percent. With reference to recent carbonates, Gehman (1962) found the organic content of lime muds from the Gulf of Batabano, Cuba, to be about 3 percent. In other words, the original organic content of carbonate muds and clay appears to be about the same; but when these sediments become limestone and shale, analyses show that the limestones have lost a considerably larger percentage of their original organic content than have the shales. Hunt (1962) notes that the organic content of limestone in the Mississippian Madison Formation of Montana is 0.56 percent; and Ronov (1958) records the average content in marine Devonian limestones from the USSR as 0.39 percent. The organic content of clay shales commonly ranges up to 1.5 percent, indicating that the original organic matter in clay shales has been largely retained, probably because of adsorbtion by clay minerals, whereas that of carbonate muds has been largely lost. The significance of this conclusion, with reference to source beds for petroleum, lies in the fact that during the course of burial of equal masses of calcilutite and shale the former will yield a much larger weight of organic matter that may become the precursor for petroleum. An obvious reservation to this conclusion stems from lack of knowledge as to the percentages of organic matter that may be lost at different ranges of depth equated with ranges of temperature and pressure. Much of the organic matter may be flushed from the limestone by formation water at depths of less than 1000 m, within which range probably no oil and only insignificant amounts of gas are generated.

With reference to the optimum depth range for the generation of oil, it is of interest to compare the section of the Niger River Delta shown in Fig. 6-5 with the distribution of oil fields on the delta shown in Fig. 7-1. The latter figure shows a distribution which is interpreted by Dailly (1976) to indicate a belt of subsidence. The section, as drawn by Evamy *et al.* (1978), certainly shows the area of greatest subsidence to be well inland from the coast. Potential reservoir beds in shoreline and distributary sand bodies are distributed throughout the delta and are probably no more abundant within the belt than without. The explanation for such a prolific belt, as described by Dailly, may be that the source beds within this subsiding belt have been subjected for a longer period of time to temperature and pressure conditions more favorable to the generation of oil. The possible surficial effects of subsidence on the distribution of potential source and reservoir beds have also been noted by Dailly.

Reservoir Rock Potential of Basins

The hydrocarbon-generation potential of a sedimentary basin depends on the quality and volume of the source beds it contains; the hydrocarbon-yield capacity depends in part on the presence within the basin of porous and permeable beds which locally form stratigraphic or structural traps. Trapping conditions require a seal of relatively impermeable rock, such as shale, to cap the updip pinchout edge or the structural closure of a potential reservoir bed. The presence of the latter only, even where associated with good hydrocarbon source beds, is no guarantee of the existence within a basin of accumulations of oil and gas. For example, in the region of the offshore Northwest Shelf of Australia, porous and permeable Tertiary limestone beds overlie a Mesozoic sequence (Fig. 2-41) known to have generated hydrocarbons; but prospects for entrapment within the limestone do not appear to be favorable because of the lack of suitable cap rocks.

Reservoir rocks are commonly sandstones and carbonates but can be any body of rock that has suitable porosity, permeability, structural configuration, and closure provided by a cap rock. Oil has been obtained from basic volcanic agglomerate associated with basaltic flows intercalated with Late Cretaceous sediments in the interior coastal plain region of Texas, from fractured Precambrian quartzite in central Kansas, from weathered basement granite in western Venezuela, and from fractured Devonian argillites in southeastern Queensland, Australia. But by far the largest part of world reserves of oil and gas is in carbonate and sandstone reservoirs, in approximately equal proportions, as estimated by Warman (1971) and illustrated in Fig. 7-3. Production in limestone, dolomitic limestone, and dolostone constituting organic bioherms, biostromes, and other permeable zones, results from stratigraphically controlled pin point and regular porosity, or from fracture porosity caused by structural deformation.

As shown in Fig. 7-3, by far the largest potential resources lie in beds deposited on ancient continental shelves. Deep-sea sediments are far less important as petroleum producers, although Late Pliocene deep-sea sands deposited as turbidites on submarine fans are oil producing in the Ventura Basin of California (Natland and Kuenen, 1951) and in basins off the east coast of Brazil (Ponte *et al.,* 1977). The main producing reservoirs were deposited as carbonate shoals and reefs and as sand sheets and bars within the inner to middle neritic zones of ancient continental shelves that have existed since the Jurassic. It is also of interest to note that the world's oil reserves are mainly in structural traps. These include the giant oil fields of the Middle East which lie mainly in fold belts of Mesozoic and Tertiary

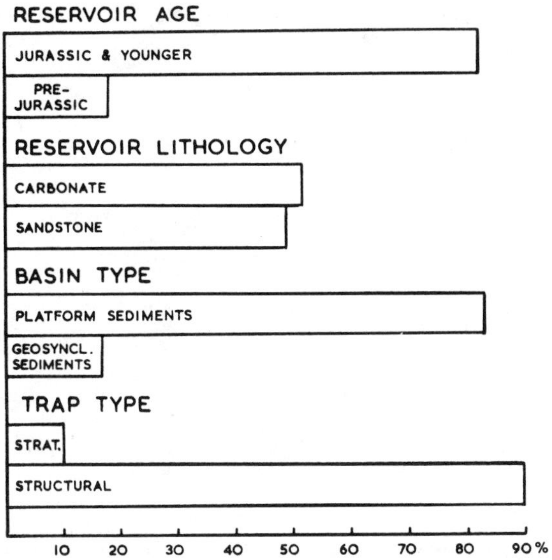

RESERVOIR AGE

JURASSIC & YOUNGER	
PRE- JURASSIC	

RESERVOIR LITHOLOGY

CARBONATE	
SANDSTONE	

BASIN TYPE

PLATFORM SEDIMENTS	
GEOSYNCL. SEDIMENTS	

TRAP TYPE

STRAT.	
STRUCTURAL	

10 20 30 40 50 60 70 80 90%

Fig. 7-3.   Basic parameters of world oil reserves. [After Warman (1971).]

limestones. Stratigraphic traps are significant, and a few, such as the East Texas oil field which produces from the Late Cretaceous Woodbine Sandstone, are of world stature.

Bioherms from which petroleum is produced are commonly regarded as stratigraphic traps, although structural elements such as regional tilting may control the ratio of oil and gas in individual but interconnected reservoirs (Gussow, 1953). On the other hand, hydrocarbon accumulations in limestone beds draped over reefs, such as those of the dolomitic Nisku Member overlying the Late Devonian Leduc reef trend in Alberta, are in structural traps. The search for reefs formed by corals, algae, stromatoparoids, polyzoans, crinoids, and other organisms obviously is confined to sediments deposited in shallower parts of a continental shelf somewhat removed from areas of muddy water adjacent to estuaries or deltas. This implies some degree of dependence on lithofacies and biofacies maps in the initial selection of areas in which to conduct a specific program of exploration. A common procedure is to employ seismic methods which may show reflections from the upper reef surface or from beds draped over the reef. In some areas the reef margin merges into surrounding calcarious shales to such an extent that there is insufficient velocity contrast to give a clearcut reflection. Having located an anomaly, it must then be drilled to ascertain its nature.

The literature on modern and ancient bioherms and biostromes is volu-

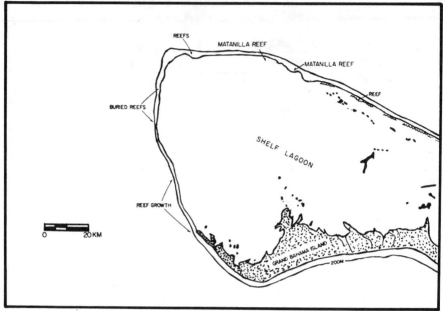

(a)

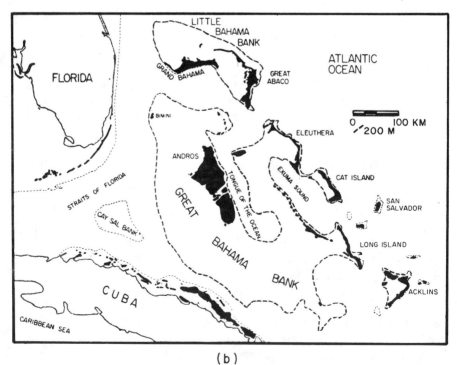

(b)

minous. Barrier reefs commonly develop as trends along the outer margin of a carbonate shelf or bank, as described by Hine and Neumann (1977) and illustrated in Fig. 7-4. The carbonates within this shelf include patch reefs and consist of organic-rich calcarenites and calcilutites. The paralic facies of the shelf become dolomitic, and may include thin beds of evaporites, in sediments deposited within the intertidal zones. The distribution of fauna is also related to the pattern of reef growth and proximity of the shoreline. Lithofacies and biofacies maps constructed on the basis of scattered control points may be sufficiently accurate to portray the salient paleogeographic and sedimentologic features of an area but are rarely definitive with respect to the delineation of probable reef trends. The best that can be hoped for is to outline a belt within which it is possible that a reef trend may exist. On this basis mineral and land rights may be obtained and seismic surveys undertaken.

Dolomitized zones in limestone beds are commonly more porous than the enclosing limestone, possibly because conversion of calcite to magnesian-rich carbonate results in a reduction of volume which may be reflected in enlargement or other modifications to the original pinpoint and regular porosity. Where dolomitized zones contain oil or gas, they may be purely stratigraphic traps, or stratigraphic-structural traps. The controlling factors in the development of secondary dolomitization in stratigraphic zones are not well understood but appear to be related to both fracturing and textural composition of the original limestone. Where control is sufficiently close and where spatial relationships can be demonstrated between dolomitized zones and particular lithofacies, it may be possible by means of extrapolation or association to predict the direction of trend of known dolomitized zones or to designate areas where such zones may exist.

The distinction between primary and secondary dolomitization cannot be sharply drawn. The former includes dolomitization of aragonite and the precipitation of dolomite in a sabkha environment (Folk and Land, 1975). It may also include dolomitization that takes place during an early stage of burial as a result of the influx of meteoric water. Land et al. (1975, p. 1602) say:

> The case for dolomitization by meteoric-sea water mixing is strengthened by the discovery of dolomite forming today, replacing

Fig. 7-4. (a) Map of Grand Bahama Island and adjacent carbonate shelf of Little Bahama Bank fringed with coralgal reefs. Note proximity of shelf edge and reef trend to 200-m bathymetric contour. (b) Map of Little Bahama Bank and Great Bahama Bank, showing areas lying within 200-m bathymetric contours, much of which is in depths of less than 10 m. [Modified after Hine and Neumann (1977).]

Jamaican Pleistocene biolithites at the modern water table (Land, 1973).

Land points out that magnesium is supplied by the saline water but that precipitation of dolomite takes place only where there is sufficient dilution with fresh water. Similar reactions are believed to occur in sabkha environments and restricted lagoons where highly saline water is periodically freshened by the influx of rainfall. Dolomitization may also occur at a much later stage of burial at depths of hundreds of meters, where the influx of fresh water under hydrodynamic conditions flushes the saline connate water and precipitates dolomite in existing cavities of the limestone. This process tends to decrease the porosity.

Secondary dolomitization involving the replacement of calcium carbonate by magnesium-rich carbonate also depends on the circulation of water at depth, but little is known of the controlling factors or the reactions that take place. Presumably, the original permeability is an important factor in the extent and direction of invasion of water, and this factor may be a function of the original texture and composition of various limestone lenses, or of the distribution of local zones of fractures. The dolomitizing solutions may also precipitate sulfides of lead and zinc, seen in traces within the Late Devonian Leduc reef trend of Alberta, and in large deposits within the Pine Point reef complex south of Great Slave Lake in the Northwest Territories of Canada. Little is known about the possible stratigraphic or structural conditions that may influence or control dolomitization and mineralization of this type. The Pine Point deposits are biostromal, well bedded, dolomitic, crystalline, show abundant evidence of replaced skeletal debris, and are associated with solution breccia formed by the removal of evaporites. Beales and Oldershaw (1969, p. 509) say:

> Closely spaced drilling by Pine Point Mines Ltd. has shown that the area was probably the site of an axis of reduced subsidence in Middle Devonian time. Along this axis a reef-controlled carbonate shoal developed and at times built up into a supratidal barrier.

It is of interest to note that some pore spaces associated with ore deposits in the brecciated biostromal dolostone are filled with bitumen and dead oil, suggesting that the beds may at one time, prior to uplift and erosion, have been a reservoir for hydrocarbons.

Fractured zones within otherwise poorly permeable carbonates can be prolific producers of oil and gas, particularly where the fracturing is within a thick sequence of fold carbonate beds having structural closure. There are many examples, including some of the world's giant oil fields producing from the Oligocene–Miocene Asmari Formation of Iran, where the range of porosity in the limestone is fair but the permeability is extremely low. Fracturing extends along the folds for tens of kilometers, resulting in continuity of fluid pressure from one well to another. Another example of hydrocarbon production from hard fractured carbonates is that from the Pincher Creek and other gas fields in folded Mississippian strata of the foothills belt in southwestern Alberta.

Fractures of this type (Fig. 7-5) may have been formed in the Asmari Formation and in thick carbonate beds elsewhere by folding of previously fractured beds (McQuillan, 1973). McQuillan is of the opinion that the orientation of fracture patterns in the Asmari indicates that they are related to local irregularities and not to the folding process, suggesting that they may have been initially formed by shock waves. Although this particular hypothesis may not have widespread implications, it is probable that features such as joints, fractures, and minor faults formed at various stages of burial of a thick carbonate sequence will be exploited by subsequent stresses during folding.

Hydrocarbons in noncarbonate reservoirs are commonly in sandstones but may occur in beds of fractured shale or claystone, and rarely in lenses of volcanic detritus, or fractured and weathered bedrock. Production is also obtained from siltstones and silty mudstones such as those of the Early Cretaceous Windalia Formation in the Barrow Island oil field of Western Australia. These tight beds must be fractured, and the natural water drive reinforced by water-injection wells. Fine-grained sandstones, such as those of the Late Cretaceous Cardium Formation in the Pembina oil field of Alberta, may also require fracturing.

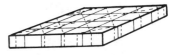
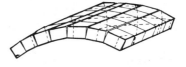

Fig. 7-5. Fracture pattern resulting from folding of preexisting fractures. [After McQuillan (1973).]

Sandstone reservoirs can be placed in two major categories with respect to the lithology of the host rock. They are contained in either lithic sandstone or quartzose sandstone, the division being somewhat arbitrary. Lithic sandstones are commonly of nonmarine origin, whereas quartzose sands are commonly of marine origin. Nonmarine sands deposited as granite wash on basement, alluvial fans, and river point bars are commonly lithic, but some estuarine, lacustrine, and aeolian sands are quartzose. Marine beach, bar, and other offshore sands are commonly quartzose but may be lithic. Deep-sea turbidites are commonly lithic. The composition of a sand depends on the nature of its provenance, the processes of recycling it may have undergone, and (with respect to river sands but not to turbidites) the distance of transportation from its source. With these reservations, it can be stated that as marine sands are the end product of sand transportation, they are generally more quartzose, finer grained, and of a more uniform grain size than river sands. Other features within a body of sandstone or within adjacent bodies of sediment, such as sedimentary structures, grain gradation, and fossils may be diagnostic of its origin.

From the point of view of exploration for oil and gas, once an accumulation has been found in a sandstone reservoir, it is necessary to determine the depositional environment of the sandstone in order to evaluate further prospects and possible trends. The significance of depositional trends and geomorphology of sandstone bodies to the entrapment of petroleum has been discussed in numerous papers, many of which have been summarized by Conybeare (1976). The application of sedimentary basin analysis to petroleum exploration, with reference to potential sandstone reservoirs, lies not only in determining the structural features within any part of a basin but also in the interpretation of depositional and stratigraphic relationships on the basis of reconstructing various facies and isopach maps with stratigraphic sections. Many case histories have been recorded of hydrocarbon accumulations in alluvial sand deposits (the deposition of which was controlled by the underlying topography), of accumulations in river distributaries and in offshore sands of a delta complex. In many such examples the control is both stratigraphic and structural, as in the multiple pay zones of sandstone bodies in the Mississippi and Niger deltas (Le Blanc, 1972; Weber, 1971). Other examples may include reservoirs in which the controlling features are not only local structures, depositional trends, and geomorphology but also unconformities. Fig. 7-6 illustrates an example where structural deformation combined with an erosional surface overlain by relatively impermeable claystone has resulted in multiple pay zones within the sandstone beds of a delta complex. It shows a section across the Kingfish oil field in the

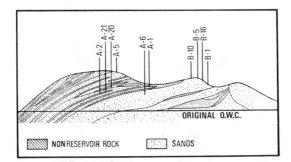

Fig. 7-6. Structural section across the offshore Kingfish oil field in the Gippsland Basin of Victoria, Australia, showing multiple pay zones within the reservoir. Producing beds are deltaic sandstones of the Eocene Latrobe Formation. O.W.C., oil–water contact. [After Khurana (1976).]

offshore area of the Gippsland Basin in Victoria, Australia. The producing sandstone beds are within the Eocene Latrobe Formation and are capped by marine shales and claystones of the Oligocene Lakes Entrance Formation. The oil occurs in distributary and marine shoreline sands locally interbedded with seams of coal. The oil–water contact is common to all the pay zones, but each has its individual three-dimensional configuration. Of particular interest is the fact that reservoir pressure in the offshore Kingfish field shows slight sinusoidal oscillations which coincide with the tidal frequency (Khurana, 1976). This reiationship suggests that there is hydraulic communication between the sea and the reservoir.

The importance of recognizing the sedimentologic and geomorphic individuality of sandstone bodies within a sandstone zone is pointed out by Seal and Gilreath (1976, p. 38) who say:

> Sands deposited in a deltaic environment may contain numerous isolated reservoirs in what appears to be a blanket sand. If multiple reservoir situations are not recognized and appropriate production plans formulated, significant hydrocarbon reserves will be left in untapped reservoirs.

They are referring to situations where hydrocarbon-bearing sandstone bodies are not in hydraulic communication and have differences in production characteristics, original pressures, and oil–water contacts. The study was based on producing sandstone zones in the Tertiary of Vermilion Block 16 oil field, offshore Louisiana. From analyses of dipmeter and electric logs, the 25 producing zones were divided into 34 correlative sandstone units comprising 108 separate sandstone bodies, thereby demonstrating that zones which appear to be blanket sandstones are in fact complex units consisting of several sandstone bodies (Fig. 7-7) each of which has its individual geomorphic, petrophysical, and reservoir characteristics. Studies in this detail are possible only in field areas where there is abundant well control; but insofar as they elucidate the stratigraphic relationships and depositional environments of a small part of a sedimen-

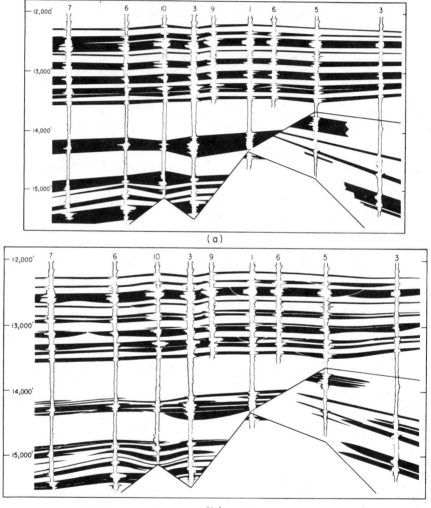

(a)

(b)

Fig. 7-7. (a) Generalized section along the axis of a structure in the offshore Vermilion Block 16 oil field, Louisiana, showing the Tertiary producing sandstones as blanket zones. (b) Reinterpretation of section above showing delineation of individual sandstone bodies deposited as delta distributaries and offshore sands. [After Seal and Gilreath (1976).]

tary basin, they throw some light on the genesis of the basin as a whole. In this way such studies serve not only to find additional accumulations of oil in known producing areas but to guide exploration in other parts of a basin where similar stratigraphic and structural conditions may have trapped oil in undiscovered sandstone reservoirs.

Structural and Stratigraphic Traps

The application of lithostratigraphic and structural analyses of sedimentary basins to the search for stratigraphic and structural traps is fundamental to exploration for oil and gas. The process of application normally follows a sequence similar to that illustrated by Forgotson and Stark (1972) in Fig. 7-8. Evaluation of a basin, or of a large part of a basin, is carried out as an overview in order to interpret the various stratigraphic and structural parameters in context with the development of the whole basin. Such an evaluation may be assisted by the use of computerized data in constructing a geostatistical model of possible stratigraphic-structural conditions in which hydrocarbons may be trapped in particular parts of a basin, as described by Forgotson and Stark (1972) and Abry (1975). With reference to the delineation of belts or zones within which stratigraphic traps may occur in sandstone or reef trends, Passega (1972, p. 2440) says:

> Environmental analysis and paleogeographic reconstruction are used increasingly in petroleum exploration to evaluate distribution of

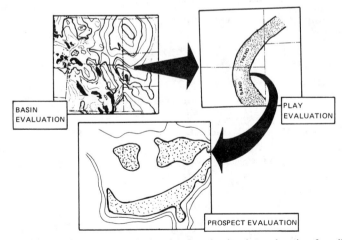

Fig. 7-8. Diagram illustrating three levels of evaluation in exploration for oil and gas. [Modified after Forgotson and Stark (1972).]

source and reservoir rocks in a basin and to delimit areas that may include stratigraphic traps.

Having outlined such a belt and mapped the possible or probable depositional trends within it, the search narrows to finding and evaluating a stratigraphic or structural-stratigraphic trap within a sandstone or carbonate body. Here the use of seismic survey methods becomes essential, particularly in view of the continual developments in technology that improve the means to solve problems relating to velocity–structure, the elimination of noise after stacking, and the direct detection of hydrocarbons, as discussed by Flowers (1976). Among these, the development of the "hot spot" technique, whereby seismic reflections from an oil–water interface can be detected, is certainly paramount. These three main levels of evaluation are interdependent and commonly assessed in the order shown in Fig. 7-8, although the third level of prospect evaluation may be undertaken by seismic surveys and drilling without prior knowledge of the two preceding levels provided a discovery has already been made on the basis of someone else's evaluation or lucky strike. Commonly the general stratigraphic and structural parameters of a sedimentary basin are known to all the operators exploring for hydrocarbons within a basin, but interpretations relating to depositional trends and possible prospects for stratigraphic and structural traps are privy to individual operators.

Initial evaluation of a sedimentary basin includes a broad overview that considers the basic features of its stratigraphic and tectonic framework, as outlined by Haun *et al.* (1970) in Figs. 7-9 and 7-10. These basic features include the volume, maximum depth of burial, and structural configuration of stratigraphic sequences which contain source and potential reservoir rocks and in which stratigraphic and structural traps may occur. Evaluations of this type, based on broad and general parameters, can be programmed to indicate and grade on a statistical basis those parts of a basin likely to contain accumulations of oil and gas. Such evaluations are included in the first level of basin evaluation as illustrated in Fig. 7-8. Further evaluation proceeds to the second level in which the search is narrowed to one or more particular stratigraphic intervals which may include depositional trends of particular interest. The boundary between the second and third levels of evaluation is not always well defined. Some features of both are listed by Halbouty (1972, p. 537) who says:

> Most basins contain facies changes, unconformities with resulting truncated beds, and buried erosional or constructive surfaces such as reefs, hills, channels, barrier sand bars, and other such phenomena— which form the basic requirements for the creation of subtle traps.

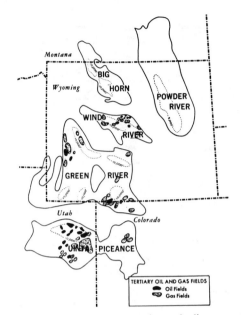

Fig. 7-9. Outline of sedimentary basins in Wyoming and adjacent regions showing maximum thickness of Paleocene–Eocene sections containing oil and gas fields. [Modified after Haun *et al.* (1970).]

If folding, normal faulting, thrusting, and the formation of salt ridges and domes are added to the picture of an evolving but continuously filling basin, the resultant structural and stratigraphic patterns become much more complex.

The third level of evaluation depends on detailed geological and geophysical studies of particular stratigraphic intervals. These studies commonly include isopach maps of lithostratigraphic units, structural maps of the tops of these units, stratigraphic and structural sections illustrating features shown on the maps, lithofacies and biofacies maps, isopach and structural seismic maps, and seismic profiles. The combined use of these maps and sections forms the basis for evaluating the petroleum prospects of a particular stratigraphic interval over the entire basin or part of the basin and for delineating anomalies that may be reservoirs for oil and gas. In the final analysis drilling is required to prove the potential of prospective anomalies although in some reservoirs seismic reflections from the horizontal oil–water interface can be recorded by means of the "bright-spot" technique. Where a stratigraphic unit is distributed throughout the entire area of an uplifted and eroded basin, the petroleum

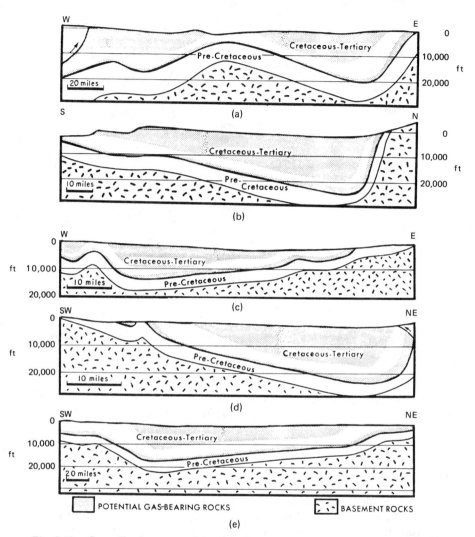

Fig. 7-10. Generalized cross sections of sedimentary basins illustrated in Fig. 7-9: (a) Green River Basin; (b) Uinta Basin; (c) Big Horn Basin; (d) Wind River Basin; (e) Powder River Basin. Unstippled part of Cretaceous–Tertiary section is predominantly marine shale. [Modified after Haun *et al.* (1970).]

accumulations are commonly found in the updip zones fringing the erosional edge of the basin, as illustrated by Lawson and Smith (1966) in Fig. 7-11. This figure shows the structure of the top of the Late Pennsylvanian Tensleep Sandstone in the Big Horn Basin of Wyoming and marks the

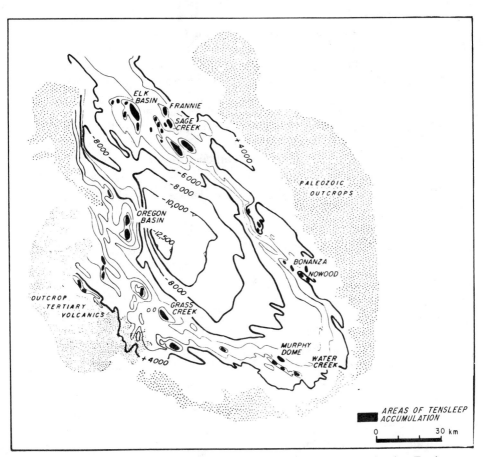

Fig. 7-11. Formline structure map of the top of the Late Pennsylvanian Tensleep Sandstone, Big Horn Basin, Wyoming, showing distribution of petroleum accumulations in the Tensleep. Contour interval in feet. [After Lawson and Smith (1966).]

location of known hydrocarbon accumulations within the Tensleep. The distribution of these accumulations within the area bordering the periphery of the basin is evident, but local control depends on several factors affecting variations in permeability, as outlined by Morgan *et al.* (1977).

The search for prospects in the third level of evaluation is frought with economic constraints which determine the economic feasibility of exploring for and producing from certain types and sizes of oil and gas accumulations. These constraints are ephemeral and quite apart from geological considerations which depend on factors that do not change. With rising well-head prices for oil and gas, it may become profitable to search for and produce from comparatively small accumulations that hitherto were not attractive. Many of these accumulations are purely stratigraphic traps, and some are controlled by the topography of unconformities. An interesting example is described in a comprehensive paper by Dolly and Busch (1972). This example, illustrated in Fig. 7-12, shows the topography of the Knox unconformity underlying the Early Ordovician Chazy Group in central Ohio. Oil accumulations occur in porous and permeable zones within erosional ridges or buried hills of the gently folded Cambrian Copper Ridge Dolomite where it underlies the unconformity. Commonly the synclinal folds of the Copper Ridge underlie erosional ridges, whereas the anticlinal folds underlie erosional valleys. Dolly and Busch (1972, p. 2366) say:

> Hydrocarbons within a ridge appear to migrate from closure to closure in a regionally updip direction after the lower reservoir is filled to its spillpoint. Adjacent ridges also are paths along which hydrocarbons have migrated, but these bear no relation to hydrocarbons in other ridges.

Elsewhere, unconformities marked by buried ridges and valleys are commonly overlain by a sandstone and shale sequence. In these situations, where the valleys are filled with sandstone and overlain by shale, they may locally form stratigraphic-structural traps for oil and gas. Several examples are described by Conybeare (1976).

As future exploration for large oil and gas accumulations moves farther out to sea in water depths of up to 2000 m overlying the outer areas of continental shelves and slopes, determination of the stratigraphic and structural framework of the underlying sedimentary pile will initially depend entirely on the interpretation of seismic records. Evaluation of petroleum prospects will, in the early stages of exploration, be based on these interpretations which place a basin within a particular stratigraphic-structural category or combination of categories such as those illustrated in Fig. 7-13 and discussed by Hedberg (1970). Trapping situations for oil and gas will depend on a number of stratigraphic and structural factors controlled or influenced by penecontemporaneous deposition, compaction, and faulting, by uplift and beveling, and by subsequent reburial. Also of prime importance from an economic viewpoint is the determination of possible and probable oil or gas provinces in offshore regions and their relationships to a number of sedimentological, physicochemical, and structural factors.

Of considerable relevance to the evaluation of potential oil and gas resources in a sedimentary basin is the problem of estimating reserves.

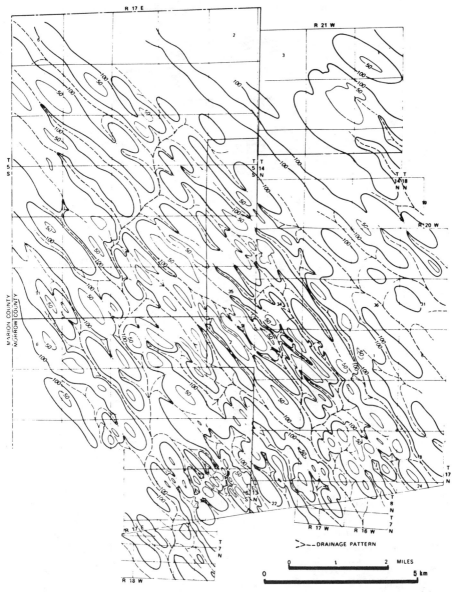

Fig. 7-12. Paleotopography of the Knox unconformity as shown by an isopach map of the combined Early Ordovician Middle and Lower Chazy Dolomite which overlies the unconformity. Contour interval in feet. [After Dolly and Busch (1972).]

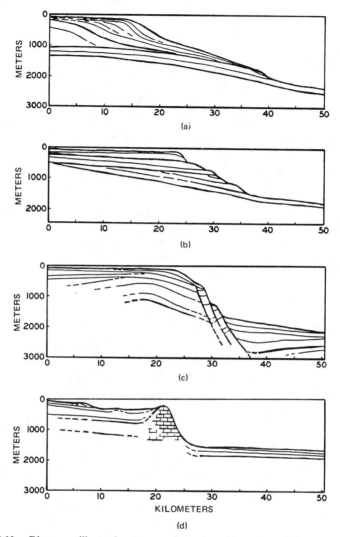

Fig. 7-13. Diagrams illustrating types of stratigraphic-structural framework within sedimentary sequences underlying continental slopes: (a) prograded; (b) erosional; (c) faulted; and (d) reef growth. [After Hedberg (1970).]

This involves a number of geological, technological, and economic factors, some of which are unknown. From a geological point of view, an estimate can be made of the possible metric tonnage of oil and equivalent gas that may have accumulated in stratigraphic and structural traps. Such an estimate can be based on a comparison of the sediment volume and lithology in one basin with respect to another basin in which the oil and gas potential is assumed to be known. This is a very approximate and probably unreliable method. Estimates can also be based on calculations of the original organic content buried with the sediment, of the percentage converted to kerogen and precursors of petroleum, of the degree of maturation, of the amounts lost by erosion or other processes within the basin, and of the tonnage that may have accumulated in large and small traps. These estimates are tentative. Expressed as tonnages, they may be very wide of the mark, but nevertheless are useful as a guide to any survey of potential petroleum resources. From an exploration point of view, such estimates are of value in grading the potential of various sedimentary basins, particularly with regard to the priority of exploration programs in different regions.

On the assumption that a sedimentary basin may contain an estimated minimum tonnage of petroleum in accumulations, or on the assumption that petroleum accumulations possibly or probably exist and may be found, an exploration program is mounted to locate prospective anomalies that may be stratigraphic or structural traps. In the event that an accumulation is found, and following sufficient seismic exploration and drilling, an estimate can be made of the tonnage of oil or volume of gas in place. This figure does not constitute a reserve, although a percentage of it may be a reserve. In fact, at the time of discovery none of it may be a reserve. The qualifying factors here are technological and economic. To date it has not been possible to extract 100% of the oil from any accumulation. In fact, the yield expected from some fields may be as low as 20%. In part, these low yields may be due to technological factors where water flooding, or fracturing of the reservoir rock, is required; and in part they may be due to geological factors such as variable permeability and interbedded layers of shale. Economic factors are of prime significance. Where the accumulation is too small, or the potential yield too low, it may be uneconomic at a current price to produce oil and gas from particular traps. In such circumstances, many potential small fields may lie idle. Interest in their development will only be generated when price rises or tax structure allow development of the field. Technological advances are also a factor in this respect, as improved or new methods of secondary or tertiary flushing may lower the cost of production.

Undiscovered accumulations of oil and gas are not reserves. A sedimentary basin may have a petroleum potential which can be expressed qualitatively or tentatively as a possible quantity. Many of the estimates of reserves of basins, countries, or of the world are in fact not estimates of reserves but estimates of possible or potential reserves. Such estimates serve a purpose but can be misunderstood and misused.

The reserves of discovered accumulations can, at any point in time, be expressed as proven, probable, or possible with reference to the current costs of production, price of oil and gas, technological capabilities, and known geological features of the trap. Any of these factors can be changed or modified. Ultimately, it is likely that technological advances in petroleum engineering and world demand for energy will support the economic viability of maximum cumulative production from both large and small stratigraphic and structural traps.

Hydrodynamic Conditions

Evaluation of the petroleum potential of a sedimentary basin must take into consideration its tectonic evolution which is a prime factor in the migration and accumulation of hydrocarbons and in changes of the flow of water through aquifers in the basin. The movement of water carrying hydrocarbons in solution has been proposed as a mechanism in the accumulation of oil and gas in some reservoirs where freshwater drive is operative (Hubbert, 1953; Hodgson et al., 1964; Meinhold (1968), Conybeare, 1970; and Price, 1976). Hydrodynamic conditions may also have been responsible for flushing hydrocarbons from large areas of a basin, as suggested by Coustau et al. (1975) with reference to eastern Algeria.

It has been established by many workers that the solubility of hydrocarbons is greater in fresh water than in saline water. Conybeare (1970, p. 668) says:

> The solubility of hydrocarbons in water of different salinity, and of varying pressures and temperatures has been given considerable attention by chemists, particularly with reference to the gases of the paraffin series: methane, ethane, propane and butane. In particular, investigations were carried out by McKetta and Katz (1948), and by Reamer et al. (1952). The main data have been summarized by McKetta and Wehe (1962). Geologists have tended to ignore the solubility of hydrocarbons in water as a factor in the accumulation of oil and gas deposits; consequently they make little mention of the possibility of moving petroleum over large distances within a sedimentary basin by means of formation water under hydrodynamic conditions. Nutting (1926) was aware of the possibility, and Dodson and Standing (1944) applied data on solubility and possible rates of flow of formation water to quantitative evaluation of rates of accumulation of petroleum.

It is suggested that a significant mechanism in the accumulation of oil in the Moonie field of Queensland, Australia may have been a drop in pressure as comparatively fresh water, carrying hydrocarbons in solution, moved updip along the southern margin of the Surat Basin. The oil–water contact in the Moonie field is underlain by fresh to brackish water having a salinity of approximately 2000 ppm (2000 mg per liter).

It has been pointed out by Price (1976) that an increase in temperature of formation water causes an increase in the solubility of hydrocarbons that constitute petroleum and that salt solutions having salinities in excess of 150,000 ppm show a marked decrease in solubility of hydrocarbons. From this it follows that decrease of pressure, decrease of temperature, and increase of salinity are factors tending to produce exsolution of hydrocarbons in water. The hydrocarbons so released would then migrate laterally and upward along more permeable zones in the strata to stratigraphic and structural traps. Price observes that in many sedimentary basins of the world there are deeply buried, thick, undercompacted, high-porosity shale sections deposited as prodelta clays. He suggests that compaction of these shale sections forces hot water into structurally higher zones where exsolution and entrapment of the hydrocarbons can take place. Such a mechanism could enable the transportation of hydrocarbons for long distances within a basin.

The solution of hydrocarbons disseminated or concentrated in small accumulations within a stratigraphic unit is a form of flushing. Certain areas may thus be underlain by a stratigraphic sequence barren of petroleum, whereas other areas underlain by the same sequence or its correlatives may be potentially productive. Flushing is also accomplished by the physical movement of petroleum in a hydrodynamic environment, as described in the classic paper by Hubbert (1953). In such cases the petroleum is moved from one structural position to another. Examples are provided in some oil fields having tilted oil–water contacts, where separate and structurally lower accumulations have been found in noses which flank the main structure and lie on the downward slope of the oil–water contact. The existence of a tilted oil–water contact is not always readily demonstrated, particularly in the absence of close control, as the differences of elevation at various boreholes may be caused by faulting or by lithostratigraphic relationships.

The flushing processes described above are essentially different in that the former involves the expulsion of connate water whereas the later involves the movement of water derived in part from surficial sources. An example of the infiltration of meteoric water into aquifers within a sedimentary basin is illustrated in Fig. 7-14 which shows a structural section based on boreholes drilled in the Sahara of eastern Algeria. Surface water is shown to infiltrate to depths of up to 3500 m. The presence of fresh-to-brackish water within permeable beds in a sedimentary basin indicates invasion of surficial water and hydrodynamic conditions. Such

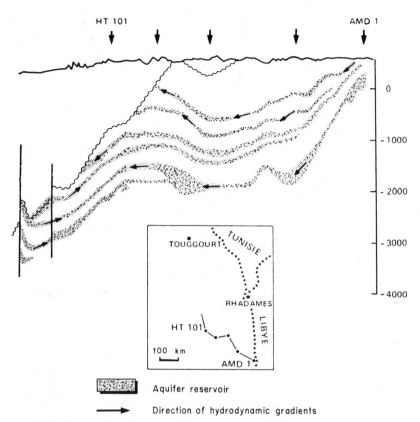

Fig. 7-14. Structural section based on boreholes drilled in the Sahara of eastern Algeria, showing movement of meteoric water in aquifers to depths of 3500 m. [Modified after Coustau *et al.* (1975).]

water may flow in any direction, from higher to lower, or from lower to higher pressures within the structural configuration of the aquifer (Hubbert, 1953), but will always flow from a higher to a lower potential. Consequently, structural closures within the aquifer are potential traps for oil and gas.

EXPLORATION FOR COAL

Depositional Environments of Basins

The search for additional reserves of coal has in recent years concentrated on regions where known or prospective coal-bearing strata are gently dipping in areas where several square kilometers can be mined by stripping and open cut methods. The selection of prospective areas depends on geological information on the stratigraphic sequence and on the results of seismic surveys. Testing and evaluation are carried out by a program of borehole drilling. It is consequently essential, particularly in regions where coal has not been discovered, to select stratigraphic intervals which are not only structurally amenable to mining but were deposited in environments suitable for the accumulation of vast deposits of organic matter. Such deposits are associated with lacustrine, alluvial, and deltaic sediments, as are those in Colorado and Wyoming described by Landis (1977). Coal may have originated as musking or bog vegetation that encroached upon and filled numerous interconnected or adjacent shallow bodies of water, as macerated organic matter, pollen, and spores washed into a lake as river backswamp vegetation including tree trunks and roots in situ or as coastal marsh on a delta. The broad river floodplains and marshy deltas of a low-lying landmass are favorable environments for the accumulation of vegetation. During a period of thousands of years, deposits of peaty vegetation will compact and accumulate to thicknesses of tens of meters. Carbonization certainly begins to take place within a few hundred years, as evidenced by the carbon dating of partly carbonized tree trunks and roots buried in preexisting backswamp areas of the ancestral Mississippi River.

The criteria for determination of the depositional environment of a stratigraphic sequence have been discussed in previous chapters, and it is only necessary here to mention the association of fluvial deposits and coal seams formed in floodplain environments and of marine-to-brackish water microfossils commonly found in shale and mudstone layers interbedded with coal seams formed as coastal marsh in a deltaic environment. An example of the first type of association is illustrated in Fig. 7-15 which shows the course of a stream channel, marked by a channel-fill sandstone body, through a swampy area in which accumulated vegetation formed a coal seam within the Pennsylvanian Carbondale Formation of southeastern Illinois, described by Wanless et al. (1970). An example of the second type of association is described by Partridge (1976) with reference to planktonic foraminiferal zones in shaly layers within coal-bearing beds of the Eocene Latrobe Formation of the Gippsland Basin in Victoria, Australia. Some of the sandstone beds associated with these coal seams are reservoirs for oil and gas. Sequences deposited in a marine environment (other than sequences comprising both marine and paralic sediments) and sequences deposited in particular nonmarine environments such as glacial outwash plains, alluvial fans, deserts, and volcanic terrain are not suitable hosts for substantial seams of coal. Consequently, a knowledge of the

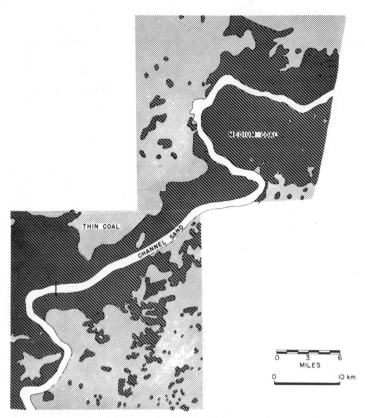

Fig. 7-15. Map showing thickness distribution of No.5 coal seam, and of a channel-fill sandstone body cutting the seam, in the Pennsylvanian Carbondale Formation of southeastern Illinois. Medium coal is more than 1.5 m thick. [After Wanless *et al.* (1970); Vail (1965); and Trescott (1964).]

regional stratigraphy and of the paleogeographic setting and depositional environments of the stratigraphic sequences to be explored is essential to any exploration program. This is not to say that coal cannot be found by prospectors without the benefit of such information or by a systematic sequence of boreholes drilled in a possible prospective area on a speculative basis. Many coal deposits were found in just such a manner. But the most effective exploration is well planned and based on all the geological, technical, and other pertinent information that is available.

Structural Configuration of Basins

The structural configuration of a sedimentary basin reflects its tectonic evolution which is also reflected by its stratigraphic framework. A comparatively rapidly rising landmass, possibly pushed up by the thrust of impinging continental masses along a zone of subduction, may yield by erosion a far greater mass of sediment than a landmass resulting from the comparatively slow and periodic upwelling of a Precambrian crator. Sediments from the former type of landmass are swept by river systems to the coast where they form deltas in which great thicknesses of sediment may accumulate. Sediments derived from Precambrian and other igneous-metamorphic cratons are commonly arenaceous and form comparatively thin sequences where they are preserved in alluvial and paralic environments. Sedimentary basins, such as the Mesozoic Alberta Basin of western Canada, may be bounded on the one side by sediments derived from a craton and on the other side by sediments derived from a landmass underlain by a tectonically active fold belt. Coal beds are not characteristic of the former but are commonly found in the latter. There may be several contributing reasons for these relationships. Granitic cratons are well jointed and fractured, and the topographic depressions are filled with permeable sandy and gravelly material, all of which factors allow for good drainage and scarcity of swampy ground. The nature of the terrain and of the lithology may also exert constraints on the growth of vegetation. Most important, cratonic regions are regions of uplift, and the paralic margins bordering the landmass of a craton are unable to accumulate a thick sedimentary sequence. The subsiding basements of major deltas, on the other hand, can readily accommodate the influx of fine sediment and nutrients necessary for the prolific growth of vegetation in backswamps and coastal marshes. In this respect it is of interest to note that brown coal beds in the Eocene Latrobe Formation of Victoria, Australia, have a cumulative thickness of more than 300 m and that one seam is up to 160 m thick (Brown *et al.*, 1968). On the basis that 1 m of brown coal is formed from 10 m of peat, one arrives at a figure of 3 km of peat formed in the Latrobe Delta. Continuous compaction probably resulted in a compositional range from carbonized matter at the bottom of the sequence to peat at the top. Compaction and tectonic subsidence resulting from block faulting are both factors in the mechanics of accumulating thick piles of sediment underlying large deltas.

Such regions of basement instability, overlain by rapidly accumulating sedimentary piles deposited by deltas, commonly become mobile belts in the later stages of their tectonic history. The association of major deep-seated thrust fault zones with thick coal measures has been noted by Pogrebnov (1974), with particular reference to the Pechora coal measures of the Pechora Basin lying on the western flank of the northern extremity of the Ural Mountains, USSR. The maximum thickness of these coal measures is 3 km. Similar structural-stratigraphic relationships obtain in other parts of the world. In some regions the stratigraphic zones in which these coal beds occur extends beyond the fold belt to relatively stable parts of a sedimentary basin where the beds are comparatively flat-lying or gently dipping. Within these correlative beds the coal seams are commonly thinner, of the order of a few meters thick, but may lie close to the surface where they are amenable to strip mining. Norris (1977) has pointed out the increasing interest in such coal seams underlying the plains of Alberta and Saskatchewan. Similar deposits underlie the plains of the western United States where the problems of mining include considerations of conservation and the environment. Assessment of coal resources and of prospective areas, based on evaluation of the structural configuration of coal-bearing strata, is locally constrained not only by requirements to preserve the natural environment but by legislation that prohibits mining in certain areas designated as preserves or national parks in many countries. The same limitations do not necessarily apply to exploration for oil and gas, the successful completion of which results in a grid of widely separated wellhead installations which are not conspicuous in the landscape.

EXPLORATION FOR MINERAL DEPOSITS

Uranium and Vanadium Ores

The search for uranium and vanadium ores in sandstone bodies of the Colorado Plateau in the United States, for pitchblende deposits in northern Canada, and for uranium ores in other parts of the world depended primarily on traditional methods of prospecting during the 1940s and early 1950s. As geological information was gathered by mapping and petrologic investigations, hypotheses began to emerge as to the origin of the uranium and the mode of emplacement of the ore bodies. Early explorations invoked hydrothermal solutions, but various workers in the field were not always in agreement as to the sources of the solutions or the structural controls. As relationships between ore bodies and the stratigraphy and structure of a region became more apparent, the evidence from many ore bodies supported an origin from meteoric waters. Langford (1977, p. 28) says:

> Although surficial concepts long have been proposed for some uranium deposits, none ever has been accepted widely. With the discovery, in Western Australia, of large uranium deposits which are unequivocally of near-surface supergene origin (Landford, 1974), serious consideration must be given to surficial and related processes as a possible mechanism for the formation of many uranium deposits.

During the first few years following the Second World War, geologists exploring for uranium deposits in the area north of Lake Athabasca, Saskatchewan, noted that the distribution of pitchblende was in fractures commonly adjacent to faults or shear zones. This spatial relationship suggested that the mineralization was hydrothermal in nature and related to a late stage of structural dislocation. Later observations demonstrated that many of the mineralized fractures extended downward from a majority unconformity. Others were within basalt layers that lay interbedded with alluvial arkosic sandstone and siltstone above the unconformity. These relationships provided evidence for supporting a meteoric origin for the mineralizing solutions. It is now thought by many geologists that drainage from the Precambrian terrain, carrying uranium in solution in an oxidizing environment, locally filtered downward through fractures in the basement rock to zones where reducing conditions brought about the precipitation of uranium minerals. In eastern Montana and the Dakotas, uranium deposits are found within the Late Cretaceous to Eocene lignitic beds in the updip parts of these tilted beds where they directly underlie an unconformity overlain by the Oligocene and Miocene sediments (Denson and Gill, 1956).

The association of uranium and vanadium minerals with carbonized organic matter and hydrocarbons is well known. Examples include the uranium content in the kerogen-rich black shales of the Cambrian Kolm Formation in Sweden, the association of uranium minerals with carbonized logs and other organic matter in sandstones of the Late Jurassic Morrison Formation in the western United States, the intergrowth of uraninite and thucholite to form uraniferous carbon in the gold-bearing conglomerate reefs of the Precambrian Witwatersrand System in South Africa (Feather and Koen, 1975), and the uranium and vanadium content in many deposits of asphalt throughout the world. The apparent genetic association of uranium and vanadium mineralization with carbon and hydrocarbons is believed to arise from the fact that in an oxidizing environment organic matter is not preserved, and uranium and vanadium are carried in solution by meteoric waters; but where these solutions penetrate fractures or enter stagnant bodies of water in a reducing environment, precipitation of uranium minerals occurs.

This association with reducing conditions is common but cannot be regarded as a rule. Landford (1977, p. 36) points out that

> The uranium-bearing calcretes in Western Australia (Langford, 1974) demonstrate that fluvial processes can produce economic uranium deposits. The main difference between the Australian deposits and those in Colorado seems to be that the Australian streams were ero-

sive, and did not accumulate alluvium with organic remains. As a result, the uranium there was precipitated by combining with vanadium in an oxidizing environment. When the Morrison and similar formations were forming in the Western United States, the streams were aggrading and much organic material was being buried. This created reducing environments in the alluvium which promoted precipitation of the uranium and the formation of primary, tabular orebodies.

Langford suggests that these tabular ore bodies were originally point bar sands and gravels (Fig. 7-16) in river channels. Such sands and gravels are scour-and-fill deposit, having subparallel but local crosscutting relationships with the enclosing strata. These permeable beds offered channels for the circulation of meteoric waters which deposited uranium and vanadium minerals in zones of reduction. Subsequently, partial dissolution and redeposition of the minerals resulted in the present relationships where the ore bodies form "rolls" cutting across the bedding. Langford says (p. 36):

> Thus it is important to differentiate between the "tabular" orebodies which are related closely to surficial processes and represent early precipitation and the "boundary-roll" deposits, most of which are remobilized tabular deposits and are secondary.

The general stratiform but partial crosscutting configuration of a crescentic ore "roll" is illustrated in Fig. 7-17 which shows sections of the ore

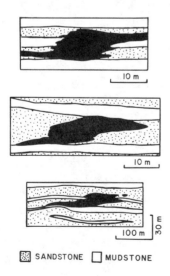

Fig. 7-16. Configurations of tabular uranium orebodies in Mesozoic fluvial sediments of the Grants region, New Mexico. [Modified after Langford (1977); Kelley et al. (1968); Harmon and Taylor (1963); and Clary et al. (1963).]

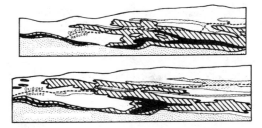

Fig. 7-17. Sections showing configuration of uranium orebody (black) within weakly mineralized zone (hatched) in sandstone and claystone beds of the Miocene Oakville Sandstone, Felder mine, southeastern Texas. [Modified after Eargle *et al.* (1975); and Kloen and Pickens (1970).]

body in the Felder uranium mine, Texas. This ore body, described by Kloen and Pickens (1970), occurs in sandstone beds within the interbedded sequence of fluvial sandstone and claystone of the Miocene Oakville Sandstone which overlies the Miocene Catahoula Tuff. The distinction between original tabular deposits and remobilized "roll" deposits is not always clearly defined in the field as with most deposits some degree of remobilization has taken place. As pointed out by Langford (1977), this makes radiometric dating of primary mineralization subject to error.

Uranium deposits are widely distributed throughout the world. Many of these are of very low grade and may, at any point of time, be considered as uneconomic. The constraints bearing on the economic viability of mining may involve any combination of geological or technical problems, geography, and the world market. The cost structure of mining, refining, and transportation of ore concentrates is subject to many variables, some which cannot be foreseen. Consequently, uranium resources are similar to oil and gas resources in that they must be expressed as reserves available only in terms of required prices needed to offset the current costs of production. Gabelman (1976, p. 2002) says:

> The supply of untapped uranium resources is not so limited as the exploration record of the last few years would suggest. Rather, the supply available to society at any time is determined by the need–cost balance, to which available supply can be adjusted by sliding economics.

Exploration for medium to high grade deposits of uranium ore must be based on a working hypothesis of ore genesis. It must also be based on the application of sedimentary basin analysis to the search for areas in which stratigraphic and structural relationships favor the deposition of ore minerals under conditions envisaged by the working hypothesis. For example, if the rationale of exploration depends on the hypothesis that uranium ore is formed in situations where meteoric water in an oxidizing environment carries uranium in solution to zones of reduction where uranium minerals are precipitated, then the search centers on finding geological relationships that indicate the possible preexistence of such conditions. These conditions could obtain where meteoric waters are draining a weathered regolith underlain by igneous rocks containing uranium. These waters, carrying uranium in solution, could penetrate fractures in the bedrock to depths of reduction, follow permeable channels in buried fluvial sediments to zones of reduction where organic matter has been carbonized but not destroyed, penetrate permeable beds beneath an unconformity to zones of reduction, or be washed into standing bodies of water such as marshes where peat is accumulating in reducing conditions. Given such a rationale, exploration depends on the recognition and delineation of unconformities with which such sedimentary processes or geological relationships may be associated. An example is seen in the Jabiluka uranium ore bodies in northeastern Northern Territory, Australia. These ore bodies lie within a sequence of schist, phyllite, quartzite, chert, and crystalline dolomite of the early Proterozoic Cahill Formation which is steeply folded and separated by a major unconformity from flat-lying sandstone and siltstone beds of the Late Proterozoic Kombolgie Formation. The ore bodies extend downward from this unconformity.

Manganese and Iron Ores

Sedimentary iron and manganese ores occur in many parts of the world, but as pointed out by Park and MacDiarmid (1970), they are most abundant in the regions of the Atlantic and Indian Oceans where they include deposits in Gabon, Liberia, Brazil, Venezuela, Canada, United States, India, and Australia. Most large sedimentary iron ore deposits are derived by supergene enrichment of Proterozoic iron formation which is referred to in various parts of the world as taconite, jaspilite, itabirite, and banded or cherty ironstone. Many sedimentary manganese deposits were also formed by supergene enrichment of Proterozoic manganiferous sediments. An exception is the Groote Eylandt manganese deposit of northern Australia which was formed during the Early Cretaceous as a pisolitic sedimentary bed deposited in shallow water. The manganese content is believed (Frazer and Belcher, 1975) to have been derived from erosion of the manganiferous sedimentary beds in the Proterozoic McArthur Group. Manganese deposits in the Pilbara region of Western Australia, described by Denholm (1977) and illustrated in Fig. 7-18, were formed during the

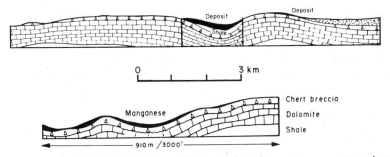

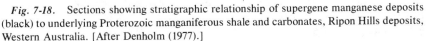

Fig. 7-18. Sections showing stratigraphic relationship of supergene manganese deposits (black) to underlying Proterozoic manganiferous shale and carbonates, Ripon Hills deposits, Western Australia. [After Denholm (1977).]

Tertiary by supergene enrichment of Proterozoic manganiferous shale. The ore consists mainly of fine-grained pyrolusite associated with hematite in a kaolinitic matrix.

The main iron formations of Australia have been described by Trendall (1973). These are all Proterozoic, although local supergene enrichment occurred during the Tertiary. Iron formations of the Yilgara and Pilbara regions of Western Australia are associated with minor beds of carbonates and thin chert layers. There is no evidence of cyclic sedimentation, and the sequence is relatively thin and laterally restricted in its distribution. Magnetite and quartz are dominant minerals in the iron formation which is dated within the range 2600–3000 million years old. The banded iron formation of the Hamersley Group in Western Australia lies in an intracratonic basin enclosed by the Archaean basement (Fig. 2-3) and is associated with beds of carbonates and cherty bands showing cyclic layering. The hematite-rich formation has minor amounts of magnetite and quartz, and the sequence is thicker and laterally more extensive than the Yilgara and Pilbara iron formations. The age of these beds is approximately 2000 million years.

MacLeod (1966), Ayres (1971), and others have described the Hamersley iron ore deposits. Hematite–goethite ore formed in situ by supergene processes involving the leaching of silica and carbonate and the redeposition of iron oxides. These are associated with conglomeratic deposits formed as iron-cemented scree. Other ores are pisolitic limonite and hematite deposits formed along fossil drainage channels. Campana (1966) expressed the view that formation of the ore deposits depended not only on stratigraphic and structural factors but to some extent on the configuration of the Miocene erosional surface.

Other Australian iron formations include the hematite-rich clastics of Yampi Sound, Western Australia, dated as 1800 million years old, three formations in South Australia, the Cleve, Holowilena and Braemar, and the Roper Bar and Constance Range beds of the McArthur Basin in the Northern Territory. The Cleve forms linear outcrops in the Middleback Range and consists of highly folded iron formation that was metamorphosed about 1800 million years ago and may be 2000 million years old. The Holowilena and Braemar iron formations are poorly banded sedimentary deposits that contain fine-grained hematite and magnetite and are associated with glacial deposits. These beds are about 750 million years old. The Roper Bar and Constance Range iron formations of the McArthur Basin in the Northern Territory are also poorly banded sedimentary deposits containing pisolitic accumulations of hematite, siderite, and chamosite. The age of these deposits is about 1400 million years.

Of particular relevance to the distribution of iron and manganese ore deposits is their common association with unconformities underlain by Proterozoic iron formation or manganiferous sediments. This association is seen in Tertiary or older supergene deposits overlying Precambrian iron and manganese-bearing beds or in sedimentary deposits containing abundant iron and manganese oxides derived from the erosion of iron formations or manganiferous beds. These sedimentary deposits may themselves be further enriched by supergene processes. Supergene enrichment in situ depends on the movement of meteoric water beneath the surface in areas underlain by iron formation and manganiferous beds. This movement in turn depends or is influenced not only by the intake from rainfall but by stratigraphic and structural factors that control the movement of water through selective channels such as fractured and other more permeable zones. The deposition of pisolitic and other sedimentary deposits in lacustrine or floodplain environments depends on the dissolution of iron and manganese from a weathered regolith underlain by iron formation or manganiferous beds and on the configuration of the drainage system which is related to the topography of the erosional surface or unconformity.

Manganese and iron ore are both soluble in the reduced state, but their oxides are comparatively insoluble. Iron in the ferrous state is more readily oxidized than manganese and, consequently, tends to be more readily fixed as deposits in the ferric state. This difference in solubility may account for the segregation of many iron and manganese sedimentary deposits. Garrels and Mackenzie (1971) point out that during weathering of iron-rich silicate minerals or iron formation, the ferrous iron released is oxidized by oxygen in the meteoric water and converted to hydrated ferric oxides that are only slightly soluble under oxygenated conditions. In lacustrine environments the role of bacteria and algae in causing the precipitation of iron and manganese may also have been of significance.

The possible origins of iron formations and manganiferous sediments have been debated and discussed in voluminous literature on the subject.

Some iron formations and manganiferous sediments have probably formed as a result of sedimentation and precipitation of recycled products derived from the erosion of preexisting iron formations and manganiferous sediments. Others may be of primary origin. Beds of both categories may be metamorphosed to hematite-rich, highly contorted jaspilite or to schistose beds containing abundant manganiferous garnet. Later supergene enrichment of such metamorphic beds may form iron ore, as in the Mesabi district of Minnesota in the United States, or as in the manganese deposits mined in southern Ghana. The formation of manganiferous nodules on the sea floor, at abyssal depths in zones where rates of sedimentation are of the order of a few millimeters in a thousand years, suggests that some manganiferous beds may have originated in this way. Others undoubtedly formed in shallow bodies of water within intracratonic basins. The origin of manganese in deep-sea nodules is not known. It may have been derived from seawater or from water expelled upward through the layers of compacting clays. In either case the syngenetic origin for such manganese deposits is unquestionable. Stratigraphic relationships and mineral distribution in iron formations and manganiferous beds also support the interpretation that these are low grade syngenetic deposits.

The application of sedimentary basin analysis to exploration for iron and manganese deposits centers primarily on the search for suitable areas where Proterozoic sediments form cratonic basins in regions of a Precambrian shield, as illustrated in Figs. 2-3 and 2-4. Where sediments contain unusually high contents of iron and manganese, there may be local deposits of iron and manganese ore formed by supergene enrichment in situ. Erosion of such sediments may also result in the formation of sedimentary deposits in standing bodies of water or in subsurface zones of moving ground water within drainage channels. Analysis of the possible or probable configuration of the unconformity overlying the iron formation or manganiferous beds may indicate area for selective diamond drilling. In many areas erosion will have removed much of the sedimentary record above the unconformity, leaving only patches of outcrops, some of which may be of ore grade and readily discovered by surface mapping and sampling.

Copper, Lead, and Zinc Ores

The paragenesis of mineral assemblages in stratified ore deposits within sedimentary beds has been much debated, and the arguments raised have been used to support hypotheses on the origin of the ore. In particular, discussion has centered on the question of whether many ore bodies are of syngenetic or epigenetic origin. Many exhibit features that support both hypotheses, and in some examples minerals originally formed in the sediments may have been altered or dissolved and reprecipitated by epigenetic processes. In either case the ore bodies are stratigraphically controlled, with particular relevance to processes of sedimentation and depositional environments in the formation of syngenetic deposits, and to sediment type, porosity, and permeability with reference to epigenetic deposits. General discussions on the various aspects of sedimentary ore deposits and on the contending views of their origins are summarized by Park and MacDiarmid (1970) and by Stanton (1972).

A classic example of an ore deposit that is generally regarded as syngenetic in origin is the Permian Kupferschiefer of northern Europe. This cupriferous bed, which underlies an area of at least 20,000 km^2, is commonly less than 1 m thick and consists of bituminous and calcareous blackish shale containing sulfides of copper, lead, and zinc. This mineralized shale bed is overlain by limestone and underlain by conglomerate and conglomeratic sandstone. The sandstone is believed to have been deposited in a desert environment (Glennie, 1972), as outwash from alluvial fans flanking a regressive escarpment of Carboniferous beds. The outwash extended to a sabkha-type shoreline on the fringes of a large, very shallow, and restricted body of water in which the Kupfershiefer shale was deposited and the sulfides precipitated. The desert coastal areas, which may have been geomorphologically similar to present-day deserts illustrated in satellite photographs and described by McKee *et al.* (1977), were transgressed by this shallow body of water and subsequently by a more open arm of the sea in which the limestone bed was deposited.

The Kupferschiefer has been described by Dunham (1964). Copper content over large areas is commonly up to 3%, and the zinc content ranges up to 1%. The association of finely disseminated sulfide minerals with bituminous matter suggests that the sulfides were precipitated in a reducing environment in which kerogen-rich organic matter was accumulating.

Another example of cupriferous shale is the Permian Flowerpot Shale of northern Texas and southwestern Oklahoma (Fig. 7-19). The ore bodies, which are flat-lying and 15–45 cm thick, extend over an area of about 30 km^2. Grades range up to 2% copper. The host rock is a gray silty shale interbedded with red-bed clastics and evaporites which are believed to have been deposited in very shallow water by a transgressive and restricted arm of the sea. Johnson (1974, p. 384) says:

> Epeirogenic movements caused slow but continual sinking of the crust beneath the inland sea and permitted accumulation of thick sequences of reddish-brown shales and interbedded evaporites.

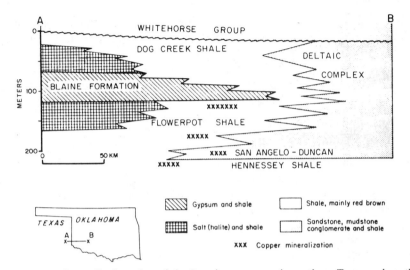

Fig. 7-19. Generalized section of the Permian sequence in northern Texas and south-western Oklahoma, showing zones of sedimentary copper deposits. [After Johnson (1974).]

The cupriferous shales were deposited with halite and gypsum in shallow coastal areas of this sea. Chalcocite is the primary copper sulfide, but whether it is of syngenetic or epigenetic origin has not been determined.

Other interesting occurrences of sedimentary copper deposits are found in several beds of the Proterozoic Belt Series in northwestern United States. This sequence comprises a sedimentary pile of fine-grained clastics and minor carbonates, up to 20 km thick, deposited in an epicontinental basin fringing the western margin of the Precambrian shield. The lower third of the sequence comprises lithic arenites and graphitic shales; the upper two-thirds are of red beds, black shales, and carbonates. The beds were deposited in shallow water, as indicated by salt casts, mud cracks, and stromatolites. Within the mineralized zones the copper minerals are mainly chalcocite, bornite, and chalcopyrite, the same assemblage as in the Kupferschiefer. Harrison (1974, p. 353) says:

> Copper of ore grade and quantity has as yet been found only in what are now impermeable white quartzites and siltites of the Revett Formation, in a zone about 30 km wide and 100 km long in northwestern Montana. The structural position of the zone on a post-Revett dome suggests that copper was reconcentrated epigenetically in permeable strata of a structural-stratigraphic trap prior to or during regional metamorphism of the formation.

This explanation presupposes some degree of structural deformation at shallow depths accompanied or followed by secondary enrichment of the primary copper content within the Revett.

Classic examples of sedimentary copper deposits are found in some of the Precambrian carbonaceous shales and argillaceous arenites of Zambia (Fig. 7-20). These ore bodies are distributed over an area of more than 1000 km². With reference to them, Jacobsen (1975, p. 348) says:

> Although the copper ores are stratabound, the ore zones are found in different lithological units within the succession and in places transgress from one stratigraphic horizon to another. Major ore formation, however, is concentrated in the arenites and calcareous argillites and siltstones of the Lower Roan, frequently intermixed with aeolian sandstones.

The origin of these copper ores, which consist mainly of chalcopyrite, bornite, and chalcocite, has been vigorously debated in numerous papers. Jacobsen says further (p. 348):

> In very general terms, the syngenetic theory, as applied to the Zambian and Zaire copper belt, considers that copper was obtained by degradation of basement rock in close proximity to the Copper Province. River and groundwater, which discharged into the near shore part of the basin, contained dissolved components that were subsequently precipitated by biological agents or in some instances concentrated mechanically as heavy sands in the foresets and bottomsets of crossbeds, in arenaceous rocks, and in sedimentary lamination in the argillites and siltstones.

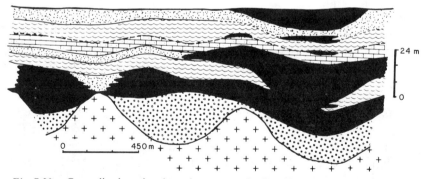

Fig. 7-20. Generalized section through copper orebodies (black) in Precambrian arenites (stippled), argillaceous rocks (wavy), and dolomite overlying igneous–metamorphic basement in Zambia. [Modified after Jacobsen (1975).]

The apparent relationships between lithofacies and ore distribution were pointed out by Garlick (1953, 1961) and by Garlick and Fleischer (1972). These suggest that within the more argillaceous facies deposited seaward from the paralic arenites and dolomites, the ore bodies show a decreasing content of chalcocite and an increasing content of chalcopyrite away from paralic lithofacies. Garlick was of the opinion that the ore bodies of the copper belt were consequently distributed in a zoned arrangement parallel to an ancient shoreline. Although the copper content of these beds probably originated during sedimentation, the present stratigraphic relationships of the ore bodies suggest that epigenesis has been an important factor in reconcentrating the minerals.

Other well-known ore bodies that are considered by many geologists to be of sedimentary origin are the fine-grained banded sulfide ores of the Rammelsberg district (Schneiderhöhn, 1953) in Germany, similar ores of the Mt. Isa district (Solomon, 1965) in Queensland, and those of the Proterozoic McArthur Group in the Gulf of Carpentaria district, Australia. The latter is an extensive deposit but so very fine-grained that it presents problems in milling. The Rammelsberg and Mount Isa banded galena–sphalerite–pyrite ores are remarkably similar in appearance. The latter occurs in the Proterozoic Mount Isa Shale of the Carpentarian System, a unit comprising argillite, siltstone, and dolomite, some beds in which exhibit small-scale cross stratification, convolute bedding, and slump breccias (Carter et al., 1961). These beds were deposited within a sedimentary pile known as the Mount Isa Geosyncline, in what is believed to have been a shallow-water to paralic environment (Brown et al., 1968). It is thought that widespread sedimentation proceeded at a comparatively slow rate for a very long period of time, as suggested by the lithologic similarity of the Carpentarian sedimentary rocks over large areas of northern Australia.

Of special relevance to the possible genesis of some stratiform ore bodies is the deposition of sulfide-rich sediments in deep parts of the Red Sea where hot brines pour on to the ocean floor from fissures along the zone of crustal parting (Fig. 7-21). The regions bounding both sides of the Red Sea are underlain by Precambrian granitic rocks (Fig. 7-22) that are highly fissured as shown in Fig. 7-21. The floor of the Red Sea, beneath which the Saudi Arabian and African crustal plates are separating, is underlain by basaltic basement and marls rich in nannofossils and pelagic microfossils. Iron-rich beds containing goethite, manganite, and manganosiderite and sulfide-rich beds containing sphalerite, chalcopyrite, and pyrite are concentrated along the central part of the Red Sea in deeper areas overlying the rift. The zinc content of these metalliferous sediments is locally as much as 20% (Mitchell and Garson, 1976). Miocene man-

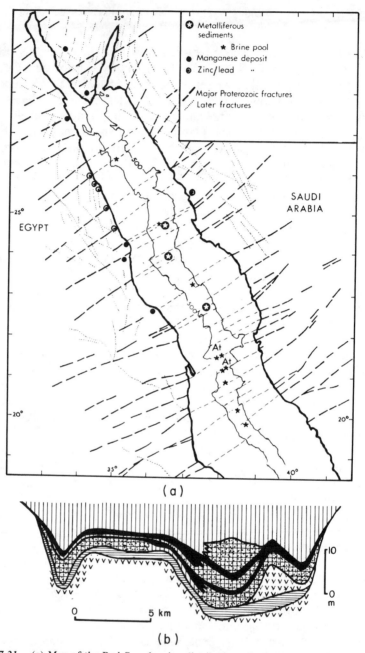

Fig. 7-21. (a) Map of the Red Sea showing distribution of metalliferous sediments, hot brine pools, and fractures in the Precambrian basement. (b) Section beneath sea floor in Atlantis II Deep (At) showing vertical distribution of metalliferous sediments: sulfides (black), goethite (stippled), overlain by montmorillonite clay (vertical), underlain by carbonates (horizontal). Basement is basalt. [Modified after Mitchell and Garson (1976); and Hackett and Bischoff (1973).]

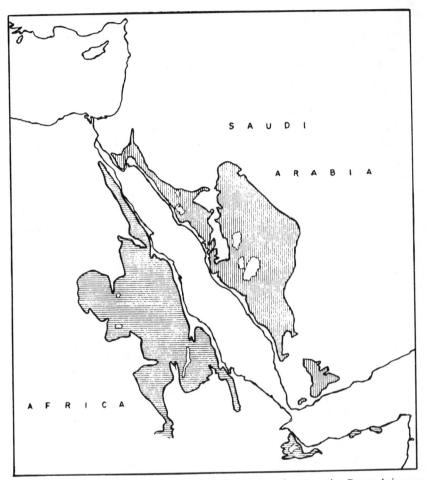

Fig. 7-22.   Map of Red Sea region showing distribution of outcropping Precambrian granitic rocks. [After Chow (1968).]

ganese and zinc–lead deposits occur along the fringes of the Red Sea shore. The manganiferous layers are interbedded with clay and are thought to be syngenetic; the zinc–lead content may originally have been deposited with the sediment but has been reconcentrated by epigenetic processes as secondary zinc and lead minerals.

The application of sedimentary basin analysis to exploration for deposits of copper, lead, and zinc is relevant only to syngenetic deposits and to those ore bodies formed by epigenetic processes from syngenetic deposits. Lithostratigraphic relationships indicate that many of these deposits are in sedimentary beds deposited in shallow neritic to paralic environments and that the sediments are predominantly argillites or argillaceous arenites. The Poterozoic Belt sequence of northwestern United States and the Mount Isa sequence of Australia were both deposited around the fringe of an Archean shield. Deposition of these beds appears to be associated with the erosion of low-lying landmasses, commonly of granitic and metamorphic rocks, and with flat coastal regions characterized by restricted circulation, the precipitation of evaporites, and local sedimentation in a reducing environment.

Many syngenetic deposits of copper, lead, and zinc are believed to have been derived by dissolution of minerals in the products of erosion and by the precipitation of minerals from these solutions in fine-grained sediments deposited in shallow-water environments. This process is particularly favored in regions where the regolith has formed largely in situ by the weathering of mineralized rocks and where most of the erosional material transported in the drainage system is silt and clay. The relatively large volume of sulfide bands compared to the volume of intercalated and admixed sediment appears to present a problem to any hypothesis of penecontemporaneous deposition of sulfides and sediment. But in local areas where rates of sedimentation are slow and where compaction of clayey sediments forces upward connate water containing minerals in solution, it may happen that precipitation of sulfides occurs at the sediment–water interface. The prime requirements for this type of depositional environment are a flat regolith overlying weathered igneous or metamorphic rocks, an extensive and shallow intracratonic body of water, good drainage from a landmass, slow accumulation of a sequence consisting largely of argillaceous sediments, and deposition in shallow water having a restricted circulation and a reducing environment.

Depth of water is by no means a limiting factor, and the most important condition is probably a reducing environment. Precipitation of metallic sulfides in the Red Sea, at depths of more than 500 m in local areas of supersaline brine near seabed hot springs, has been mentioned. The tectonic setting of these mineral deposits is the antithesis of shallow-water deposition in an intracratonic basin overlying a stable continental crust. These sulfide deposits of the Red Sea overlie and are intercalated with flows of basalt ejected from fissures along a central rift marking the pulling apart of the Arabian and African crustal plates. If there are other examples in the geological record of syngenetic sulfide deposits formed in a similar tectonic environment, then it can be expected that they will also be associated with basaltic flows and other volcanics.

Phosphate Rock

Phosphate rock has a wide geographic and time range throughout the world, but the areas in which beds are sufficiently phosphatic to constitute phosphorite are limited. Of these only a few have a high enough grade, a large enough volume, and a favorable geographic location to be considered as possible prospects for mining operations. Other considerations pertaining to changing world demand, prices, and costs of production are paramount.

Beds of pelletal phosphorite occur in carbonaceous sediments of the Proterozoic Belt Series in Montana (Gulbrandsen, 1966). This occurrence is somewhat exceptional in that most Proterozoic phosphorites are of collophane lutites. Phosphate rock has also been found in Cambrian, Ordovician, Permian, and Cretaceous shelf sediments in Australia (Howard, 1972); but with the exception of Cambrian deposits in the Duchess and other areas of the Mount Isa region in Queensland, the known occurrences are not of economic significance. Phosphatic zones have been found in Late Ordovician rocks of central Tennessee (Wilson, 1962). This sequence of rocks comprises a western deeper water facies of laminated argillaceous limestone and an eastern shallower water facies of silty limestone. Granular phosphorite occurs between the two facies, associated with very fossiliferous limestone composed of reef detritus that accumulated along the seaward margin of a shallow carbonate shelf. Phosphatic beds are also well known in the Permian Phosphoria Formation of Wyoming, Idaho, and Utah where they occur as dark marine shales associated with cherty limestone. The Phosphoria was deposited in a shallow marine environment and grades laterally to the east into a paralic to terrestrial red bed sequence. Cretaceous phosphate beds are widespread throughout the Middle East and North Africa where they were deposited as shallow water marine shelf sediments. Local phosphatic layers are also present in the Late Cretaceous chalk beds of southern England. These beds contain cherty layers and overlie a limestone which in turn overlies a transgressive and glauconitic unit, the Late Cretaceous Greensand.

The phosphate content of these and other phosphorites is commonly in the form of collophane, an amorphous and optically isotropic mass, having a conchoidal fracture, and composed essentially of hydrated calcium phosphate, commonly with abundant acicular crystallites of apatite. The genesis of collophane may, in some regions, have involved the diffusion of calcium and phosphate ions through silica gel formed in sea-floor muds (Russell and Trueman, 1971). Phosphorites associated with black shales commonly contain uranium, probably as a uranium phosphate. Vanadium is also present in some phosphatic rocks in which it is associated with carbonaceous matter or kerogen.

The phosphorites of northern Queensland, Australia (Fig. 7-23), are found in the Georgina Basin south of the Gulf of Carpentaria (Fig. 7-24). A thick sequence of Cambrian dolomite occupies the central area of the

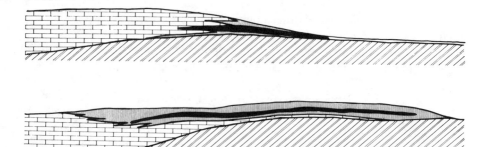

Fig. 7-23. Diagrammatic sections of northern phosphate deposits, Gulf of Carpenteria region, Australia, showing limestone, phosphorite (black), and associated bedded cherts and siliceous phosphatic siltstones (vertical hachures) overlying Proterozoic beds. [After Thomson and Russell (1971).]

basin, whereas much thinner sequences overlie the basin margins. Cook (1972, p. 1193) says:

> Thin-bedded limestones occur around the periphery, and these in turn pass into mudstones and cherts near the margins of the basin.

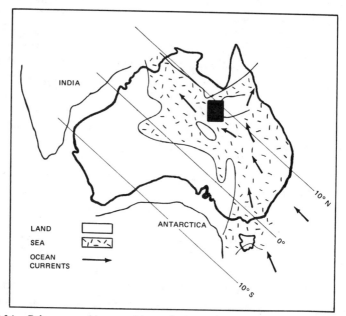

Fig. 7-24. Paleogeographic map of Australia during the late Early Cambrian, showing the probable positions of India and Antarctica with respect to paleolatitudes. Area of phosphorite deposits (black). [Modified after Howard (1972).]

Although several Cambrian units are slightly phosphatic, the thick high grade phosphorites are restricted to the marginal mudstone-chert facies and in particular to the Middle Cambrian Beetle Creek Formation and its lateral equivalents.

The Beetle Creek Formation, which commonly exhibits thin uniform bedding but locally includes sedimentary breccia, comprises calcareous and cherty phosphatic siltstone, chert, fetid dolomite, phosphatic limestone, and phosphorite. The phosphorites contain varieties of apatite, most of which occur as submicroscopic crystallites comprising collophane. Most of the collophane occurs as pelletal phosphorite, but some occurs as collophane mudstone and as cement in sedimentary breccia. The main gangue constituents of the phosphorite are chert and carbonate minerals.

In the Queensland deposits the association of carbonaceous black (unweathered) chert, fetid limestone, and phosphorite is similar to that of beds in the Permian Phosphoria Formation of the northwestern United States and of the Late Cretaceous phosphate deposits of the Middle East. Deposition appears to have been in shallow marine to paralic environments on a broad continental shelf bounded by a landmass to the north and south (Fig. 7-24). This landmass was predominantly of Precambrian crystalline and metasedimentary rocks. The influx of terrigenous sediments onto this shelf was minimal, and these were derived mainly from a source to the west, which indicates to Russell and Trueman (1971) that the borderlands were of low relief and the climate was probably arid. Sedimentation in this environment proceeded slowly, with penecontemporaneous development of the collophane in both pelletal and mudstone phosphorites. Cook (1972, p. 1193) says:

> The pelletal type is composed predominantly of ovules which formed under shallow marine conditions partly by the phosphatization of rounded fossil fragments and partly by phosphatization below the sediment-water interface. There are two varieties of the nonpelletal type: a collophane mudstone believed to have been deposited in the intertidal or supratidal zone and subject to subaerial processes, and phoscretes which are solely the product of subaerial weathering.

It has been pointed out by Howard (1972, p. 1180) that

> The discoveries of phosphorite in the lower Middle Cambrian sediments of the Georgina Basin resulted from systematic Australia-wide exploration programs by several companies. The exploration involved the application of concepts relating basin stratigraphy, lithologic associations, tectonics and phosphate occurrences to paleogeography, in particular paleolatitude, and areas of supposed

oceanic upwelling. Lithologic and gamma-ray logs, magnetic, gravity and seismic survey data were derived from work done by oil companies and the Bureau of Mineral Resources. In terms of technique, these data and their interpretations were integrated with the results of examination, chemical testing and selected analyses of cores and cuttings, together with university rock and palaeontological collections. This was followed by the field examination of favourable suites of sediments and/or established phosphogenic sections, the gamma-ray logging of abandoned water bores, and, finally, rotary drilling.

This summarizes succinctly the main geological considerations, techniques, and procedures followed during the course of exploration for phosphorites in northern Australia.

Mention has been made of the possible role of upwelling ocean currents. It is postulated that where cold deep-water currents, rich in phosphorous, move up a continental slope onto a shallow and extensive continental shelf, the phosphorous is precipitated. As shown by the Fig. 7-24, the inferred directions of movements of ocean currents during the late Early Cambrian suggest that such a mechanism may have been instrumental in the development of phosphorite deposits in northern Australia. Exchange of water between the ocean and shallow shelf embayments may have been restricted by sea-floor topography similar to that of the Santo Domingo embayment on the west coast of Baja California (D'Anglejan, 1967). The embayment, in which phosphates occur, is bounded on the seaward side by submarine escarpments resulting from faulting and on the east by a shoreline with low relief.

The application of techniques, procedures, and concepts involved in sedimentary basin analysis to the search for phosphorites in Australia has been pointed out by Howard (1972). Phosphorite deposits are known to have formed in various parts of the world during intervals of time from the Proterozoic to the Neogene, including the Miocene deposits of Florida which were at one time thought to have formed by the leaching of guano or bone beds, and the precipitation of phosphates in the underlying limestone. Most of these deposits are associated with dark cherty shales and carbonates which were apparently deposited in shallow marine embayments or on shelves where the rates of sedimentation were slow, circulation of seawater somewhat restricted, and the salinity higher than normal. This type of depositional environment suggests proximity to a landmass of low relief and also to a situation of marine transgression over large areas of coastal lowlands. Paucity of sedimentation suggests also that many phosphorite deposits occur in comparatively thin stratigraphic sequences of carbonates, cherts, and shales that overlie an unconformity or paracontinuity. A further requirement may be the periodic incursion of phosphorus-rich marine waters such as those of the Arctic and Antarctic regions. Where currents well up along escarpments or continental slopes, they may spill over into embayments or the shallows of carbonate banks during periods of storms or high tides. Whether or not it is necessary to invoke an influx of cold upwelling seawater is a matter of conjecture; furthermore, the geological criteria that might indicate such a situation are problematical.

EXPLORATION FOR AQUIFERS AND STORAGE RESERVOIRS

Aquifers

The demand and consequent search for fresh water in many parts of the world is acute. The requirements of water for hydroelectric power plants, industry, domestic use, and recreational purposes are becoming increasingly urgent. Most of these requirements depend on surficial water such as rivers and lakes, and on ground water in sand, gravel, or fractured bedrock at depths commonly up to a few hundred meters below the surface. These water resources are not included within the compass of this book which deals only with fresh to brackish water in the strata of sedimentary basins. Most porous and permeable beds contain salty connate water which has been locked into the original sediment, or squeezed from adjacent sediments by compaction. This water is not of commercial value. The following discussion will concern only those sedimentary beds known as aquifers, in which meteoric water moves downward along the bedding plane to depths within a sedimentary basin. This situation requires a porous and permeable bed, commonly but not exclusively a sandstone, or a jointed and vugular limestone that crops out at a low angle over an extensive area around the erosional edge of a sedimentary basin in a region of ample rainfall.

Aquifers are of considerable importance in many rural areas of Australia where water supply from such beds is essential for the survival of livestock during the dry season. Water from bores drilled to these aquifers ranges in salinity from potable for human consumption to brackish but acceptable to cattle, a range including a few hundred to a few thousand milligrams per liter. In the Great Artesian Basin of eastern Australia, the most prolific aquifers are sandstone beds within the Late Jurassic to Early Cretaceous Blythesdale Formation. This sequence includes fluvial quartzose and feldspathic sandstones, silty shales, and thin seams of coal in the lower part. The lowermost member and main aquifer is the Mooga Sandstone, a quartzose sandstone that is locally conglomeratic at the base. Beneath the Blythesdale is a thick sequence of shaly, coal-bearing beds. Referring to the stratigraphy of the eastern portion of the Great Artesian Basin, Hind and Helby (1969, p. 481) say:

> These coaly strata are restricted to the eastern portion of the basin in New South Wales. The upper limit of these formations is sharply defined by the occurrence of massive quartzose sands, representing the major intake beds of the basin. In the north these sands are commonly referred to as the "Blythesdale Group", whereas in the southeast portion of the basin (Coonamble Lobe) they are known as the Pilliga Sandstone. The sandstones of the "Blythesdale Group" are overlain by a marine shale representing the lower limits of the Rolling Downs Group.

There is no marked unconformity between the Jurassic and Cretaceous in this region, although a disconformity or paracontinuity may exist. The Blythesdale is essentially nonmarine, its Late Jurassic basal member comprises fluvial sandstone, and its uppermost member includes an early Aptian marine fauna buried during the widespread Early Cretaceous marine transgression.

Hind and Helby (1969) report that there are some 3000 water bores within the New South Wales region of the Great Artesian Basin, the main production being from aquifers in the lower part of the Blythesdale. Of these bores nearly 700 are artesian wells having a combined unrestricted rate of flow of approximately 12 million liters/hr. The remaining 2300 wells have flow rates ranging from a few thousand to 150,000 liters/hr. The salinity of water obtained from aquifers in the lower part of the Blythesdale is commonly less than 700 mgm/liter. The main intake for artesian wells in New South Wales is in outcrops and subcrops of the Pilliga Sandstone and other sandstones of the Blythesdale which extends in an arc for many tens of kilometers around the erosional edge of the basin. Hind and Helby state that rates of flow within the aquifers are in the range 1.5–10 million/year and that the rate of withdrawal exceeds the rate of recharge to such an extent that serious shrinkage of the flow area and lowering of water levels in the bores have occurred.

In the Carnarvon Basin of Western Australia the main aquifer is the Early Cretaceous Birdrong Sandstone which rests unconformably on Permian beds and is overlain by the relatively impermeable Muderong Shale. According to Condon (1954), the Birdrong consists mainly of quartzose and glauconitic sandstone with good porosity and permeability.

The lower part of the Birdrong contains abundant fragments of fossil wood, and the basal part is coarse to very coarse. Total thickness ranges to 30 m. The Birdrong is a marine transgressive sandstone that covers a wide area in the subsurface and crops out as escarpments. Of particular interest is the fact that it formed the reservoir for Rough Range No.1, Australia's first oil well drilled in 1952 and abandoned as economically unproductive. The water in many bores is hot, about 50–60°C, and brackish. Some bores produce up to 300,000 liters/hr, the rates of flow increasing away from the sources of intake.

The main aquifer of the Canning Basin in Western Australia is the sandstone at the base of the Mesozoic sequence, which unconformably overlies the Permian. This coarse and locally conglomeratic bed was probably deposited during the Late Jurassic (Veevers and Wells, 1961). The sandstone, which is friable and has good porosity and permeability, contains fragments of fossilized wood. Locally, it is interbedded with thin lenses of carbonaceous to lignitic siltstone and is evidently of fluvial to lacustrine origin. Production of potable water is obtained mainly from bores in the coastal area.

In the Amadeus Basin of central Australia the most prolific yields of water are obtained from the Late Silurian–Devonian Mereenie Sandstone, the Early Permian Crown Point Formation, and the Jurassic (?) De Souza Sandstone (Wells *et al.*, 1970). The Mereenie, which crops out over an area of some 2500 km² around the edge of the basin and underlies approximately 25,000 km² within the basin, has a thickness of up to 1000 m. The formation consists of white to pale brown, cross-bedded, quartzose sandstone with local conglomeratic lenses at the base. The grains are fine, well sorted, and well rounded. Vertical tubes of *Scolithus* and indeterminate trace fossils indicate that in part at least the Mereenie was deposited in a beach environment. Other sections may be of eolian origin, possibly deposited as coastal sand dunes.

The Mereenie is a diachronous stratigraphic unit deposited during a period of marine transgression. In the northern part of the Amadeus Basin it lies unconformably on the Cambro–Ordovician Stokes Formation, in which area the base of the Mereenie is marked by conglomerate lenses. In the southern part of the basin, the relationship of the Mereenie with underlying beds is conformable.

Porosity and permeability in the Mereenie are excellent, the range of the former being 18–25% and that of the latter 80–1500 millidarcys (Wells *et al.*, 1970). Where the water is not contaminated by brines moving across faulted contacts from other beds, it has a salinity of a few hundred milligrams of total dissolved salts per liter. This is within the acceptable range for potable water.

The Early Permian Crown Point Formation consists of conglomerate, tillite, sandstone, and siltstone, with thin lenses of lignite in the upper part of the sequence which is about 60 m thick. In the southeastern part of the Amadeus Basin the Crown Point unconformably overlies the Devono–Carboniferous Finke Group and is unconformably overlain by the Jurassic (?) De Souza Sandstone. The Crown Point consists largely of sediments derived from the erosion of a landmass during a period of glaciation and is probably entirely nonmarine. The main aquifer within the Crown Point is a coarse-grained, conglomeratic sandstone which yields water having a salinity range similar to that of the Mereenie.

The De Souza Sandstone lies unconformably on both the Finke Group and the Crown Point. The formation has a maximum thickness of 90 m of cross-bedded sandstone that is friable, quartzose, micaceous, and very porous and permeable. The fluvial character of the beds suggests deposition in estuarine and river floodplain environments. The overlying Early Cretaceous Rumbalara Shale contains *Inoceramus* and probably represents lagoon and bay sediments deposited during an early phase of marine transgression. The De Souza occurs only where the Mesozoic sequence of the Great Artesian Basin overlaps the southeastern margin of the Amadeus Basin. This sandstone is one of the chief aquifers of the Great Artesian Basin and yields excellent rates of flow of water having a salinity range of 500–3000 mgm/liter.

Of the numerous aquifers in many other parts of the world (Fig. 7-14) which could be discussed, only a few will be mentioned. In northeastern Florida an important aquifer is the Eocene Ocala Group (Meisburger and Field, 1976) which consists largely of shelf carbonates deposited disconformably on the Cretaceous and overlain by relatively impermeable Miocene beds. These beds dip gently and extend under the sea where subcrops have been invaded by seawater. During the period 1904–1946 the inland encroachment of salt water, resulting from excessive withdrawal of fresh water, locally exceeded five miles (Martinez, 1971).

In the Gulf Coast region south of Houston, Texas, freshwater aquifers occur in a Pliocene–Pleistocene deltaic sequence (Kreitler *et al.*, 1977). The sandstone geometry of these aquifers, as determined by sandstone percentage maps, is shown to be that of distributary channels. These aquifers are extensively cut by growth faults, resulting in thicker sand development on the downthrow or coastward side of the faults. Aquifers of this type can occur in a sequence including coal seams (Fig. 7-15) and where the coal is being mined, particularly by underground methods, can cause problems of flooding.

An unusual type of freshwater-bearing bed is the Honolulu aquifer (Dale, 1974) which consists of fractured and rubbly basalt overlain by relatively impermeable Pleistocene sediments in the coastal area of the island. This aquifer is extremely permeable and has a hydraulic conductivity of about 3000 m/day. Downdip movement of water is locally impeded by erosional valleys which cut into the aquifer and are filled with Pleistocene sediments.

The stratigraphic position and depositional environment of many aquifers suggest that they have commonly been deposited as marine transgressive units of quartzose sand, shelf carbonates, and as fluvial and eolian deposits on a land surface. In all such cases their close spatial and genetic relationships to an angular unconformity, disconformity, or paracontinuity are evident. The application of sedimentary basin analysis to the search for aquifers depends in large part on the determination of these relationships, with particular reference to the presence of relatively impermeable beds of shale overlying the potential aquifers. In this regard it is emphasized (Chapter 6) that a sedimentary basin is not a hydrodynamically homogeneous unit but may include one or more aquifers, each of which has its own hydrodynamic properties and potentiometric surface.

Storage for Water and Natural Gas

The available supply of fresh water to the storage facilities of an urban or industrial area commonly fluctuates to an alarming extent. In recent years considerable attention has been directed in various parts of the world to exploration for suitable subsurface reservoirs that would cushion the effects of water consumption during periods of drought and consequent drawdown on water levels in the storage facilities. Kimbler *et al.* (1973, p. 1598) say:

> In the advanced industrial countries the most favourable, least expensive sites for surface reservoirs are already in use or the land already is preempted for other uses and is unavailable for the storage of water. In addition, there are many flat areas in coastal zones, also underlain by saline aquifers, that are unsuitable for water storage although a surplus of fresh water is available in such areas at certain times of the year. The lack of a reliable, year-round supply of water has been a major factor in preventing commercial and residential development in these areas.

Kimbler *et al.* state that empirical experiments on the storage of fresh water in slightly saline aquifers, as well as studies based on laboratory experiments and mathematical models, show that it is feasible to store fresh water in subsurface reservoirs.

In selecting a suitable reservoir, certain structural, petrophysical, and petrochemical factors must be considered. The reservoir must be readily accessible in that it must lie at a comparatively shallow depth. The structural configuration and stratigraphic relationships of the reservoir are also of prime concern. Most water-bearing beds have some degree of dip, and in many there is a measurable flow of slightly saline formation water. Injected fresh water, having a lower density than the formation water, forms a lens that lies above the saline formation water and moves in response to the structural and hydrodynamic conditions of the reservoir. That is, it tends to move updip and also in the direction of flow of the formation water. The relative effects of these two vectors determine the amount and direction of displacement of the fresh water from its point of injection within a given period of time. For practical purposes this displacement may not be significant. Another obvious factor is the permeability of the reservoir, and this will in many cases be affected by petrochemical factors. For example, during tests in Virginia of the effects of injecting fresh water into and withdrawing it from a confined aquifer containing brackish water, it was found by Brown and Silvey (1973) that both physical and chemical changes occurred in the reservoir. Data indicated that as fresh water was injected, calcium and magnesium replaced sodium in clays contained in the aquifer. The effect was to decrease very slightly the tendency of the clays to disperse during injection of fresh water. During withdrawal of fresh water a reverse cation exchange took place as the sodium concentration of the mixed fresh and saline formation water increased. Flushing with a solution of calcium chloride prior to injection of fresh water eliminated this problem. It is of interest to note that of 100 million liters of fresh water injected and retained in the aquifer for six weeks, 85% of potable water could be recovered.

Other problems arising during the injection of fresh water into a storage reservoir have been enumerated by Signor (1973, p. 1605) who says:

> Artificial groundwater recharge, by any method, is subject to limitations caused by some mechanism degrading the hydraulic conductivity of the porous media through which the recharge water is being infiltered or injected. Reduction of hydraulic conductivity may be caused by suspended solids, bacterial growth, chemical reactions of dissolved solids with the porous media or existing native water, and air entrapment or dissolution of gases in interstices of the porous medium.

Suitable storage reservoirs for fresh water include ancient coastal sand bodies such as beaches, bars, and barrier islands buried inland and also the buried channel sand bodies of ancient river distributaries. In coastal areas such Pleistocene or Tertiary bodies are commonly of poorly con-solidated porous and permeable quartzose sandstone. Where these bodies are capped by relatively impermeable shale, they can form satisfactory reservoirs for the temporary storage of fresh water. Porous limestone beds, particularly those that are cavernous, jointed, and underlie an un-conformity marked by an ancient karst topography, commonly provide channels along which there is a rapid flow of water. Such beds are subject to the incursion of contaminated meteoric water and may not be suitable as freshwater reservoirs.

The storage of natural gas may be required for a number of reasons. In many fields producing oil, condensate, and natural gas, most of the gas produced with the liquid fractions is pumped back into the field because the marketing of it is not economically viable or because it is necessary to retard the decline of reservoir pressure. Both reasons may preclude the flaring off of appreciable volumes of gas. This procedure of pumping gas back into a field which basically produces oil or condensate is not primar-ily a method of gas storage. On the other hand, a field such as the one in the Po Basin on the outskirts of Naples, Italy, that not only produces gas but has gas pumped into it, is considered as a storage reservoir. This field was originally a producing field only, but as production declined and reservoir pressure dropped, it was repressurized with gas from other fields and is used as a reserve for periods of peak gas consumption in the industrial city of Naples. Similarly, the Bow Island gas field situated southeast of Calgary, Alberta, was discovered in 1909, supplied gas to Calgary from 1912 (Belyea et al., 1968), and many years later was used as a storage reservoir for surplus gas from the Turner Valley field. For years, during the 1920s and 1930s, the condensate from Turner Valley gas was stripped off and the gas flared at rates exceeding 3 million cubic meters a day. On overcast nights the reddish glow from the flares could be seen 60 km away. Another example of structural traps that have served as both producing and storage reservoirs are the Little Capon and Augusta fields in West Virginia. These reservoirs are in the Oriskany Sandstone capped by shales of the Early Devonian Onondaga Formation. Porosity results from fractures, and the two fields have estimated total recoverable re-serves from storage of 400 million cubic meters of natural gas (Jacobeen and Kanes, 1974).

Storage reservoirs may be used to supplement the gas load during pe-riods of peak demand, to save unwanted gas after the liquid fractions have been stripped off, or to store large quantities of gas for situations of national emergency. A requirement of the first and last contingency is proximity to the area of gas consumption, as long gas pipelines are more difficult to maintain and safeguard than are short ones. Ideally, a storage reservoir should be situated close to an urban or industrial area. For this reason some consideration has been given to the search for potential storage reservoirs near Sydney in the Sydney Basin, Australia, and near cities in other parts of the world.

The search for such reservoirs depends initially on the application of sedimentary basin analysis to the problems of assessing the structural and stratigraphic framework of the region searched and on the determination of lithologic sequences and petrophysical characteristics of potential reservoir rocks. This means that basin analysis must indicate not only whether the strata are folded or faulted, and to what extent, but also the nature, thickness, and general deposition environment with reference to the predominant molasse, flysch, carbonate, or volcanic character of the sedimentary pile. Further determinations of particular structures, geometry of stratigraphic bodies, lithologic sequences, and petrophysical characteristics depend on seismic surveys and drilling. The requirements of any suitable storage reservoir are those of any producing reservoir, in other words, structural closure, an impermeable caprock, adequate volume, porosity, and permeability. A further requirement for a storage reservoir that is desirable but not essential to a producing reservoir is proximity to the area of gas consumption.

In cases where depleted fields are used for storage, their capacity to hold hydrocarbons has been demonstrated. These fields may be structural, structural-stratigraphic, or purely stratigraphic traps. Where it is necessary to search for a potential reservoir, priority should be given to structural features or to bioherms. In particular, structures which are indicated by seismic surveys to have prospects for closure should be investigated. These may be antiforms or domes deforming a sandstone or carbonate sequence, and they may have resulted from tectonic stress or the draping effects of compaction over buried hills, bioherms, or salt domes. Seismic surveys may also indicate the presence of a potential caprock. Determinations of fluid content, porosity, and permeability depend primarily on drilling and testing operations. In this regard, with respect to petrophysical characteristics, it is seldom helpful to examine the outcrops of potential reservoir rocks because surficial weathering has altered the porosity and permeability.

Storage for Radioactive and Chemical Waste

The disposal of liquid wastes from cities, industrial plants, other chemical or power installations, and mines poses an evergrowing problem. In particular, the problems of disposal of high-level radioactive wastes, organohalogens, mercury and cadmium compounds, and certain products from the manufacture of biological and chemical warfare substances are so acute as to arouse worldwide controversy. The disposal of other liquid wastes, including acids and alkalis, organosilicones, cyanides, fluorides, pesticides, compounds or arsenic, lead, copper, and zinc, and low-level radioactive material is less critical but becoming an increasing burden. In the following discussion it is not proposed to consider the various procedures or methods of disposal advocated but only to refer to the possibility of storage in sedimentary basins and the relevant application of sedimentary basin analysis.

Problems of disposal of chemical, and in particular of radioactive liquid wastes in deep impermeable beds within sedimentary basins, have occupied the attention of many people, some of the early investigators including Theis (1955), Birch (1958), De Laguna (1959), Roedder (1959), Skibitzke (1961), and Nace (1961). A few years later, Drescher (1965) concluded that the present state of knowledge indicated that although low-level radioactive wastes could physically be injected into deep saltwater aquifers in large sedimentary basins, it was not possible to accurately predict the effects on the hydrology of the aquifer or the possible extent or rate of movement of the waste-contaminated water. He advocated programs of hydrologic studies to monitor the movement of water in selected basins. Balter (1971, p. 2081) reported that the U.S. Atomic Energy Commission required all high-level radioactive liquid wastes from fuel reprocessing plants to be solidified and shipped to a national repository, possibly a salt mine. He concluded that:

> Because of a general requirement for adequate monitoring to assure the safe and effective operation of a deep well injection system, this method has generally not been used for disposal of radioactive wastes. It appears that the injection into deep permeable formations may be a practical solution for the disposal of large quantities of tritium-bearing wastes from water reactors and nuclear fuel reprocessing plants in the future. Additional research is also required on the potential deep disposal of noble gases such as krypton-85 from reactor and reprocessing plant off-gas streams.

In the Federal Republic of Germany radioactive waste from nuclear research and power production is stored in the Asse salt mine (Kuhn, 1973).

The need for additional research on the possible reactions of waste constituents with the subsurface formation into which they are injected was also pointed out by Donaldson (1971, p. 2083) who said:

> The safety hazards of underground waste disposal should receive careful consideration in the planning stages of a waste-disposal well.

Such wells may be drilled to depths exceeding a thousand meters in permeable sandstone, fractured or vugular limestone, fractured shale, or jointed basement rock. Problems other than those inherent in the petrophysical nature of the rock or sedimentary basin can be introduced, as outlined by Ehrlich (1971), who states that biologic activity, particularly by certain anaerobic bacteria, can clog waste injection wells. Another problem has been mentioned by Hanshaw (1971) who points out that, although shales are commonly regarded as good aquicludes because of their poor permeability, the clay mineral content of a shale bed may cause it to behave as a semipermeable membrane, resulting in the invasion and transfer across a shale bed of chemical constituents in the waste. This effect may be a problem only where the aquiclude is thin. Monitoring of more than 15,000 million liters of acid industrial waste injected into a cavernous limestone aquifer near Pensacola, Florida, showed no invasion of contaminants or pressure change in the aquifer overlying but separated from the injection bed by some 50 meters of claystone (Goolsby, 1971). Within the injection bed, waste material moved more than 1.5 km from the injection site during a period of less than a year, indicating the necessity to understand details of the stratigraphy and hydrology of any sedimentary stratum considered as a potential reservoir for waste disposal. Another example of storage of industrial and municipal waste in highly permeable cavernous limestone, at depths of up to 1500 m in southern Florida, is cited by Kaufman *et al.* (1973). At one location, since 1966, more than 3000 million liters of highly organic and acidic effluent from a sugar mill have been injected into this limestone. Part of the acid component of the waste is quickly neutralized by dissolution of limestone, and much of the organic matter is degraded by the action of anaerobic bacteria.

Apart from the lithostratigraphic and hydrologic characteristics of the proposed storage reservoir, the nature of the waste is of prime consideration. Radioactive waste could not be stored in a limestone bed such as those mentioned above. This was emphasized by Barlow (1973, p. 1591) who said:

> Deep well disposal is not advanced as a cure-all for problems related to waste-liquid disposal. Its use is relatively limited considering the wide divergence in chemical composition of wastes.

A further warning was sounded by Lofgren (1973) who pointed out that withdrawal of fluids over a period of many years has already affected the hydraulic systems of many sedimentary basins and that in California alone probably 18,000 km² have subsided by up to 8 m in a few places. Injection of wastes into sedimentary beds that have subsided may decrease the rate of subsidence but result in contamination of freshwater aquifers by upward migration of fluids from the injected beds, particularly along the planes of faults caused by subsidence. The storage of liquid low- to medium-level radioactive wastes is a particularly pressing problem, and if underground storage in a sedimentary basin is sought, it must obviously be in a thick section of impermeable rock that can be shattered by hydraulic fracturing through boreholes or reached by sinking a shaft. A Triassic sedimentary basin South Carolina that appears to offer such possibilities has been described by Marine (1974). Seismic, gravity, and magnetic surveys, in addition to drilling, have shown the basin to be about 50 km long, 10 km wide, and 1500 m thick. The stratigraphic sequence is predominantly of mudstone with very minor lenses of poorly sorted sandstone. The water yield and permeability in all wells drilled is extremely low. Other basins will be studied in the future with particular regard to the type and volume of waste to be disposed.

Evaluation of sedimentary basins as sites for the disposal of wastes requires the combined efforts of scientists and engineers working in a variety of fields including geology, hydrology, chemistry, and physics. Mistakes that might have unforseen consequences years ahead could be costly in terms of their environmental or socioeconomic impact. The onus is on the present to take care of the future.

References

AAPG (1972). Carbonate Rocks II: Porosity and Classification of Reservoir Rocks. Am. Assoc. Petrol. Geologists Reprint Ser. 5.

Abry, C. G. (1975). Geostatic model for predicting oil: Tatum Basin, New Mexico, *Am. Assoc. Pet. Geol. Bull.* **59,** 2111–2122.

Adams, J. E. (1969). Semi-cyclicity in the Castile Evaporite, *In* "Cyclic Sedimentation in the Permian Basin" (J. G. Elam and S. Chuber, eds.), pp. 197–203. West Texas Geol. Soc. Publ.

Adams, W. L. (1964). Diagenetic aspects of Lower Morrowan, Pennsylvanian, sandstones, northwestern Oklahoma, *Am. Assoc. Pet. Geol. Bull.* **48,** 1568–1580.

Ager, D. V. (1973). "The Nature of the Stratigraphical Record." Macmillan, New York.

AGI (1976). "Dictionary of Geological Terms." Am. Geol. Inst. Publ.

Agrell, H. (1974). Glaciation and deglaciation in the Sommen-Asunden region, southeastern Sweden, *Bull. Geol. Inst. Univ. Uppsala* **4,** 125–188.

Akpati, B. N. (1978). Geological structure and evolution of the Keta basin, Ghana, West Africa, *Geol. Soc. Am. Bull.* **89,** 124–132.

Alam, M. (1972). Tectonic classification of Bengal Basin, *Geol. Soc. Am. Bull.* **83,** 519–522.

Alcock, F. J. (1936). "Geology of Lake Athabaska Region, Saskatchewan." Geol. Surv. Can. Mem. 196.

Allan, J. R., and Mathews, R. K. (1977). Carbon and oxygen isotopes as diagenetic and stratigraphic tools: surface and subsurface data, Barbados, West Indies. *Geology* **5,** 16–20.

Allen, J. R. L. (1970). "Physical Processes of Sedimentation." Allen and Unwin, London.

Alling, H. L., and Briggs, L. I. (1961). Stratigraphy of Upper Silurian Cayugan evaporites, *Am. Assoc. Pet. Geol. Bull.* **45,** 515–547.

Alvarez, W. (1976). A former continuation of the Alps, *Geol. Soc. Am. Bull.* **87,** 891–896.

American Committee on Stratigraphic Nomenclature (1961). Code of stratigraphic nomenclature, *Am. Assoc. Pet. Geol. Bull.* **45,** 645–665.

Amstutz, G. C. (ed.) (1964). Sedimentology and ore genesis, *Dev. Sedimentol.* **2.**

Amstutz, G. C., and Bernard, A. J. (eds.) (1973). "Ores in Sediments." Springer-Verlag, Berlin and New York.

Anderson, R. Y., Dean, W. E., Kirkland, D. W., and Snider, H. I. (1972). Permian Castile varved evaporite sequence, West Texas and New Mexico, *Geol. Soc. Am. Bull.* **83,** 59–86.

Andresen, M. J. (1962). Paleodrainage patterns: their mapping from subsurface data, and their paleogeographic value, *Am. Assoc. Pet. Geol. Bull.* **46,** 398–405.

Anhaeusser, C. R., Mason, R., Viljoen, M. J., and Viljoen, R. P. (1969). A reappraisal of some aspects of Precambrian Shield geology, *Geol. Soc. Am. Bull.* **80,** 2175–2200.

Archie, G. E. (1952). Classification of carbonate reservoir rocks and petrophysical considerations, *Am. Assoc. Pet. Geol. Bull.* **36,** 278–298.

Armstrong, A. K. (1974). Carboniferous carbonate depositional models, preliminary lithofacies and paleotectonic maps, Arctic Alaska, *Am. Assoc. Pet. Geol. Bull.* **58,** 621–645.

Arrhenius, G. O. S. (1963). Pelagic sediments, *In* "The Sea," Vol. III, pp. 655–727. Wiley (Interscience), New York.

Asquith, D. O. (1974). Sedimentary models, cycles, and deltas, Upper Cretaceous, Wyoming, *Am. Assoc. Pet. Geol. Bull.* **58,** 2274–2283.

Assoc. for Geol. Collab. of Japan (1965). "The Geological Development of the Japanese Islands." Tsukiji Shokan, Tokyo.

Aubouin, J. (1965). Geosynclines, *In* "Developments in Geotectonics," Vol. I. Elsevier, Amsterdam.

Austin, G. H. (1973). Regional geology of Eastern Canada offshore, *Am. Assoc. Pet. Geol. Bull.* **57,** 1250–1275.

Avbovbo, A. A. (1978). Tertiary lithostratigraphy of Niger Delta, *Am. Assoc. Pet. Geol. Bull.* **62,** 295–306.

Ayres, D. E. (1971). The hematite ores of Mount Tom Price and Mount Whaleback, Hamersley iron province, *Aust. Inst. Min. Metall.* **238,** 47–58.

Bagnold, R. A. (1941). "The Physics of Blown Sand and Desert Dunes." Methuen, London.

Bailey, N. J. L., Evans, C. R., and Milner, C. W. D. (1974). Applying petroleum geochemistry to search for oil: examples from Western Canada Basin, *Am. Assoc. Pet. Geol. Bull.* **58,** 2284–2294.

Baker, D. R., and Claypool, G. E. (1970). Effects of incipient metamorphism on organic matter in mudrock, *Am. Assoc. Pet. Geol. Bull.* **54,** 456–468.

Balkwill, H. R. (1978). Evolution of Sverdrup Basin, Arctic Canada. *Am. Assoc. Pet. Geol. Bull.* **62,** 1004–1028.

Ballard, J. A., and Feden, R. H. (1970). Diapiric structures on the Campeche shelf and slope, western Gulf of Mexico, *Geol. Soc. Am. Bull.* **81,** 505–512.

Barberi, F., Borsi, S., Ferrara, G., Marinelli, G., Santacroce, R., Tazieff, H., and Varet, J. (1972). Evolution of the Danakil Depression (Afar, Ethiopia) in light of radiometric age determinations, *J. Geol.* **80,** 720–729.

Barlow, A. C. (1973). Philosophy of deep-well disposal (abstract), *Am. Assoc. Pet. Geol. Bull.* **57,** 1591.

Barlow, J. A., and Haun, J. D. (1966). Regional stratigraphy of Frontier Formation and relation to Salt Creek Field, Wyoming, *Am. Assoc. Pet. Geol. Bull.* **50,** 2185–2196.

Barr, S. M. (1974). Structure and tectonics of the continental slope west of southern Vancouver Island, *Can. J. Earth Sci.* **11,** 1187–1199.

Barrata, M. (1910). "La catastrofe sismica Calabro Messine (28 Dicembre 1908)." Società Geografica Italiana.

Bartholomé, P., De Magneé, I., Evrard P., and Moreau, J. (eds.) (1974). "Gisements stratiformes et provinces cuprifères." Société Géologique de Belqique.

Baumgartner, T. R., and Van Andell, T. H. (1971). Diapirs of the continental margin of Angola, Africa, *Geol. Soc. Am. Bull.* **82**, 793–802.

Beales, F. W., and Oldershaw, A. E. (1969). Evaporite-solution brecciation and Devonian carbonate reservoir porosity in western Canada, *Am. Assoc. Pet. Geol. Bull.* **53**, 503–512.

Beard, D. C., and Weyl, P. K. (1973). Influence of texture on porosity and permeability of unconsolidated sand, *Am. Assoc. Pet. Geol. Bull.* **57**, 349–369.

Beaumont, E. C., Dane, C. H., and Sears, J. D. (1956). Revised nomenclature of Mesaverde Group in San Juan Basin, New Mexico, *Am. Assoc. Pet. Geol. Bull.* **40**, 2149–2162.

Beck, R. H., and Lehner, P. (1974). Oceans, new frontiers in exploration, *Am. Assoc. Pet. Geol. Bull.* **58**, 376–395.

Behrendt, J. C., Popenoe, P., and Mattick, R. E. (1969). A geophysical study of North Park and the surrounding ranges, Colorado, *Geol. Soc. Am. Bull.* **80**, 1523–1538.

Belter, W. (1971). Deep disposal systems for radioactive wastes (abstract), *Am. Assoc. Pet. Geol. Bull.* **55**, 2081.

Belyea, H. R., Sanford, B. V., and Hood, P. J. (1968). The geological survey of Canada and the oil industry, *In* "Dusters and Gushers: The Canadian Oil and Gas Industry," pp. 99–104. Pitt Publ., Toronto.

Bennacef, A., Beuf, S., Biju-Duval, B., De Charpal, O., Gabriel, O., and Rognon, P. (1971). Example of cratonic sedimentation: Lower Paleozoic of Algerian Sahara, *Am. Assoc. Pet. Geol. Bull.* **55**, 2225–2245.

Berg, R. R. (1975). Capillary pressures in stratigraphic traps, *Am. Assoc. Pet. Geol. Bull.* **59**, 939–956.

Berg, R. R. (1975a). Depositional environment of Upper Cretaceous Sussex Sandstone, House Creek Field, Wyoming, *Am. Assoc. Pet. Geol. Bull.* **59**, 2099–2110.

Berg, R. R., and Davies, D. K. (1968). Origin of Lower Cretaceous Muddy Sandstone at Bell Creek Field, Montana, *Am. Assoc. Pet. Geol. Bull.* **52**, 1888–1898.

Berger, J., Blanchard, J. E., Keen, M. J., McAllister, R. E., and Tsong, C. F. (1965). Geophysical observations on sediments and basement structure underlying Sable Island, Nova Scotia, *Am. Assoc. Pet. Geol. Bull.* **49**, 959–965.

Berger, W. H. (1970). Biogenous deep-sea sediments: fractionation by deep-sea circulation, *Geol. Soc. Am. Bull.* **81**, 1385–1402.

Berger, W. H., and Mayer, L. A. (1978). Deep-sea carbonates: acoustic reflectors and lysocline fluctuations, *Geology* **6**, 11–15.

Bernard, H. A., and Le Blanc, R. J. (1965). Resume of Quaternary geology of the north-western Gulf of Mexico province, *In* "The Quaternary of the United States," pp. 137–185. Princeton Univ. Press, Princeton, New Jersey.

Bernard, H. A., Major, C. F., and Parrott, B. S. (1959). The Galveston barrier island and environs: a model for predicting reservoir occurrence and trend, *Trans. Gulf Coast Assoc. Geol. Soc.* **9**, 221–224.

Bernard, H. A., Major C. F., Parrott, B. S., and Le Blanc, R. J. (1970). "Recent Sediments of Southeast Texas, a Field Guide to the Brazos Alluvial and Deltaic Plains and the Galveston Barrier Island Complex." Bureau Econ. Geol. Guidebook 11, Univ. of Texas, Austin, Texas.

Bigarella, J. J. (1972). Eolian environments: their characteristics, recognition, and importance, *In* "Recognition of Ancient Sedimentary Environments," pp. 12–62. Soc. Econ. Paleontologists Mineralogists Spec. Publ. 16.

Birch, F. (1958). "Thermal Considerations in Deep Disposal of Radioactive Waste." Nat. Research Council Publ. 588, National Academy Science, Washington, D.C.

p, M. S. (1960). "Subsurface Mapping." Wiley, New York.

e, M. C. *et al.* (1978). Neogene Basin formation in relation to plate-tectonic evolution of San Andreas fault system, California, *Am. Assoc. Pet. Geol. Bull.* **62**, 344–372.

att, H., Middleton, G., and Murray, R. (1972). "Origin of Sedimentary Rocks." Prentice-Hall, Englewood Cliffs, New Jersey.

Bloomer, R. R. (1977). Depositional environments of a reservoir sandstone in west-central Texas, *Am. Assoc. Pet. Geol. Bull.* **62**, 344–372.

Boggs, S. (1974). Sand-wave fields in Taiwan Strait, *Geology* **2**, 251–253.

Bond, G. C. (1973). A Late Paleozoic volcanic arc in the eastern Alaska Range, Alaska, *J. Geol.* **81**, 557–575.

Bonnefous, J. (1972). Geology of the quartzitic "Gargaf Formation" in the Sirte Basin, Libya, *Centre Rech. Pau-SNPA Bull.* **6**, 225–261.

Borden, J. F., and Brant, R. A. (1941). East Tuskega Pool, Creek County, Oklahoma, *In* "Stratigraphic Type Oil Fields," *Am. Assoc. Pet. Geol. Spec. Publ.* 436–455.

Bott, M. H. P. (1971). Evolution of young continental margins and formation of shelf basins, *Tectonophysics* **11**, 319–327.

Bott, M. H. P. (ed.) (1976). Sedimentary basins of continental margins and cratons, *Tectonophysics (Spec. Issue)* **36**.

Botvinkina, L. N., and Yablokov, V. S. (1964). Specific features of deltaic deposits in coal-bearing and cupriferous formations, *In* "Deltaic and Shallow Marine Deposits," pp. 39–47. Elsevier, Amsterdam.

Boucot, A. (1975). "Evolution and Extinction Rate Control." Elsevier, Amsterdam.

Bouma, A. H. (1968). Distribution of minor structures in Gulf of Mexico sediments, *Trans. Gulf Coast Assoc. Geol. Soc.* **18**, 26–33.

Bouma, A. H., and Brouwer, A. (eds.) (1964). Turbidites, *In* "Developments in Sedimentology," Vol. 3. Elsevier, Amsterdam.

Braunstein J. (ed.) (1973). *Symp. Underground Waste Management Artificial Recharge* **1**, 2. Am. Assoc. Pet. Geol., U.S. Geol. Surv., Int. Assoc. of Hyrdrol. Sci., George Banta Press.

Braunstein, J. (ed.) (1974). "Facies and the Reconstruction of Environments." Am. Assoc. Pet. Geol. Reprint Ser. 10.

Braunstein J. (ed.) (1976). "North American Oil and Gas Fields." Am. Assoc. Pet. Geol. Mem. 24.

Bray, E. E., and Evans, E. D. (1965). Hydrocarbons in non-reservoir rock source beds, *Am. Assoc. Pet. Geol. Bull.* **49**, 248–257.

Brenner, R. L., and Davies, D. K. (1974). Oxfordian sedimentation in western interior United States, *Am. Assoc. Pet. Geol. Bull.* **58**, 407–428.

Bricker, O. P. (ed.) (1971). "Carbonate Cements." Johns Hopkins Press, Baltimore, Maryland.

Brink, A. H. (1974). Petroleum geology of Gabon Basin, *Am. Assoc. Pet. Geol. Bull.* **58**, 216–235.

Brognon, G. P., and Verrier, G. R. (1966). Oil and geology in Cuanza Basin of Angola, *Am. Assoc. Pet. Geol. Bull.* **50**, 108–158.

Broin, T. L., and Schenck, D. R. (1977). Computer approach to vertical-variability mapping (abstract), *Am. Assoc. Pet. Geol. Bull.* **61**, 771.

Brooks, B. T. (1938). The chemical and geochemical aspects of the origin of petroleum, *In* "The Science of Petroleum," pp. 46–53. Oxford Univ. Press., London and New York.

Brown, D. A. Campbell, K. S. W., and Crook, K. A. W. (1968). "The Geological Evolution of Australia and New Zealand." Pergamon, Oxford.

Brown, D. L., and Silvey, W. D. (1973). Underground storage and retrieval of fresh water from brackish-water aquifer (abstract), *Am. Assoc. Pet. Geol. Bull.* **57**, 1592.

Brown, L. F., McGowen, J. H., Seals, M. J., Waller, T. H., and Ray, . compaction in development of geometry of superposed elongate (abstract), *Am. Assoc. Pet. Geol. Bull.* **51**, 455–456.

Bruce, C. H. (1973). Pressured shale and related sediment deformation: n. development of regional contemporaneous faults, *Am. Assoc. Pet. Geol. Bu.* 886.

Bryan, G. M., and Simpson, E. S. W. (1971). Seismic refraction measurements continental shelf between the Orange River and Cape Town, *In* "The Geology c East Atlantic Continental Margin" (F. M. Delany, ed.), Vol. 4, Africa, pp. 187–1. Inst. Geol. Sci. Rpt. 70/16.

Bryant, W. R., Meyerhoff, A. A., Brown, N. K., Furrer, M. A., Pyle, T. E., and Antoine, J. W. (1969). Escarpments, reef trends, and diapiric structures, eastern Gulf of Mexico, *Am. Assoc. Pet. Geol. Bull.* **53**, 2506–2542.

Bull, W. B. (1972). Recognition of alluvial-fan deposits in the stratigraphic record, *In* "Recognition of Ancient Sedimentary Environments" (J. K. Rigby and W. K. Hamblin, eds.), pp. 63–83. Soc. Econ. Paleontol. Mineral. Spec. Publ. 16.

Bunce, E. T., Phillips, J. D., and Chase, R. L. (1974). Geophysical study of Antilles Outer Ridge, Puerto Rico Trench, and northeast margin of Caribbean Sea, *Am. Assoc. Pet. Geol. Bull.* **58**, 106–123.

Burch, S. H. (1968). Tectonic emplacement of the Burro Mountain ultramafic body, Santa Lucia Range, California, *Geol. Soc. Am. Bull.* **79**, 527–544.

Burchfield, B. C., and Davis, G. A. (1975). Nature and controls of Cordilleran orogenesis, western United States: extensions of an earlier thesis, *Am. J. Sci.* **275-A**, 397–436.

Burke, D. B. (1967). Aerial photograph survey of Dixie Valley, Nevada, *In* "Geophysical Study of Dixie Valley Region, Nevada." U.S. Air Force Cambridge Research Labs. Sci. Rep. AFCRL-66-848, Part IV.

Busch, D. A. (1959). Prospecting for stratigraphic traps, *Am. Assoc. Pet. Geol. Bull.* **43**, 2829–2843.

Busch, D. A. (1971). Genetic units in delta prospecting, *Am. Assoc. Pet. Geol. Bull.* **55**, 1137–1154.

Busch, D. A. (1974). "Stratigraphic Traps in Sandstone—Exploration Techniques." Am. Assoc. Pet. Geol. Mem. 21.

Busch, D. A. (1975). Influence of growth faulting on sedimentation and prospect evaluation, *Am. Assoc. Pet. Geol. Bull.* **59**, 217–230.

Busch, D. A. (1977). Review: Geomorphology of oil and gas fields in sandstone bodies (C. E. B. Conybeare), *Am. Assoc. Pet. Geol. Bull.* **61**, 3, 450–453.

Campana, B. (1966). Stratigraphic-structural-palaeoclimatic controls of the newly discovered iron ore deposits of Western Australia, *Mineral. Deposita* **1**, 53–59.

Carlson, C. E. (1968). Triassic-Jurassic of Alberta, Saskatchewan, Manitoba, Montana, and North Dakota, *Am. Assoc. Pet. Geol. Bull.* **52**, 1969–1983.

Carlson, C. G., and Anderson, S. B. (1965). Sedimentary and tectonic history of North Dakota part of Williston Basin, *Am. Assoc. Pet. Geol. Bull.* **49**, 1833–1846.

Carozzi, A. V. (1960). "Microscope Sedimentary Petrology." Wiley, New York.

Carozzi, A. V. (ed.) (1974). Sedimentary rocks: concepts and history, "Benchmark Papers in Geology," Vol. 12. Dowden, Hutchinson, and Ross, Stroudsburg, Pennsylvania.

Carter, D. J., Audley-Charles, M. G., and Barber, A. J. (1976). Stratigraphic analysis of island arc-continental margin collision in eastern Indonesia, *J. Geol. Soc.* **132**, 179–198.

Carter, E. K., Brooks, J. H., and Walker, K. R. (1961). "The Precambrian Mineral Belt of North-Western Queensland." Bur. Mineral Res., Canberra, Bull. 51.

Carver, R. E. (1968). Differential compaction as a cause of regional contemporaneous faults, *Am. Assoc. Pet. Geol. Bull.* **52**, 414–419.

Castaño, J. R., and Sparks, D. M. (1974). Interpretation of vitrinite reflectance measurements in sedimentary rocks, and determination of burial history using vitrinite reflectance and authigenic minerals, *Geol. Soc. Am. Spec. Pap.* **153**, 31–52.

Cebull, S. E., Shurbet, D. H., Keller, G. P., and Russell, L. R. (1976). Possible role of transform faults in the development of apparent offsets in the Ouachite–Southern Appalachian tectonic belt, *J. Geol.* **84**, 107–114.

Chase, T. E., Menard, H. W., and Mammerickx, J. (1971). Topography of the North Pacific. La Jolle Inst. Marine Res. Rep. TR-17.

Chilingarian, G. V., and Wolf, K. H. (1975). Compaction of coarse-grained sediments, *In* "Developments in Sedimentology," Vol. 18A. Elsevier, Amsterdam.

Choquette, P. W., and Pray, L. C. (1970). Geologic nomenclature and classification of porosity in sedimentary carbonames, *Am. Assoc. Pet. Geol. Bull.* **54**, 207–250.

Chow, T. J. (1968). Lead isotopes of the Red Sea region, *Earth Planet Sci. Lett.* **5**, 143–147.

Christie, R. L. (1967). "Reconnaissance of the Surficial Geology of Northeastern Ellesmere Island, Arctic Archipelago." Geol. Surv. Can. Bull. 138.

Chuber, S. (1962). Late Mesozoic stratigraphy of the Sacramento Valley, *San Joaquin Geol. Soc. Trans.* **1**, 3–16.

Claerbout, J. F. (1976). "Fundamentals of Geophysical Data Processing." McGraw-Hill, New York.

Clark, R. H., and Rouse, J. T. (1971). A closed system for generation and entrapment of hydrocarbons in Cenozoic deltas, Louisiana Gulf Coast, *Am. Assoc. Pet. Geol. Bull.* **55**, 1170–1178.

Clary, T. A., Moberley, C. M., and Moulton, G. E. (1963). Geological setting of an anomalous ore deposit in the Section 30 mine, Ambrosia Lake area, *In* "Geology and Technology of the Grants Region," pp. 72–79. New Mexico Bur. Mines Mineral Res. Mem. 15.

Claypool, G. E., Love, A. H., and Maughan, E. K. (1978). Organic geochemistry, incipient metamorphism, and oil generation in black shale members of Phosphoria Formation, western interior United States, *Am. Assoc. Pet. Geol. Bull.* **62**, 98–120.

Cobbing, E. J. (1976). The geosynclinal pair at the continental margin of Peru, *Tectonphysics* **36**, 157–165.

Cohen, Z. (1976). Early Cretaceous buried canyon: influence on accumulation of hydrocarbons in Helez Field, Isral, *Am. Assoc. Pet. Geol. Bull.* **60**, 108–114.

Combs, E. J. (ed.) (1967). Geology of selected sedimentary basins: introduction, *Am. Assoc. Pet. Geol. Bull.* **51**, 648–650.

Conant, L. C. and Goudarzi, G. H. (1967). Stratigraphic and tectonic framework of Libya. *Am. Assoc. Pet. Geol. Bull.* **51**, 719–730.

Condon, M. A. (1954). "Progress Report on the Stratigraphy and Structure of the Carnarvon Basin, Western Australia." Bur. Mineral Res., Canberra, Rep. 15.

Conkin, J. E., and Conkin, B. M. (1975). The Devonian–Mississippian and Kinderhookian–Osagean boundaries in the east-central United States are paracontinuities, *Univ. Lousville, Stud. in Paleon. Strat.* **4**.

Conolly, J. R., and Ferm, J. C. (1971). Permo-Triassic sedimentation patterns, Sydney Basin, Australia, *Am. Assoc. Pet. Geol. Bull.* **55**, 2018–2032.

Conybeare, C. E. B. (1949). Genesis and structure of the Athabasca Series, Goldfields, Saskatchewan, *Northwest Sci.* **23**, 165–174.

Conybeare, C. E. B. (1965). Hydrocarbon-generation potential and hydrocarbon-yield capacity of sedimentary basins, *Bull. Can. Pet. Geol.* **13**, 509–528.

Conybeare, C. E. B. (1966). Origin of Athabasca oil sands: a review, *Bull. Can. Pet. Geol.* **14**, 145–163.

Conybeare, C. E. B. (1967). Influence of compaction on stratigraphic analysis, *Bull. Can. Pet. Geol.* **15**, 331–345.

Conybeare, C. E. B. (1970). Solubility and mobility of petroleum under hydrodynamic conditions, Surat Basin, Queensland, *J. Geol. Soc. Aust.* **16**, 667–681.

Conybeare, C. E. B. (1972). Petroleum accumulations in ancient river sands: a paleogeographic view, *Trans. Int. Geol. Congr., 24th* **5**, 59–74.

Conybeare, C. E. B. (1976). Geomorphology of oil and gas fields in sandstone bodies, *In* "Developments in Petroleum Science," Vol. 4. Elsevier, Amsterdam.

Conybeare, C. E. B., and Crook, K. A. W. (1968). "Manual of Sedimentary Structures." Bur. Mineral Res., Canberra, Bull. 102.

Conybeare, C. E. B., and Jessop, R. G. C. (1972). Exploration for oil-bearing sand trends in the Fly River area, western Papua, *Aust. Pet. Exp. Assoc. J.* **12**, 69–73.

Coogan, A. H., Bebout, D. G., and Maggio, C. (1972). Depositional environments and geologic history of Golden Lane and Poza Rica trend, Mexico, an alternative view, *Am. Assoc. Pet. Geol. Bull.* **56**, 1419–1447.

Cook, H. E. (ed.) (1977). "Deep-water Carbonate Environments." Soc. Econ. Paleontologists Mineralogists Spec. Publ. 25.

Cook, P. J. (1972). Petrology and geochemistry of the phosphate deposits of northwest Queensland, Australia, *Econ. Geol.* **67**, 1193–1213.

Cook, T. D. (ed.) (1972). "Underground Waste Management and Environmental Implications." Am. Assoc. Pet. Geol. Mem. 18.

Cook, K. L., and Hardman, E. (1967). Regional gravity survey of the Hurricane Fault area and Iron Springs district, Utah, *Geol. Soc. Am. Bull.* **78**, 1063–1076.

Corliss, J. B., and Ballard, R. D. (1977). Oases of life in the cold abyss, *Nat. Geog. Mag.* **152**, 441–453.

Cotter, E. (1971). Paleoflow characteristics of a Late Cretaceous river in Utah from analysis of sedimentary structures in the Ferron Sandstone, *J. Sediment Pet.* **41**, 129–138.

Coustau, H., Rumeau, J., Sourisse, C., Chiarelli, A., and Tison, J. (1975). Classification hydrodynamique des bassins sedimentaires utilisation combinee avec d'autres methodes pour rationaliser l'exploration dans des bassins non-productifs, *Proc. World Pet. Congr., 9th* **2**, 105–119.

Crawford, A. R., and Campbell, K. S. W. (1973). Large-scale horizontal displacement within Australo-Antarctica in the Ordovician, *Nature (London) Phys. Sci.* **241**, 11–14.

Crimes, T. P., and Harper, J. C. (eds.) (1970). "Trace Fossils." Seel House Press, Liverpool.

Crosby, E. J. (1972). Classification of sedimentary environments, *In* "Recognition of Sedimentary Environments" (J. K. Rigby and W. K. Hamblin, eds.). pp. 4–11. Soc. Econ. Paleontolog. Mineral. Spec. Publ. 16.

Crowder, D. F., McKee, E. H., Ross, D. C., and Krauskopf, K. B. (1973). Granitic rocks of the White Mountains area, California–Nevada: age and regional significance. *Geol. Soc. Am. Bull.* **84**, 285–296.

Cross, A. T., Maxfield, E. B., Cotter, E., and Cross, C. C. (1975). Field guide and road log to the Western Book Cliffs, Castle Valley, and parts of the Wasatch Plateau, *Brigham Young Univ. Geol. Stud.* **22**.

Curray, J. R. (1964). Transgressions and regressions, *In* "Papers in Marine Geology" (R. L. Miller, ed.), pp. 175–203. Macmillan, New York.

Curray, J. R., and Moore, D. G. (1971). Growth of the Bengal deep-sea fan and denudation in the Himalayas, *Geol. Soc. Am. Bull.* **82**, 563–572.

Curray, J. R., Dickinson, W. R., Dow, W. G., Emery, K. O., Seely, D. R., Vail, P. R., and Yarborough, H. (1977). "Geology of Continental Margins." Am. Assoc. Pet. Geol. Cont. Educ. Cat. 879.

Currie, J. B. (1977). Significant geologic processes in development of fracture porosity, *Am. Assoc. Pet. Geol. Bull.* **61**, 1086–1089.

Currie, J. B., and Nwachukwu, S. O. (1974). Evidence of incipient fracture porosity in reservoir rocks at depth, *Bull. Can. Pet. Geol.* **22**, 42–58.

Curry, W. H. (1977). Teapot Dome—past, present, and future, *Am. Assoc. Pet. Geol. Bull.* **61**, 671–697.

Dale, R. H. (1974). Simulation model of Honolulu basal aquifer (abstract), *Am. Assoc. Pet. Geol. Bull.* **58**, 1435.

D'Anglejan, B. J. (1967). Origin of marine phosphorites off Baja California, *Marine Geol.* **5**, 15–44.

Dahlstrom, C. D. A. (1969). Balanced cross sections, *Can. J. Earth Sci.* **6**, 743–757.

Dailly, G. C. (1976). Pendulum effect and Niger Delta prolific belt, *Am. Assoc. Pet. Geol. Bull.* **60**, 1543–1575.

Dalziel, I. W. D., and Dott, R. H. (1970). "Geology of the Baraboo District, Wisconsin." Wisconsin Geol. Nat. Hist. Surv. Inf. Circ. 14.

Davies, G. R. (1975). Devonian Reef Complexes of Canada, 1, Rainbow, Swan Hills. Can. Soc. Pet. Geol. Reprint Series 1.

Davies, G. R. (1977). Former magnesian calcite and aragonite submarine cements in upper Paleozoic reefs of the Canadian Arctic: a summary, *Geology* **5**, 11–15.

Davies, D. K., Ethridge, F. G., and Berg, R. R. (1971). Recognition of barrier environments, *Am. Assoc. Pet. Geol. Bull.* **55**, 550–565.

Davies, D. K., and Ethridge, F. G. (1975). Sandstone composition and depositional environment, *Am. Assoc. Pet. Geol. Bull.* **59**, 239–264.

Davis, J. C. (1975). Training geologists in geomathematics and use of computers, *Am. Assoc. Pet. Geol. Bull.* **59**, 159–160.

Davies, P. J. (1975). Shallow seismic structure of the continental shelf, southeast Australia, *J. Geol. Soc. Aust.* **22**, 345–359.

Davis, R. A., and Fox, W. T. (1972). Coastal processes and nearshore sand bars, *J. Sediment. Pet.* **42**, 401–412.

Day, R. W. (1969). The Lower Cretaceous of the Great Artesian Basin, *In* "Stratigraphy and Palaeontology" (K. S. W. Campbell, ed.), pp. 140–173. Aust. Nat. Univ. Press.

De Almeida, F. F. M. (1953). Botucatú, a Triassic desert of South America, *Trans. Intl. Geol. Congr., 19th* **7**, 9–24.

Degens, E. T. (1965). "Geochemistry of Sediments: A Brief Survey." Prentice-Hall, Englewood Cliffs, New Jersey.

de Jong, K. A. (1974). Melange (Olistostrome) near Lago Titicaca, Peru, *Am. Assoc. Pet. Geol. Bull.* **58**, 729–741.

De Laguna, W. (1959). What is safe waste disposal? *At. Sci. Bull.* **15**, 34–43.

Demenitskaya, R. M. (1967). "Kora i mantiya Zemli." Izdat. Nedra, Moscow.

De Mille, G., Shouldice, J. R., and Nelson, H. W. (1964). Collapse structures related to evaporites of the Prairie Formation, Saskatchewan, *Geol. Soc. Am. Bull.* **75**, 307–316.

Denholm, L. S. (1977). Investigation of the ferruginous manganese deposits at Ripon Hills, Pilbara manganese province, Western Australia, *Aust. Inst. Min. Metall.* **264**, 9–17.

Denson, N. M., and Gill, J. R. (1956). Uranium-bearing lignite and its relation to volcanic tuffs in eastern Montana and the Dakotas, *Proc. Int. Conf. Peaceful Uses Energy, Geneva, 1955* **6**, 464–467.

Derr, M. (1974). Sedimentary structure and depositional environment of paleochannels in the Jurassic Morrison Formation near Green River, Utah, *Brigham Young Univ. Geol. Stud.* **21**, 3–39.

De Villiers, J., and Simpson, E. S. W. (1974). Late Precambrian tectonic patterns in southwestern Africa, *Univ. Cape Town Precambrian Res. Unit Bull.* **15**, 141–152.

Devine, S. B., Ferrell, R. E., and Billings, G. K. (1973). Mineral distribution patterns, deep Gulf of Mexico, *Am. Assoc. Pet. Geol. Bull.* **57**, 28–41.

Dickas, A. B., and Payne, J. L. (1967). Upper Paleocene buried channel in Sacramento Valley, California, *Am. Assoc. Pet. Geol. Bull.* **51**, 873–882.

Dickey, P. A. (1968). Contemporary nonmarine sedimentation in Soviet Central Asia, *Am. Assoc. Pet. Geol. Bull.* **52**, 2396–2421.

Dickey, P. A. (1972). Geology of oil fields of West Siberian Lowlands, *Am. Assoc. Pet. Geol. Bull.* **56**, 454–471.

Dickinson, K. A., Berryhill, H. L., and Holmes, C. W. (1972). Criteria for recognizing ancient barrier coastlines, *In* "Recognition of Ancient Sedimentary Environments," pp. 192–214. Soc. Econ. Paleontol. Mineral. Spec. Publ. 16.

Dickinson, W. R. (ed.) (1972). "Tectonics and Sedimentation." Soc. Econ. Paleontol. Mineral. Spec. Publ. 22.

Dietz, R. S., and Holden, J. C. (1966). Deep-sea deposits in but not on the continents, *Am. Assoc. Pet. Geol. Bull.* **50**, 351–362.

Dietz, R. S., and Holden, J. C. (1966). Miogeoclines (miogeosynclines) in space and time, *J. Geol.* **74**, 566–583.

Dimitrijević, M. D., and Dimitrijević, M. N. (1973). Olistostrome mélange in the Yugoslavian Dinarides and Late Mesozoic plate tectonics, *J. Geol.* **81**, 328–340.

Dingle, R. V., and Scrutton, R. A. (1974). Continental breakup and the development of post-Paleozoic sedimentary basins around southern Africa, *Geol. Soc. Am. Bull.* **85**, 1467–1474.

Dobrin, M. B. (1976). "Introduction to Geophysical Prospecting," 3rd ed. McGraw-Hill, New York.

Dodson, C. R., and Standing, M. B. (1944). Pressure-volume-temperature and solubility relations for natural gas-water mixtures, *In* "Drilling and Petroleum Practices," pp. 173–179. Am. Petrol. Inst. Publ.

Dolly, E. D., and Busch, D. A. (1972). Stratigraphic, structural, and geomorphologic factors controlling oil accumulation in Upper Cambrian strata of central Ohio, *Am. Assoc. Pet. Geol. Bull.* **56**, 2335–2368.

Donaldson, E. C. (1971). Injection wells and operations today (abstract), *Am. Assoc. Pet. Geol. Bull.* **5**, 2082.

Dössegger, Von R., and Müller, W. H. (1976). Die sedimentserien der Engadiner Dolomiten und ihre lithostratigraphische gliederung, *Ecol. Geol. Helv.* **69**, 229–238.

Dott, R. H. (1963). Dynamics of subaqueous gravity depositional processes, *Am. Assoc. Pet. Geol. Bull.* **47**, 104–128.

Dott, R. H. (1974). Cambrian tropical storm waves in Wisconsin, *Geology* **2**, 243–246.

Dott, R. H., and Shaver, R. H. (eds.) (1974). "Modern and Ancient Geosynclinal Sedimentation." Soc. Econ. Paleontol. Mineral. Spec. Publ. 19.

Drake, C. L., Ewing, W. M., and Sutton, G. H. (1959). Continental margins and geosynclines—the east coast of North America north of Cape Hatteras, *In* "Physics and Chemistry of the Earth," Vol. 3, pp. 110–198. Pergamon, Oxford.

Drescher, W. V. (1965). Hydrology of deep-well disposal of radioactive liquid wastes, *In* "Fluids in Subsurface Environments," pp. 399–406. Am. Assoc. Pet. Geol. Mem. 4.

Dunbar, C. O., and Rodgers, J. (1957). "Principles of Stratigraphy." Wiley, New York.

Dunham, K. C. (1964). Neptunist concepts in ore genesis, *Econ. Geol.* **59**, 1–21.

Dzulynski, S., and Walton, E. K. (1965). Sedimentary features of flysch and greywackes, *Dev. Sedimen.* **7**.

Eargle, D. H., Dickinson, K. A., and Davis, B. O. (1975). South Texas uranium deposits, *Am. Assoc. Pet. Geol. Bull.* **59**, 766–779.

Edgerley, D. W., and Crist, R. P. (1974). Salt and diapiric anomalies in the southeast Bonaparte Gulf Basin, *Aust. Pet. Exp. Assoc. J.* **14**, 85–94.

Ehrlich, G. G. (1971). Role of biota in underground waste injection and storage (abstract), *Am. Assoc. Pet. Geol. Bull.* **5**, 2083.

Elam. J. G., and Chuber, S. (eds.) (1969). "Cyclic Sedimentation in the Permian Basin." West Texas Geol. Soc. Publ.

Elston, D. P., and Shoemaker, E. M. (1963). Salt anticlines of the Paradox Basin, Colorado and Utah, *In Symp. Salt,* pp. 131–146. Northern Ohio Geol. Soc. Publ.

Embley, R. W. (1976). New evidence for occurrence of debris flow deposits in the deep sea, *Geology* **4**, 371–374.

Embry, A. F., and Klovan, J. E. (1971). A Late Devonian reef tract on northeastern Banks Island, N.W.T., *Am. Assoc. Pet. Geol. Bull.* **19**, 730–781.

Emery, K. O., and Bray, E. E. (1962). Radiocarbon dating of California basin sediments, *Am. Assoc. Pet. Geol. Bull.* **46**, 1839–1856.

Erdman, J. G. (1961). Some chemical aspects of petroleum genesis as related to the problem of source bed recognition, *Geochim. Cosmochim. Acta* **22**, 16–36.

Evamy, B. D., Harembource, J., Kamerling, P., Knaap, W. A., Molloy, F. A., and Rowlands, P. H. (1978). Hydrocarbon habitat of Tertiary Niger Delta, *Am. Assoc. Pet. Geol. Bull.* **62**, 1–39.

Evans, G. (1970). Coastal and nearshore sedimentation: a comparison of clastic and carbonate deposition, *Proc. Geol. Assoc.* **81**, 493–508.

Evans, R. (1978). Origin and significance of evaporites in basins around Atlantic margin, *Am. Assoc. Pet. Geol. Bull.* **62**, 223–234.

Evans, W. E. (1970). Imbricate linear sandstone bodies in Viking Formation in Dodsland-Hoosier area of southwestern Saskatchewan, Canada, *Am. Assoc. Pet. Geol. Bull.* **54**, 469–486.

Exon, N. F., and Willcox, J. B. (1978). Geology and petroleum potential of Exmouth Plateau area off Western Australia, *Am. Assoc. Pet. Geol. Bull.* **62**, 40–72.

Feather, C. E., and Koen, G. M. (1975). The mineralogy of the Witwatersrand reefs, *Minerals Sci. Eng.* **7**, 189–224.

Feden, R. H., Fleming, H. S., and Perry, R. K. (1975). The Mid-Atlantic Ridge at 33°N: the Hayes Fracture Zone, *Earth Planet. Sci. Lett.* **26**, 292–298.

Fisher, A. G., and Heezen, B. C. (1971). "Initial Reports of Deep Sea Drilling Project," Vol. 6, U. S. Govt. Printing Office, Washington, D.C.

Fisher, G. W., Pettijohn, F. J., Reed, J. C., and Weaver, K. N. (eds.) (1970). "Studies of Appalachian Geology: Central and Southern." Wiley (Interscience), New York.

Fisher, J. H. (ed.) (1977). Reefs and evaporites: concepts and depositional models, *Am. Assoc. Pet. Geol. Stud. Geol.* **5**.

Fisher, R. L. (1961). Middle America Trench: topography and structure, *Geol. Soc. Am. Bull.* **72**, 703–720.

Fisher, R. L., and Engel, C. G. (1969). Ultramafic and basaltic rocks dredged from the nearshore flank of the Tonga Trench, *Geol. Soc. Am. Bull.* **80**, 1373–1378.

Fisher, W. L., and Brown, L. F. (1972). "Clastic Depositional Systems—A Genetic Approach to Facies Analysis: Annotated Outline and Bibliography." Bur. Econ. Geol. Publ., Univ. Texas, Austin, Texas.

Fisher, W. L., and McGowen, J. H. (1969). Depositional systems in Wilcox Group (Eocene) of Texas and their relation to occurrence of oil and gas, *Am. Assoc. Pet. Geol. Bull.* **53**, 30–54.

Fisher, W. L., and Rodda, P. V. (1969). Edwards Formation (Lower Cretaceous), Texas: dolomitization in a carbonate platform system, *Am. Assoc. Pet. Geol. Bull.* **53**, 55–72.

Fisk, H. N., McFarlan, E., Kolb, C. R., and Wilbert, L. J. (1954). Sedimentary framework of the modern Mississippi delta, *J. Sediment. Pet.* **24**, 76–99.

Flawn, P. T. (1965). Basement—not the bottom but the beginning, *Am. Assoc. Pet. Geol. Bull.* **49**, 883–886.

Flawn, P. T., Goldstein, A., King, P. B., and Weaver, C. E. (eds.) (1961). "The Ouachita System." Univ. Texas. Publ. 6120.

Flowers, B. S. (1976). Overview of exploration geophysics—recent breakthroughs and challenging new problems, *Am. Assoc. Pet. Geol. Bull.* **60**, 3–11.

Folk, R. L., and Land, L. S. (1975). Mg/Ca ratio and salinity: two controls over crystallization of dolomite, *Am. Assoc. Pet. Geol. Bull.* **59**, 60–68.

Forgotson, J. M. (1957). Nature, usage and definition of marker-defined vertically segregated rock units, *Am. Assoc. Pet. Geol. Bull.* **41**, 2108–2113.

Forgotson, J. M. (1960). Review and classification of quantitative mapping techniques, *Am. Assoc. Pet. Geol. Bull.* **44**, 83–100.

Forgotson, J. M., and Inglehart, C. F. (1967). Current uses of computers by exploration geologists, *Am. Assoc. Pet. Geol. Bull.* **51**, 1202–1224.

Forgotson, J. M., and Stark, P. H. (1972). Well-data files and the computer, a case history from northern Rocky Mountains, *Am. Assoc. Pet. Geol. Bull.* **56**, 1114–1127.

Forgotson, J. M., Statler, A. T., and David, M. (1966). Influence of regional tectonics and local structure on deposition of Morrow Formation in western Anadarko Basin. *Am. Assoc. Pet. Geol.* **50**, 518–532.

Foster, M. W. (1974). Recent Antarctic and sub-Antarctic brachiopods. *Am. Geophys. Union,* Antarctic Research Ser. **21**, 189.

Fox, P. J., Schreiber, E., and Heezen, B. C. (1971). The geology of the Caribbean crust: Tertiary sediments, granitic and basic rocks from Aves Ridge, *Tectonophysics* **12**, 89–109.

Franklin, E. H., and Clifton, B. B. (1971). Halibut field, southeastern Australia, *Am. Assoc. Pet. Geol. Bull.* **55**, 1262–1279.

Fraser, G. S. (1976). Sedimentology of a Middle Ordovician quartz arenite-carbonate transition in the upper Mississippi Valley, *Geol. Soc. Am. Bull.* **87**, 833–845.

Frazer, F. W., and Belcher, C. B. (1975). Mineralogical studies of the Groote Eylandt manganese ore deposits, *Aust. Inst. Min. Metall. Proc.* **254**, 29–36.

Freeman, E. W. (1902). The awful doom of St. Pierre, *Pearson's Mag.* 313–325.

Frey, R. (ed.) (1975). "The Study of Trace Fossils." Springer-Verlag, Berlin and New York.

Friedman, G. M. (1961). Distinction between dune, beach and river sands from their textural characteristics, *J. Sediment. Pet.* **31**, 514–529.

Füchtbauer, H., and Müller, G. (1970). "Sedimente und Sedimentgesteine." Schweizerbartsche Verlagsbuchhandlung, Stuttgart.

Fuller, J. G. C. M., and Porter, J. W. (1969). Evaporite formations with petroleum reservoirs in Devonian and Mississippian of Alberta Saskatchewan, and North Dakota, *Am. Assoc. Pet. Geol. Bull.* **53**, 909–926.

Fuller, M. L. (1920). Carbon ratios in Carboniferous coals of Oklahoma, and their relation to petroleum, *Econ. Geol.* **15**, 226–227.

Gabelman, J. W. (1976). Expectations from uranium exploration, *Am. Assoc. Pet. Geol. Bull.* **60**, 1993–2004.

Gagliano, S. M., and McIntire, W. G. (1968). "Delta Components in Recent and Ancient Deltaic Sedimentation: A Comparison." Louisiana State Univ. Coastal Studies Inst. Publ.

Galehouse, J. S. (1967). Provenance and paleocurrents of the Paso Robles Formation, California, *Geol. Soc. Am. Bull.* **78**, 951–978.

Galley, J. E. (1958). "Oil and Geology in the Permian Basin of Texas and New Mexico." Am. Assoc. Petrol. Geol. Spec. Publ.

Galley, J. E. (ed) (1968). "Subsurface Disposal in Geologic Basins—A Study of Reservoir Strata." Am. Assoc. Pet. Geol. Mem. 10.

Galloway, W. E., and Brown, L. F. (1973). Depositional systems and shelf-slope relations on cratonic basin margin, uppermost Pennsylvanian of north-central Texas, *Am. Assoc. Pet. Geol. Bull.* **57**, 1185–1218.

Garland, G. D. (1971). "Introduction to Geophysics." Saunders, Philadelphia, Pennsylvania.

Garlick, W. G. (1953). Reflections on prospecting and ore genesis in Northern Rhodesia, *Trans. Inst. Min. Metall. (London)* **63**, 9–20.

Garlick, W. G. (1961). The syngenetic theory, *In* "The geology of the Northern Rhodesian Copperbelt" (F. Mendelsohn, ed.), pp. 146–165. Macdonald, London.

Garlick, W. G., and Fleischer, A. (1972). Sedimentary environment of Zambian copper deposition, *Geol. Mijnbouw* **51**, 277–298.

Garrels, R. M., and MacKenzie, F. T. (1971). "Evolution of Sedimentary Rocks." Norton, New York.

Gaskell, T. F., Hill, M. N., and Swallow, J. C. (1959). Seismic measurements made by H.M.S. Challenger in the Atlantic, Pacific and Indian Oceans and in the Mediterranean Sea, *R. Soc. London Phil. Trans. Ser. A* **251**, 23–84.

Gatehouse, C. G. (1972). Formations of the Gidgealpa Group in the Cooper Basin, *Aust. Oil Gas Rev.* **18**, 10–15.

Gehman, H. M. (1962). Organic matter in limestone, *Geochim. Cosmochim. Acta* **26**, 885–897.

Gibbs, R. J. (1976). Amazon River sediment transport in the Atlantic Ocean, *Geology* **4**, 45–48.

Gilluly, J., Reed, J. C., and Cady, W. (1970). Sedimentary volumes and their significance, *Geol. Soc. Am. Bull.* **81**, 353–375.

Ginsburg, R. N. (ed.) (1973). "Evolving Concepts in Sedimentology." Johns Hopkins Univ. Press, Baltimore, Maryland.

Glasby, G. P. (ed.) (1977). "Marine Manganese Deposits." Elsevier, Amsterdam.

Glennie, K. W. (1970). Desert sedimentary environments, *Dev. Sediment.* **14**.

Glennie, K. W. (1972). Permian Rotliegendes of northwest Europe interpreted in light of modern desert sedimentation studies, *Am. Assoc. Pet. Geol. Bull.* **56**, 1048–1071.

Goldbery, R., and Holland, W. N. (1973). Stratigraphy and sedimentation of redbed facies in Narrabeen Group of Sydney Basin, Australia, *Am. Assoc. Pet. Geol. Bull.* **57**, 1314–1334.

Goolsby, D. A. (1971). Geochemical effects and movement of injected industrial waste in a limestone aquifer (abstract), *Am. Assoc. Pet. Geol. Bull.* **55**, 2084.

Gostin, V. A., and Herbert, C. (1973). Stratigraphy of the Upper Carboniferous and Lower Permian sequence, southern Sydney Basin, *J. Geol. Soc. Aust.* **20**, 49–70.

Grabau, A. W. (1960). "Principles of Stratigraphy." Dover, New York.

Gradstein, F. M., and Van Gelder, A. (1971). Prograding clastic fans and transition from a fluviatile to a marine environment in Neogene deposits of eastern Crete, *Geol. Mijnbouw* **50**, 383–392.

Grender, G. C., Rapoport, L. A., and Segers, R. G. (1974). Experiment in quantitative geological modeling, *Am. Assoc. Pet. Geol. Bull.* **58**, 488–498.

Griffin, G. M. (1962). Regional clay-mineral facies—products of weathering intensity and current distribution in the northeastern Gulf of Mexico, *Geol. Soc. Am. Bull.* **73**, 737–768.

Griffith, B. R., and Hodgson, E. A. (1971). Offshore Gippsland Basin fields, *Aust. Pet. Exp. Assoc. J.* **11**, 85–89.

Griffiths, J. C. (1967). "Scientific Method in Analysis of Sediments." McGraw-Hill, New York.

Grim, R. E., and Johns, W. D. (1954). Clay-mineral investigations of sediments in the northern Gulf of Mexico, *In* "Clays and Clay Minerals," pp. 81–103. Nat. Res. Council Publ. 327.

Geol. Soc. Am. Coal Geol. Div. (1975). "Field Guide and Road Log to the Western Book Cliffs, Castle Valley, and Parts of the Wasatch Plateau." Brigham Young Univ. Geol. Stud. 22.

Gulbrandsen, R. A. (1966). "Pre-Cambrian Phosphorite in the Belt Series in Montana," pp. 199–202. U.S. Geol. Surv. Prof. Paper 550-D.

Gussow, W. C. (1953). Differential trapping of hydrocarbons, *J. Alberta Soc. Pet. Geol.* **1**, 4–5.

Gutenberg, B., and Richter, C. F. (1954). "Seismicity of the Earth and Associated Phenomena." Princeton Univ. Press, Princeton, New Jersey.

Hackett, J. P., and Bischoff, J. L. (1973). New data on the stratigraphy, extent and geologic history of the Red Sea geothermal deposits, *Econ. Geol.* **68**, 533–564.

Hageman, Ir. B. P. (1969). Development of the western part of the Netherlands during the Holocene, *Geol. Mijnbouw* **48**, 373–388.

Halbouty, M. T. (1969). Hidden trends and subtle traps in Gulf Coast, *Am. Assoc. Pet. Geol. Bull.* **53**, 3–29.

Halbouty, M. T. (1972). Rationale for deliberate pursuit of stratigraphic, unconformity, and paleogeomorphic traps, *Am. Assoc. Pet. Geol. Bull.* **56**, 537–541.

Hall, D. H., and Hajnal, Z. (1962). The gravimeter in studies of buried valleys, *Geophysics* **27**, 939–951.

Hall, R. (1976). Ophiolite emplacement and the evolution of the Taurus suture zone, southeastern Turkey, *Geol. Soc. Am. Bull.* **87**, 1078–1088.

Hallam, A., and Sellwood, B. W. (1976). Middle Mesozoic sedimentation in relation to tectonics in the British area, *J. Geol.* **84**, 301–321.

Ham, W. E. (ed.) (1962). "Classification of Carbonate Rocks: a Symposium." Am. Assoc. Pet. Geol. Mem. 1.

Hamblin, W. K. (1965). Origin of "Reverse Drag" on the downthrow side of normal faults, *Geol. Soc. Am. Bull.* **76**, 1145–1164.

Hamilton, E. L. (1973). Marine geology of the Aleutian abyssal plain, *Mar. Geol.* **14**, 295–325.

Hamilton, R. M., and Gale, A. W. (1968). Seismicity and structure of North Island, New Zealand, *J. Geophys. Res.* **73**, 3859.

Hamilton, W. (1970). The Uralides and the motion of the Russian and Siberian platforms, *Geol. Soc. Am. Bull.* **81**, 2553–2576.

Hancock, N. J., Hurst, J. M., and Fürsich, F. T. (1974). The depths inhabited by Silurian brachiopod communities, *Q. J. Geol. Soc. London* **130**, 151–156.

Hansen, A. R. (1966). Reef trends of Mississippian Ratcliffe Zone, northeast Montana and northwest North Dakota, *Am. Assoc. Pet. Geol. Bull.* **50**, 2260–2268.

Hanshaw, B. B. (1971). Natural membrane phenomena and subsurface waste emplacement (abstract), *Am. Assoc. Pet. Geol. Bull.* **5**, 2085.

Harbaugh, J. W., and Bonham-Carter, G. (1970). "Computer Simulation in Geology." Wiley (Interscience), New York.

Harmon, G. F., and Taylor, P. S. (1963). Geology and ore deposits of the Sandstone mine, southeastern Ambrosia Lake area, *In* "Geology and Technology of the Grants Uranium Region," pp. 102–107. New Mexico Bur. Mines Mineral Resour. Mem. 15.

Harms, J. C. (1966). Stratigraphic traps in a valley fill, Western Nebraska, *Am. Assoc. Pet. Geol. Bull.* **50**, 2119–2149.

Harrington, H. J. (1965). Space, things, time and events—an essay on stratigraphy, *Am. Assoc. Pet. Geol. Bull.* **49**, 1601–1646.

Harris, L. D. (1976). Thin-skinned tectonics and potential hydrocarbon traps—illustrated by a seismic profile in the valley and ridge province of Tennessee, *U.S. Geol. Surv. J. Res.* **4**, 379–386.

Harrison, J. E. (1974). Copper mineralization in miogeosynclinal clastics of the Belt Super-group, northwestern United States, *In* "Gisements Stratiformes et Provinces Cupri-fères," pp. 353–366. Société Géologique de Belgique.

Harwood, R. J. (1977). Oil and gas generation by laboratory pyrolysis of kerogen, *Am. Assoc. Pet. Geol. Bull.* **61**, 2082–2102.

Hastings, D. A. (1974). Proposed origin for Guianian diamonds: comment, *Geology* **2**, 475–476.

Hatch, F. H., and Rastall, R. H. (1965). "Petrology of the Sedimentary Rocks" 4th ed., revised by J. T. Greensmith. Murby, London.

Haun, J. D. (1962). Guidebook, *Ann. Field Conf., 17th* pp. 19–22. Wyoming Geol. Assoc. Publ.

Haun, J. D. (1963). Stratigraphy of Dakota Group and relationship to petroleum occurrence, northern Denver Basin, *In* "Guidebook," *Ann. Field Conf., 14th* pp. 119–134. Rocky Mt. Assoc. Geol. Publ.

Haun, J. D., Barlow, J. A., and Hallinger, D. E. (1970). Natural gas resources, Rocky Mountain Region, *Am. Assoc. Pet. Geol. Bull.* **54**, 1706–1718.

Hawkings, T. J., Hatlelid, W. G., Bowerman, J. N., and Coffman, R. C. (1976). Taglu gas field, Beaufort Basin, Northwest Territories, *In* "North American Oil and Gas Fields" (J. Braunstein, ed.), pp. 51–57. Am. Assoc. Pet. Geol. Mem. 24.

Hay, J. T. C. (1973). Control methods in subsurface map construction, *Am. Assoc. Pet. Geol. Bull.* **57**, 2386–2395.

Hay, W. H. (ed.) (1974). "Studies in Paleo-Oceanography," Soc. Econ. Paleontol. Mineral. Spec. Publ. 20.

Hazzard, J. C. (ed.) (1965). Rocky Mountain sedimentary basins, *Am. Assoc. Pet. Geol. Bull.* **49**, 11.

Heckel, P. H. (1977). Origin of phosphatic black shale facies in Pennsylvanian cyclothems of mid-continent North America, *Am. Assoc. Pet. Geol. Bull.* **61**, 1045–1068.

Hedberg, H. D. (1964). Geologic aspects of origin of petroleum, *Am. Assoc. Pet. Geol. Bull.* **48**, 1755–1803.

Hedberg, H. D. (1968). Significance of high-wax oils with respect to genesis of petroleum, *Am. Assoc. Pet. Geol. Bull.* **52**, 736–750.

Hedberg, H. D. (1970). Continental margins from viewpoint of the petroleum geologist, *Am. Assoc. Pet. Geol. Bull.* **54**, 3–43.

Hedberg, H. D. (ed.) (1976). "International Stratigraphic Guide: A Guide to Stratigraphic Classification, Terminology, and Procedure." Wiley, New York.

Heezen, B. C. (ed.) (1977). Influence of abyssal circulation on sedimentary accumulations in space and time, *Dev. Sediment.* **23**.

Heezen, B. C., and Hollister, C. D. (1971). "The Face of the Deep." Oxford Univ. Press, London and New York.

Heezen, B. C., Johnson, G. L. and Hollister, C. D. (1969). The Northwest Atlantic Mid-Ocean Canyon, *Can. Earth Sci.* **6**, 1441–1453.

Hesse, R. (1974). Long-distance continuity of turbidites: possible evidence for an Early Cretaceous trench—abyssal plain in the East Alps, *Geol. Soc. Am. Bull.* **85**, 859–870.

Hiltabrand, R. R., Ferrell, R. E., and Billings, G. K. (1973). Experimental diagenesis of Gulf Coast argillaceous sediment, *Am. Assoc. Pet. Geol. Bull.* **57**, 338–348.

Hind, M. C., and Helby, R. J. (1969). The Great Artesian Basin in New South Wales, in the geology of New South Wales, *J. Geol. Soc. Aust.* **16**, 481–497.

Hine, A. C., and Neumann, A. C. (1977). Shallow carbonate-bank-margin growth and structure, Little Bahama Bank, Bahamas, *Am. Assoc. Pet. Geol. Bull.* **61**, 376–406.

Hodgson, G. W., Hitchon, B., and Taguchi, K. (1964). The water and hydrocarbon cycles in the formation of oil accumulations, *In* "Recent Researches in the Fields of Hydrosphere, Atmosphere and Nuclear Geochemistry" (Sugawara Festival Vol.), pp. 217–242. Maruzen, Tokyo.

Hoffman, P. (1974). Shallow and deepwater stromatolites in Lower Proterzoic platform-to-basin facies change, Great Slave Lake, Canada, *Am. Assoc. Pet. Geol. Bull.* **58**, 856–867.

Holeman, J. N. (1968). The sediment yield of the major rivers of the world, *Water Resources Res.* **4**, 737–747.

Honea, J. W. (1956). Sam Fordyce-Vanderbilt fault system of southwest Texas, *Trans. Gulf Coast Assoc. Geol. Soc.* **6**, 51–52.

Hood, A., and Castaño, J. R. (1974). "Organic Metamorphism: Its Relationship to Petroleum Generation and Application to Studies of Authigenic Minerals," pp. 85–118. United Nations ESCAP CCOP Tech. Bull. 8.

Hoots, H. W., Bear, T. L., and Kleinpell, W. D. (1954). Stratigraphic traps for oil and gas in San Joaquin Valley, *In* "Geology of Southern California," pp. 29–32. California Div. Mines Bull. 170.

Hoskins, C. W. (1964). Molluscan biofacies in calcareous sediments, Gulf of Batabano, Cuba, *Am. Assoc. Pet. Geol. Bull.* **48**, 1680–1704.

Houbolt, J. J. H. C. (1968). Recent sediments in the southern bight of the North Sea, *Geol. Mijnbouw* **47**, 245–273.

Hough, J. L., and Menard, H. W. (eds.) (1956). "Finding Ancient Shorelines—a Symposium." Soc. Econ. Paleontol. Mineral. Spec. Publ. 3.

Houlik, C. W. (1973). Interpretation of carbonate-detrital silicate transitions in the Carboniferous of western Wyoming, *Am. Assoc. Pet. Geol. Bull.* **57**, 498–509.

Howard, P. F. (1972). Exploration for phosphorite—a case history, *Econ. Geol.* **67**, 1180–1192.

Hower, J., Eslinger, E. V., Hower, M. E., and Perry, E. A. (1976). Mechamism of burial metamorphism of argillaceous sediment: 1, mineralogical and chemical evidence, *Geol. Soc. Am. Bull.* **87**, 725–737.

Hoyt, J. H. (1959). Erosional channel in the Middle Wilcox near Yoakum, Lavaca County, Texas, *Trans. Gulf Coast Geol. Soc.* **9**, 41–50.

Hoyt, J. H. (1969). Chenier *versus* barrier, genetic and stratigraphic distinction, *Am. Assoc. Pet. Geol. Bull.* **53**, 299–306.

Hoyt, J. H., and Weimer, R. J. (1963). Comparison of modern and ancient beaches, central Georgia coast, *Am. Assoc. Pet. Geol. Bull.* **47**, 529–531.

Hoyt, J. H., and Henry, V. J. (1965). Significance of inlet sedimentation in recognition of ancient barrier islands, Guidebook, *Ann. Field Conf., 19th*, pp. 190–194. Wyoming Geol. Assoc. Publ.

Hriskevich, M. E. (1970). Middle Devonian reef production, Rainbow area, Alberta, Canada, *Am. Assoc. Pet. Geol. Bull.* **54**, 2260–2281.

Hsü, K. J. (1972). Origin of saline giants: a critical review after the discovery of the Mediterranean evaporite, *Earth Sci. Rev.* **8**, 371–396.

Hsü, K. J., and Jenkyns, H. C. (eds.) (1974). "Pelagic Sediments: On Land and Under the

Sea," Int. Assoc. Sedimentol. Spec. Publ. 1. Oxford Univ. Press (Blackwell), London and New York.

Huang, T. (1972). "Distribution of Planktonic Foraminifers in the Surface Sediments of Taiwan Strait," pp. 31–73. United Nations ECAFE CCOP Tech. Bull. 6.

Hubbert, M. K. (1953). Entrapment of petroleum under hydrodynamic conditions. *Am. Assoc. Pet. Geol. Bull.* **37**, 1954–2026.

Hunt, J. M. (1961). Distribution of hydrocarbons in sedimentary rocks, *Geochim. Cosmochim. Acta* **22**, 37–49.

Hunt, J. M. (1962). Geochemical data on organic matter in limestones, *Proc. Int. Sci. Oil Conf., Budapest.*

Huntoon, P. W., and Sears, J. W. (1975). Bright Angel and Eminence Faults, eastern Grand Canyon, Arizona, *Geol. Soc. Am. Bull.* **86**, 465–472.

Illing, L. V., Wells, A. V., and Taylor, J. C. M. (1965). Penecontemporaneous dolomite in the Persian Gulf, *In* "Dolomitization and limestone diagenesis: a symposium," Soc. Econ. Paleontol. Mineral. Spec. Paper **13**, 89–111.

Inderbitzen, A. L. (1974). "Deep-Sea Sediments." Plenum Press, New York.

Ion, D. C. (ed.) (1965). Salt Basins Around Africa: A Symposium. Inst. Pet. Publ., London.

Irwin, M. L. (1965). General theory of epeiric clear water sedimentation, *Am. Assoc. Pet. Geol. Bull.* **49**, 445–459.

Ishiwada, Y. (1971). "Analysis of Petroleum Source Rocks from the Philippines," pp. 83–91. United Nations ECAFE CCOP Tech. Bull. 4.

Ishiwatari, R., Rohrback, B. G., and Kaplan, I. R. (1978). Hydrocarbon generation by thermal alteration of kerogen from different sediments, *Am. Assoc. Pet. Geol. Bull.* **62**, 687–692.

Jackson, W. E. (1964). Depositional topography and cyclic deposition in west-central Texas, *Am. Assoc. Pet. Geol. Bull.* **48**, 317–328.

Jacobeen, F., and Kanes, W. H. (1974). Structure of Broadtop synclinorium and its implications for Appalachian structural style, *Am. Assoc. Pet. Geol. Bull.* **58**, 362–375.

Jacobsen, J. B. E. (1975). Copper deposits in time and space, *Minerals Sci. Eng.* **7**, 337–371.

Jenik, A. J., and Lerbekmo, J. F. (1968). Facies and geometry of Swan Hills reef member of Beaverhill Lake Formation (Upper Devonian), Goose River field, Alberta, Canada, *Am. Assoc. Pet. Geol. Bull.* **52**, 21–56.

Johns, W. D., and Shimoyama, A. (1972). Clay minerals and petroleum-forming reactions during burial and diagenesis, *Am. Assoc. Pet. Geol. Bull.* **56**, 2160–2167.

Johnson, D. W. (1959). "Shore Processes and Shoreline Development." Wiley, New York.

Johnson, J. G. (1971). Timing and coordination or orogenic, epeirogenic, and eustatic events, *Geol. Soc. Am. Bull.* **82**, 3262–3298.

Johnson, J. G., and Potter, E. C. (1976). The depths inhabited by Silurian brachiopod communities: a reply, *Geology* **4**, 189–191.

Johnson, K. S. (1974). Permian copper shales of southwestern United States, *In* "Gisements Stratiformes et Provinces Cupriferes," pp. 383–393. Soc. Géol. de Belgique Publ.

Jones, O. A., and Endean, R. (eds.) (1977). "Biology and Geology of Coral Reefs," Vols. 1 and 4, Geology. Academic Press, New York.

Jopling, A. V., and McDonald, B. C. (eds.) (1975). "Glaciofluvial and Glaciolacustrine Sedimentation," Soc. Econ. Paleontol. Mineral. Spec. Publ. 23.

Jöreskog, K. G., Klovan, J. E., and Reyment, R. A. (1976). Geological factor analysis, *In* "Methods in Geomathematics," Vol. 1. Elsevier, Amsterdam.

Junner, N. R. (1943). The diamond deposits of the Gold Coast, *Gold Coast Geol. Surv. Bull.* **12**, 3–49.

Kahle, C. F. (ed.) (1974). "Plate Tectonics: Assessments and Reassessments." Am. Assoc. Pet. Geol. Mem. 23.

Kamen-Kaye, M. (1967). "Basin subsidence and hypersubsidence. *Am. Assoc. Pet. Geol. Bull.* **51**, 1833–1842.

Kamen-Kaye, M. (1970). Geology and productivity of Persian Gulf synclinorium, *Am. Assoc. Pet. Geol. Bull.* **54**, 2371–2394.

Katich, P. J. (1958). Stratigraphy of the Eagle evaporites, *Symp. Pennsylvanian Rocks of Colorado Adjacent Areas,* pp. 106–110. Rocky Mountain Assoc. Geol. Publ., Denver, Colorado.

Kaufman, M. I., Goolsby, D. A., and Faulkner, G. L. (1973). Injection of acidic industrial waste into saline carbonate aquifer—geochemical aspects (abstract), *Am. Assoc. Pet. Geol. Bull.* **57**, 1597.

Kay, G. M. (1942). Ottawa–Bonnechere graben and Lake Ontario homocline, *Geol. Soc. Am. Bull.* **53**, 585–646.

Kay, G. M. (1951). "North American Geosynclines." Geol. Soc. Am. Mem. 48.

Kay, G. M. (ed.) (1969). "North Atlantic—geology and continental drift." Am. Assoc. Pet. Geol. Mem. 12.

Kaye, P., Edmond, G. M., and Challinor, A. (1972). The Rankin trend Northwest Shelf, Western Australia. *Aust. Pet. Exp. Assoc. J.* **12**, 3–8.

Kelley, V. C., Kittel, D. F., and Melancon, P. E. (1968). Uranium deposits of the Grants region, *In* "Ore deposits of the United States" (J. D. Ridge, ed.), *Am. Inst. Min. Metall. Eng. Bull.* **1**, 748–766.

Kendall, C. G. St. C., and Skipwith, Sir Patrick A. D. (1969). Holocene shallow-water carbonate and evaporite sediments of Khor al Bazam, Abu Dhabi, southwest Persian Gulf, *Am. Assoc. Pet. Geol. Bull.* **53**, 841–869.

Kennedy, W. J., and Garrison, R. E. (1975). Morphology and genesis of nodular chalks and hardgrounds in the Upper Cretaceous of southern England, *Sedimentology* **22**, 311–386.

Kennedy, W. Q. (1964). "The Structural Differentiation of Africa in the Pan-African (±500 m.y.) Tectonic Episode," pp. 48–49. Res. Inst. Afr. Geol., Univ. Leeds, 8th Ann. Rep.

Kent, P. E. (1967). Progress of exploration in North Sea, *Am. Assoc. Pet. Geol. Bull.* **51**, 731–741.

Kent, P. E., Satterthwaite, G. E., and Spencer, A. M. (1969). "Time and Place in Orogeny." Geol. Soc. London Spec. Publ. 3.

Kesler, S. E., Kienle, C. F., and Bateson, J. H. (1974). Tectonic significance of intrusive rocks in Maya Mountains, British Honduras, *Geol. Soc. Am. Bull.* **85**, 549–552.

Kharaka, Y. K., and Smalley, W. C. (1976). Flow of water and solutes through compacted clays, *Am. Assoc. Pet. Geol. Bull.* **60**, 973–980.

Khurana, A. K. (1976). Influence of tidal phenomena on interpretation of pressure build-up and pulse tests, *Aust. Pet. Exp. J.* **16**, 99–105.

Kimbler, O. K., Kazmann, R. G., and Whitehead, W. R. (1973). Saline aquifers—future storage reservoirs for fresh water ? (abstract), *Am. Assoc. Pet. Geol. Bull.* **57**, 1598.

King, P. B. (1961). The subsurface Ouachita structural belt of the Ouachita Mountains, *In* "The Ouachita System" (P. T. Flawn *et al.,* eds.), pp. 83–98. Univ. Texas. Publ. 6120.

King, R. E. (ed.) (1972). "Stratigraphic Oil and Gas Fields—Classification, Exploration Methods, and Case Histories." Am. Assoc. Pet. Geol. Mem. 16.

Kirkland, D. W., and Evans, R. (eds.) (1973). Marine evaporites: origin, diagenesis, and geochemistry, "Benchmark Papers in Geology," Vol. 7. Dowden, Hutchinson, and Ross, Stroudsburg, Pennsylvania.

Klein, G. de V. (1974). Estimating water depths from analysis of barrier island and deltaic sedimentary sequences, *Geology* **2**, 409–412.

Klenova, M. V. *et al.* (1956). "Contemporary Sediments of the Caspian Sea" (transl.). Akad. Nauk., Moscow.

Kloen, M. L., and Pickens, W. R. (1970). "Geology of the Felder Uranium Mine." Live Oak County, Texas. Soc. Min. Eng., Preprint 70-1-38.

Klovan, J. E. (1974). Development of western Canadian Devonian reefs and comparison with Holocene analogues, *Am. Assoc. Pet. Geol. Bull.* **58**, 787–799.

Knebel, G. M. (1946). The transformation of organic material into petroleum, *Am. Assoc. Pet. Geol. Bull.* **30**, 1935–1954.

Koch, G. S., and Link, R. F. (1970). "Statistical Analysis of Geological Data," Vol. 1, p. 2. Wiley, New York.

Komar, P. D. (1976). "Beach Processes and Sedimentation." Prentice-Hall. Englewood Cliffs, New Jersey.

Kottlowski, F. E. (1965). "Measuring Stratigraphic Sections." Holt, New York.

Kottlowski, F. E. (ed.) (1972). "Continental Shelves—Origin and Significance." Am. Assoc. Pet. Geol. Reprint Ser. 3.

Kozlova, O. G., and Mukhina, V. V. (1967). Diatoms and silicoflagellates in suspension and floor sediments of the Pacific Ocean, *Int. Geol. Rev.* **9**, 1322–1342.

Kraft, J. C. (1971). Sedimentary environment facies patterns and geologic history of a Holocene marine transgression, *Geol. Soc. Am. Bull.* **82**, 2131–2158.

Kraft, J. C., Sheridan, R. E., and Maisano, M. (1971). Time-stratigraphic units and petroleum entrapment-models in Baltimore Canyon Basin and Atlantic margin geosynclines, *Am. Assoc. Pet. Geol. Bull.* **55**, 658–679.

Kranzler, I. (1966). Origin of oil in lower member of Tyler Formation of central Montana, *Am. Assoc. Pet. Geol. Bull.* **50**, 2245–2259.

Kreitler, C. W., Guevera, E. H., Granata, G., and McKalips, D. G. (1977). Geometry of the Gulf Coast aquifers, Houston–Galveston, Texas (abstract), *Am. Assoc. Pet. Geol. Bull.* **61**, 1539.

Krueger, W. C. (1968). Depositional environments of sandstones as interpreted from electrical measurements—an introduction, *Trans. Gulf Coast Assoc. Geol. Soc.* **18**, 226–241.

Krumbein, W. C. (1948). Lithofacies maps and regional sedimentary-stratigraphic analysis, *Am. Assoc. Pet. Geol. Bull.* **32**, 1909–1923.

Krumbein, W. C., and Sloss, L. L. (1963). "Stratigraphy and Sedimentation," 2nd ed. Freeman, San Francisco, California.

Krumbein, W. C., and Graybill, F. A. (1965). "An Introduction to Statistical Models in Geology." McGraw Hill, New York.

Krumbein, W. C., and Monk, G. D. (1942). "Permeability as a Function of the Size Parameters of Unconsolidated Sand." Am. Inst. Min. Metall. Eng. Tech. Publ. 1492. Also *Pet. Tech.* **5**, 1–11.

Kühn, K. (1973). Asse saltmine, Federal Republic of Germany—operating facility for underground disposal of radioactive waste (abstract), *Am. Assoc. Pet. Geol. Bull.* **57**, 1598–1599.

Kukal, Z. (1971). "Geology of Recent Sediments." Academic Press, New York.

Kulm, L. D., Scheidegger, K. F., Prince, R. A., Dymond, J., and Moore, T. C. (1973). Tholeiitic basalt ridge in the Peru trench, *Geology* **1**, 11–14.

Kulm, L. D., Roush, R. C., Harlett, J. C., Neudeck, R. H., Chambers, D. M., and Runge, E. J. (1975). Oregon continental shelf sedimentation: interrelationships of facies distribution and sedimentary processes, *J. Geol.* **83**, 145–175.

Lajoie, J. (ed.) (1970). "Flysch Sedimentology in North America." Geol. Assoc. Can. Spec. Paper 7.

Land, L. S. (1973). Holocene meteoric dolomitization of Pleistocene limestones, north Jamacia, *Sedimentology* **20**, 411–424.

Land, L. S., Salem, M. R. I., and Morrow, D. W. (1975). Paleohydrology of ancient dolomites: geochemical evidence, *Am. Assoc. Pet. Geol. Bull.* **59**, 1602–1625.

Landis, E. R. (1977). Coal resources of Colorado and Wyoming (abstract), *Am. Assoc. Pet. Geol. Bull.* **6**, 806.

Langford, F. F. (1974). A supergene origin for vein-type uranium ores in the light of Western Australian calcrete-carnotite deposits, *Econ. Geol.* **69**, 516–526.

Langford, F. F. (1977). Surficial origin of North American pitchblende and related uranium deposits, *Am. Assoc. Pet. Geol. Bull.* **61**, 28–42.

Laporte, L. F. (1968). "Ancient Environments." Prentice-Hall, Englewood Cliffs, New Jersey.

Laporte, L. F. (ed.) (1974). "Reefs in Time and Space." Soc. Econ. Paleontol. Mineral. Spec. Publ. 18.

Larskaya, Y. S., and Zhabrev, D. V. (1964). Effect of stratal temperature and pressure on the composition of dispersed organic matter, *Dokl. Akad. Nauk SSSR* **157**, 897–900.

Lawson, D. E., and Smith, J. R. (1966). Pennsylvanian and Permian influence on Tensleep oil accumulation, Big Horn Basin, Wyoming, *Am. Assoc. Pet. Geol. Bull.* **50**, 2197–2220.

Leavitt, E. M. (1968). Petrology, palaeontology, Carson Creek North Reef Complex, Alberta, *Bull. Can. Pet. Geol.* **16**, 298–413.

Le Blanc, R. J. (1972). Geometry of sandstone reservoir bodies, In "Underground Waste Management and Environmental Implications," pp. 133–190. Am. Assoc. Pet. Geol. Mem. 18.

Lehner, P. (1969). Salt tectonics and Pleistocene stratigraphy on continental slope of northern Gulf of Mexico, *Am. Assoc. Pet. Geol. Bull.* **53**, 2431–2479.

Leighton, M. W., and Pendexter, C. (1962). Carbonate rock types, In "Classification of Carbonate Rocks," pp. 33–61. Am. Assoc. Pet. Geol. Mem. 1.

Leipper, D. F. (1954). Physical oceanography of the Gulf of Mexico, In "Gulf of Mexico, Its Origin, Waters, and Marine Life." U. S. Dept. Interior Fishery Bull. 89.

Leopold, L. B., Wolman, M. G., and Miller, J. P. (1964). "Fluvial Processes in Geomorphology." Freeman, San Francisco, California.

Levorsen, A. I. (1967). "Geology of Petroleum," 2nd ed. Freeman, San Francisco, California.

Lewis, K. B. (1971). Slumping on a continental slope inclined at $1°–4°$, *Sedimentology* **16**, 97–110.

Lewis, T. L., and Schwietering, J. F. (1971). Distribution of the Cleveland Black Shale in Ohio, *Geol. Soc. Am. Bull.* **82**, 3477–3482.

Leyden, R., Bryan, G., and Ewing, M. (1972). Geophysical reconnaissance of African shelf: 2, Margin sediments from Gulf of Guinea to Walvis Ridge, *Am. Assoc. Pet. Geol. Bull.* **56**, 682–693.

Lineback, J. A. (1969). Illinois Basin—starved-sediment during Mississippian, *Am. Assoc. Pet. Geol. Bull.* **53**, 112–126.

Link, T. A. (1949). Interpretation of foothills structure, Alberta, Canada, *Am. Assoc. Pet. Geol. Bull.* **33**, 1481.

Lisitzin, A. P. (1972). "Sedimentation in the World Ocean" (K. S. Rodolfo, ed.). Soc. Econ. Paleontol. Mineral. Spec. Publ. 17.

Lofgren, B. E. (1973). Hazards of waste disposal in groundwater basins (abstract), *Am. Assoc. Pet. Geol. Bull.* **57**, 1600.

Logan, B. W. *et al.* (1969). "Carbonate Sediments and Reefs, Yucatan shelf, Mexico," pp. 1–196. Am. Assoc. Pet. Geol. Mem. 11.

Lohmann, H. H. (1972). Salt dissolution in subsurface of British North Sea as interpreted from seismograms, *Am. Assoc. Pet. Geol. Bull.* **56**, 472–479.

Lombard, A. (1972). "Séries Sédimentaires, Genesè-évolution." Masson, Paris.

Longwell, C. R. (ed.) (1949). "Sedimentary Facies in Geologic History: A Symposium." Geol. Soc. Am. Mem. 39.

Louis, M., and Tissot, B. (1967). Influence de la temperature et de la pression sur la formation des hydrocarbures dans les argiles a kerogene. *Proc. World Pet. Congr., 7th* **2**, 47–60.

Lowrie, A. (1978). Buried trench south of the Gulf of Panama, *Geology* **6**, 434–436.

Luyendyk, B. P., and MacDonald, K. C. (1976). Spreading centre terms and concepts, *Geology* **4**, 369–370.

Lyons, P. L., and Dobrin, M. B. (1972). Seismic exploration for stratigraphic traps, *In* "Stratigraphic Oil and Gas Fields," pp. 225–243. Am. Assoc. Pet. Geol. Mem. 16.

Machatschek, F. (1969). "Geomorphology," 9th ed. Oliver and Boyd, Edinburgh.

MacKenzie, D. B. (1972). Primary stratigraphic traps in sandstone, *In* "Stratigraphic Oil and Gas Fields," pp. 47–63. Am. Assoc. Pet. Geol. Mem. 16.

MacLeod, W. N. (1966). "The Geology and Iron Deposits of the Hamersley Range Area, Western Australia." West Aust. Geol. Surv. Bull. 117.

Magara, K. (1976). Thickness of removed sedimentary rocks, paleopore pressure, and paleotemperature, southwestern part of Western Canadian Basin, *Am. Assoc. Pet. Geol. Bull.* **60**, 554–565.

Maher, J. C. (1953). Permian and Pennsylvanian rocks of southeastern Colorado, *Am. Assoc. Pet. Geol. Bull.* **37**, 913–939.

Maksimov, V. L., Rudakov, G. V., Filanovskiy, V. Y., and Shevaldin, I. E. (1968). "Characteristics of Oil Recovery from Fields in West Siberia" (transl.). Nedra, Moscow.

Maldonado, A., and Stanley, D. J. (1977). Lithofacies as a function of depth in the Strait of Sicily, *Geology* **5**, 111–117.

Malek-Aslani, M. (1970). Lower Wolfcampian reef in Kemnitz Field, Lea County, New Mexico, *Am. Assoc. Pet. Geol. Bull.* **54**, 2317–2335.

Mallory, W. W. (1960). Outline of Pennsylvanian stratigraphy of Colorado, *In* "Guide to the Geology of Colorado," pp. 23–33. Rocky Mountain Assoc. Geol. Publ.

Marine, I. W. (1974). Geohydrology of buried Triassic basin at Savannah River Plant, South Carolina, *Am. Assoc. Pet. Geol. Bull.* **58**, 1825–1837.

Martin, B. D. (1963). Rosedale channel—evidence for Late Miocene submarine erosion in Great Valley of California, *Am. Assoc. Pet. Geol. Bull.* **47**, 441–456.

Martin, R. (1966). Paleogeomorphology and its application to exploration for oil and gas (with examples from Western Canada), *Am. Assoc. Pet. Geol. Bull.* **50**, 2277–2311.

Martinez, J. D. (1971). Environmental significance of salt, *Am. Assoc. Pet. Geol. Bull.* **55**, 810–825.

Mascle, J. R., Bornhold, B. D., and Renard, V. (1973). Diapiric structures off Niger Delta, *Am. Assoc. Pet. Geol. Bull.* **57**, 1672–1678.

Masters, C. D. (1967). Use of sedimentary structures in determination of depositional environments, Mesaverde Formation, Williams Fork Mountains, Colorado, *Am. Assoc. Pet. Geol. Bull.* **51**, 2033–2043.

Matuszczak, R. A. (1973). Wattenberg Field, Denver Basin, Colorado, *Mountain Geol.* **10**, 99–105.

Maxwell, J. C. (1964). Influence of depth, temperature, and geologic age on porosity of quartzose sandstone, *Am. Assoc. Pet. Geol. Bull.* **48**, 697–709.

McBride, E. F., and Kimberly, J. E. (1963). Sedimentology of Smithwick Shale (Pennsylvanian), eastern Llano region, Texas, *Am. Assoc. Pet. Geol. Bull.* **47,** 1840–1854.

McCamis, J. G., and Griffith, L. S. (1968). Middle Devonian facies relations, Zama area, Alberta, *Am. Assoc. Pet. Geol. Bull.* **52,** 1899–1924.

McCrossan, R. G., and Porter, J. W. (1973). The geology and petroleum potential of the Canadian sedimentary basins—a synthesis, *In* "The Future Petroleum Provinces of Canada—Their Geology and Potential," pp. 589–720. Can. Soc. Pet. Geol. Mem. 1.

McCubbin, D. G. (1969). Cretaceous strike-valley sandstone reservoirs, northwestern New Mexico, *Am. Assoc. Pet. Geol. Bull.* **53,** 2114–2140.

McDaniel, G. A. (1968). Application of sedimentary directional features and scalar properties to hydrocarbon exploration, *Am. Assoc. Pet. Geol. Bull.* **52,** 1689–99.

McGinnis, L. D., and Ervin, C. P. (1974). Earthquake and block tectonics in the Illinois Basin, *Geology* **2,** 517–519.

McGlasson, E. H. (1968). The Siluro-Devonian of West Texas and Southeast New Mexico, *In* "Delaware Basin Exploration" (Field Trip Guidebook), pp. 35–44. West Texas Geol. Soc. Publ.

McGregor, A. A., and Biggs, C. A. (1968). Bell Creek Field, Montana: a rich stratigraphic trap, *Am. Assoc. Pet. Geol. Bull.* **52,** 1869–1887.

McKee, E. D. (1966). Structure of sand dunes at White Sands National Monument, New Mexico, *Sedimentology* **7,** 1–69.

McKee, E. D. (1969). Paleozoic rocks of Grand Canyon, *In* "Geology and Natural History of the Grand Canyon Region," pp. 78–90. Four Corners Geol. Soc. Spec. Publ.

McKee, E. D., Breed, C. S., and Fryberger, S. G. (1977). Desert sand seas, *In* "Skylab Explores the Earth," Vol. 2, pp. 5–48. NASA Publ. SP-380.

McKee, E. M. (1963). Paleogeomorphology, a practical exploration technique, *Oil Gas J.* **61,** 140–143.

McKetta, J. J., and Katz, D. L. (1948). Methane–*n*-butane–water system in two- and three-phase regions, *Ind. Eng. Chem.* **40,** 853–863.

McKetta, J. J., and Wehe, A. H. (1962). Hydrocarbon-water and formation water correlations, *In* "Petroleum Production Handbook" (T. C. Frick and R. W. Taylor, eds.), Vol. 2, p. 22. McGraw-Hill, New York.

McKirdy, D. M. (1976). Biochemical markers in stromatolites, *Dev. Sediment.* **20,** 163–191.

McKirdy, D. M., and Powell, T. G. (1974). Metamorphic alteration of carbon isotopic composition in ancient sedimentary organic matter: new evidence from Australia and South Africa, *Geology* **2,** 591–595.

McKirdy, D. M., Sumartojo, J., Tucker, D. H., and Gostin, V. (1975). Organic, mineralogic and magnetic indications of metamorphism in the Tapley Hill Formation, Adelaide Geosyncline, *Precambrian Res.* **2,** 345–373.

McMaster, R. L., and Lachance, T. P. (1968). Seismic reflectivity studies on northwestern African continental shelf: Strait of Gibraltar to Mauritania, *Am. Assoc. Pet. Geol. Bull.* **52,** 2387–2395.

McMaster, R. L., and Ashraf, A. (1973). Extent and formation of deeply buried channels on the continental shelf off southern New England, *J. Geol.* **81,** 374–379.

McQuillan, H. (1973). Small-scale fracture density in Asmari Formation of southwest Iran and its relation to bed thickness and structural setting, *Am. Assoc. Pet. Geol. Bull.* **57,** 2367–2385.

Meckel, L. D. (1967). Origin of Pottsville Conglomerates (Pennsylvanian) in the central Appalachians, *Geol. Soc. Am. Bull.* **78,** 223–258.

Meinhold, R. (1968). Über den Zusammenhang geothermischer, hydrodynamischer und geochemischer Anomalien und deren Bedeutung für die Klärung der Entstehung von Erdöllagerstätten, *Z. Ang. Geol.* **14**, 233–240.

Meinzer, O. E., and Wenzel, L. K. (1942). Hydrology, *In* "Physics of the Earth" (O. E. Meinzer, ed.). McGraw-Hill, New York.

Meisburger, E. P., and Field, M. E. (1976). Neogene sediments of Atlantic inner continental shelf off northern Florida, *Am. Assoc. Pet. Geol. Bull.* **60**, 2019–2037.

Mellen, F. F. (1974). Patterson Sandstone, Ordovician or Silurian? *Am. Assoc. Pet. Geol. Bull.* **58**, 143–148.

Mendelsohn, F. (1961). Ore genesis; summary of the evidence, *In* "The Geology of the Northern Rhodesian Copper Belt" (F. Mendelsohn, ed.), pp. 130–146. Macdonald, London.

Merki, P. J. (1972). Structural geology of the Cenozoic Niger delta, *Proc. Conf. Afr. Geol.* pp. 635–646. Ibadan Univ. Press, Ibadan, Nigeria.

Merriam, D. F. (ed.) (1972). "Mathematical Models of Sedimentary Processes." Plenum Press, New York.

Miall, A. D. (1970). Continental marine transition in the Devonian of Prince of Wales Island, Northwest Territories, *Can. J. Earth Sci.* **7**, 125–144.

Miall, A. D. (1976). Devonian geology of Banks Island, Arctic Canada, and its bearing on the tectonic development of the circum-Arctic region, *Geol. Soc. Am. Bull.* **87**, 1599–1608.

Middleton, G. V. (ed.) (1965). "Primary Sedimentary Structures and Their Hydrodynamic Interpretation." Soc. Econ. Paleontol. Mineral. Spec. Paper 12.

Middleton, G. V. (1970). "Experimental Studies Related to Problems of Flysch Sedimentation," pp. 253–272. Geol. Assoc. Can. Spec. Paper 7.

Milliman, J. D. (1974). Marine carbonates, "Recent Sedimentary Carbonates," Vol. 1. Springer-Verlag, Berlin and New York.

Mitchell, A. H. G. (1970). Facies of an Early Miocene volcanic arc, Malekula Island, New Hebrides, *Sedimentology* **14**, 201–243.

Mitchell, A. H. G., and Garson, M. S. (1976). Mineralization at plate boundaries, *Minerals Sci. Eng.* **8**, 129–169.

Molenaar, C. M. (ed.) (1969). "Geology and Natural History of the Grand Canyon Region" (a symposium). Four Corners Geol. Soc. Publ.

Mollan, R. G., Craig, R. W., and Lofting, M. J. W. (1970). Geologic framework of continental shelf off northwest Australia, *Am. Assoc. Pet. Geol. Bull.* **54**, 583–600.

Moore, C. A. (1964). "Handbook of Subsurface Geology." Harper, New York.

Moore, G. F., and Karig, D. E. (1976). Development of sedimentary basins on the lower trench slope, *Geology* **4**, 693–697.

Moore, G. T., and Asquith, D. O. (1971). Delta: term and concept, *Geol. Soc. Am. Bull.* **82**, 2563–2568.

Moore, H. B. (1965). "Bottom Temperatures Between Miami and Bimini." Miami Univ. Inst. Mar. Sci. Tech. Rep.

Morgan, J. P., and Shaver, R. H. (eds.) (1970). "Deltaic Sedimentation, Modern and Ancient." Soc. Econ. Paleontol. Mineral. Spec. Publ. 15.

Morgan, J. T., Cordiner, F. S., and Livingstone, A. R. (1977). Tensleep reservoir study, Oregon Basin field, Wyoming (abstract), *Am. Assoc. Pet. Geol. Bull.* **6**, 816.

Morris, R. C. (1971). Stratigraphy and sedimentology of Jackfork Group, Arkansas, *Am. Assoc. Pet. Geol. Bull.* **55**, 387–402.

Motts, W. S. (1968). The control of ground-water occurrence by lithofacies in the Guadalupian Reef Complex near Carlsbad, New Mexico, *Geol. Soc. Am. Bull.* **79**, 283–298.

Müller, G. (1967). "Methods in Sedimentary Petrology" (Transl. by H. U. Schmincke). Hafner, New York.

Müller, G., and Friedman, G. M. (eds.) (1968). "Recent Developments in Carbonate Sedimentology in Central Europe." Springer-Verlag, Berlin and New York.

Murchison, D., and Westoll, T. S. (eds.) (1968). "Coal and Coal-Bearing Strata." Oliver and Boyd, Edinburgh.

Murray, G. E. (1966). Salt structures of Gulf of Mexico Basin—a review, *Am. Assoc. Pet. Geol. Bull.* **50**, 439–478.

Mutti, E. (1974). Examples of ancient deep-sea fan deposits from circum-Mediterranean geosynclines, *In* "Modern and Ancient Geosynclinal Sedimentation" (R. H. Dott and R. H. Shaver, eds.) pp. 92–105. Soc. Econ. Paleontol. Mineral. Spec. Publ. 19.

Nace, R. L. (1961). Underground storage of radioactive waste, *Interstate Oil Compact Commun. Bull.* **3**, 1.

Namy, J. N. (1974). Origin of a Lower Pennsylvanian depositional cycle, *J. Sediment Pet.* **44**, 795–805.

Nanz, R. H. (1960). Exploration and Earth Formations Associated with Petroleum Deposits. U.S. Patent 2,963,641.

Natland, M. L., and Kuenen, P. H. (1951). "Sedimentary History of the Ventura Basin, California, and the Action of Turbidity Currents," pp. 76–107. Soc. Econ. Paleontol. Mineral. Spec. Publ. 2.

Naylor, D., and Mounteney, S. N. (1975). "Geology of the North-West European Continental Shelf." Graham Trotman Dudley, London.

Neev, D., and Emery, K. O. (1967). "The Dead Sea: Depositional Processes and Environment of Evaporites," Israel Geol. Surv. Bull. 41.

Negi, B. S. *et al.* (1968). Tectonic Map of India. Indian Oil Nat. Gas. Comm. Publ.

Nelson, B. W. (ed.) (1972). Environmental Framework of Coastal Plain Estuaries. Geol. Soc. Am. Mem. 133.

Nelson, R. A., and Handin, J. (1977). Experimental study of fracture permeability in porous rock, *Am. Assoc. Pet. Geol. Bull.* **61**, 227–236.

Neprochnov, I. P., and Korylin, V. M. (1964). The results of seismic research on the structure of the earth crust and sedimentary mass in the Indian Ocean, *Int. Geol. Congr., Rep. Sov. Geol., 22nd* pp. 52–61.

Nettleton, L. L. (1962). Gravity and magnetics for geologists and seismologists, *Am. Assoc. Pet. Geol. Bull.* **46**, 1815–1838.

Neumann, A. C., Kofoed, J. W., and Keller, G. H. (1977). Lithoherms in the Straits of Florida, *Geology* **5**, 4–10.

Nicolini, O. (1970). "Gîtologie des Concentrations Minérales Stratiformes." Gauthier-Villars, Paris.

Normark, W. R. (1970). Growth pattern of deep-sea fans, *Am. Assoc. Pet. Geol. Bull.* **54**, 2170–2195.

Norris, D. K. (1977). Old and new horizons in coal potential in Western and Arctic Canada (abstract), *Am. Assoc. Pet. Geol. Bull.* **6**, 818.

Nutting, P. G. (1926). Geothermal relations between petroleum, silica and water, *Econ. Geol.* **21**, 234–242.

Ocamb, R. D. (1961). Growth faults of south Louisiana, *Trans. Gulf Coast Assoc. Geol. Soc.* **11**, 139–176.

O'Connor, M. J., and Gretener, P. E. (1974). Quantitative modelling of the processes of differential compaction, *Bull. Can. Pet. Geol.* **22**, 241–268.

Off, T. (1963). Rhythmic linear sand bodies caused by tidal currents, *Am. Assoc. Pet. Geol. Bull.* **47**, 324–341.

Officer, C. B., and Ewing, W. M. (1954). Continental shelf, continental slope and continental rise south of Nova Scotia, *In* "Geophysical Investigations in Emerged and Submerged Atlantic Coastal Plain," *Geol. Soc. Am. Bull.* **65**, 653–669.

Ohlen, H. R., and McIntyre, L. B. (1965). Stratigraphy and tectonic features of Paradox Basin, Four Corners Area, *Am. Assoc. Pet. Geol. Bull.* **49**, 2020–2040.

Ojakangas, R. W. (1968). Cretaceous sedimentation Sacramento Valley, California, *Geol. Soc. Am. Bull.* **79**, 973–1008.

Olenin, V. B. (1967). The principles of classification of oil and gas basins, *Aust. Oil Gas. J.* **13**, 40–46.

Oomkens, E. (1974). Lithofacies relations in the Late Quaternary Niger Delta complex, *Sedimentology* **21**, 195–222.

Oxburgh, E. R. (1974). The plain man's guide to plate tectonics, *Proc. Geol. Assoc.* **85**, 299–357.

Oxburgh, E. R., and Turcotte, D. L. (1970). Thermal structure of island arcs, *Geol. Soc. Am. Bull.* **81**, 1665–1688.

Padula, V. T. (1969). Oil shale of Permian Irati Formation, Brazil, *Am. Assoc. Pet. Geol. Bull.* **53**, 591–602.

Page, B. M. (1963). Gravity tectonics near Passo Della Cisa, northern Apennines, Italy, *Geol. Soc. Am. Bull.* **74**, 655–672.

Park, C. F., and MacDiarmid, R. A. (1970). "Ore Deposits," 2nd ed. Freeman, San Francisco, California.

Parke, M. L., Emery, K. O., Szymankiewicz, R., and Reynolds, L. M. (1971). "Structural Framework of the Continental Margin in the South China Sea," pp. 103–142. United Nations ECAFE, CCOP Tech. Bull. 4.

Partridge, A. D. (1976). The geological expression of eustacy in the Early Tertiary of the Gippsland Basin, *Aust. Pet. Exp. Assoc. J.* **16**, 73–79.

Passega, R. (1972). Sediment sorting related to basin mobility and environment, *Am. Assoc. Pet. Geol. Bull.* **56**, 2440–2450.

Payne, M. W. (1976). Basinal sandstone facies, Delaware Basin, West Texas and Southeast New Mexico, *Am. Assoc. Pet. Geol. Bull.* **60**, 517–527.

Payton, C. E. (1977). "Seismic Stratigraphy—Applications to Hydrocarbon Exploration." Am. Assoc. Pet. Geol. Mem. 26.

Pegrum, R. M., and Mounteney, N. (1978). Rift basins flanking North Atlantic Ocean and their relation to North Sea area, *Am. Assoc. Pet. Geol. Bull.* **62**, 419–441.

Pelletier, B. R. (1958). Pocono paleocurrents in Pennsylvania and Maryland, *Geol. Soc. Am. Bull.* **69**, 1033–1064.

Peters, K. E., Ishiwatari, R., and Kaplan, I. R. (1977). Color of kerogen as index of organic maturity, *Am. Assoc. Pet. Geol. Bull.* **61**, 504–510.

Peterson, J. A., and Osmond, J. C. (ed.) (1961). "Geometry of Sandstone Bodies: A Symposium." Am. Assoc. Pet. Geol., Tulsa, Oklahoma.

Peterson, J. A., and Hite, R. J. (1969). Pennsylvanian evaporite-carbonate cycles and their relation to petroleum occurrence, southern Rocky Mountains, *Am. Assoc. Pet. Geol. Bull.* **53**, 884–908.

Petters, S. W. (1978). Mid-Cretaceous paleoenvironments and biostratigraphy of the Benue Trough, Nigeria, *Geol. Soc. Am. Bull.* **89**, 151–154.

Pettijohn, F. J. (1962). Palecurrents and paleogeography, *Am. Assoc. Pet. Geol. Bull.* **46**, 1468–1493.

Pettijohn, F. J. (1975). "Sedimentary Rocks," 3rd ed. Harper, New York.

Pettijohn, F. J., Potter, P. E., and Siever, R. (1972). "Sand and Sandstone." Springer-Verlag, Berlin and New York.

Philippi, G. T. (1956). Identification of oil source beds by chemical means, *Proc. Int. Geol. Congr., 20th, Mexico* pp. 43–44.

Philippi, G. T. (1965). On the depth, time and mechanism of petroleum generation, *Geochim. Cosmochim. Acta* **29**, 1021–1049.

Picard, M. D., and High, L. R. (1973). Sedimentary structures of ephemeral streams, *Dev. Sediment.* **17.**

Pirson, S. J. (1970). "Geologic Well Log Analysis." Gulf Publ., Houston, Texas.

Pitcher, M. G. (ed.) (1973). "Arctic Geology." Am. Assoc. Pet. Geol. Mem. 19.

Pittman, E. D. (1974). Porosity and permeability changes during diagenesis of Pleistocene corals, Barbados, West Indies, *Geol. Soc. Am. Bull.* **85,** 1811–1820.

Pogrebnov, N. I. (1974). Restriction of major coal basins to zones of regional faults in the earth's crust (transl.), *Int. Geol. Rev.* **17,** 1007–1012; *Sov. Geol.* **1,** 120–125.

Ponte, F. C., Fonseca, J. dos R., and Morales, R. G. (1977). Petroleum geology of eastern Brazilian continental margin. *Am. Assoc. Pet. Geol. Bull.* **61,** 1470–1482.

Potter, P. E. (1962). "Late Mississippian Sandstones of Illinois," Illinois Geol. Surv. Circ. 340.

Potter, P. E. (1962a). Regional distribution patterns of Pennsylvanian sandstones in Illinois Basin, *Am. Assoc. Pet. Geol. Bull.* **46,** 1890–1911.

Potter, P. E. (1967). Sand bodies and sedimentary environments: a review, *Am. Assoc. Pet. Geol. Bull.* **51,** 337–365.

Potter, P. E. *et al.* (1958). Chester cross-bedding and sandstone trends in Illinois Basin, *Am. Assoc. Pet. Geol. Bull.* **42,** 1013–1048.

Potter, P. E., and Simon, J. A. (1961). "Anvil Rock Sandstone and Channel Cutouts of Herrin (No. 6) Coal in West-central Illinois." Illinois Geol. Surv. Circ. 314.

Potter, P. E., and Pettijohn, F. J. (1963). "Paleocurrents and Basin Analysis." Springer-Verlag, Berlin and New York.

Potter, P. E., Heling, D., Shimp, N. F., and Van Wie. (1975). Clay mineralogy of modern alluvial muds of the Mississippi River Basin, *Bull. Centre Rech.* **Pau-SNPA 9,** 353–389.

Powell, D. E. (1976). The geological evolution of the continental margin off northwest Australia, *Aust. Pet. Exp. Assoc. J.* **16,** 13–23.

Powers, M. C. (1967). Fluid-release mechanisms in compacting marine mudrocks and their importance in oil exploration, *Am. Assoc. Pet. Geol. Bull.* **51,** 1240–1254.

Pray, L. C., and Murray, R. C. (eds.) (1965). "Dolomitization and Limestone Diagenesis." Soc. Econ. Paleontol. Mineral. Spec. Publ. 13.

Price, L. C. (1976). Aqueous solubility of petroleum as applied to its origin and primary migration, *Am. Assoc. Pet. Geol. Bull.* **60,** 213–244.

Price, R. A., and Douglas, R. J. W. (eds.) (1972). "Variations in Tectonic Styles in Canada." Geol. Assoc. Can. Spec. Paper 11.

Proshlyakov, B. K. (1960). Reservoir properties of rocks as a function of their depth and lithology, *Geol. Neft. Gazeta* **4,** 24–29.

Pryor, W. A. (1973). Permeability-porosity patterns and variations in some Holocene sand-bodies, *Am. Assoc. Pet. Geol. Bull.* **57,** 162–189.

Pusey, W. C. (1973). Paleotemperatures in Gulf Coast using ESR-kerogen method (abstract), *Am. Assoc. Pet. Geol. Bull.* **57,** 1836.

Quigley, M. D. (1965). Geologic history of Piceance Creek-Eagle Basins, *Am. Assoc. Pet. Geol. Bull.* **49,** 1974–1996.

Raasch, G. O. (1958). Baraboo monadnock and paleowind direction, *J. Alberta Soc. Pet. Geol.* **6,** 183–187.

Rad, U. von (1968). Comparison of sedimentation in the Bavarian Flysch (Cretaceous) and Recent San Diego Trough, *J. Sediment Pet.* **38,** 1120–1154.

Rao, M. S. (1960). Organic matter in marine sediments off east coast of India, *Am. Assoc. Pet. Geol. Bull.* **44,** 1705–1713.

Reamer, H. H., Sage, B. H., and Lacey, W. N. (1952). Phase equilibrium in hydrocarbon system; *n*-butane–water system in the two phase region, *Ind. Eng. Chem.* **44,** 609–615.

Rees, F. B. (1972). Methods of mapping and illustrating stratigraphic traps, *In* "Stratigraphic Oil and Gas Fields—Classification, Exploration Methods, and Case Histories" (R. E. King, ed.), pp. 168–221. Am. Assoc. Pet. Geol. Mem. 16.

Reid, A. R. (1974). Proposed origin for Guianian diamonds, *Geology* **2**, 67–68.

Reid, A. R. (1974a). Proposed origin for Guianian diamonds: reply, *Geology* **2**, 476.

Reid, J. L. (1962). On circulation, phosphate–phosphorus content, and zooplankton volumes in the upper part of the Pacific Ocean, *Limnol. Oceanogr.* **7**, 287–306.

Reineck, H. E. (1974). Present and ancient shallow marine deposits, *Bull. Centre Rech Pau-SNPA* **8**, 295–304.

Reineck, H. E., and Singh, I. B. (1973). "Depositional Sedimentary Environments." Springer-Verlag, Berlin and New York.

Revelle, R., Bramlette, M. N., Arrhenius, G., and Goldberg, E. D. (1955). "Pelagic Sediments." Geol. Soc. Am. Spec. Paper 62.

Reyment, R. A. (1971). "Introduction to Quantitative Paleoecology." Elsevier, Amsterdam.

Reyre, D. (1966). Particularités géologique des bassins côtiers de l' Oueste Africain, 1: Littoral Atlantique, *In* "Bassins Sedimentaires du Littoral Africain," pp. 253–301. Assoc. Afr. Geol. Surv., Paris.

Ricateau, R., and Villemin, J. (1973). Evolution au crétacé supérieur de la pente séparant le domaine de plate-forme du sillon sous-pyrénéen en Aquitaine meridionale, *Bull. Soc. Géol. Fr.* **7**, 30–39.

Rich, J. L. (1935). Graphical methods for eliminating regional dip, *Am. Assoc. Pet. Geol. Bull.* **19**, 1538–1540.

Riedel, W. R. (1959). Siliceous organic remains in pelagic sediments, *In* "Silica in Sediments," pp. 80–91. Soc. Econ. Paleontol. Mineral. Spec. Publ. 7.

Rieke, H. H., and Chilingarian, G. V. (1974). Compaction of argillaceous sediments, *Dev. Sediment.* **16**.

Rigby, J. K., and Hamblin, W. K. (eds.) (1972). Recognition of Ancient Sedimentary Environments. Soc. Econ. Paleontol. Mineral. Spec. Publ. 16.

Rittenhouse, G. (1943). A visual method of estimating two-dimensional sphericity, *J. Sediment. Pet.* **13**, 79–81.

Rittenhouse, G. (1971). Pore-space reduction by solution and cementation, *Am. Assoc. Pet. Geol. Bull.* **55**, 80–91.

Roberts, D. G., and Caston, V. N. D. (1975). Petroleum potential of the deep Atlantic Ocean, *Proc. World Pet. Congr., 9th* **2**, 281–298.

Roberts, R. J. (1972). Evolution of the Cordilleran Fold Belt, *Geol. Soc. Am. Bull.* **83**, 1989–2004.

Robinson, R. B. (1966). Classification of reservoir rocks by surface texture, *Am. Assoc. Pet. Geol. Bull.* **50**, 547–559.

Robinson, J. E., Charlesworth, H. A. K., and Ellis, M. J. (1969). Structural analysis using spatial filtering in interior plains of south-central Alberta, *Am. Assoc. Pet. Geol. Bull.* **53**, 2341–2367.

Robson, D. A. (1971). The structure of the Gulf of Suez (Clysmic) rift, with special reference to the eastern side, *Q. J. Geol. Soc. London* **127**, 247–276.

Roedder, E. (1959). "Problems in the Disposal of Acid Aluminum Nitrate High-Level Radioactive Waste Solutions by Injection into Deep-Lying Permeable Formations." U.S. Geol. Surv. Bull. 1088.

Roeder, D. H. (1967). "Rocky Mountains: der geologische aufbau des Kanadischen felsengebirges." Gebrüder borntraeger, Berlin.

Ronov, A. B. (1958). Organic carbon in sedimentary rocks, *Geochemistry* **5**, 510–536.

Roobol, M. J., and Smith, A. L. (1976). Mount Pelée, Martinique: a pattern of alternating eruptive styles, *Geology* **4**, 521–524.

Roof, J. G., and Rutherford, W. M. (1958). Rate of migration of petroleum by proposed mechanisms, *Am. Assoc. Pet. Geol. Bull.* **42**, 963–980.

Rose, P. R. (1976). Mississippian carbonate shelf margins, western United States, *U.S. Geol. Surv. J. Res.* **4**, 449–466.

Rosholt, J. N., and Bartel, A. J. (1969). Uranium, thorium, and lead systematics in Granite Mountains, Wyoming, *Earth Planetary Sci. Lett.* **7**, 141–147.

Ross, C. A. (ed.) (1974). "Paleogeographic Provinces and Provinciality". Soc. Econ. Paleontol. Mineral. Spec. Publ. 21.

Russell, R. T., and Trueman, N. A. (1971). The geology of the Duchess phosphate deposits, northwestern Queensland, Australia, *Econ. Geol.* **66**, 1186–1214.

Ryan, W. B., and Heezen, B. C. (1965). Ionian Sea submarine canyons and the 1908 Messina Turbidity current, *Geol. Soc. Am. Bull.* **76**, 915–932.

Ryder, R. T., Fouch, T. D., and Elison, J. H. (1976). Early Tertiary sedimentation in the western Uinta Basin, Utah, *Geol. Soc. Am. Bull.* **87**, 496–512.

Sabins, F. F. (1963). Anatomy of stratigraphic trap, Bisti field, New Mexico, *Am. Assoc. Pet. Geol. Bull.* **47**, 193–228.

Sabins, F. F. (1964). Symmetry, stratigraphy, and petrography of cyclic Cretaceous deposits in San Juan Basin, *Am. Assoc. Pet. Geol. Bull.* **48**, 292–316.

Safanov, A. (1962). "The Challenge of the Sacramento Valley, California," pp. 77–97. California Div. Mines and Geol. Bull. 181.

Sanderson, I. D. (1974). Sedimentary structures and their environmental significance in the Navajo Sandstone, San Rafael Swell, Utah, *Brigham Young Univ. Geol. Stud.* **21**, 215–246.

Sanford, R. M., and Lange, F. W. (1960). Basin-study approach to oil evaluation of Paraná miogeosyncline, south Brazil, *Am. Assoc. Pet. Geol. Bull.* **44**, 1316–1370.

Sastri, V. V., Sinha, R. N., Singh, G., and Murti, K. V. S. (1973). Stratigraphy and tectonics of sedimentary basins on east coast of peninsular India, *Am. Assoc. Pet. Geol. Bull.* **57**, 655–678.

Schäfer, W. (1972). "Ecology and Palaeoecology of Marine Environments" (I. Oertel, transl.; G. Y. Craig, ed.). Oliver and Boyd, Edinburgh.

Schlager, W., and Schlager, M. (1973). Clastic sediments associated with radiolarites (Tauglboden-Schichten, Upper Jurassic, eastern Alps), *Sedimentology* **20**, 65–89.

Schlee, J. S., and Moench, R. H. (1961). Properties and genesis of "Jackpile" Sandstone, Laguna, New Mexico, *In* "Geometry of Sandstone Bodies," pp. 134–150. Am. Assoc. Pet. Geol. Publ.

Schlee, J., Behrendt, J. C., Grow, J. A., Robb, J. M., Mattick, R. E., Taylor, P. T., and Lawson, B. J. (1976). Regional geologic framework off northeastern United States, *Am. Assoc. Pet. Geol. Bull.* **60**, 926–951.

Schlumberger Ltd. (1970). "Fundamentals of Dipmeter Interpretations." Schlumberger, New York.

Schmalz, R. F. (1970). Environment of marine evaporite deposition, *Minerals Ind.* **35**, 1–7.

Schmitt, G. T. (1953). Regional stratigraphic analysis of Middle and Upper marine Jurassic in northern Rocky Mountains–Great Plains, *Am. Assoc. Pet. Geol. Bull.* **37**, 355–393.

Schneiderhöhn, H. (1953). Konvergenzerscheinungen zwischen magmatischen und sedimentären Lagerstätten, *Geol. Rundsch.* **42**, 34–43.

Scholle, P. A. (1978). "Carbonate Rock Constituents, Textures, Cements, and Porosities." Am. Assoc. Pet. Geol. Mem. 27.

Schönberger, H., and De Roever, E. W. F. (1974). Possible origin of diamonds in the Guiana shield: comment, *Geology* **2**, 474–475.

Schramm, M. W. (1964). Paleogeologic and quantitative lithofacies analysis, Simpson Group, Oklahoma, *Am. Assoc. Pet. Geol. Bull.* **48,** 1164–1195.

Schramm, M. W. (1968). Application of trend analysis to pre-Morrow surface, southeastern Hugoton embayment area, Texas, Oklahoma, and Kansas, *Am. Assoc. Pet. Geol. Bull.* **52,** 1655–1661.

Schreiber, B. C., and Schreiber, E. (1977). The salt that was, *Geology* **5,** 527–528.

Schultz, L. K., and Grover, R. L. (1974). Geology of Georges Bank basin, *Am. Assoc. Pet. Geol. Bull.* **58,** 1159–1168.

Schwartz, M. L. (ed.) (1973). "Barrier Islands." Dowden, Hutchinson, and Ross, Stroudsburg, Pennsylvania.

Schwartz, M. L. (ed.) (1972). "Spits and Bars." Dowden, Hutchinson, and Ross, Stroudsburg, Pennsylvania.

Schwartzacher, W. (1961). Petrology and structure of some Lower Carboniferous reefs in northwestern Ireland, *Am. Assoc. Pet. Geol. Bull.* **45,** 1481–1503.

Schwarzacher, W. (1975). Sedimentation models and quantitative stratigraphy, *Dev. Sediment.* **19.**

Scott, J., Collins, G. A., and Hodgson, G. W. (1954). Trace metals in the McMurray oil sands and other Cretaceous reservoirs in Alberta, *Trans. Can. Inst. Mining and Metall.* **57,** 34–40.

Scott, K. M. (1966). Sedimentology and dispersal pattern of a Cretaceous flysch sequence, Patagonian Andes, southern Chile, *Am. Assoc. Pet. Geol. Bull.* **50,** 72–107.

Scrutton, R. A. (1973). Structure and evolution of the sea floor south of South Africa, *Earth Planetary Sci. Lett.* **19,** 250–256.

Scrutton, R. A. (1973a). Gravity results from the continental margin of south-western Africa, *Mar. Geophys. Res.* **2,** 11–22.

Scrutton, R. A., and Dingle, R. V. (1974). Basement control over sedimentation on the continental margin west of southern Africa, *Trans. Geol. Soc. Afr.* **77,** 253–260.

Seal, W. L., and Gilreath, J. A. (1976). How to get the most gas from noncontinuous reservoirs, *World Oil* **182,** 38–42.

Seglund, J. A. (1974). Collapse-fault systems of Louisiana Gulf Coast, *Am. Assoc. Pet. Geol. Bull.* **58,** 2389–2397.

Seilacher, A. (1964). Biogenic sedimentary structures, *In* "Approaches to Paleoecology" (J. Imbrie and N. D. Newell, eds.), pp. 296–316. Wiley, New York.

Seilacher, A. (1967). Bathymetry of trace fossils, *Mar. Geol.* **5,** 413–428.

Selley, R. C. (1965). Diagnostic characters of fluviatile sediments of the Torridonian Formation (Precambrian) of North-West Scotland, *J. Sediment Pet.* **35,** 366–380.

Selley, R. C. (1970). "Ancient Sedimentary Environments." Chapman and Hall, London.

Sengupta, S. (1966). Geological and geophysical studies in western part of Bengal Basin, India, *Am. Assoc. Pet. Geol. Bull.* **50,** 1001–1017.

Shabica, S. V., and Boucot, A. V. (1976). The depth inhabited by Silurian brachiopod communities: comment and reply, *Geology* **4,** 132.

Shannon, J. P., and Dahl, A. R. (1971). Deltaic stratigraphic traps in West Tuscola Field, Taylor County, Texas, *Am. Assoc. Pet. Geol. Bull.* **55,** 1194–1205.

Shelton, J. W. (1965). Trend and genesis of lowermost sandstone unit of Eagle Sandstone at Billings, Montana, *Am. Assoc. Pet. Geol. Bull.* **49,** 1385–1397.

Shelton, J. W. (1967). Stratigraphic models and general criteria for recognition of alluvial, barrier-bar, and turbidity-current sand deposits, *Am. Assoc. Pet. Geol. Bull.* **51,** 2441–2461.

Shelton, J. W. (1968). Role of contemporaneous faulting during basinal subsidence, *Am. Assoc. Pet. Geol. Bull.* **52,** 399–413.

Shelton, J. W. (1973). Dielectric anisotropy and grain orientation: discussion, *Am. Assoc. Pet. Geol. Bull.* **59**, 1787–1788.

Shelton, J. W. (1973a). "Models of Sand and Sandstone Deposits: A Methodology for Determining Sand Genesis and Trend." Oklahoma Geol. Surv. Bull. 118.

Shelton, J. W., and Mack, D. E. (1970). Grain orientation in determination of paleocurrents and sandstone trends, *Am. Assoc. Pet. Geol. Bull.* **54**, 1108–1119.

Shelton, J. W., Terrell, D. W., and Karvelot, M. D. (1972). "Genesis and Geometry of Sandstones." Oklahoma City Geol. Soc. Publ.

Shenhav, H. (1971). Lower Cretaceous sandstone reservoirs, Israel: petrography, porosity, permeability, *Am. Assoc. Pet. Geol. Bull.* **55**, 2194–2224.

Shepard, F. P. (1973). "Submarine Geology," 3rd ed. Harper, New York.

Shepard, F. P., and Dill, R. F. (1966). "Submarine Canyons and Other Sea Valleys." Rand McNally, Chicago Illinois.

Sheriff, R. E. (1976). Inferring stratigraphy from seismic data, *Am. Assoc. Pet. Geol. Bull.* **60**, 528–542.

Short, K. C., and Stäuble, A. U. (1967). Outline of geology of Niger Delta, *Am. Assoc. Pet. Geol. Bull.* **51**, 761–779.

Shideler, G. L. (1970). Provenance of Johns Valley Boulders in Late Paleozoic Ouachita facies, southeastern Oklahoma and southwestern Arkansas, *Am. Assoc. Pet. Geol. Bull.* **54**, 789–806.

Shier, D. E. (1969). Vermetid reefs and coastal development in the Ten Thousand Islands, Southwest Florida, *Geol. Soc. Am. Bull.* **80**, 485–508.

Shinn, E. A., Halley, R. B., Hudson, J. H., and Lidz, B. H. (1977). Limestone compaction: an enigma, *Geology* **5**, 21–24.

Shirley, M. L., and Ragsdale, J. A. (eds.) (1966). "Deltas in Their Geological Framework." Houston Geol. Soc. Publ.

Short, K. C., and Stäuble, A. J. (1967). Outline of geology of Niger Delta, *Am. Assoc. Pet. Geol. Bull.* **51**, 761–779.

Signor, D. C. (1973). Laboratory studies related to artificial recharge (abstract), *Am. Assoc. Pet. Geol.,* **57**, 1605–1606.

Skibitzke, H. E. (1961). "Temperature Rise Within Radioactive Liquid Wastes Injected into Deep Formations." U.S. Geol. Surv., Prof. Paper 386-A.

Skinner, H. C. W. (1963). Precipitation of calcian dolomites and magnesium calcites in the southeast of South Australia, *Am. J. Sci.* **261**, 449–472.

Skorniakova, N. S., and Petelin, V. P. (1967). Sediments of the central part of the southern Pacific, *Okeanologiia* **7**, 1005–1019.

Sloss, L. L., and Scherer, W. (1975). Geometry of sedimentary basins: applications to Devonian of North America and Europe, *In* "Quantitative Studies in the Geological Sciences" (E. H. T. Whitten, ed.), pp. 71–88. Geol. Soc. Am. Mem. 142.

Smith, F. G. (1966). "Geological Data Processing." Harper, New York.

Smith, P. V. (1954). Studies on the origin of petroleum: occurrence of hydrocarbons in recent sediments, *Am. Assoc. Pet. Geol. Bull.* **38**, 377–404.

Sokolov, V. A. (1957). Possibility of formation and migration of oil in young sedimentary deposits, *In* "Problems of Oil Migration and Formation of Oil and Gas Deposits" *(Proc. Lvov Conf., 1957)*, pp. 59–63. Gostoptekhizdat, Moscow, 1959.

Sokolowski, S. (ed.) (1970). "Geology of Poland, 1, Pre-Cambrian and Palaeozoic." Geol. Inst. Poland.

Solomon, P. J. (1965). Investigations into sulphide mineralization at Mt. Isa, Queensland, *Econ. Geol.* **60**, 737–765.

Spencer, A. M. (ed.) (1974). "Mesozoic–Cenozoic Orogenic Belts." Geol. Soc. London Spec. Publ. 4, Scottish Academic Press.

Spjeldnoes, N. (1961). Ordovician climatic zones, Norsk. Geol. Tidsskr. 41, 45–77.

Spjeldnoes, N. (1973). Moraine stratigraphy, with examples from the basal Cambrian ("Eocambrian") and Ordovician glaciations, Bull. Geol. Inst. Univ. Uppsala 5, 165–171.

Sprunt, E. S., and Nur, A. (1976). Reduction of porosity by pressure solution: experimental verification, Geology 4, 463–466.

Stanley, D. J. (1967). Comparing patterns of sedimentation in some modern and ancient submarine canyons, Earth Planetary Sci. Lett. 3, 371–380.

Stanton, R. L. (1972). "Ore Petrology." McGraw Hill, New York.

Stearns, D. W. (1971). Mechanisms of drape folding in the Wyoming Province, In "Guidebook," Ann. Florida Conf., 23rd pp. 125–143. Wyoming Geol. Assoc. Publ.

Steele, R. J. (1976). Some concepts of seismic stratigraphy with application to the Gippsland Basin, Aust. Pet. Exp. Assoc. J. 16, 67–71.

Stehli, F. G., Creath, W. B., Upshaw, C. F., and Forgotson, J. M. (1972). Depositional history of Gulfian Cretaceous of East Texas Embayment, Am. Assoc. Pet. Geol. Bull. 56, 38–67.

Steiner, J. (1977). An expanding Earth on the basis of sea-floor spreading and subduction rates, Geology 5, 313–318.

Stephenson, L. P. (1977). Porosity dependence on temperature: limits on maximum possible effect, Am. Assoc. Pet. Geol. Bull. 61, 407–415.

Stetson, H. C., and Trask, P. D. (1953). "The Sediments of the Western Gulf of Mexico." MIT and Woods Hole Ocean. Inst. Publ. 12.

Stewart, A. J. (1967). An interpretation of the structure of the Blatherskite Nappe, Alice Springs, Northern Territory, J. Geol. Soc. Aust. 14, 175–184.

Stewart, J. H. (1971). Basin and range structure: a system of horsts and grabens produced by deep-seated extension, Geol. Soc. Am. Bull. 82, 1019–1044.

Stewart, J. H. (1972). Initial deposits in the Cordilleran Geosyncline: evidence of a Late Precambrian (<850 m.y.) continental separation, Geol. Soc. Am. Bull. 83, 1345–1360.

Stewart, R. J. (1976). Turbidites of the Aleutian abyssal plain: mineralogy, provenance, and constraints for Cenozoic motion of the Pacific plate, Geol. Soc. Am. Bull. 87, 793–808.

St. John, V. P. (1970). The gravity field and structure of Papua and New Guinea, Aust. Pet. Exp. Assoc. J. 10, 41–55.

Stokes, W. L. (1961). Fluvial and eolian sandstone bodies in Colorado Plateau. In "Geometry of Sandstone Bodies," pp. 151–178. Am. Assoc. Pet. Geol. Spec. Publ.

Stone, W. D. (1972). Stratigraphy and exploration of the Lower Cretaceous Muddy Formation, northern Powder River Basin, Wyoming and Montana, Mountain Geol. 9, 355–378.

Stoneley, R. (1967). The structural development of the Gulf of Alaska sedimentary province in southern Alaska, Q. J. Geol. Soc. London 123, 25–57.

Stride, A. H. (1963). Current-swept sea floors near the southern half of Great Britain, Q. J. Geol. Soc. London 119, 175–199.

Suttner, L. J. (1969). Stratigraphic and petrographic analysis of Upper Jurassic–Lower Cretaceous Morrison and Kootenai Formations, southwest Montana, Am. Assoc. Pet. Geol. Bull. 53, 1391–1410.

Sutton, R. G. (1959). Use of flute casts in stratigraphic correlation, Am. Assoc. Pet. Geol. Bull. 43, 230–237.

Swann, D. H. (1964). Late Mississippian rhythmic sediments of Mississippi Valley, Am. Assoc. Pet. Geol. Bull. 48, 637–658.

Swann, D. H., and Atherton, E. (1948). Subsurface correlation of lower Chester strata of the Eastern Interior Basin, *J. Geol.* **56**, 269–287.

Swift, D. J. P., and Lyall, A. K. (1968). Reconnaissance of bedrock geology by sub-bottom profiler, Bay of Fundy, *Geol. Soc. Am. Bull.* **79**, 639–646.

Swift, D. J. P., and McMullen, R. M. (1968). Preliminary studies of intertidal sand bodies in the Minas Basin, Bay of Fundy, Nova Scotia, *Can. J. Earth Sci.* **5**, 175–183.

Szumilas, D. (1977). Using computer for time-to-depth conversion and structure mapping in complexly faulted areas. (abstract), *Am. Assoc. Pet. Geol. Bull.* **61**, 834.

Tanner, S. F. (1965). Upper Jurassic paleogeography of the Four Corners region, *J. Sediment. Pet.* **35**, 564–574.

Tazieff, H., Varet, J., Barberi, F., and Giglia, G. (1972). Tectonic significance of the Afar (or Danakil) Depression, *Nature (London)* **235**, 144–147.

Tellen, K. E. (1977). Paleocurrents in part of the Franciscan complex, California, *Geology* **5**, 49–51.

Temple, P. G. (1968). Mechanics of large-scale gravity sliding in the Greek Peleponnesos, *Geol. Soc. Am. Bull.* **79**, 687–700.

Theis, C. V. (1955). Geologic and hydrologic factors in ground disposal of waste, *Proc. Sanit. Eng. Conf., Baltimore, Maryland, 1954* pp. 261–283. U.S. Atomic Energy Comm. Publ. 275.

Thode, H. G., Monster, J., and Dunford, H. B. (1958). Sulphur isotope abundance in petroleum and associated materials, *Am. Assoc. Pet. Geol. Bull.* **42**, 2619–2641.

Thomas, B. M., and Smith, D. N. (1974). A summary of the petroleum geology of the Carnarvon Basin, *Aust. Pet. Exp. Assoc. J.* **14**, 66–76.

Thomas, R. L., Kemp, A. L., and Lewis, C. F. M. (1973). The surficial sediments of Lake Huron, *Can. J. Earth Sci.* **10**, 226–271.

Thomas, W. A. (1973). Southwestern Appalachian structural system beneath the Gulf Coastal Plain, *Am. J. Sci.* **273-A**, 372–390.

Thompson, T. L. (1976). Plate tectonics in oil and gas exploration of continental margins, *Am. Assoc. Pet. Geol. Bull.* **60**, 1463–1501.

Thomson, L. D., and Russell, R. T. (1971). Discovery, exploration, and investigations of phosphate deposits in Queensland, *Proc. Aust. Inst. Min. Metall.* **240**, 1–14.

Tillman, R. W. (1971). Petrology and paleoenvironments, Robinson Member, Minturn Formation (Desmoinesian), Eagle Basin, Colorado, *Am. Assoc. Pet. Geol. Bull.* **55**, 593–620.

Tissot, B., Durand, B., Espitalié, J., and Combaz, A. (1974). Influence of nature and diagenesis of organic matter in formation of petroleum, *Am. Assoc. Pet. Geol. Bull.* **58**, 499–506.

Tixier, M. P., and Forsythe, R. L. (1951). Application of electric logging in Canada, *Bull. Can. Inst. Min. Metall.* September, 358–369.

Tjhin, K. T. (1976). Trobriand basin exploration, Papua New Guinea, *Aust. Pet. Exp. Assoc. J.* **16**, 81–90.

Todd, R. G. (1976). Oolite-Bar progradation, San Andres Formation, Midland Basin, Texas, *Am. Assoc. Pet. Geol. Bull.* **60**, 907–925.

Todd, T. W. (1964). Petrology of Pennsylvanian rocks, Bighorn Basin, Wyoming, *Am. Assoc. Pet. Geol. Bull.* **48**, 1063–1090.

Trask, P. D. (1939). Organic content of recent marine sediments, *In* "Recent Marine Sediments." Soc. Econ. Paleontol. Mineral. Publ.

Trendall, A. F. (1973). Precambrian iron-formations of Australia, *Econ. Geol.* **68**, 1023–1034.

Trescott, P. C. (1964). The Influence of Preceding Sedimentary Patterns on the Thickness of a Pennsylvanian Coal. M. Sc. thesis, Univ. Illinois.

Trettin, H. P. (1969). "Geology in Ordovician to Pennsylvanian Rocks, M'Clintock Inlet, North Coast of Ellesmere Island, Canadian Arctic Archipelago." Geol. Surv. Can. Bull. 183.

Trusheim, F. (1960). Mechanism of salt migration in northern Germany, Am. Assoc. Pet. Geol. Bull. **44**, 1519–1540.

Tuke, M. F. (1968). Autochthonous and allochthonous rocks in the Pistolet Bay area in northernmost Newfoundland, Can. J. Earth Sci. 501–513.

Uchupi, E., Emery, K. O., Bowin, C. O., and Phillips, J. D. (1976). Continental margin off western Africa: Senegal to Portugal, Am. Assoc. Pet. Geol. Bull. **60**, 809–878.

UNESCO (1973). Genesis of Precambrian iron and manganese deposits, Proc. Kiev. Symp. Earth Sci. **9**.

Vail, P. R., Mitchum, R. M., Sangree, J. B., and Thompson, S. (1975). Eustatic cycles from seismic sequence analysis, Am. Assoc. Pet. Geol. Symp., Dallas, Texas.

Vail, R. S. (1965). Environmental Mapping of Some Lower Pennsylvanian Strata of a Small Area in Southeastern Illinois. B.Sc. thesis, Univ. Illinois.

Van Der Lingen, G. J. (ed.) (1977). "Diagenesis of Deep-Sea Biogenic Sediments." Benchmark Papers in Geol. 40, Dowden, Hutchinson, and Ross, Stroudsburg, Pennsylvania.

Van Hoorn, B. (1969). Submarine canyon and fan deposits in the Upper Cretaceous of the south-central Pyrenees, Spain, Geol. Mijnbouw **48**, 67–72.

Van Houten, F. B. (1973). Meaning of molasse, Geol. Soc. Am. Bull. **84**, 1973–1975.

Van Houton, F. B. (ed.) (1977). "Ancient Continental Deposits." Benchmark Papers in Geol. 43, Dowden, Hutchinson, and Ross, Stroudsburg, Pennsylvania.

Van Straaten, L. M. J. U. (ed.) (1964). Deltaic and shallow marine deposits, Dev. Sediment. **1**.

Vasil'yev, V. G. (ed.) (1968). "Geology of Oil." Ministerstvo Geologii SSSR.

Vassoyevich, N. B., Korchagina, Y. I., Lopatin, N. V., and Chernyshev, V. V. (1969). Principal phase of oil formation, Moskov. Univ. Vestn. **6**, 3–27 [English transl.: Int. Geol. Rev. **12**, 1276–1296 (1970)].

Veevers, J. J. and Wells, A. T. (1961). The Geology of the Canning Basin, Western Australia. Bur. Mineral Resources (Canberra), Bull. **60**, 323 p.

Vincelette, R. R., and Soeparjadi, R. A. (1976). Oil-bearing reefs in Salwati Basin of Irian Jaya, Indonesia, Am. Assoc. Pet. Geol. Bull. **60**, 1448–1462.

Visher, G. S. (1965). Use of vertical profile in environmental reconstruction, Am. Assoc. Pet. Geol. Bull. **49**, 41–61.

Von Der Borch, C. C. (1967). Marginal plateaus: their relationship to continental shelf sedimentary basins in southern Australia, J. Geol. Soc. Aust. **14**, 309–316.

Wageman, J. M., Hilde, T. W. C., and Emery, K. O. (1970). Structural framework of East China Sea and Yellow Sea, Am. Assoc. Pet. Geol. Bull. **54**, 1611–1643.

Walker, C. T. (1964). Paleosalinity in Upper Visean Yoredale Formation of England—geochemical method for locating porosity, Am. Assoc. Pet. Geol. Bull. **48**, 207–220.

Walker, C. T. (1968). Evaluation of boron as a paleosalinity indicator and its application to offshore prospects, Am. Assoc. Pet. Geol. Bull. **52**, 751–766.

Walker, R. G. (1978). Deep-water sandstone facies and ancient submarine fans: models for exploration for stratigraphic traps, Am. Assoc. Pet. Geol. Bull. **62**, 932–966.

Wanless, H. R. (1969). Eustatic shifts in sea level during the deposition of Late Paleozoic sediments in the central United States, In "Cyclic Sedimentation in the Permian Basin" (J. G. Elam and S. Chuber, eds.), pp. 41–54. West Texas Geol. Soc. Publ.

Wanless, H. R., Baroffio, J. R., Gamble, J. C., Horne, J. C., Orlopp, D. R., Rocha-Campos,

A., Souter, J. E., Trescott, P. C., Vail, R. F., and Wright, C. R. (1970). Late Paleozoic deltas in the central and eastern United States, *In* "Deltaic Sedimentation, Modern and Ancient." Soc. Econ. Paleontol. Mineral. Spec. Publ. 15.

Wardlaw, N. C. (1976). Pore geometry of carbonate rocks as revealed by pore casts and capillary pressure, *Am. Assoc. Pet. Geol. Bull.* **60**, 245–257.

Warman, H. R. (1971). Why explore for oil and where? *Aust. Pet. Exp. Assoc. J.* **11**, 9–13.

Warren, P. S., and Stelck, C. R. (1958). The Nikanassin–Luscar hiatus in the Canadian Rockies, *Trans. R. Soc. Can., 3rd.* **52**, 55–62.

Watson, J. M., and Swanson, C. A. (1975). North Sea—major petroleum province, *Am. Assoc. Pet. Geol. Bull.* **59**, 1098–1112.

Weaver, C. E. (1967). Potassium, illite and the ocean, *Geochim. Cosmochim. Acta* **31**, 2181–2196.

Weaver, C. E. (1968). Geochemical study of a reef complex, *Am. Assoc. Pet. Geol. Bull.* **52**, 2153–2169.

Weaver, C. E., and Beck, K. C. (1977). Miocene of the S. E. United States, *Dev. Sediment.* **22**.

Webb, J. E. (1974). Relation of oil migration to secondary clay cementation, Cretaceous sandstones, Wyoming, *Am. Assoc. Pet. Geol. Bull.* **58**, 2245–2249.

Weber, K. J. (1971). Sedimentological aspects of oil fields in the Niger Delta, *Geol. Mijnbouw* **50**, 559–576.

Weber, K. J., and Daukoru, E. (1975). Petroleum geology of the Niger Delta, *Proc. World Pet. Congr., 9th* **2**, 209–221.

Weeks, L. G. (1952). Factors of sedimentary basin development that control oil occurrence, *Am. Assoc. Pet. Geol. Bull.* **36**, 2071–2124.

Weeks, L. G. (1961). Origin, migration, and occurrence of petroleum, *In* "Petroleum Exploration Handbook." McGraw-Hill, New York.

Weeks, L. G., and Hopkins, B. M. (1967). Geology and exploration of three Bass Strait basins, Australia, *Am. Assoc. Pet. Geol. Bull.* **51**, 742–760.

Weimer, R. J. (1966). Time-stratigraphic analysis and petroleum accumulations, Patrick Draw field, Sweetwater County, Wyoming, *Am. Assoc. Pet. Geol. Bull.* **50**, 2150–2175.

Weimer, R. J. (1970). Rates of deltaic sedimentation and intrabasin deformation, Upper Cretaceous of Rocky Mountain region, *In* "Deltaic Sedimentation, Modern and Ancient," pp. 270–292. Soc. Econ. Paleontol. Mineral. Spec. Publ. 15.

Weimer, R. J. (1977). "Deltaic and Shallow Marine Sandstones: Sedimentation, Tectonics, and Petroleum Occurrences." Am. Assoc. Pet. Geol. Cont. Educ. Cat. 876.

Wells, A. T., Forman, D. J., Ranford, L. C., and Cook, P. J. (1970). "Geology of the Amadeus Basin, Central Australia." Bur. Mineral Resources, Canberra Bull. 100.

Wermund, E. G., and Jenkins, W. A. (1970). Recognition of deltas by fitting trend surfaces to Upper Pennsylvanian sandstones in north-central Texas, *In* "Deltaic Sedimentation, Modern and Ancient," pp. 156–269. Soc. Econ. Paleontol. Mineral. Spec. Publ. 15.

West, I. M. (1973). Carbonate cementation of some Pleistocene temperate marine sediments, *Sedimentology* **20**, 229–249.

White, D. (1935). Metamorphism of organic sediments and derived oil, *Am. Assoc. Pet. Geol. Bull.* **19**, 589–617.

White, W. A. (1972). Deep erosion by continental ice sheets, *Geol. Soc. Am. Bull.* **83**, 1037–1056.

Whittaker, A. (1975). A postulated post-Hercynian rift valley system in southern Britain, *Geol. Mag.* **112**, 137–149.

Whitten, E. H. T. (ed.) (1975). "Quantitative Studies in the Geological Sciences." Geol. Soc. Am. Mem. 142.

accumulation, continental

s models for the Ouachita

consolidation of kaolinitic

d around Gulf of St. Law-

history, *Aust. Pet. Exp.*

iddle Ordovician rocks of

of Ordovician and younger *Assoc. Pet. Geol. Bull.* **47,**

eru, *Am. Assoc. Pet. Geol.*

ins, *Am. Assoc. Pet. Geol.*

pringer-Verlag, Berlin and

ientation, *Am. Assoc. Pet.*

ientation: reply to John W.

iques des sillons aquitains,

nian Red Fork Sandstone in *Geol. Bull.* **52,** 1638–1654.

ntal Shelf of North-West ritain.

nbayment, South Australia,

W. H. (1973). Pliocene and

and Texas, *Am. Assoc. Pet.*

drift, and plate tectonics.

on, and Ross, Stroudsburg,

iver Basin, *Am. Assoc. Pet.*

Dating of Paleosols." Israel

urrents and paleogeography 5—1540.

s tectonic framework, *Am.*

rifting in northern Nevada: *Geology* **6,** 111–116.

Subject Index

A

Abyssal environments, 9–10
  geomorphology of, 275–282
Abyssal plain deposits, geomorphology of, 278–282
Adige River, sediment from, in Po Delta, 200
Afghanistan, Amu Darya Delta of, 224
Africa
  alluvial fan deposits in, 222
  Atlas orogenic belt of, 74
  basins of, 97
  cratonic nuclei in, 46
  crustal plate separation near, 479
  debris flows offshore of, 136
  Gabon Basin of, 81
  growth faults in, 188
  salt bed offshore of, 104
  salt extraction by natives of, 241, 243
  South, see South Africa
  southeast, sedimentary basin near, 83
  West, see West Africa
Agbada Formation, in Niger Delta, 192
Agulhas basin, continental slope of, 67–68
Alaska
  Cordilleran Geosyncline of, 39
  glaciomarine beds in, 219
Alaskan plate, Pacific crustal plate against, 72
Alberta (Canada)
  basin superposition in, 183
  Belly River Formation in, 147, 427
  Black Creek Basin of, 358, 361
  coal seams of, 468
  Drumheller Valley in, 211
  early barrier reefs of, 266

  Elk Point Basin of, 358, 361
  Ireton Formation of, 413
  Leduc reef trend in, 445, 448
  marine beds in, 153
  McMurray Formation in, 338
  miogeosyncline of, 27
  Nisku Member of, 445
  Ostracod Limestone of, 166
  Paleogeomorphic maps of, 429
  Pembina oil field of, 332, 449
  Pincher Creek gas fields in, 449
  Pleistocene lakes in, 17
  Prairie Formation in, 110
  Rocky Mountains of, 171, 259
  Stettler Formation in, 244
  Swan Hills Formation of, 266
  Turner Valley Field of, 352, 368
  Viking Formation in, 163
  Waterton Lakes area, 338
  Dina Formation of, 326
  structural configuration of, 467
Aleutian Abyssal Plain geomorphology of, 279
Aleutian Trench, abyssal plain south of, 279
Algeria
  hydrodynamic conditions of Sahara in, 463–464
  paleocurrent studies on, 431
Allochthon, movement of, 74
Alluvial fans, description and occurrence of, 219–220
Alpine orogenic belt, olistostromes in, 72
Alps, macrostratigraphic studies of, 310
Amadeus Basin, aquifers in, 489
Amazon cone, debris flows in, 136
Amazon Delta
  coal seam origination in, 289

mud sheets of, 254
sediment of, 152, 202, 248–249
sedimentation cycles in, 144
swamps of, 207
course of, 214
sediments of, 248–249
deposition, 395
Amsden Foundation, stratigraphic sections
    of, 376
Amu Darya Delta, depositional environment
    of, 224
Amur Delta, swamps of, 209
Anadarko Basin
    hydrocarbon-bearing sandstones in, 347
    Red Fork Sandstone of, 215, 217
Andaman Islands, Bengal Fan near, 114
Andaman Sea, mud sheets of, 254
Andes mountains, carbonate-quartz sand
    facies in, 286–287
Andros Island, carbonate banks of, 259
Angola
    Cuanza Basin of, 116
    salt bed offshore of, 104
Antarctica
    biofacies of, 157
    phosphate-rich waters of, 487
Apennine belt, formation of, 74
Appalachians
    miogeosyncline–eugeosyncline couple in,
        112
    Pocono Sandstone in, 314
    strata offset of, 97
Aquifers, exploration for, 487–491
Aquitaine Basin, flysch sequence in, 284
Arabie, alluvial fan deposits in, 222
Arabian crustal plate, splitting of, 66
Arabian Precambrian Shield, as probable al-
    luvial fan, 222
Aral Sea, desert area near, 224
Archean basement, of Western Australia, 44
Arctic and arctic islands (Canadian)
    glaciomarine beds of, 218–219
    Gondwanaland and, 219
    paleogeographic maps of, 422
    Paleozoic reefs of, 344
    Peel Sound Formation of, 395
    Uralides and, 64
    Weatherall Formation of, 269, 422
Arctic Circle
    muskeg of, 208

phosphorus-rich waters of, 487
Argentina, Colorado and Negro rivers of,
    214
Arkansas
    Johns Valley Formation of, 408–409
    macrostratigraphic studies in, 310
    Ouachita Mountains of, 97
Asiatic plate, Indian crustal plate thrusting
    under, 60
Asmari Formation, oil fields of, 449
Astrobleme, in sedimentary basins, 100
Athabasca (lake)
    provenance interpretations of, 328
    uranium deposits near, 469
Athabasca Sandstone, provenance interpre-
    tations of, 328
Atlantic area, structural framework of, 80
Atlantic Ocean
    ancient breakthrough to Mediterranean,
        155, 244
    effect on Brazilian deltas, 201
    sedimentary iron and manganese ores in,
        472
Atlantic Ridge, from continental mass
    movement, 29–30
Atlas orogenic belt, formation of, 74
August fields, as gas reservoir, 493
Aulacogenes, properties of, 24
Australia
    Amadeus Basin of, 489
    ancient submarine channels of, 276–277
    aquifers in, 487–490
    Archean basement of, 44
    banded sulfide ores of, 479
    basement faults in, 185
    basin superposition in, 182
    Bass Basin of, 21
    Blatherskite Nappe in, 368
    Bonaparte Gulf Basin of, 83
    Bowen Basin in, 24
    Exmouth Plateau of, 367
    glaciomarine beds in, 218
    glacial period in, 219
    Great Artesian Basin of, 424, 487
    Great Barrier Reef of, 47, 259
    Hamersley Basin of, 44
    hydrocarbon source rock analysis in, 340
    Lake Eyre of, 241
    Latrobe Formation in, 18, 233, 288–289
    MacDonnell Rangers of, 368–369

Narrabeen Group in, 312
oir reservoirs in, 444
Otway Basin of, 128
phosphate rock in, 483
  origin, 486
sedimentary iron and manganese ores in,
  472
submarine plateau of, 49
Sydney Basin of, 161, 494
Surat Basin of, 179, 389–390
Toolebuc Member of, 163
Western
  Barrow Island oil field of, 449
  basement faults in, 185
  Canning Basin of, 489
  Carnarvon Basin of, 488
  Hamersley Group in, 473
  iron deposits in, 140, 472, 473
  Northwest Shelf of, 88
  Pilbara manganese deposits in, 472–473
  uranium deposits in, 468–469
  Windalia Radiolarite of, 167
  Yampi Sound of, 473–474
Australian crust plate, direction of, 69

B

Bahamas
  argonite precipitation in, 139–140
  carbonate sediments of, 12, 101, 164, 236,
    259, 264, 286, 324
  depositional environment of, 43, 153
  reef growth in, 46
Baja California, phosphate beds in, 486
Balearic Basin, evaporitic sequence of, 245
Baltimore Canyon Trough, sedimentary
  rocks of, 121
Banda Sea, sedimentary basins of, 61
Bangkok, sedimentary basins of, 61
Bangladesh
  Bengal Fan near, 114
  depositional environment of, 46
Banks Island, Weatherall Formation of, 269
Barbados, porosity of limestone in, 336
Barium, in petroleum exploration, 345–346
Barrier islands, formation of, 249–252
Barrow Island oil field, rock types in, 449
Basement faults, in sedimentary basins,
  183–186
Basin hingelines, definition and descriptions
  of, 180–182

Basin studies group, functions and organization of, 30–33
Bass Basin, 21
  description of, 128
Bathyal environments, geomorphology of,
  275–282
Bavarian Alps, macrostratigraphic studies
  of, 311
Bay of Bengal, delta complex emptying into,
  46
Beachport Plateau, basement faults in,
  185–186
Beaufort Basin, sediments in, 92
Beetle Creek Formation, phosphorites in,
  485
Belden Formation, stratigraphic relationships of, 414
Bell Creek Field, paleogeomorphic map of,
  429, 430
Belly River Formation
  paleogeography of, 427
  regressive sequences in, 147
  rock types in, 153, 326, 332
Belt series
  ore deposits in, 482
    copper, 477
Bengal Fan, characteristics of, 9, 59–60,
  114
Benin Formation, in Niger Delta, 192
Big Horn Basin
  stratigraphic sections of, 376
  Tensleep Sandstone of, 456–457
Biofacies
  of barrier reefs, 266–267, 273
  maps of, 412–414
  of sedimentary basin, 155–157
Biogenic structures, examples of, 307–320
Bioherms
  development of, 100–101
  in sedimentary basins, 77
    effects, 40
  as stratigraphic traps, 445
Biostratigraphic analysis, 7
Bioturbation
  disturbance of lagoons by, 236
  effect on sheet sands, 249
Birdrong Sandstone, aquifers in, 488
Birrimian sequence, Tarkwaian sequence
  derivation from, 328
Bitlis Massif, mélanges in, 72

Bitter Springs Formation, structural features
   of, 369
Black Creek Basin, development of, 358,
   361
Black Sea
   euxinic environment in, 287
   swamps of, 209
Blatherskite Nappe, structural features of,
   367–368
Blythesdale Formation, aquifers in, 487–488
Bonaparte Gulf Basin, development of, 83
Booch Sandstone, as ancient delta distributary, 232
Borehole geophysical surveys, in subsurface
   mapping, 329
Boron, as water salinity indicator, 345
Botucatú Formation, as fossil sand dune,
   223
Bow Island gas field, as storage reservoir,
   493
Bowen Basin
   of Queensland, Australia, 24
   superposition on Surat Basin, 182
Brazil
   basins of, 97
   Botucatú Formation of, 223
   diamond-bearing sediments in, 328
   Doce Delta of, 200–201
   early swamps of, 214
   oil reservoirs in, 444
   Paraiba Delta of, 200–201
   sedimentary iron and manganese ores in,
      472
British Columbia
   basin superposition in, 183
   Cache Creek Series of, 153, 326–327, 427
   Columbia River basalts of, 225
   fossil reefs of, 274
   miogeosyncline of, 27
   Rocky Mountains of, 259
   Stephan shale of, 288
Brunei-Saigon Basin, sedimentation in, 61
Burgos Basin, growth faults in, 194
Burial metamorphism
   diagenesis from, 96
   of sedimentary rock, clay mineral analysis
      for, 342
Burma
   Irrawaddy Delta of, 239
   mud sheets of, 254

C

Cache Creek Series
   paleogeography of, 427
   rock types in, 153, 326–327
Cahill Formation, in Jabiluka uranium ore
   bodies, 472
Calcasieu Lake, collapse–fault structures in,
   109
California
   basement faults in, 185
   Death Valley of, 241
   exposed granitic rocks in, 297
   Nicaragua Rise of, 363
   offshore region of, 8
   Paso Robles Formation in, 327–328
   Sacramento Valley Basin of, 357, 359
   Salton Sea of, 206
   sedimentary basin changes in, 496
   Ventura Basin of, 444
Canada
   Arctic Islands, see Arctic Islands, Canada
   Alberta Basin of, 467
   Beaufort Basin in, 92
   Cardium Sandstone of, 347
   Cordilleran region of, 262–263
   early stromatoporoid reefs of, 269
   Exshaw Shale of, 146
   glaciomarine beds of, 218–219
   Inoceramus fossils in, 323
   Gulf of St. Lawrence in, 101
   hydrocarbon source rock analysis in, 340
   Northwest Territories of, 219
   Paleozoic reefs of, 344
   Precambrian basement in, 186
   reef sections of, 259
   Rocky Mountains of, 307
   Stettler Formation of, 244
   trace mineral analysis of oil-bearing rock
      in, 345–346
   uranium exploration in, 468, 469
   Weatherall Formation of, 269
Canary Islands Ridge, effect on Moroccan
   salt diapirs, 123
Cannel coal
   deposits of, 207
   derivation of, 288
Cape Hatteras, depositional margins off, 94
Cape Town, sea-floor movement near, 67–68
Capitan Reef, mineral deposits in, 141

Capricorn Group, of Great Barrier Reef, 262
Carbonate banks, in Bahamas, 12, 101, 164, 236, 259, 264, 286, 324
Carbonate muds and sands, description and occurrence of, 258–275
Carbonate rocks, analysis of, 344–346
Carbonate–quartz sand facies, description and occurrence of, 284–287
Carbonate sequences, of sedimentary basins, 12–15.
Carbondale Formation, coal seam in, 465
Cardium Sandstone, paleogeothermal gradients in, 347
Caribbean Sea
  basins underlying, 361
  Magdalena Delta of, 209
Carnarvon Basin, aquifers in, 488
Cascadia Basin, sediments overlying, 123
Caspar Formation, as fossil sand dune, 224
Caspian Sea, size of, 206
Castile Formation
  in Capitan Reef, 141
  depositional environment of, 245
Catastrophic events, effect on lithofacies, 155
Catahoula Tuff, of Felder uranium mine, 471
Catatumba Delta, meandering pattern of, 228
Catskill Delta, paleogeographic framework of, 420
Cauvery Basin, structure of, 87–88
Cayugan Series, salt deposits in, 245
Cedar Mesa Formation, of Cutler Group, 56
Celebes Sea, sedimentary basins of, 61
Central America, crustal rifts in, 67
Cerro Toro Formation, flysch deposits in, 311–312
Chalk beds, of England, 319–320
Charles Formation, algal pelletoid reefs of, 262
Chattanooga Shale
  paleogeologic maps of, 383
  as transgressive stratigraphic unit, 146
Chazy Group, Knox unconformity under, 458
Chemical waste, storage for, exploration, 494–497
Chester sands, as tidal current ridges, 256
Chester Series, size and development of, 361
Chile, Cerro Toro Formation in, 311–312

China, activated platform basins in, 22
China Coastal Stream, effect on biofacies, 157
Cincinnati Arch, Richmond Groups on, 383–384
Cinder cones of Bass Basin, 128
Cisco Series, fossil reefs of, 274
Clay minerals, rock analysis for, 342–344
Clay sediments, diagenesis of, 172
Cleveland Shale, paleogeographic framework of, 419–420
Clore Formation, underlying Degonia Formation, 397
Coal
  depositional environments of, 464–466
  deposits, from swamplands, 205
  exploration for, 464–468
Coal beds
  occurrence of, 18, 233, 288–289
  structural configuration for, 466–468
  vitrinite in, reflectance of, 441
Coal distribution maps, construction and uses of, 405–407
Coal facies, description and occurrence of, 288–290
Coast Range, batholiths of, 27
Coastal swamps, as depositional environments, 233
Collapse features, of salt beds, 107
Colombia, Magdalena Delta of, 209
Colorado
  coal deposits in, 465
  coarse clastics in, 409–410
  Denver Basin of, 364
  J Sandstone of, 425
  Mesaverde Group in, 148–149
  Navajo Sandstone of, 223
  paleogeography of, 415–416, 425
  Paradox Basin of, 56
  San Juan Basin of, 391
  Uranium deposits in, 469
Colorado Plateau
  growth faults in, 189
  uranium and vanadium ore exploration in, 468
  course of, 214
  depositional environment of, 224
  Grand Canyon of, 307
Columbia Basin, development of, 361

Columbia River
  basalts of, 225
  sediment movement in, 202
Compaction, sediment consolidation by,
  167–173
Compaction rates, in sedimentary basins,
  131
Computer
  use in geological data evaluation, 434–437
  use in paleotopographic map interpreta-
    tion, 385
Configurations, of sedimentary basins,
  351–391
Continental drift, theory of, 63
Copper ores, exploration for, 475–482
Copper Ridge Dolomite, oil accumulations
  in, 458
Cordilleran region
  cyclical sequences of, 262–263
  sedimentary deposits in, 38
    microstratigraphy, 326–327
Corsica, orogeny in, 72
Cratonic nuclei, of Archean basement, 46
Cross bedding, in sedimentary basins, 17–18
Cryptoexplosions, effect on sedimentary ba-
  sins, 100
Cuanza Basin, continental basin of, 116
Cuba, Gulf of Batabano in, 157, 443
Cupriferous shale, example of, 476
Cutbank Formation, sandstone bodies in,
  397
Cutler Group, stratigraphic framework of,
  56–57
Cyclic deposition, of sandstone in Mancos
  Shale, 377
Cyclical sequences
  in delta formation, 48, 149–150
  with limestone, 262–263
Cyclic stratification, of lakes, 207
Cyclic varves, in sabkhas, 241

D

Dampier Archipelago, Northwest Shelf of,
  88
Danakil Depression, rifting in, 66–67
Danakil Desert, evaporite deposits of, 241
Danube River, delta swamps of, 209
Data, marker bed use in selection of, 162–
  163

De Chelly Formation, of Cutler Group, 56
De Souza Sandstone, aquifers in, 489, 490
Dead Sea
  evaporite deposits of, 241
  size of, 206
Death Valley, evaporite deposits of, 241
Debris flows, description of, 136–137
Deeper-water facies, of Cordilleran
  Geosyncline, 39
Degonia Formation, isopach map of, 397,
  398
Delaware Basin, Delaware Mountain Group
  of, 330
Delaware Mountain Group, radioactivity
  and sonic characteristics of, 330
Delta lobes
  interfingering of, 148
  progradation of, 173
Deltic distributaries, paralic environments
  of, 228–253
Denmark, Tollund Man discovered in, 289
Denver Basin, structural maps of, 364, 365
Depocenters
  definition of, 116–117
  independent development of, 99
  in progradation of delta lobes, 173
  of sedimentary basins, 92–93
  sedimentation rate in, 131
  shift of, 77
Depositional edges, of sedimentary basins,
  41, 94–95
Depositional environments, 197–293
  abyssal, 9–10
  bathyal and abyssal as, 275–282
  classification of, 2–10
  of coal beds, 464–466
  criteria for recognition of, 290–293
  description of, 41–61
  geomorphic and sedimentary associations
    in, 197–282
  intracratonic–epicontinental, 7–9
  neritic, 254–275
  neritic–bathyal, 8
  paralic, 228–253
  of rock-time units, 414–420
Deserts, as depositional environment, 222–
  228
Devonian carbonate reefs, of North America, 46
Devonian flysch, 27

Diachronism, of sedimentary basins, 159–163
Diachronous unit(s)
    Exshaw Shale as, 146
    lithostratigraphic aspects of, 151
Diagenesis
    from burial metamorphism, 96
    of carbonate rocks, 344
    of sediments, 170–173
    compaction and, 346–349
Diamictite, of Cordilleran Geosyncline, 39
Diamonds
    in conglomeratic beds, 98
    sea-floor spreading and, 328–329
Diapirs
    formation of, 99
    in sedimentary basins, 77
Dielectric anisotropy, of sandstones, 349–350
Dina Formation, lithic sediments of, 326
Dinosaur skeletons, in Edmonton Formation, 211–212
Disconformities, in sedimentary basins, 177–180
Displaced margins, of sedimentary basins, 96–99
Dissolution features
    description of, 105–107
    of sedimentary basins, 99–111
Doce Delta, subaerial configurations of, 200–201
Dolomitized zones, in limestone beds, 447
Drill cuttings
    analysis of, 294–350
    by lithology, 321–324
    macrofossil identification in, 364
    porosity and permeability of, 334
Drumheller Valley, Edmonton Formation in, 211
Durban, sea-floor movement near, 67–68
Duvernay Formation, trace mineral analysis of, 345–346

E

Eagle Ford Formation, rock-type ratio map of, 402
Eagle Sandstone
    grain orientation in, 350
    as regressive sequence, 147

Eagle Valley, stratigraphic relationships of, 414–415
Earthquake epicenters, causes of, 63
East China Sea, sedimentary basin of, 60
East Siberian Sea, glacial outwash plain under, 219
East Texas Embayment, Eagle Ford Formation of, 402
East Texas oil field, production from Woodbine Sandstone, 383, 445
Economic geology, basin analysis application to, 438–497
Ecuador, Galapagos Rift near, 288
Edmonton Formation, deposition of, 211
Edwards Formation, rudist reefs of, 270
El Salvador, Middle America trench near, 52
Electric logs
    of cross bedding, 18
    in interpretation of time-stratigraphic units, 160
    limestone beds on, 166
    rock porosity studies by, 337
    in sedimentation rate studies, 134
    self-potential curves of, 331–332
    sections, of Red Fork Sandstone, 217, 218
    in studies of
        barrier islands, 250
        Powder River Basin, 174
        sheet sands, 250
    use in microstratigraphy, 320
    of Viking Formation, 163
Elk Point Basin, Black Creek Basin underlying, 358, 361
Elk Point Group, Prairie Formation in, 111
Ellesmerides, Uralides and, 64
Ellesmere Islands, carbonates in, 64
England
    calcareous quartz sandstones of, 324
    chalk beds of, 319–320, 483
    phosphate beds in, 483
    Yoredale Formation of, 345
Engleford Shale, of East Texas oil field, 383
Entrada Formation, as fossil sand dunes, 223
Epigenetic origin, of mineral ores, 476
Erickson Sandstone, paleogeographic map of, 425, 427
Eritrea, Danakil Desert of, 241
Eroded margins, of sedimentary basins, 95–96

Erosion cycles, in provenance interpretations, 328
Ethiopia, Danakil Desert of, 241
Eugeosyncline
  in Appalachian region, 112
  characteristics of, 26–27
  of lava flow sheets, 252
  of submarine trenches, 53
Eugeosyncline basins, tectonic control of, 68
Euphrates river, delta of, 224
Eureka and Avon Hills oil and gas fields, Viking Formation in, 332
Euxinic facies, description and occurrence of, 287–288
Evaporites, in carbonate–salt mud sheets, 239, 241
Everglades of Florida
  coal seal origination in, 289
  swamps of, 207
Evolution, of sedimentary basins, 37–129
Excelsior Field (Alberta), Viking Formation in, 163
Exmouth Plateau, structural maps of, 367
Exploration, for natural resources, 36
Exshaw Shale, as transgressive stratigraphic unit, 146

F

Faults, displaced margins from, 96–97
Felder uranium mine, ore body sections in, 471
Field notes
  on biogenic structures, 308
  on crop distribution, 294–295
  on lithological studies, 303
Finke Group, in Amadeus Basin, 490
Fiji Basin, sedimentation in, 53
Flippen Limestone, description of, 275
Florida
  coast, carbonate sequences of, 12
  limestone waste reservoirs in, 496
  Ocala Group in, 490
  Ten Thousand Islands of, 259–260
Florida Strait, 259
  depositional environment of, 153
  Jordan Knoll of, 104
Flowerpot Shale, as cupriferous shale, 476
Flysch facies

of Cordilleran Geosyncline, 39
description of, 283–284
Fossil biota, biofacies maps and, 412
Fossil reefs, modern reefs compared to, 274
Fossil sand dunes, occurrence of, 223
Fossil soils, detection of, 178
Fossils
  trace type
    in turbidite sequences, 20
    types, 309
Fountain Formation, stratigraphic relationships of, 415
Fracture porosity, of sedimentary rock, 338
France, Pyrenees of, 284
Franciscan Formation, surface mapping of, 297–298
Frio Formation, of Burgos Basin, 195–196

G

Gabon
  Ogooue Delta of, 234
  salt bed offshore of, 104
  sedimentary iron and manganese ores in, 472
Gabon Basin developmental stages of, 81
Galapagos Rift
  hydrogen sulfide from, 288
  organisms in, 200
Gallup Formation, paleogeomorphic map of, 430
Gamma radiation, of rocks, measurement of, 163
Ganges–Brahmaputra Delta, 114
  sedimentation of, 60
  rate, 131
Ganges–Brahmaputra river system, depositional environment of, 46
Gas Draw Sandstone, shoestring sandstone bodies of, 397, 399
Gas reservoirs, see also Oil
  ancient river channels as, 312
  basin potential for, 444–453
  computer use in estimation of, 437
  estimation of, 458, 460
  in fractured zones, 448–449
  in rivers and floodplains, 209, 211
  rock porosity importance in, 336
  sand sheets as, 230

sedimentary basin configuration studies for, 352
source rock potential for, 439–443
structural and stratigraphic traps for, 453–462
Gault Formation, turbidites in, 311
Geology, economic, *see* Economic geology
Geomagnetic surveys, sea-floor spreading from, 62–63
Geomorphic features, of depositional environments, 197–282
Geophysical data, use in oil explorations, 352–353
Georges Bank Basin, sedimentary strata of, 121
Georgina Basin, phosphorites in, 483–484
Geothermal gradients, for oil formation, 439
Germany
    banded sulfide ores of, 479
    radioactive waste storage in, 495
Gevaram Canyon, as submarine channel, 277
Gevaram Formation, as submarine channel, 277–278
Ghana
    beach environment in, 42
    Tarkwaian sequence in, 328
Gibraltar, Atlantic Ocean breakthrough at, 155
Gila Desert, Gulf of California near, 224
Gillen Member, structural features of, 369
Gippsland Basin
    basement faults in, 185
    isopach map and sections of, 125, 126
    Kingfish oil field in, 450–451
    Latrobe Formation in, 18, 233, 288–289, 465
    Marlin Channel of, 277
    oil and gas deposits in, 233
Glacial outwash, occurrence of, 219
Glaciomarine sediments, deposition of, 218
Glenwood Formation, description of, 145–146
Godavari river, distributaries of, 228
Gondwanaland
    as early continent, 219
    glaciation of, 150
Goyllarisquisga Formation, lithofacies distribution of, 286–287

Graded bedding
    in flysch sequences, 284
    in Sacramento Valley Basin, 358
    in submarine channels, 276
    of turbidite sequences, 20
Grain gradation
    measurement of, 18
    in sand bodies, 150
Grain orientation, in sedimentary rocks, 349–350
Grand Banks
    sections across, 118
    submarine canyon near, 52
Grand Canyon, stratigraphic features of, 307
Gravitational slumping, of sediment, 49
Gravity surveys, of basement faults, 355
Great Artesian Basin
    aquifers in, 487, 490
    paleogeographic maps of, 424
Great Australian Bight, basement faults in, 186
Great Barrier Reef
    carbonate sedimentation in, 259, 260
    growth of, 47
    reefs resembling, 267
Great Lakes
    Cayugan series near, 245
    size of, 206
Great Salt Lake, size of, 206
Great Slave Lake
    dolomitizing solutions in, 448
    Precambrian basement in, 186
Greenland, Uralides and, 64
Greensand, phosphate beds in, 483
Greenstone belts, of Precambrian shields, 25–26, 37
Groote Eylandt manganese deposit, formation of, 472
Growth, of sedimentary basins, 99–111
Growth faults, in sedimentary basins, 187–196
Guatemala, Middle America trench near, 52
Guiana current, effect on Amazon sediments, 249
Gulf of Aden, sedimentary basin of, 66
Gulf of Alaska
    abyssal plain of, 279
    structural section of, 71–72
Gulf of Batabano
    biofacies of, 157

organic content of lime muds of, 443
Gulf of California
    early desert areas of, 224
    paleogeography of, 416
Gulf of Carpentaria
    banded sulfide ores of, 479
    phosphorites in, 483–484
Gulf of Korea, tidal current ridges of, 256
Gulf of Mexico
    Amazon cone in, 136
    barrier island formation in, 154
    continental slopes of, 200
    faulting of, 26
    Florida Strait of, 104
    mineral distribution map of, 405, 406
    organic content of clays in, 443
    physiography of, 8
    sedimentation rate in, 131
Gulf of St. Lawrence, salt domes in, 101
Gulf of Thailand Basin, sedimentation in, 61
Gunsight Limestone, description of, 275
Guyana, diamond-bearing sediments in, 328

H

Halgaito Formation of Cutler Group, 56
Hamersley Basin
    of Western Australia, 44
    iron deposits in, 473
Heavitree Quartzite, structural features of,
    368–369
Heavy minerals, of Paso Robles Formation,
    327–328
Helez Formation
    porosity and permeability of, 336
    as submarine channel, 278
Himalayan orogenic belt, delta formation
    from, 46
Himalayas, denudation rate of, 60
Honolulu aquifer, description of, 490–491
Huang Ho River, sediments from, 61
Hume Formation, paleogeographic maps of,
    422, 423
Hydrocarbons, See also Gas, Oil
    diagenesis in indications of, 173
    exploration for, sedimentary basin config-
        uration studies in, 352
    generation in sedimentary basins, kerogen
        role in, 347
    productive trends of, 118
reservoirs and traps
    formation, 89
    hydrodynamic conditions affecting,
        462–464
    near basin hingelines, 180–181
    in Niger Delta, 193
    truncated structures and, 384
    rock analysis for, 339–341
Hydrodynamic conditions, of hydrocarbon
    reservoirs, 462–464
Hydrodynamic maps, uses and construction
    of, 388–391

I

Iceland, volcanoes of, 252
Idaho, Phosphoria Formation in, 483
Illinois
    Carbondale Formation in, 465
    Degonia Formation in, 397
Illinois Basin
    Chester sands of, 256, 361
    stratigraphic framework of, 353–355
    Valmeyeran Series of, 132
India
    Bengal Fan of, 114
    Cauvery Basin of, 87–88
    Godavari river of, 228
    organic content of coastal clay muds of,
        443
    Kistna river of, 228
    sedimentary iron and manganese ores in,
        472
Indian crustal plate, northward thrusting of,
    60
Indian Ocean, sedimentary iron and man-
    ganese ores in, 472
Indonesia, Salawati Basin of, 366
Inoceramus fossils, in Western Canada, 323
Intracratonic–epicontinental environments,
    7–9
Ionian Sea, submarine channels of, 276
Iran
    Euphrates and Tigris rivers of, 224
    oil fields in Asmari Formation in, 449
    stratigraphic features of, 118, 307
Iraq
    Euphrates and Tigris rivers of, 224
    stratigraphic features of, 307
Ireton Formation

Tentaculites in, 413
trace mineral analysis of, 345–346
Irian Jaya, Salawati Basin of, 366
Iron ores
 exploration for, 472–475
 possible origin of, 474–476
Irrawaddy Delta, mud sheet of, 239
Isocarb maps, construction and uses of, 405–407
Isograde maps, construction and uses of, 408–409
Isolith maps, construction and uses of, 395–401
Isopach maps, of sedimentary basins, 369–374
Israel
 ancient submarine channels of, 276–277
 Dead Sea of, 241
 Helez Formation of, 276–277, 336
Italy
 Apennine belt of, 74
 orogeny in, 72
 Po Basin in, 493

J

J Sandstone, paleogeographic framework of, 417–419
Jabiluka uranium ore bodies, geology of, 472
Jackfork Group, flysch deposits of, 311
Jackpile Sandstone, as ancient river channel, 312
Japan, 61
 Sakhalin in, Amur Delta near, 209
Java Sea, sedimentary basins of, 61
Java Trench
 Bengal Fan and, 114
 crust structure beneath, 69
Johns Valley Formation, isograde maps of, 408–409
Jordon, Dead Sea of, 241
Jordan Knoll, seismic profile of, 104

K

Kansas, oil reservoirs in, 445
Kentucky, Palestine Sandstone of, 256
Kenya, salt saturation in Lake Turkana of, 142
Kerogen
 determination of, 339

in sedimentary basins, 96
hydrocarbon formation and, 347, 440
Kingfish oil field, sandstone reservoirs of, 450–451
Kistna river, distributaries of, 228
Knox unconformity, under Chazy Group, topography of, 458
Kolm Formation, uranium in, 469
Kombolgie Formation, in Jabiluka uranium ore bodies, 472
Kootenai Formation, sandstone channesl in, 397
Korea, coastal tidal current ridges of, 256
Kupferschiefer shales, formation of, 141, 476
Kuroshio Current, effect on biofacies, 157
Kuwait, marshy delta near, 224

L

Labette Formation, stratigraphic relationships in, 150
Labrador, submarine canyon of, 51–52
Lago Titicaco, rock masses of, 138
Lagoons, as depositional environments, 236
Laguna Madre, description of, 236
Lake Eyre, sediments in, 241
Lake Huron, sedimentation patterns of, 206–207
Lake Maracaibo, drainage patterns of, 228
Lakes, as depositional environments, 204–209
Lakes Entrance marine mudstones
 of Latrobe Formation, 277
 oil fields of, 451
Latrobe Formation
 in Australia, 18
 coal deposits in, 233, 288–289, 465, 467
 geomorphology of, 277
 oil fields in, 451
Lava flow sheets, as depositional environment, 252–253
Leduc reef trend
 in Alberta, Canada, 445
 dolomitizing solutions in, 448
Lead ores, exploration for, 475–482
Lewis Shale, delta lobes interfingering of, 149–150
Liberia, sedimentary iron and manganese ores in, 472

Libya
  alluvial fan deposits in, 222
  Sirte Basin of, 78
Limestone beds, dolomitized zones in, 447
Lithification, of sediments, 170–173
Lithofacies
  of sedimentary basins, 150–155
  stratigraphic associations and, 282–290
Lithofacies maps, types and uses of, 391–411
Lithogenesis, of sedimentary beds, 321–324
Lithology, of sedimentary basins, 10–20, 302–307, 321–324
Lithostratigraphic units, diachronism in, 164
Logging of sequences, in an outcrop, 378
Long Island, barrier island formation along, 154
Longshore currents
  effect on
    Amazon sediments, 249
    grain orientation, 349
  sediment movement by, 202, 238–239
Los Angeles Basin, deposition rate in, 363
Louisiana
  collapse–fault structures in, 109
  sedimentation cycles offshore of, 144–145
  Vermilian Block 16 oil field in, 451–452
Loves Creek Member, structural features of, 369

M

Macrostratigraphy of sedimentary basins, 294–320
Madison Formation, organic content of limestones in, 443
Magdalena Delta, swamps of, 209
Magnetic anomalies, in sedimentary basins, 64
Magnetic surveys, of basement configuration, 355
Malagasy, sedimentary basin near, 83
Malekula Island, paleogeography of, 53
Mancos Shale
  cyclic deposition of deltaic sediments in, 377
  delta lobes interfingering of, 149–150
Manganese ores
  exploration for, 472–475
  possible origin of, 474–476
Maps and sections

collation and representation of, 414–437
  of sedimentary basins, 351–437
    structural features, 364, 366
  vertical exagerration of, 353
Maracaibo Basin, biofacies maps of, 414
Marathon Mountains, strata offset in, 97
Marble Falls Group, early algal reef of, 266
Marker beds, in sedimentary basins, 162–163
Margins
  depositional, 94–95
  displaced, 96–99
  eroded, 95–96
  of sedimentary basins, 92–99
Marine seismic surveys, of abyssal environments, 9
Maritime Alps, submarine channels in, 278
Maroon Formation, stratigraphic relationships of, 415, 416
Martinique, Mont Pelée explosion on, 252–253
Maryland, stratigraphic accumulations near, 120–121
Mass sedimentation, characteristics of, 134–135
Maturation, of organic matter, in shales and limestones, 439
McArthur Group
  banded sulfide ores of, 479
  manganese derived from, 472
McDonnell Ranges, Heavitree Quartzite in, 368
McMurray Formation
  lithic sediments of, 326
  tarry oils from, 338
Mediterranean Basin
  Atlantic Ocean effects on, 155
  salt deposits under, 244
Mediterranean region, Alpine orogenic belt of, 72
Mediterranean Sea
  Balearic Basin as precursor of, 245
  debris flows on floor of, 137
Mekong Basin, sedimentation in, 61
Mélange
  debris flows and, 138
  in Dinaric region of Yugoslavia, 376
  from subduction, 72
Mereenic Sandstone, aquifers in, 489–490
Mesaverde Formation paleogeographic map of, 425, 427

Mesaverde Group, stratigraphic relation-
  ships in, 148–149
Mesozoic Age, Surat Basin of, 24
Messina Cone, submarine channels of, 276
Metallic sulfides, in Red Sea basin, 66
Mexico
  Burgos Basin of, 194
  carbonate banks of, 259
  Cordilleran Geosyncline of, 39
  Frio Formation in, 195–196
  reef growth in, 46
  Rio Grande Delta in, 236
  southern swamps in,
    209
  Tamabra Limestones of, 271
  Yucatan region of, 153
Microflora, in biofacies, 413–414
Microstratigraphy, of sedimentary base-
  ments, 320–333
Mid-Atlantic Ridge, sea-floor profile of, 66
Middle America trench, description of,
  52–53
Middle East
  Dead Sea of, 206
  oil fields of, 444–445
  phosphate beds in, 483, 485
Middleback Range, iron deposits in, 474
Midland Basin, ancient submarine fan of,
  276
Mineral deposits, exploration for, 468–487
Mineral distribution maps, construction and
  uses of, 404–405
Minturn Formation, stratigraphic relationships of, 414
Miogeosyncline
  in Appalachian region, 112
  occurrence of, 26
Misener Sandstone, in Oklahoma, 147
Mississippi Basin, clay mineral content of,
  342
Mississippi Delta
  as bird-foot delta, 200, 228
  broad shelf of, 8
  compaction rates in, 131
  hydrocarbon reservoirs in, 450
  regional environment of, 43
  river-shift effects on, 174
  sedimentation in, 41–42
  stratigraphic framework of, 359
  subsurface beds of, 48
  swamps of, 209

Mississippi River
  ancestral, carbon dating of tree trunks in,
    465
  illite in, 405
Mississippi Valley, St. Peter Sandstone of,
  145
Molasse facies and sequences
  of Cordilleran Geosyncline, 39
  description of, 282–283
Mont Pelée, explosion of, 252
Montana
  Bell Creek Field in, 430
  Cutbank Formation of, 397
  Eagle Sandstone of, 147, 350
  Madison Formation of, 443
  Morrison Formation of, 421
  Muddy Sandstone of, 430
  Powder River Basin, 174, 337, 429, 430
  Revett Formation in, 477
  Rierdon Formation of, 421
  Swift Formation of, 421
  Tyler Formation of, 214–215
  Williston Basin in, 262
Mooga Sandstone, aquifer in, 488
Moonie oil field, hydrodynamic conditions
  in, 462–463
Morocco, salt diapirs on coast of, 121, 123
Morrison Formation
  paleogeographic maps of, 421, 422
  uranium deposits in, 469
Morrow Formation
  hydrocarbon-bearing sandstones in, 347
  paleotopographic map of, 385–386
Mount Isa Shale, ore bodies of, 479
Mozambique
  sedimentary basin near, 83
  Zambeze Delta of, 234
Muddy Formation, shoestring sandstone
  bodies of, 397, 399
Muddy Sandstone
  paleogeographic framework of, 417–419
  map, 429, 430
Muderong Shale, Birdrong Sandstone over-
  lay by, 488–489

N

Nacimiento Fault, surface mapping of,
  297–298
Narrabeen Group, cross bedding in, 312
Nashville Dome, paleogeologic maps of, 383

Natural gas, storage for, 491–494
Navajo Sandstone, as fossil sand dune, 223
Nebraska, Denver Basin of, 364
Negro River, course of, 214
Neritic–bathyal environments, 8
Neritic environments, characteristics of, 254–275
Netherlands
    mud and salt flats of, 239
    sedimentation patterns in, 250–251
Nevada, exposed granitic rocks in, 297
New Caledonia Basin, sedimentation in, 53
New England, stratigraphic accumulations near, 120–121
New Guinea Basin, stratigraphic framework of, 355, 357
New Hampshire, Appalachian fold belt in, 112
New Hebrides Trench, sedimentation in, 53
New Mexico
    Capitan Reef of, 141
    Delaware Basin in, 245
    Delaware Mountain Group of, 330
    Entrada Formation of, 223
    Jackpile Sandstone of, 312
    Marble Falls Group of, 266
    San Juan Basin of, 391, 430
New South Wales
    aquifers in, 488
    glaciomarine beds in, 218
    Narrabeen Group in, 312
New York, Appalachian fold belt in, 112
New Zealand, North Island in, 188
Newcastle Sandstone, paleogeographic framework of, 417–419
Newfoundland Ridge, submarine canyon of, 51–52
Nicaragua Rise, size of, 361, 363
Niger Basin, stratigraphic framework of, 358
Niger Delta
    formation of, 47–48, 248, 358
    growth faults in, 192
    oil fields in, 443, 451
    paleogeographic map of, 425, 426
Nigeria
    beach environment in, 42
    Niger Delta of, 192, 248, 425, 426
    salt domes offshore of, 103
Nile Delta, island chains in, 154
Nile River, swamps of, 207

Nisku Member, oil traps in, 445
Noncarbonate sequences, of sedimentary basins, 15–20
Nonvolcanic sequences, of noncarbonate rocks, 17
North Africa
    Ordovician glaciation in, 302
    phosphate beds in, 483
North Dakota
    Prairie Formation in, 110
    Stettler Formation in, 244
    Williston Basin in, 262
North German Basin, Permian salt diapirs in, 101
North Island (New Zealand), growth faults in, 188
North Pole, Northwest Territories and, 219
North Sea
    Rotliegendes Formation of, 224
    salt diapirs in, 107, 109
    tidal current ridges of, 256
    Viking Graben of, 77, 78
    Zechstein sequence of, 245
Northern Territory (Australia)
    Blatherskite Nappe in, 368
    Jabiluka uranium ore bodies in, 472
Northwest Atlantic Mid-Ocean Canyon, description of, 51–52
Northwest Borneo Geosyncline, basins in, 61
Northwest Shelf, of Western Australia, 88
Northwest Territories (Canada)
    Beaufort Basin in, 92
    Gondwanaland compared to, 219
    Pine Point reef complex, 448
    Prairie Formation in, 110
Norway, Viking Graben near, 77
Nova Scotia, basement ridge of, 57
Nubial Sandstone, as possible alluvial fan, 222
Nubian crustal plate, splitting of, 66

O

Oakville Sandstone, of Felder uranium mine, 471
Ocala Group, aquifers in, 490
Ocean currents, effect on biofacies, 156
Oceanic crust
    of Red Sea, 66

in sedimentary basins, 83
Oceanic basement, of Aleutian Abyssal
    Plain, 279
Ogooue Delta, coastal swamps of, 234
Ohio
    Chazy Group in, 458
    Cleveland Shale of, 419–420
Oil, *see also* Gas reservoirs, Hydrocarbons
    basin potential for, 444–453
    clay minerals as catalyst in formation of,
        343
    reefs producing, 266–267, 271
Oil reservoirs
    estimation of, 458, 460
        computer use, 437
    exploration for, 439–464
    hydrodynamic conditions of, 462–464
    in fractured zones, 448–449
    in rivers and floodplains, 209, 211, 312
    rock porosity importance in, 336–337
    sand sheets as, 230
    sedimentary basin configuration studies
        for, 352
    source rock potential for, 439–443
    strontium and barium contents of, 345
    tarry, 338
    structural and stratigraphic traps for,
        453–462
Oil sands, characteristics of, 338
Oklahoma
    Anadarko Basin of, 215, 217, 347
    Booch Sandstone of, 232
    Flowerpot Shale in, 476
    Johns Valley Formation of, 408–409
    limestone bed characteristics of, 331
    Misener Sandstone of, 147
    Morrow Formation of, 385–386
    Red Fork Sandstone of, 374
    Viola Formation in, 383
Oksut Formation, paleogeographic maps of,
    422, 423
Olistostromes
    definition of, 72
    recognition of, 138
Onondaga Formation, as gas reservoir, 493
Orange River, in Africa, sedimentary basins
    of, 87
Organ Rock Formation, of Cutler Group, 56
Organic matter
    in lagoons, 236

of shales and limestones, 439
Oriskany Sandstone, as gas reservoir, 493
Orogenic belt
    erosional material of, 47
    from subduction, 63–64
Ostracod Limestone, strata of, 166
Otway Basin, description of, 128
Ouachita Mountains, strata offset of, 97
Outcrops, analysis of, 294–350
Oxygen isotope ratios, in carbonate rocks,
    345

P

Pacific Ocean
    abyssal plains of, 279
    sedimentary basins of, 61
Paleocurrent and paleowind pattern maps,
    interpretations and use of, 431–432
Paleocurrents, direction indicators of, 349
Paleogeographic framework, composite rep-
    resentations of, 417
Paleogeographic maps, of rock-time units,
    420–427
Paleogeologic maps, construction and uses
    of, 379–384
Paleogeomorphic units, of rock-time units,
    427–430
Paleopressure, in sedimentary basins, 346
Paleothermal gradients, in sedimentary ba-
    sins, 346–349
Paleotopographic maps, uses and construc-
    tion of, 384–388
Palestine Sandstone, as tidal current ridge,
    256
Palinspastic adjustment procedure, 433–434
    uses of, 99
Palinspastic reconstruction, of a sedimen-
    tary basin, 55
Papua, Early Creataceous bars in, 332
Papua Basin, stratigraphic framework of,
    357
Papua–New Guinea
    gravity map of, 355, 357
    Trobriand Basin of, 90–91
Paracontinuities
    in prograded delta lobes, 174–175
    of sedimentary basins, 93–94, 128–129,
        177–180

in sedimentary piles, 118
superposition of, 143–144
Paradox Basin, stratigraphic framework of, 56
Paraiba Delta, subaerial delta of, 200–201
Paralic environments, of sedimentary basins, 228–253
Parkman sandstone, paleogeographic map of, 425, 427
Paso Robles Formation, heavy minerals in, 327–328
Passo Della Cisa, melange in, 74
Pecatonica Formation, deposition of, 146
Pechora coal measures, thickness of, 467
Pedimentation, alluvial fans from, 222
Peel Sound Formation, rock-type distribution map of, 394–395
Pelagic deposits, over Pacific abyssal plains, 279
Pembina oil field
  rock types in, 449
  stratigraphic relationships in, 147, 332
Peoples Republic of China, 60
Permeability
  clay mineral effects on, 342
  measurement of, 18
  of sedimentary rocks, 333–339
Permian Age, Bowen Basin of, 24
Persian Gulf
  carbonate sediments of, 101, 259
  reef growth in, 46
  sedimentation patterns in, 244, 250–251
  stratigraphic accumulations in, 118
Peru
  carbonate–quartz sand facies in basin in, 286
  coastline submarine trench of, 27
  Lago Titicaco in, 138
Petroleum, see Oil
Petrophysical characteristics
  of sedimentary beds, 321
    integration, 330–331
Phosphate rock
  exploration for, 482–487
  in sedimentary basins, 140
Phosphoria Formation, phosphate beds in, 483, 485
Pierre Shale
  paleogeography of, 425, 427

prograding lobes of, 174
Pilbara region manganese deposits in, 472
Pilliga Sandstone aquifers in, 488
Pincher Creek, gas fields of, 449
Pine Point reef complex, dolomitizing solutions in, 448
Pleistocene Age, continental shelf exposure during, 19
Pleistocene lakes, in Alberta, Canada, 17
Po Basin, gas reservoir in, 493
Po Delta, configuration of, 200
Pocono Sandstone, pebble gradation in, 314
Point Lookout Sandstone, hydrodynamic maps of, 391
Porosity
  clay mineral effects on, 342
  of sedimentary rocks, 333–339
Potentiometric maps, uses of, 391
Powder River Basin
  Muddy Sandstone of, 429, 430
  Pierre Shale of, 174
  porosity of Sussex Sandstone of, 337
Prairie Formation salt removal from, 110
Precambrian basement
  in Canada, 186
  of Grand Canyon, 307
Precambrian Hoggar basement, paleocurrent studies of, 431–432
Precambrian shield
  copper deposits in, 477
  greenstone belts of, 25–26, 37
  occurrence of, 25
Precipice Sandstone, hydrodynamic maps of, 389–390
Pressure–solution, effects on sandstone porosity, 347, 349
Prince Edward Island, salt domes offshore of, 101
Prince of Wales Island, Peel Sound Formation on, 395
Prodelta mud sheets, description and occurrence of, 254–255
Provenance
  of lithic sediments, 324–329
  of sedimentary rock, clay mineral analysis for, 342
Pueblo Formation, fossil reefs of, 274
Pyrenees, flysch sequences in, 284

Q

Queensland (Australia)
banded sulfide ores of, 479
basin superposition in, 182
Bowen Basin of, 24
Great Barrier Reef of, 260
hydrocarbon source rock analysis in, 340
Moonie oil field of, 462–463
oil reservoirs in, 444
phosphate rock in, 483
Surat Basin of, 179, 389–390, 463
Toolebuc Member of, 163

R

Radioactive waste, storage for, exploration, 494–497
Radiolarite, occurrence of, 20
Rammelsberg district (Germany), banded sulfide ores of, 479
Rankin Platform, Exmouth Plateau and, 367
Reconstruction, of basin growth, 432–433
Recycling, in sandstones, 96
Red beds, as climatic indicators, 154
Red bed–evaporite facies, fossil sand dunes with, 224
Red Fork Sandstone
ancient river system of, 215, 217
isopach map of, 373, 374
Red Sea
metallic sulfides in, 140–141
sedimentary basin of, 66
sulfide-rich sediments of, 479
Reefs
development of, in carbonate banks, 259
oil-producing type, 266
Regional tilt adjustment procedure, 433–434
Regressive cycles, of sedimentation, 144–150
Reservoir beds, potential of, 444
Revett Formation, copper deposits in, 477
Rhine Delta, mud and salt flats of, 239
Rhine Graben, structure of, 77, 78
Richmond Groups, paleogeologic maps of, 383
Rierdon Formation
paleogeographic maps of, 421, 422
sandstone channels in, 397

Rift valley, of Mid-Atlantic Ridge, 66
Rio Grande Delta
coastal region south of, 236
illite in, 405
Riverdale beds, salt domes in, 103
Rivers, floodplain deposits of, 209, 211
Rock samples
geological mappings of, 294
lithology of, 302–307
Rock-stratigraphic units, time-stratigraphic units compared to, 159
Rock-time units
depositional environments of, 414
paleogeographic maps of, 420–427
paleogeomorphic maps of, 427–430
Rock-type distribution maps, construction and uses of, 392–395
Rock-type percentage maps, construction and uses of, 403–404
Rock-type ratio maps, construction and use of, 401–403
Rocky Mountains
badland topography of, 211
limestone of, 259
lithification in, 171
stratigraphic features of, 307
smectite-bearing mud near, 342
Rolling Downs Group, in Blythesdale Group, 488
Rotliegendes Formation, as fossil sand dune, 224
Rudist reefs, examples of, 270–271
Rumbalara Shale, 490
Russia, see USSR
Russian Platform, paleomagnetic studies of, 64

S

Sabkhas
environment of, 29
intertidal zones with, 241
mineral deposits in, 141, 244
Sable Island, basement ridge of, 57
Sacajawea Formation, stratigraphic sections of, 376
Sacramento Valley Basin, stratigraphic framework of, 357–358

Sahara, hydrodynamic conditions under, 463–464
Sakhalin, Amur Delta near, 209
Salado Formation, depositional environment of, 245
Salawati Basin, oil-bearing reefs of, 366
Salt deposits and beds
  formation of, 243–244
  salt removal from, 107
Salt diapirs
  on Moroccan coast, 121, 123
  seismic profiles of, 104
Salt domes, effect on sedimentary basins, 40
Salton Sea
  early desert areas of, 224
  size of, 206
San Andreas Fault, surface mapping of, 297–298
San Juan Basin paleogeomorphic map of, 430
  Point Lookout Sandstone of, 391
Sand dunes development of, 223
Sand sheets, description and occurrence of, 247–249
Sandstone bodies, recognition of, for oil reservoirs, 451–452
San Francisco River, delta configuration of, 201
Saskatchewan
  coal seams in, 468
  paleogeomorphic maps of, 429
  Rosetown collapse feature in, 109–110
  Viking Formation in, 147, 332
Saudi Arabia
  crustal plate separation near, 479
  strategraphic accumulations in, 118
Scotland, alluvial fan deposits of, 222
Scott Reef, structure of, 89
Sea-floor spreading, 141
  in provenance interpretations, 328–329
  theory of, 63
Sediment
  consolidation of, 167–173
  transport of, 135–136
Sediment-starved basins, description of 132, 134
Sedimentary basins
  analysis of, application to economic geology, 438–497
  aquifer exploration in, 487–491

basement faults in, 183–186
basin studies group of, 30–33
biofacies of, 155–157
  maps, 412–414
carbonate sequences of, 12–15
  rock analysis of, 344–346
classification of, 1–36
clay mineral content analysis of, 342–344
coal bed exploration in, 464–468
coal distribution maps of, 405–407
compaction of, 346–349
  rates, 131
configurations of, 351–391
depositional environments of, 197–293
diagenesis of, 170–173
  compaction and, 346–349
dissolution features of, 99–111
evolution and development of, 37–129
grain orientation in rocks of, 349–350
growth of, 99–111
  features, 100–111
  reconstruction, 432–433
growth faults in, 187–196
hingelines of, 180–182
hydrocarbon source analysis of, 339–341
hydrodynamic maps of, 388–391
isocarb maps of, 405–407
isograde maps of, 408–410
isolith maps of, 395–401
isopach maps of, 369–374
lithification of, 170–173
lithofacies of, 150–155
  maps of, 391–141
lithology of, 10–20
macrostratigraphy of, 294–320
maps and sections of, 351–437
  collation, 414–437
margins of, 92–99
marker beds of, 162–163
mass sedimentation in, 134–135
microstratigraphy of, 320–333
mineral deposits in
  distribution maps of, 404–405
  exploration for, 468–472
noncarbonate sequences of, 15
oil and gas exploration in, 439–464
paleogeographic maps of, 420–427
paleogeomorphic maps of, 427–430
paleotopographic maps of, 384–388
palinspastic adjustment studies of, 55,

433–434
physicochemical properties of rock in, 333–350
porosity and permeability, 333–339
physicochemical sedimentation processes in, 139–143
reservoir rock potential of, 444–453
rock-type distribution maps of, 392–395
rock-type percentage maps of, 403–404
rock-type ratio maps of, 401–403
sediment consolidation in, 167–173
sediment-starved, 132–134
sediment transport in, 135–136
sedimentation rate in, 131–139
shale color maps of, 410–411
solid geometry of, 112–129
strata correlation in, 163–167
stratigraphic concepts of, 130–196
framework, 173–183, 353–363
stratigraphic sections of, 374–379
structural concepts of, 130–196
computer analysis, 364, 366
framework, 353–363
structural framework of, 20–25, 61–92, 183–196
structural maps and sections of, 364–369
subsurface mapping of, 329–333
superposition of, 182–183
strata in, 142–144
surface mapping of, 295–302
tectonic control of 25–30, 61–92, 358, 361
terrestrial environments of, 204–228
time-stratigraphic units of, 159–163
transgressive and regressive cycles in, 144–150
Sedimentary troughs, definition and description of, 78, 80
Sedimentation rate, in sedimentary basins, 130–139
Seismic profiles, of marker beds, 163
Seismic refraction profiles, of Rhine and Viking Grabens, 77
Seismic surveys
of Nova Scotia basin, 57
of stratigraphic layers, 165
Shale color maps, construction and uses of, 410–411
Shetland Islands, Viking Graben near, 77
Siberia
Amur Delta of, 209

glaciomarine beds in, 219
Siberian Platform, paleomagnetic studies of, 64
Sicily
coastal submarine channels of, 276
orogeny in, 72
Silica gel, precipitation in basins, 140
Silled basin, Bass Basin as, 128
Simpson Group, paleogeologic maps of, 383
Singapore, sedimentary basins of, 61
Sirte Basin, structure of, 78
Sole faults, displaced margins from, 96–97
Solid geometry, of sedimentary basins, 112–129, 351
Solution breccia, definition of, 245
Sonic surveys, of debris flows, 137
Source rock analysis, of sedimentary basins, 439–443
South Africa
crustal rifts in sea floor of, 67
sedimentary basins near, 83–84, 116
Witwatersrand System in, 469
South America, diamond-bearing sediments in, 98, 328
South Carolina, possible basin for waste storage in, 497
South China Sea, sedimentary basins of, 61
South Korea, 60
South Pole, Gondwanaland and, 219
Spanish Sahara, debris flows offshore of, 136–137
Sri Lanka, Cauvery Basin of, 87–88
St. Lawrence River, submerged valley of, 52
St. Peter Sandstone
porosity of, 336
Tonti Member of, 178
transgressive sequence in, 145, 151
Starved Rock Member, of St. Peter Sandstone, 145
Stephan Shale, euxinic facies of, 288
Stettler Formation, facies map of, 244
Storage reservoirs, exploration for, 487–491
Strait of Gibralter, seawater passage through, 244
Straits of Florida, limestone mounds of, analysis, 345
Strata
correlation of, 163–167
superposition of, 142–144
Stratigraphic concepts, of sedimentary ba-

sins, 130–196
Stratigraphic framework
  definition of, 54–55, 74, 77
  of sedimentary basins, 173–183
Stratigraphic hingelines, structural
  hingelines compared to, 181–182
Stratigraphic markers, 374
Stratigraphic objective, selection of, 33–36
Stratigraphic traps, for oil and gas, 453–462
Stratigraphic sections, uses and interpreta-
  tions of, 374–379
Stratigraphic sequences, geomorphic fea-
  tures of, 197
Strontium, in petroleum exploration, 345–
  346
Structural framework, of sedimentary ba-
  sins, 20–25, 61–92, 183–196
Structural hingelines
  stratigraphic hingelines compared to,
  181–182
Structural maps, of sedimentary basins,
  364–369
Structural traps, for oil and gas, 453–462
Stylolites, as dissolution feature, 106
Subaerial sliding, of rock masses, 138–139
Subduction
  crustal disturbance and, 68
  definition of, 63
  in sedimentary basins, 25
  strata offsets and, 97–98
Submarine canyons, near Australia, 49
Submarine channel deposits, geomorphology of, 276–282
Submarine fans, geomorphology of, 275–276
Submarine plateau, of South Australia, 49
Submarine trenches, 40
  causes of, 63
  depositional environments of, 9–10
  near curstal plates, 68
  sedimentation in, 53
Subsurface mapping, description and uses
  of, 329–333
Sulfide(s)
  metallic, in Red Sea, 140–141
  ores, of Rammelsberg district (Germany),
  origin of, 479
  rich sediments, 479
Superposition, of sedimentary basins, 182–
  183
Surat Basin, 24

discontinuities in, 179
hydrodynamic conditions in, 463
  maps of, 389–390
  superposition on Bowen Basin, 182
Surface mapping, of sedimentary basins,
  295–302
Survey or Fracture Zone, near Aleutian
  Abyssal Plain, 279
Sussex Sandstone, porosity and permeability of, 337
Swamps, as depositional environments,
  204–209
Swan Hills Formation, early algal reef of,
  266
Sweden
  bedrock outcrops in, 300
  Kolm Formation in, 469
Swift Formation
  paleogeographic maps of, 421, 422
  sandstone channels in, 397
Sydney Basin
  lithofacies of, 161
  as possible gas reservoir, 494
Syngenetic origin, of mineral ores, 476
Syracuse Formation, salt deposits in, 245,
  247

                         T

Taiwan strait, biofacies of, 156–157
Tamabra Limestones, rudist reefs of, 271
Taphrogeny, in graben formation, 91
Tarkwaian sequence, diamondiferous con-
  glomerate of, 328
Tarry oils, from sandstone beds, 338
Tensleep Formations, sediments in, 245
Tectonic control, of sedimentary basins,
  25–30, 61–92
Tectonic framework, of sedimentary basin
  development, 358
Ten Thousand Islands, geomorphology of,
  259–260
Tennessee
  Nashville Dome of, 383
  phosphate rock in, 483
  stratigraphic sequence of Paleozoic rocks
  in, 295
Tensleep Formation
  hydrocarbon reservoirs in, 456–457
  stratigraphic sections of, 376

Texas
Capitan Reef of, 141
Delaware Basin in, 245
Delaware Mountain Group of, 330
Edwards Formation of, 270–271
Felder uranium mine in, 471
Flowerpot Shale in, 476
fossil reefs of, 274
Frio Formation in, 195–196
growth faults in, 190–191
lagoons in, 236
Marathon Mountains of, 97
Marble Falls Group of, 266
Midland Basin of, 276
oil reservoirs in, 444, 445
shale units in, 441
Tertiary barrier bars in, 332
Tholeiite, in Uralides, 64
Tidal current ridges, description and occur-
rence of, 255–257
Tidal mud flats, sediments from, 238–239
Tierra del Fuego, crustal plate movements
in, 311
Tigris river, delta environment of, 224
Time-marker, for stratigraphic sections, 34
Time-stratigraphic units
facies changes and, 151
of sedimentary basins, 159–163
in strata correlation, 164
Timor Trough, crust structure beneath, 69
Tollund Man, age of, 289
Tonga Trench, ultramafic and basaltic rocks
in, 281
Tonti Member
disconformity in, 178–179
St. Peter Sandstone of, 145
Toolebuc Member, marker beds of, 163
Trace element analyses, of carbonate rocks,
345
Transcurrent faults, displaced margins from,
96–97
Transform faults, strata offsets from, 97–98
Transgressive cycles, of sedimentation,
144–150
Transkei Basin, continental slope of, 67–68
Trobriand Basin, in Papua–New Guinea,
90–91
Trucial Coast, sedimentation patterns of,
251
Turbidites

in abyssal plain deposits, 279
in Nicaragua rise, 363
of submarine fans, 276
Turkey, Bitlis Massif in, 72
Turner Valley gas Field
hydrocarbon migration from, 352
storage in Bow Island gas field, 493
structural features of, 368
Tuscarora Quartzite, pebble gradation in,
314
Tyler Formation, ancient river system of,
214–215

U

Uncompahgre Uplift, of Paradox Basin, 56
Unconformities, in sedimentary basins,
177–180
United Arab Republic, Nile Delta in, 154–
155
United States
Belt series in, 482
Cambrian strata in Wisconsin of, 302
copper deposits in, 477
Cordilleran region of, 262–263
Illinois Basin of, 353–355
isopach map of Mesozoic Erathem of,
371, 373
macrostratigraphic studies in, 310
mineral deposit exploration in, 468–471
manganese and iron, 472
phosphate rock in, 483, 485
Pocono Sandstone in, 314
St. Peter Sandstone of, 336
Ural Mountains
paleomagnetic studies of, 64
Pechora coal measures of, 467
West Siberian Lowland near, 129
Uralides, medial eugeosyncline of, 64
Uranium ores, exploration for, 468–472
USSR
Amu Darya Delta of, 224
organic content of limestones in, 443
Pechora coal measures of, 467
sediment consilidation patterns in, 169
West Siberian Lowland of, 129
Utah
Great Salt Lake of, 206
Phosphoria Formation in, 483
Paradox Basin of, 56

V

Valmeyeran Series, isopach map of, 132, 133
Vanadium ores, exploration for, 468–472
Vancouver Island, basement topography near, 123
Velocity survey, of stratigraphic layers, 165
Venezuela
 basins of, 97
 Catatumba Delta of, 229
 oil reservoirs in, 444
 pollens in rocks of, 163
 sedimentary iron and manganese ores in, 472
Venezuela Basin, development of, 361
Ventura Basin
 deposition rate in, 363
 oil reservoirs in, 444
Vermilion Block 16 oil field, sandstone-producing bodies in, 451–452
Vermont, Appalachian fold belt in, 112
Vertical exaggeration, in structural maps, 366
Victoria (Australia)
 Gippsland Basin of, 126, 185, 233, 277, 288–289, 451, 465
 Latrobe Formation in, 18, 233, 467
 submarine plateau of, 49
Viking Formation
 marker beds in, 163
 petrophysical characteristics of, 332
 as regressive sequence, 147
Viking Graben, structure of, 77, 78
Viola Formation, paleogeologic maps of, 383
Vitrinite, in coal, reflectance measurement of, 441
Volcanic chains, causes of, 63
Volcanic flow sheets, as sedimentary association, 225–228
Volcanic sequences
 marine type, 16
 types of, 16
Volcanics, preservation of, 252
Volga River, sedimentation in Delta of, 207

W

Wadden Sea, sedimentation patterns of, 251
Washington, Columbia River basalts of, 225
Waterton Lakes, gas reservoirs of, 338

Weatherall Formation
 paleogeographic maps of, 422
 stromatoporoid reefs of, 269
West Africa
 diamond occurrence in, 98
 sea-floor spreading and, 328
West Indies, surface limestones of, 344
West Siberian Lowland, megabasin of, 128–129
Water, storage for, 491–494
West Virginia, Augusta gas fields in, 493
Wetar Strait Australian plate beneath, 69
Williams Fork Formation, paleogeographic map of, 425, 427
Williston Basin, reefs of, 262
Windalia Formation, oil reservoirs in, 449
Windalia Radiolarite, deposition of, 167
Wisconsin, quartzite boulders in, 302
Witwatersrand System, uranium in, 469
Woodbine Sandstone, oil production from, 383, 445
Wreck Island, oil exploration on, 262
Wyoming
 Big Horn Basin of, 376, 456–457
 Caspar Formation in, 224
 clay mineral content in petroleum reservoir rocks in, 343–344
 coal deposits in, 465
 Denver Basin of, 364
 depositional environments of, 245
 isopach maps of Mesozoic sequence in, 371–373
 Mancos Shale of, 377
 Muddy Formation of, 399
 paleogeographic framework of, 417
 maps of, 425, 427
 Phosphoria Formation in, 483
 Powder River Basin of, 174
 sedimentary basins in, 298
 Sussex Sandstone of, 337
 Upper Cretaceous of, 174

Y

Yampi Sound, iron deposits in, 473–474
Yangtze Kiang River, sediments from, 61
Yellow Sea, 61
 Gulf of Korea in, 256
Yoredale Formation, boron content of, 345

Yucatan
 carbonate banks of, 258–259
 depositional environment of, 153
 reef growth in, 46
Yucatan Basin, development of, 361
Yucatan Peninsula, carbonate sediments of,
 101
Yugoslavia, mélange in, 376
Yukon Delta, glaciomarine beds in, 219

Z

Zaire copper belt, origin of, 478
Zambeze Delta, 87
 coastal swamps of, 234
Zambia, copper deposits of, 478
Zechstein evaporites, 245
 fossil sand dunes and, 224
 Kupferschiefer shales and, 141
Zinc ores, exploration for, 475–482